T0228444

Contact Angle, Wettability and Adhesion
Volume 4

CONTACT ANGLE, WETTABILITY AND ADHESION

VOLUME 4

Editor:
K.L. Mittal

CRC Press
Taylor & Francis Group
Boca Raton London New York

CRC Press is an imprint of the
Taylor & Francis Group, an **informa** business

First published 2006 by VSP

Published 2019 by CRC Press
Taylor & Francis Group
6000 Broken Sound Parkway NW, Suite 300
Boca Raton, FL 33487-2742

© 2006 by Taylor & Francis Group, LLC
CRC Press is an imprint of Taylor & Francis Group, an Informa business

First issued in paperback 2019

No claim to original U.S. Government works

ISBN 13: 978-0-367-44632-1 (pbk)
ISBN 13: 978-90-6764-436-5 (hbk)

This book contains information obtained from authentic and highly regarded sources. Reasonable efforts have been made to publish reliable data and information, but the author and publisher cannot assume responsibility for the validity of all materials or the consequences of their use. The authors and publishers have attempted to trace the copyright holders of all material reproduced in this publication and apologize to copyright holders if permission to publish in this form has not been obtained. If any copyright material has not been acknowledged please write and let us know so we may rectify in any future reprint.

Except as permitted under U.S. Copyright Law, no part of this book may be reprinted, reproduced, transmitted, or utilized in any form by any electronic, mechanical, or other means, now known or hereafter invented, including photocopying, microfilming, and recording, or in any information storage or retrieval system, without written permission from the publishers.

For permission to photocopy or use material electronically from this work, please access www.copyright.com (http://www.copyright.com/) or contact the Copyright Clearance Center, Inc. (CCC), 222 Rosewood Drive, Danvers, MA 01923, 978-750-8400. CCC is a not-for-profit organization that provides licenses and registration for a variety of users. For organizations that have been granted a photocopy license by the CCC, a separate system of payment has been arranged.

Trademark Notice: Product or corporate names may be trademarks or registered trademarks, and are used only for identification and explanation without intent to infringe.

Visit the Taylor & Francis Web site at
http://www.taylorandfrancis.com

and the CRC Press Web site at
http://www.crcpress.com

Contents

Part 3: Wettability and Its Modification

Part 4: Surface Characteristics and Adhesion

Contact Angle, Wettability and Adhesion, Vol. 4, pp. ix–xi
Ed. K.L. Mittal
© VSP 2006

Preface

This volume documents the proceedings of the Fourth International Symposium on Contact Angle, Wettability and Adhesion held under the auspices of MST Conferences in Philadelphia, Pennsylvania, June 14–16, 2004. The premier symposium with the same title was held in 1992 in honor of Prof. Robert J. Good as a part of the American Chemical Society meeting in San Francisco, California, and the second and third events in this series were held in 2000 in Newark, New Jersey and 2002 in Providence, Rhode Island, respectively. The proceedings of these earlier symposia have been properly chronicled as three hard-bound books [1–3]. Because of the tremendous interest in and technological importance of this topic, the next (Fifth) symposium is planned to be held in Toronto, Canada during June 21–23, 2006.

As pointed out in the Prefaces to the previous volumes in this series, the world of wettability is very wide and it plays a crucial role in many and varied technological areas ranging from microfluidics to biomedical to agriculture to welding. By the way, these days there is a tremendous interest in superhydrophobic materials and there is a high tempo of research activity in manipulating the chemical and physical nature of surfaces to impart them such characteristic.

This symposium was organized with the following objectives in mind: to update and consolidate the information on this topic since the last symposium, to provide a forum for cross-pollination of ideas, and to highlight the moot issues within the broad purview of this topic. The technical program for the symposium comprised 67 papers covering many different aspects of this topic. The presentations ranged from critical reviews of particular subtopics to original research contributions, and the presenters hailed from academia, industry and other research organizations. This meeting was a veritable global event both in scope and spirit. There were lively and brisk discussions – both formally and informally – throughout the symposium, and on occasions quite divergent views were expressed.

Now coming to this volume, it contains a total of 31 papers covering many ramifications of contact angle, wettability and adhesion. It must be recorded that all manuscripts were rigorously peer-reviewed and revised (some twice or thrice) and properly edited before inclusion in this book. Concomitantly, this volume represents an archival publication of the highest standard. It should not be considered a proceedings volume in the usual and ordinary sense, as many so-called proceedings volumes are neither peer-reviewed nor adequately edited.

This book (called Volume 4) is divided into four parts: Part 1: General Papers; Part 2: Contact Angle Measurements/Determination and Solid Surface Free Energy; Part 3: Wettability and Its Modification; and Part 4: Surface Characteristics and Adhesion. The topics covered include: fundamental aspects of contact line region; evaporative behavior of sessile drops; various factors influencing contact angle measurements; different kinds of contact angles; various ways to measure contact angles; contact angle hysteresis; contact angle measurements on various materials (smooth, rough, porous, heterogeneous); effect of electric field on contact angle (electrowetting); wetting and spreading on heterogeneous surfaces; factors influencing wetting/spreading phenomena; determination of solid surface free energy via contact angle measurements; application of AFM in determining solid surface tension at the nano-scale; ultralyophobic surfaces; surface modification and wettability; multiphase flow dynamics in porous media; thin film coatings for textile materials; bio-fouling resistant coatings; relationship between wetting and adhesion; and relevance/importance of wetting and surface energetics in technological applications, including cleaning of flooring materials, kinetics of oil removal from coating materials, cell adhesion, and mold compound–metal adhesion in semiconductor packaging.

Apropos, this year (2005) marks the 200th anniversary of the seminal paper published by Young (Thomas Young, *Phil. Trans. Royal Soc.* **95**, 65–87 (1805)) which described the thermodynamics between a solid and a liquid in terms of their respective surface tensions and the contact angle formed between them. Concomitantly, this book finds a special place in the literature as it coincides with the 200th anniversary of the famous Young's Equation. It should be noted, however, that the reader will not find such equation in the original paper by Young (cf. supra). Apparently, some clever and brilliant scientist later expressed Young's ideas in the form of this widely used, extremely useful, and one of the prominent equations within the realm of surface chemistry.

This volume and its predecessors [1–3] containing a wealth of information on various facets of contact angle, wettability and adhesion provide a unified and comprehensive reference source, and yours truly hopes that anyone interested or involved – centrally or peripherally – in any aspects of wetting will find these volumes of great interest and value.

Acknowledgements

Now comes the pleasant task of thanking those who helped in many and varied ways. First, as usual, I am thankful to Dr. Robert H. Lacombe, a dear friend and colleague, for taking care of the organizational details of this symposium. The unsung heroes (reviewers) must be copiously thanked for their time and effort in providing many valuable comments which are a *sine qua non* in maintaining high standard of any publication. Special thanks are extended to the authors for their interest, cooperation and for providing written accounts of their presentations, without which this book would not have seen the light of day. Finally my sincere appreciation goes to the staff of Brill Academic Publishers for materializing this book.

K. L. Mittal
P.O. Box 1280
Hopewell Jct., NY 12533

December 2005

1. K. L. Mittal (Ed.), *Contact Angle, Wettability and Adhesion*, VSP, Utrecht (1993).
2. K. L. Mittal (Ed.), *Contact Angle, Wettability and Adhesion*, Vol. 2, VSP, Utrecht (2002).
3. K. L. Mittal (Ed.), *Contact Angle, Wettability and Adhesion*, Vol. 3, VSP, Utrecht (2003).

Part 1
General Papers

Contact Angle, Wettability and Adhesion, Vol. 4, pp. 3–28
Ed. K.L. Mittal
© VSP 2006

Study of contact angles, contact line dynamics and interfacial liquid slip by a mean-field free-energy lattice Boltzmann model

JUNFENG ZHANG[1] and DANIEL Y. KWOK[2,*]

[1]*Department of Biomedical Engineering, School of Medicine, Johns Hopkins University, Baltimore, MD 21205, USA*
[2]*Nanoscale Technology and Engineering Laboratory, Department of Mechanical Engineering, Schulich School of Engineering, University of Calgary, Calgary, Alberta T2N 1N4, Canada*

Abstract—We summarize here a mean-field representation of fluid free-energy to a lattice Boltzmann scheme recently proposed for interfacial studies. The interfacial behaviors obtained from this new multi-phase lattice Boltzmann model (LBM) were validated by means of the Laplace equation of capillarity and the capillary wave dispersion relation. Applications of this mean-field LBM to various interfacial studies are reviewed, including wettability on heterogeneous surfaces, self-propelled drop movement, contact line dynamics and solid–liquid interfacial slip. The mean-field LBM simulates systems with better physical reality in terms of solid–liquid interactions and could be an alternative for simulating interfacial phenomena.

Keywords: Lattice Boltzmann method; solid–fluid interactions; contact angle; contact line dynamics; Cassie's equation; heterogeneous surface; solid–fluid interfacial slip.

1. INTRODUCTION

The lattice Boltzmann method (LBM) has been extensively developed during the past decade in simulating fluid dynamics [1–5]. As an extension of the lattice gas automata (LGA), LBM describes macroscopic complex flows by dealing with the underlying particle interactions. One of the attractions of LBM is that the numerical algorithm can be easily implemented with complex solid or free boundaries for multicomponent/multiphase systems. In bulk fluid, LBM is, in fact, a Navier–Stokes solver; however, at the fluid–solid interface, the mesoscopic nature of this method becomes manifest as boundary conditions (BCs) are imposed on the particle distributions rather than directly on the fluid quantities such as velocity [6]. As LBM can be considered as a mesoscopic approach, i.e., between microscopic

*To whom correspondence should be addressed. Tel.: (1-403) 210-8428; Fax: (1-403) 282-8406; e-mail: daniel.kwok@ucalgary.ca

molecular dynamics (MD) and conventional macroscopic fluid dynamics, it would be useful when microscopic statistics and macroscopic description of flow are both important; for example, in problems involving surface tension, capillarity and phase transition in multiphase/multicomponent systems [7]. For these reasons, LBM has been widely employed in interfacial studies [7–11]. However, most of the existing multiphase schemes achieve phase separation and interface formation only phenomenologically and usually produce significant spurious currents within the interfacial region even at equilibrium [12–17]. For example, Yeomans and co-workers were among the first to employ a thermodynamic free-energy approach for the LBM scheme in modeling isothermal systems with liquid–vapor interfaces or with two mutually interacting fluids [17, 18]. Others have also employed this and similar free-energy models to simulate various hydrodynamic systems [12, 14, 19]. In Yeomans' model [17, 18], free energy is expressed by a square-gradient expression of the van der Waals theory using the Cahn–Hilliard description of non-equilibrium dynamics [12, 20]. In reality, however, the presence of an impenetrable solid boundary imposes a discontinuity on the local fluid density near the wall. Thus, the gradient expansion approximation used in deriving the square-gradient theory is inadequate for description of solid–fluid interfaces [21, 22]. As a result, meaningful fluid simulations by LBM involving solid–fluid interfaces often require adjustment of interaction parameters which might not truly reflect the physics of specific solid–fluid systems.

Since solid–fluid interactions play a vital role in interfacial phenomena for micro/nanofluidics [6, 23], dynamics of wetting [24] and thin liquid film stability [25], we have recently proposed a free-energy approach to the LBM by means of a mean-field representation [26]. Such a mean-field representation has also been employed by us and others to study various wetting and adhesion phenomena [21, 27, 28]. In this work, we summarize our recent study on the use of a mean-field LBM to examine various contact angle and interfacial phenomena, including wettability on heterogeneous surfaces [29], dynamic contact angles [30] and apparent liquid slip over a solid–fluid interface [31].

2. THE MEAN-FIELD FREE-ENERGY LBM SCHEME

According to the mean-field version of van der Waals' theory [21, 27, 28, 32–34], the total free-energy function for a fluid system can be expressed as

$$F = \int d\mathbf{r} \left\{ \psi[\rho(\mathbf{r})] + \frac{1}{2}\rho(\mathbf{r}) \int d\mathbf{r}' \phi_{ff}(\mathbf{r}' - \mathbf{r})[\rho(\mathbf{r}') - \rho(\mathbf{r})] + \rho(\mathbf{r})V(\mathbf{r}) \right\}, \quad (1)$$

where $\psi(\rho)$ is a local free-energy with respect to the bulk density ρ. The second term is a non-local term taking into account the free-energy cost due to variations in density; $\phi_{ff}(\mathbf{r}' - \mathbf{r})$ is the interaction potential between two fluid particles located at $\mathbf{r}'$ and $\mathbf{r}$. This term can be reduced to that of a square-gradient approximation when the local density varies only slightly [21, 22]. The third term represents contribution of external potential energy $V(\mathbf{r})$ to the free-energy F. Both integrations are taken over the entire space.

With this expression for free-energy, we followed the procedures described in Ref. [35] and defined a non-local pressure as

$$P(\mathbf{r}) = \rho(\mathbf{r})\psi'[\rho(\mathbf{r})] - \psi[\rho(\mathbf{r})] + \frac{1}{2}\rho(\mathbf{r})\int d\mathbf{r}'\phi_{ff}(\mathbf{r}' - \mathbf{r})[\rho(\mathbf{r}') - \rho(\mathbf{r})]. \quad (2)$$

For a bulk fluid with uniform density, the non-local integral term disappears and equation (2) reverts to the equation-of-state for the fluid

$$P = \rho\psi'(\rho) - \psi(\rho). \quad (3)$$

For the sake of clarity, here we briefly describe the implementation of these results into an LBM algorithm. In general, after discretization in time and space, the lattice Boltzmann equation (LBE) with a Bhatnagar–Gross–Krook (BGK) collision term can be written as

$$f_i(\mathbf{x} + \mathbf{e}_i, t + 1) - f_i(\mathbf{x}, t) = -\frac{1}{\tau}\left[f_i(\mathbf{x}, t) - f_i^{eq}(\mathbf{x}, t)\right], \quad (4)$$

where the distribution function $f_i(\mathbf{x}, t)$ denotes fluid particle population moving in the direction of $\mathbf{e}_i$ at a lattice site $\mathbf{x}$ and at a time step t; τ is a relaxation time and $f_i^{eq}(\mathbf{x}, t)$ is a prescribed equilibrium distribution function with

$$\begin{aligned}
f_0^{eq} &= \rho\left[d_0 - \frac{1}{c^2}\mathbf{u}^2\right], \\
f_i^{eq} &= \rho\left[\frac{1 - d_0}{l} + \frac{D}{c^2 l}\mathbf{e}_i \cdot \mathbf{u} + \frac{D(D+2)}{2c^4 l}(\mathbf{e}_i \cdot \mathbf{u})^2 - \frac{D}{2lc^2}\mathbf{u}^2\right], \\
i &= 1, \ldots, l
\end{aligned} \quad (5)$$

for a D-dimensional lattice with l links where the particle speed is $|\mathbf{e}_i| = c$. The constant d_0 is the equilibrium fraction of particles at rest [16, 36]. The macroscopic density ρ and velocity $\mathbf{u}$ can be calculated from the distribution function f_i as:

$$\rho = \sum_i f_i, \quad (6)$$

and

$$\rho\mathbf{u} = \sum_i f_i\mathbf{e}_i. \quad (7)$$

However, if an external force $\mathbf{F}(\mathbf{x}, t)$ exists, we can modify equation (7) to reflect the momentum change as

$$\rho\mathbf{u} = \sum_i f_i\mathbf{e}_i + \tau\mathbf{F} \quad (8)$$

and employ the $\mathbf{u}$ produced here to calculate the equilibrium distribution function f_i^{eq} in equation (5) [14, 16]. Redefining the fluid momentum $\rho\mathbf{v}$ to be an average of the momentum before collision $\sum_i f_i e_i$ and that after collision $(\rho u + F)$ yields

$$\rho v = \sum_i f_i e_i + \frac{1}{2} F. \tag{9}$$

Following the Chapman–Enskog procedure, a Navier–Stokes equation with the equation-of-state

$$P = \frac{c^2(1 - d_0)}{D} \rho + \Phi \tag{10}$$

can be obtained, where Φ is the potential energy field related to F by

$$F(x, t) = -\nabla \Phi(x, t). \tag{11}$$

In order to obtain the Navier–Stokes equation with a pressure term similar to that given by equation (2), we set an artificial Φ as follows

$$\Phi(x, t) = \rho(x)\psi'[\rho(x)] - \psi[\rho(x)]$$
$$+ \frac{1}{2}\rho(x) \int dx' \phi_{ff}(x' - x)[\rho(x') - \rho(x)] - \frac{c^2(1 - d_0)}{D} \rho(x). \tag{12}$$

The above equations constitute a complete LBM scheme with the mean-field free-energy function implemented. This proposed mean-field free-energy LBM has been tested against various criteria [26], including density distributions, interfacial tensions, isotropy and Galilean invariance. Our results also agree well with those from other thermodynamic and molecular dynamics (MD) studies.

3. EXAMINATION OF LIQUID–VAPOR INTERFACE BEHAVIORS

Nevertheless, before applying the mean-field LBM method to study a more complex interfacial problem, examination of the liquid–vapor interface behaviors is also necessary. In this section, we summarize the bubble and capillary wave tests from Ref. [30], which have been commonly employed to test the validity by different multi-phase/multi-component LBM models [13, 16–18, 36]. The bubble tests were carried out over a 128×128 D2Q7 lattice domain with periodic BCs in both directions, having $D = 2$, $l = 6$, $d_0 = 1/2$, $c = 1$, $e_0 = [0, 0]$ and $e_i = [\cos(i - 1)\pi/3, \sin(i - 1)\pi/3]$ $(i = 1, \ldots, 6)$ in equation (5). Following Refs [12, 17, 18, 26], we employed a van der Waals fluid model to express the free-energy of bulk fluid:

$$\psi(\rho) = \rho k_B T \ln \frac{\rho}{1 - b\rho} - a\rho^2, \tag{13}$$

where a and b are the van der Waals constants; k_B is the Boltzmann constant and T is the absolute temperature. Here, we have selected $a = 9/49$, $b = 2/21$ and

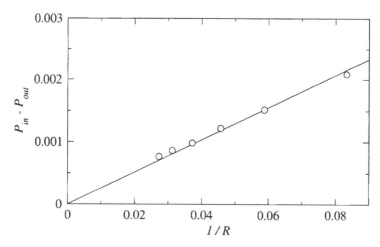

Figure 1. Pressure differences across the interface for different bubble sizes from LBM simulations (symbols). The solid line is a curve-fit according to Laplace's equation, equation (15).

the scaled temperature $k_B T = 0.52$. The fluid–fluid interaction was simplified by considering only that between nearest neighbors [16, 36]:

$$\phi_{ff}(\mathbf{x'} - \mathbf{x}) = \begin{cases} K, & |\mathbf{x'} - \mathbf{x}| = c \\ 0, & |\mathbf{x'} - \mathbf{x}| \neq c \end{cases}, \tag{14}$$

where K measures the interaction strength among the nearest neighboring particles; and the non-local integral term in equation (12) can be replaced by a summation over the neighbors of a site $\mathbf{x}$. In this work, K was set to -0.01. A negative value implies an attraction between two fluid particles and its magnitude changes the liquid–vapor interfacial tension.

The initial density distribution was assigned with a small higher density region in the center. After about 10 000 steps, the system reached equilibrium and the drop radius and densities inside and outside were measured. The pressure difference across the interface can be obtained through the equation of state (equation (3)). The results are plotted in Fig. 1 together with a fitted line according to Laplace's equation

$$P_{in} - P_{out} = \gamma / R, \tag{15}$$

where γ is the interfacial tension and R is the bubble radius. Clearly, the symbols from our LBM results follow the Laplace equation well and the interfacial tension γ was found to be 0.026 from the slope of the fitted line. By direct integration of the excess free energy across a planar interface, the interfacial tension was found to be 0.029 [26], which is, in principle, similar to the value obtained here by the bubble test simulations.

The Galilean invariance is another important property for LBM models, since the system behavior should be invariant even when the reference coordinates are

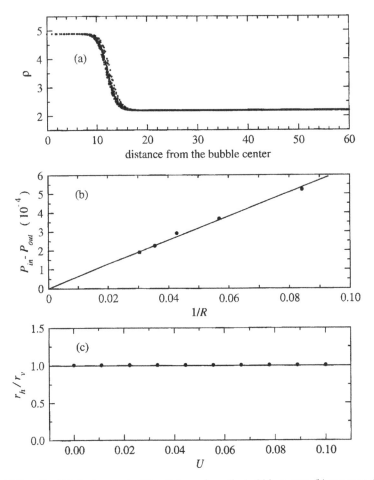

Figure 2. (a) Density distribution against the distance from the bubble center, (b) pressure differences across the bubble interface of different bubble sizes and (c) ratio of the two radii at perpendicular directions when the reference system moves at U.

being translated at any given constant velocity. The data in Fig. 2c were obtained by setting different initial horizontal velocity U for the reference coordinates and measuring the corresponding liquid bubble size, radii in horizontal r_h and vertical direction r_v. Clearly, the bubble behaves as a perfect circle and does not change with the reference velocity U. These results suggest that our model is indeed better than the one proposed by Yeomans and co-workers [18] and is comparable to the modified one [19] with respect to the Galilean invariance.

We have also calculated the dispersion relation for capillary waves [30]. In the absence of gravity, this relation for a fluid-fluid interface is given by fluid dynamics as

$$\omega^2 \sim k^3,\tag{16}$$

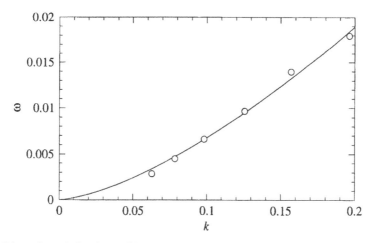

Figure 3. Dispersion relation for capillary waves on a liquid–vapor interface. The symbols are LBM results while the solid line is a curve-fit according to fluid mechanics theory, equation (16).

where ω is frequency and k is the wavenumber of capillary wave [17, 37]. Our simulations were performed over a $W \times H$ rectangular lattice domains with periodic BCs in the horizontal direction and bounce-back BCs on the top and bottom layers. The height H was selected as $10W$ to overcome the effects of finite fluid depth. Initially, the density in the upper part of the domain was set to be higher; while that of the lower to be less. After equilibrium had been established, a planar liquid–vapor interface was formed in the middle and we introduced a single sinusoidal disturbance with a wavelength W on the interface. The wave amplitude was thereafter measured to obtain frequency ω and wavenumber $k = 2\pi/W$. Our results (symbols) are displayed in Fig. 3 with a non-linear fit (solid line) of $\omega = \alpha k^\beta$. The fitted line provides a value of β as 1.48 and is nearly identical to the theoretical value of 3/2 in equation (16), validating our LBM for liquid–vapor interfacial studies.

4. WETTABILITY ON HETEROGENEOUS SURFACES

The wettability of a liquid in contact with a solid surface plays a dominant role in many industrial processes, such as painting, coating and adhesion. The simplest and most common method to evaluate wettability and solid surface tensions is believed to be contact angle measurements. Although this approach appears straightforward, the underlying physics and its interpretation are not trivial [38, 39]. The basis of contact angle measurements to determine solid surface tensions relies on the well known Young's equation [40]. For a liquid drop resting on a solid surface, the Young equation is written as

$$\gamma_{lv} \cos\theta = \gamma_{sv} - \gamma_{sl}, \tag{17}$$

where γ denotes the interfacial tension with subscripts lv, sv and sl representing the liquid–vapor, solid–vapor and solid–liquid interfaces, respectively; θ is the contact angle at a location where the tangent along a liquid–vapor interface intersects the solid surface. It is well-known that equation (17) is a mechanical equilibrium condition based on an ideal situation where the surface is perfectly smooth and homogeneous. This condition, however, does not always hold in reality.

Cassie was the first to propose a model of contact angle on a heterogeneous surface [41, 42]. According to Cassie's equation, the contact angle θ on a heterogeneous surface consisting of two materials is given by

$$\cos \theta = \alpha_1 \cos \theta_1 + \alpha_2 \cos \theta_2, \tag{18}$$

where θ_i is the contact angle on a homogeneous surface of pure material i; α_i is the fractional area of material i and we have

$$\alpha_1 + \alpha_2 = 1. \tag{19}$$

Because of experimental difficulty, computer simulations using MD [43] and Monte Carlo [44] have been employed to demonstrate the validity of Cassie's equation.

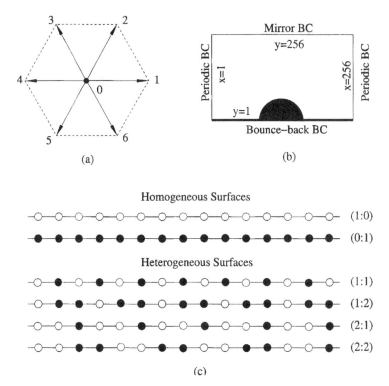

Figure 4. The simulation set-up in this study: (a) the lattice velocities of a D2Q7 lattice; (b) a 256×256 simulation domain and BCs; and (c) homogeneous and heterogeneous surface models. $(n_1 : n_2)$ represents the ratio of hydrophobic ($K_w = 0.04$) and hydrophilic ($K_w = 0.08$) patches.

This question regarding the Cassie equation has remained the subject of many contradicting studies [45–52].

In this section, we summarize the LBM results for contact angle behaviors on heterogeneous surfaces [29]. Three kinds of BCs were applied to the 256 × 256 D2Q7 lattice domain (Fig. 4a): a general bounce-back BC on the solid surface (y = 1); a mirror BC on the top (y = 256); and periodic BCs on the left and right ends (x = 1 and x = 256) (Fig. 4b). The solid–fluid interaction was also simplified as the fluid–fluid interaction described above and was implemented as an attraction force between a solid ($\mathbf{x}_s$) and a fluid ($\mathbf{x}_f$) site as

$$\mathbf{F}_s = \begin{cases} K_w \rho(\mathbf{x}_f)(\mathbf{x}_s - \mathbf{x}_f), & |\mathbf{x}_s - \mathbf{x}_f| = c \\ 0, & |\mathbf{x}_s - \mathbf{x}_f| \neq c \end{cases}. \tag{20}$$

The coefficient K_w is positive for attraction forces and by adjusting its magnitude, solid surfaces with different wettabilities (contact angles) can be easily modeled, from wetting to non-wetting [26]. A heterogeneous solid surface was modeled by assigning different interaction strengths K_w to the solid sites (Fig. 4c). A higher K_w value implies a stronger solid–liquid interaction and hence stronger adhesion. Our results in the following sections suggest that the Cassie equation is valid only when the heterogeneous patches are smaller than the thickness of a liquid–vapor interface. For heterogeneous patches larger than the interfacial thickness, the apparent contact angle deviates from the Cassie prediction, depending on the local surface feature near the contact point that the liquid–vapor interface can "see".

4.1. Liquid–vapor interface and contact angle on homogeneous surfaces

Using the above parameters, the liquid–vapor interface generated by the mean-field LBM is shown in Fig. 5. Here, the fluid densities of liquid and vapor phases are ρ_l = 5.63 and ρ_v = 1.45, respectively, with an interfacial thickness of about 6 lattice

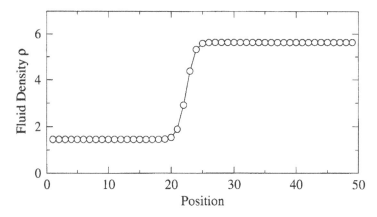

Figure 5. Calculated fluid density distribution across a typical liquid–vapor interface from the mean-field LBM.

units. Contact angles were obtained by treating the liquid–vapor interface as a curve of mean density $(\rho_l + \rho_v)/2 = 3.54$. This can be done through a geometric relation by assuming the drop profiles as circular or by direct manual measurements; and they agree well within $\pm 2°$. The contact angles on two homogeneous solid surfaces having $K_w = 0.04$ and $K_w = 0.08$ are $124.4°$ and $50.5°$, respectively. We would also like to point out that, unlike other LBM models [17, 18], contact angle values between 0 and $180°$ can be easily generated without using a less realistic repulsive solid–fluid interaction. This is consistent with MD simulations and physical reality [26].

4.2. The Cassie equation

To examine Cassie's relation (equation (18)), we modeled a heterogeneous solid surface consisting of two different patches, by assigning different attractive strengths to the solid sites. Two kinds of solid sites were used with $K_{w1} = 0.04$ and $K_{w2} = 0.08$ (displayed as white and black filled circles, respectively, in Fig. 4c. The heterogeneous patches were arranged as $n_1 \times K_{w1}$ followed by $n_2 \times K_{w2}$ sites, denoted as $(n_1 : n_2)$ hereafter; and such patterns were repeated to cover the entire surface $(y = 1)$. The contact angles obtained from these surfaces are plotted in Fig. 6 *versus* $\alpha_1 = n_1/(n_1 + n_2)$, i.e., the fractional area of K_{w1}; the symbols are LBM results while the solid line is a best-fitted line according to the Cassie relation (equation (18)). Clearly, the simulated results and those from the Cassie relation are in good agreement; the minor deviations are within a contact angle measurement error of $\pm 2°$.

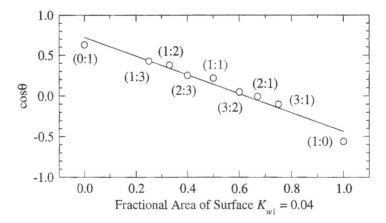

Figure 6. Contact angles obtained on heterogeneous surfaces with different hydrophobic ($K_{w1} = 0.04$) to hydrophilic ($K_{w2} = 0.08$) ratios: $(n_1:n_2)=(0:1)$, (1:4), (1:3), (1:2), (2:3), (1:1), (3:2), (2:1), (3:1), (4:1) and (1:0) (left to right).

4.3. Patch size effect

Here, we examine the effect of patch size on contact angles when the pattern ratio remains constant. We fixed this ratio by selecting $\alpha_1 = n_1/(n_1 + n_2) = \alpha_2 = n_2/(n_1 + n_2) = 0.5$ and changed its patch size from (1:1) up to (20:20). The results are shown in Fig. 7 for contact angles at the left and right contact points (Fig. 7a) and the center positions of the contact area (Fig. 7b). We had originally anticipated to observe different angles on the two sides as the local heterogeneous features may be different. However, the two measured angles turned out to be similar, or same within the measurement errors considered. Drop-profile images were also analyzed and found to be indeed symmetric about its center. When the patch size is small ($n \leqslant 3$), the angles are consistent with Cassie's prediction (81.8°, dashed line in Fig. 7a). For a larger patch size ($n > 4$), there exist many metastable contact-angle values which deviate from Cassie's prediction, even when the pattern ratio remains the same ($\alpha_1 = \alpha_2 = 0.5$). Other ratios, ($n : 2n$), ($2n : n$), ($3n : n$) and ($n : 4n$), were also studied and found to have similar results. As mentioned earlier, we found that the liquid–vapor interface has a thickness of approx. 6 lattice units

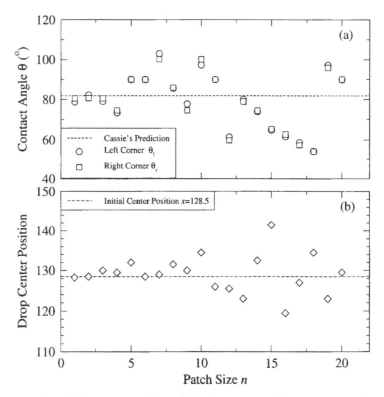

Figure 7. (a) Left and right contact angles and (b) drop-center positions on a $(n{:}n)$ heterogeneous surface consisting of 50% hydrophobic and 50% hydrophilic patches, where n represents the patch size.

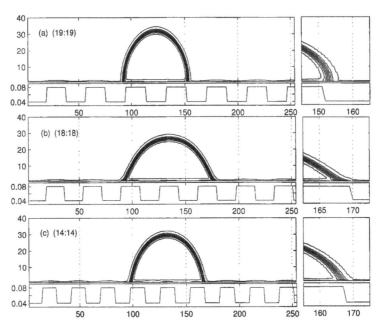

Figure 8. Fluid density contours (upper figures) and solid attraction K_w patterns (lower figures) for three surface models having (a) (19:19), (b) (18:18) and (c) (14:14) patches with the same pattern ratio 50%. The figures on the right display the local features near the right contact points.

(*cf.*, Fig. 5). It appears that the validity of the Cassie equation depends on how much the liquid–vapor interface can "see" the patches. In Fig. 7a, $n \leqslant 3$ corresponds to a patch size smaller than the thickness of the liquid–vapor interface ($n = 6$); we anticipate that the liquid–vapor interface is capable of "seeing" both patches at the same time; whereas for $n > 4$, the liquid–vapor interface with an interfacial thickness of $n = 6$ would not be able to "see" both patches completely at once. Thus, the Cassie equation appears to be valid only when the patch size is on the order of the liquid–vapor interfacial thickness. These results imply, at a macroscopic scale, that the Cassie equation is, in general, not valid as it is nearly impossible to pattern a surface having regular hydrophobic and hydrophilic patches on the order of the interfacial thickness (approx. 1 nm). These results are also in agreement with the conclusions obtained from a free-energy thermodynamic approach [39].

To further verify the above analysis, we plotted in Fig. 8 the fluid density contours (upper figures) and the solid attraction strength K_w (lower figures) for three specific cases with different patch sizes: (a) (19:19), (b) (18:18) and (c) (14:14). We selected these three cases because of their representative contact angle values: $\theta_{(19:19)} = 96°$, $\theta_{(18:18)} = 54°$ and $\theta_{(14:14)} = 74°$. The respective solid–liquid contact area ratios are $(A_{0.04}/A_{0.08})_{(19:19)}$ approx. 1/2, $(A_{0.04}/A_{0.08})_{(18:18)}$ approx. 2/3 and $(A_{0.04}/A_{0.08})_{(14:14)}$ approx. 2/3. Despite the similarity in the contact area ratios for the (18:18) and (14:14) surfaces, their contact angles are different. To be specific, the contact area of the surface (19:19) has a larger hydrophilic fraction, yet its contact angle is the largest among the three; this is due to the fact that the liquid–vapor interfacial region sits almost completely on top of the hydrophobic site, resulting in a higher contact angle. Thus, we conclude that contact angle results from a local equilibrium among the interfacial tensions at the three-phase contact line where the liquid–vapor interfacial region brackets. This is reconfirmed by enlarging the figures near the three-phase contact point shown on the right of Fig. 8; due to symmetry, the contact points on the left are not displayed. If we look closely at the local features near the contact points, the liquid–vapor interface for the (19:19) surface is sitting mainly on the hydrophobic fraction of the surface ($K_w = 0.04$) and, hence, it results in a larger contact angle. In contrast, the (18:18) interface sits on the hydrophilic fraction of the surface ($K_w = 0.08$) and, thus, produces a smaller angle. For the (14:14) surface, the interface sees both of the hydrophobic and hydrophilic patches on the surface, resulting in an intermediate contact angle.

5. SELF-PROPELLED DROP MOVEMENT

A number of recent theoretical and experimental studies [53–58] have shown that drop motion can be initiated and self-propelled by a surface energy gradient on a substrate. Such gradients can be generated either passively using surfaces with spatial variations in free energy [57, 58] or actively using surfactant-like agents that adsorb onto the contacted surface and induce localized dewetting events [53–56]. Both of these methods rely on the difference in wettability between opposite edges of a droplet to create an unbalanced surface tension force between the edges, resulting in droplet propulsion. In Fig. 7b, we see that as the patch size n increases, the drop appears to move away from its initial center position $x = 128.5$, suggesting that the drop is trying to find a more comfortable lower-energy state at which the contact angles on the left and right sides are the same, i.e., no surface tension gradient. We found that the above results might be used to describe the self-propelled drop movement experiments recently reported [53–55]. When a drop sits on a heterogeneous surface with a wettability gradient, it will self-propel and move on the surface as induced by the unbalanced surface tension force. Such a drop will keep on moving as long as the unbalanced surface tension force is maintained. To demonstrate this movement, we set the solid surface as half hydrophobic ($K_{w1} = 0.04$ for $x = 1$–128, $y = 1$) and half hydrophilic ($K_{w2} = 0.08$ for $x = 129$–256, $y = 1$). At the beginning of the simulation, a drop was placed in the center of the surface ($x = 128.5$) where the left side is set to be hydrophilic ($K_{w1} = 0.08$) and the right side to be hydrophobic $K_{w2} = 0.04$). This setting is similar to the experimental set-up reported in Ref. [55]. As the drop does not prefer to stay on the hydrophobic surface, there is a stronger surface tension gradient towards the hydrophilic surface (left) and the drop will move spontaneously towards the high energy surface. Figure 9a shows the positions of the moving contact points at the left and right with the simulation time steps; some snapshots of the drop

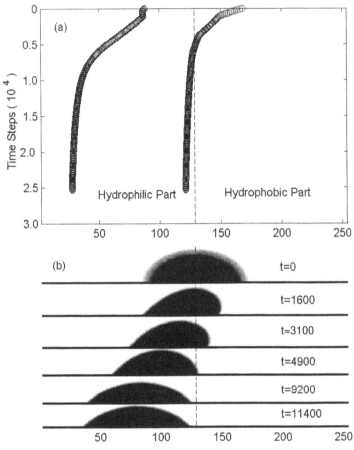

Figure 9. Self-propelled drop movement on a heterogeneous surface consisting of hydrophobic and hydrophilic patches: (a) the left and right contact point positions for different time steps, t and (b) snapshots during the self-propelled drop movement with time steps. The dashed lines indicate the boundary of the hydrophobic and hydrophilic patches.

shapes are also displayed in Fig. 9b. At $t = 0$, the drop has a wider liquid–vapor interface due to an initial density distribution, set as 5.6 for the liquid phase and 1.4 for vapor phase with an interfacial thickness of about 10 lattice units (*cf.*, Fig. 9a). At $t = 1600$, it can be seen in Fig. 9b that the contact angle on the right is larger than that on the left; the drop has started moving towards the more hydrophilic surface. For $t > 4900$, the initial surface tension gradient starts to decrease and the drop reaches its equilibrium shape at about $t = 11\,400$. It is also apparent in Fig. 9a that the velocity of the drop beyond $t > 10\,000$ has decreased significantly from its initial velocity; the drop retains its momentum due to inertia until it is equilibrated. These behaviors are similar to those previous reported experimentally [53, 54, 56].

6. CONTACT ANGLE AND CONTACT LINE DYNAMICS OF LIQUID–VAPOR INTERFACES

The moving contact line problem is important to a wide range of phenomena and industrial processes, such as coating and oil recovery [24, 59]. Classical fluid dynamics treatment on movement of contact line leads to a stress singularity and a multi-valued velocity field near the contact line [60, 61]. To overcome such difficulties, several theoretical models have been proposed [8, 60–62]; these methods, however, typically require the dynamic contact angle as a BC. Experimental studies have shown that dynamic contact angle can deviate significantly from the static value when contact line is in motion [63–67]. Yet, the nature of this phenomenon is not well understood. The general behavior of dynamic contact angles is that they increase with advancing contact line motion and decrease when the contact line is receded. In order to understand this phenomenon, several relations have been presented [60, 63, 68, 69]. The fact is that these relations always involve some unknown parameters, which can only be determined by non-linear fittings with experimental data.

On a microscopic level, MD simulations have also been employed to study this problem [70–74]. Most of them were performed over a two-component system and the velocity fields were usually difficult to resolve [71]. Recently, Fan *et al.* [8] conducted LBM simulations on a two-component system to study dynamic advancing contact angle and pressure drop across a fluid interface, using an interparticle potential model proposed by Shan and Chan [36]. The reported changes in dynamic advancing contact angles, however, were limited and increased by only about 3.5 degrees. In this section, we summarize our recent mean-field LBM study [30] on this contact line problem by focusing on the dynamic contact angles (advancing and receding) and the velocity field near the liquid–vapor interface. The results had been compared with those from other theoretical models and good agreement was found.

Similar to the procedures described in Ref. [8], simulations of moving contact lines were conducted on a 2D channel. By taking advantage of symmetry about the centerline, only half of the channel was considered here with a mirror BC applied along the centerline (Fig. 10). The simulation domain was 1000×20 with periodic BCs at the left and right ends. A general bounce-back no-slip BC was applied along the channel wall. In this section, two kinds of channel surfaces, one hydrophobic ($K_w = 0.04$) and another hydrophilic ($K_w = 0.08$), were studied. Here, we label a hydrophobic surface to be low energy (contact angle $> 90°$) and the hydrophilic one to be high energy (contact angle $< 90°$).

The density distribution in the channel at the beginning was set to the density of liquid phase, except a segment in the center which was set to that of the vapor phase, in order to simulate a vapor bubble in the channel enclosed by two liquid columns with two liquid–vapor interfaces. The densities of liquid and vapor phases were obtained as described in the previous section.

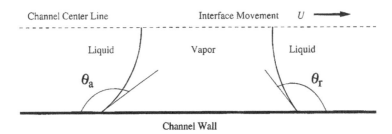

Figure 10. A schematic diagram (not to scale) of the simulation framework. θ_a and θ_r in the figure are the advancing contact angle with contact line velocity U and the receding contact angle with $-U$, respectively.

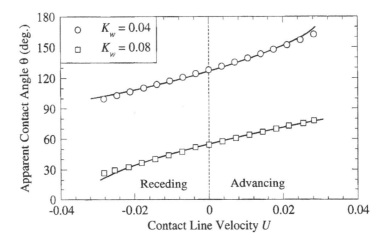

Figure 11. Dynamic contact angles obtained at different contact line velocities (symbols), with non-linear fits (solid lines) according to Blake's theory (equation (21)).

To cause the interfaces to move, we applied a body force on the fluid particles in the direction along the channel and various velocities can be easily obtained by adjusting the force magnitude. Such velocities were calculated by measuring the interface positions at different time steps. The dynamic contact angles were also measured at both ends of the enclosed bubble, corresponding, respectively, to the advancing and receding contact angles (Fig. 10). Our contact angle calculation utilizes a geometry relation which is the same as that given in Refs [8] and [75], which assumed the liquid–vapor interface as a circular arc. The liquid–vapor interface was treated as the isoline of the average density.

The dynamic contact angles at different velocities are displayed in Fig. 11 where the solid lines were determined by non-linear fits using Blake's adsorption/desorption model due to its good agreement with experimental results [63, 76]. According to Blake's theory, the relation between dynamic contact angle θ and contact line velocity U is given as

$$U = 2\kappa_w^0 \lambda \sinh \left[\frac{\gamma (\cos \theta_0 - \cos \theta)}{2nk_B T} \right],$$ (21)

where θ_0 is the static contact angle ($U = 0$). Others are parameters related to the adsorption/desorption process: n is the number of adsorption sites per unit area, λ is the average length of each particle displacement and κ_w^0 is the equilibrium frequency of particle displacement. Typically, these parameters can only be determined by curve-fitting [63]. It can be seen in Fig. 11 that our simulation results follow Blake's relation (equation (21)), very closely, except for large velocities. The difference may be caused by the fact that when the velocity becomes large, the liquid–vapor interface shape deviates from a circular shape and the contact angle obtained from the geometry relation becomes less accurate. However, manual measurements of such angles are difficult to be consistent and could cause a much larger uncertainty. Nevertheless, our LBM results demonstrate the general behaviors of dynamic contact angles. The fitting parameter values are $\theta_0 = 126.0°$, $2\kappa_w^0 \lambda = 12.62$, $2nk_B T/\gamma = 0.23$ for the $K_w = 0.04$ surface, and $\theta_0 = 55.1°$, $2\kappa_w^0 \lambda = 52.06$, $2nk_B T/\gamma = 0.67$ for the $K_w = 0.08$ surface; the latter implies more adsorption sites and higher displacement frequency on a surface with stronger attractions to the fluid particles.

Another aspect that we were interested in is the flow in the interfacial region. Figure 12 displays the velocity fields around the advancing and receding ends. The hydrophobic channel wall was modeled using $K_w = 0.04$ with an interfacial velocity of $U = 0.014$. Focusing on the three-phase contact points, the flow fields are similar to that of the picture using Blake's adsorption/desorption process: when fluid 1 moves forward to displace fluid 2 (the two fluids may either be both liquids or a liquid and a vapor), the fluid 2 particles adsorbed on the solid surface would be pushed away (desorption) and the fluid 1 particles would be attached onto the surface sites left behind by the particles of fluid 2 (adsorption) [63].

In moving contact line studies, the frame of reference is typically selected for stationary contact line and interface [61, 62, 70, 75, 77]. Thus, we plotted the flow streamlines (from Fig. 12) in Fig. 13 with the contours of constant density (circular curves in the center) in such a reference system. However, as the flow field near interfaces is complicated, errors can be introduced when integrating fluid velocities across the interfaces to produce streamlines. We note that far away from the interfaces, the flow displays the same features as the Poiseuille flow with a parabolic velocity profile. The general streamline patterns near the interfaces are, however, quite similar to the results from fluid dynamics [61, 62, 75, 77] and MD simulation [70]. The flow agrees in character with the wedge flow solution [78]. We also observed small vortexes near the contact line and beside the interface as found in Refs [61] and [75]. The reason for the mass transfer across the interface at the advancing end is not clear. It may be due to spurious current in the interfacial region [15], fluid diffusion as reported in [61], or simply the integration errors mentioned

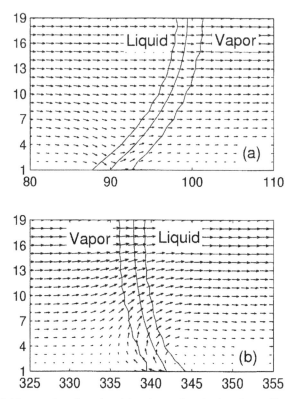

Figure 12. Flow fields near the advancing (a) and receding (b) interfaces. The curves are constant density contours. x and y axes have dimensions of lattice units

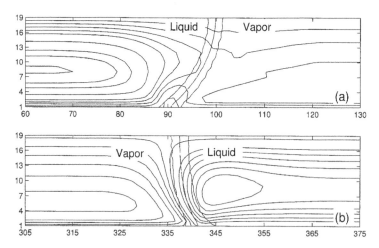

Figure 13. Streamlines and density contours (circular curves in center) in the vicinities of the advancing (a) and receding (b) interfaces. The reference frame is moving at a constant velocity with the interfaces.

above. More recently, Briant *et al.* [79] studied droplet deformation in a shear flow field and the mass transfer was explained as an evaporation/condensation process to overcome the contact line velocity singularity.

7. APPARENT SLIP OVER SOLID–LIQUID INTERFACE WITH NO-SLIP BCs

In fluid mechanics, the no-slip BC between a fluid and a solid surface has traditionally been an assumption in solving the governing Navier–Stokes (NS) equations [80]. Despite macroscopic experimental support, it still remains an assumption without physical principles. In fact, studies on fluid slip have long been an interesting subject since the pioneering work by Navier [81] and Maxwell [82]. Recent measurements, however, indicate significant slip on solid surfaces [83–87]. Due to the difficulties in direct microscopic observation near the solid–fluid interface, MD simulations have been widely used to study the relationship between fluid slip and the properties of both fluid and solid [88–91]. In general, both experimental and MD simulation results show that there is a strong relationship between the magnitude of slip and the solid–fluid interaction: the weaker the interaction, the larger is the contact angle and hence the slip.

Recently, Succi [6] has applied the LBM to study fluid slip on solid surfaces by employing a mix of bounce-back and specular reflection BCs. With respect to slip, Nie *et al.* [92] and Lim *et al.* [93] performed simulations of micro-systems by relating the LBM relaxation time to the Knudsen number. According to Ref. [94], slip velocity relates directly to the relaxation time, and thus different relaxation times would certainly produce different degrees of slippage. However, all these attempts failed to relate the slip magnitude with the solid–fluid interactions, which indeed plays a dominant role in such phenomenon.

The mean-field LBM model presented above had been employed to examine fluid slip [31] where the solid–fluid interaction was modeled as an exponentially decaying attractive force [21]

$$F_w(\mathbf{x}) = \rho(\mathbf{x}) K_w e^{-h/\alpha}, \tag{22}$$

where K_w is the interaction strength; h is a distance from point $\mathbf{x}$ to the solid surface in lattice units; and α is a parameter controlling the decaying behavior. In these simulations, we selected $a = 9/49$, $b = 2/21$ and $k_B T = 0.55$. A 40×100 D2Q7 lattice domain for the slip simulations and a 128×256 D2Q7 domain for the contact angle simulations were employed with a relaxation time $\tau = 1$. The BCs applied to the top and bottom layer nodes are the general mid-grid bounce-back BCs to simulate no-slip solid–fluid interfaces; periodic BCs were applied to the other two sides [5]. The fluid density in the slip simulations was set to be that of the liquid phase at equilibrium ($\rho_{bulk} = 4.895$).

Typical density and velocity profiles of pressure-driven Poiseuille flows are displayed in Fig. 14. The filled circles are simulation results from our mean-field LBM scheme with $K_w = 0$; for comparison purpose, we also illustrate the results from our mean-field model in the limit of a standard no-slip Poiseuille flow as open circles. All other parameters in these two simulations remain the same. Because of symmetry, only half of the profiles are shown. As the wall is located at $x = 0.5$, the right boundary (at $x = 19.5$) corresponds to the channel center line. Unlike the constant density distribution (open circles) from the general LBM, there is a dry (low-density) layer between the bulk liquid and the wall (at $x = 0.5$) from our mean-field model (filled circles). Such a dry layer reflects the specific solid–fluid interactions in the vicinity of the wall. This result is similar to those obtained from thermodynamics [27] and observed in MD simulations [95]. However, because short-range interactions are neglected in this approximation, the density profile shows no oscillatory behavior near the wall as found from other MD studies [21, 27].

In Fig. 14b, the fluid velocity at the solid–fluid interface (at $x = 0.5$) is not directly available. Extrapolating the velocity data to this interface position results in a more or less zero velocity for both the mean-field LBM simulations and that in the limit of a standard no-slip Poiseuille flow. This demonstrates that the no-slip BC is

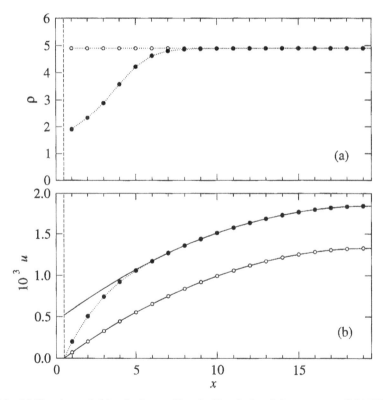

Figure 14. (a) Density and (b) velocity profiles (half) calculated from a mean-field LBM (filled circles) and that in the limit of a standard no-slip Poiseuille flow (open circles). Solid lines in (b) are parabolic fittings using only the data points (for $x \geqslant 10$) away from the solid wall; the solid wall is represented by a dashed line at $x = 0.5$.

satisfied and there is no microscopic slip at the solid–fluid interface. Thus, the slip phenomenon discussed below does not appear to be a numerical artifact described elsewhere [94]. Comparing the two velocity profiles in Fig. 14b, we found that the velocity from the mean-field LBM (solid circles) increases much faster in the dry layer ($1 \leqslant x \leqslant 6$); in the inner region (for $x \geqslant 7$), however, the variation of the two velocity profiles becomes similar. Through a parabolic fitting for the data points (for $x \geqslant 10$) in Fig. 14b, we found that the velocity data from a general LBM follow the curve exactly; whereas, those from the mean-field LBM show good agreement only for $x \geqslant 6$, where the density is approximately constant. Extrapolating these fitted profiles to zero velocity yields a slip length δ, i.e., the distance between this zero velocity point and the wall; δ is positive if this zero-velocity point is outside the channel and negative if inside [89]. The slip lengths found in this specific example are 2.78 and 0 for the mean-field and general LBM, respectively. Overall, the velocity profile from the mean-field LBM model employed here is qualitatively similar to those obtained from MD simulations [89, 91].

As a matter of fact, experimental and MD studies have shown that slip usually occurs on a hydrophobic surface. The origin of wettability and contact angle phenomena is, of course, from intermolecular interactions: the weaker the solid–fluid interaction, the more hydrophobic is the surface and, hence, the larger the

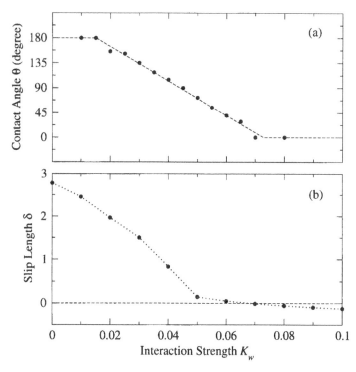

Figure 15. Variation of (a) contact angle θ and (b) slip length δ with the solid–fluid interaction strength K_w.

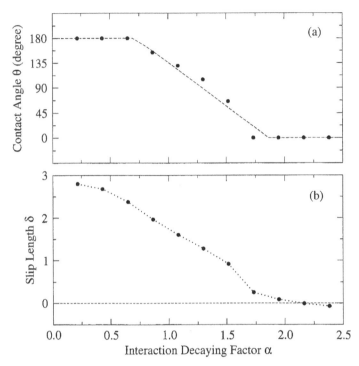

Figure 16. Variation of (a) contact angle θ and (b) slip length δ with the solid–fluid interaction decaying factor α.

contact angle. For example, the choice of $K_w = 0$ in Fig. 14 should represent a low-energy (hydrophobic) surface. In Fig. 15, the contact angle θ and slip length δ values against the solid–fluid interaction strength K_w with $\alpha = \sqrt{3}/2$ is plotted. The contact angle is found to be nearly a linear function of K_w between $\theta = 0$–$180°$ and is in agreement with those from other studies [9, 95]. We point out that, unlike other multiphase LBM models, a contact-angle value between 0 and $180°$ can be generated here without using a less realistic repulsive solid–fluid interaction. This is also consistent with physical reality and those observed in MD simulations [95].

In Fig. 15, as solid–fluid interaction strength increases, the slip length decreases quickly and becomes negative when $K_w > 0.06$. Beyond this value, the decrease in slip length becomes slower. Similar negative and small slip lengths had also been observed in MD simulations [89]. In fact, a positive slip length will produce a larger flow rate and can be considered as a wider channel (cf., Fig. 14b); a negative slip length implies a smaller flow rate which corresponds to a narrower channel. The latter case appears to be possible when the solid–fluid interaction (adhesion) is very strong and molecules near the solid wall would have less mobility; the wall can then be thought of having an extra covered layer of molecules, resulting in a narrower channel. Focusing on the contact-angle and slip-length behaviors, we see that they follow similar decreasing trends as the solid–fluid attraction increases.

Another interesting factor that may influence the apparent slip length is the decaying behavior of the solid–fluid interaction in equation (22). Thus, we plot in Fig. 16 the contact angle and slip length *versus* α. In this result, as α changes, the interaction strength K_w was adjusted to maintain the same value of F_w/ρ at $x = 1$, i.e., all the first-layer fluid particles are set to experience the same body force from the wall. As α increases, the attractive force F_w in equation (22) will decay more slowly and, thus, can attract more fluid particles further away from the wall, resulting in a smaller contact angle as well as slip length. This phenomenon is similar to that shown in Fig. 15 and a small negative slip length is also observed. We have also studied the effects of the externally applied pressure on δ [31]. However, unlike those from Refs [6, 88], our results suggest that slip length is independent of the magnitude of the driving force.

8. CONCLUSIONS

We have summarized a new multiphase lattice Boltzmann model (LBM) with the mean-field expression for fluid free-energy implemented. The fluid–solid interactions are more physical as they relate directly to the contact angle and fluid density profile near the solid. Interfacial behaviors obtained from this LBM model had been examined by means of the Laplace capillary equation and the capillary wave dispersion relation; the results are in good agreement with theoretical predictions.

By employing the mean-field LBM, several interfacial phenomena were studied. We examined the contact angle behaviors on heterogeneous surfaces with different patterns and patch sizes. The results suggest that Cassie's relation is, in general, not valid in macroscopic contact angle measurements. These findings agree well with thermodynamics analysis and provide a more physical picture near the contact point. Simulations were also conducted for self-propelled drop movement by means of a local surface tension gradient.

With an applied external force, the contact angle and contact line dynamics were also examined. Dynamic contact angles from our simulations followed the general trend observed experimentally and could be described by Blake's theory. The fitting parameters with Blake's theory indicate a strong and active adsorption process on a hydrophilic surface. The evaporation/condensation mechanism of contact line movement was also observed.

The mean-field LBM was also employed to study solid–liquid interfacial slip and represents the first attempt to relate the slip magnitude to the solid–fluid interactions *via* the LBM approach. Even with a no-slip BC, apparent slip can be observed because of the specific solid–fluid interactions. A larger slip was found on more hydrophobic surfaces. With stronger interactions, a smaller negative slip length was also observed. These results are in good agreement with other experimental and numerical studies.

In summary, our recently conducted simulations have demonstrated the potential of the mean-field free-energy LBM for future fluid–solid interfacial studies.

Acknowledgements

This work was supported by the Natural Sciences and Engineering Research Council of Canada (NSERC). J. Z. acknowledges financial support from NSERC through a postdoctoral fellowship at Johns Hopkins University.

REFERENCES

1. R. Benzi, S. Succi and M. Vergassola, *Phys. Rep.* **222**, 145 (1992).
2. S. Chen and G. D. Doolen, *Annu. Rev. Fluid Mech.* **30**, 329 (1998).
3. Y. H. Qian, D. d'Humieres and P. Lallemand, *Europhys. Lett.* **17**, 479 (1992).
4. D. A. Wolf-Gladrow, *Lattice-Gas Cellular Automata and Lattice Boltzmann Models: An Introduction*. Springer, Berlin (2000).
5. S. Succi, *The Lattice Boltzmann Equation*. Oxford University Press, Oxford (2001).
6. S. Succi, *Phys. Rev. Lett.* **89**, 064502 (2002).
7. Z. L. Yang, T. N. Dinh, R. R. Nourgaliev and B. R. Sehgal, *Int. J. Heat Mass Transfer* **44**, 195 (2001).
8. L. Fan, H. Fang and Z. Lin, *Phys. Rev. E* **63**, 051603 (2001).
9. Q. Kang, D. Zhang and S. Chen, *Phys. Fluids* **14**, 3203 (2002).
10. P. Raiskinmaki, A. Koponen, J. Merikoski and J. Timonen, *Comput. Mater. Sci.* **18**, 7 (2000).
11. P. Raiskinmaki, A. Shakib-Manesh, A. Jasberg, A. Koponen, J. Merikoski and J. Timonen, *J. Stat. Phys.* **107**, 143 (2002).
12. A. D. Angelopoulos, V. N. Paunov, V. N. Burganos and A. C. Payatakes, *Phys. Rev. E* **57**, 3237 (1998).
13. C. Appert, D. H. Rothman and S. Zaleski, *Physica D* **47**, 85 (1991).
14. J. M. Buick, Ph.D. thesis, Lattice Boltzmann Methods in Interfacial Wave Modelling, The University of Edinburgh, Edinburgh (1997).
15. S. Hou, X. Shan, Q. Zou, G. D. Doolen and W. E. Soll, *J. Comput. Phys.* **138**, 695 (1997).
16. X. Shan and H. Chen, *Phys. Rev. E* **49**, 2941 (1994).
17. M. R. Swift, W. R. Osborn and J. M. Yeomans, *Phys. Rev. Lett.* **75**, 830 (1995).
18. M. R. Swift, E. Orlandmi, W. R. Osborn and J. M. Yeomans, *Phys. Rev. E* **54**, 5041 (1996).
19. A. N. Kalarakis, V. N. Burganos and A. C. Payatakes, *Phys. Rev. E* **65**, 056702 (2002).
20. J. W. Cahn and J. E. Hilliard, *J. Chem. Phys.* **28**, 258 (1958).
21. D. E. Sullivan, *J. Chem. Phys.* **74**, 2604 (1981).
22. B. Widom, *J. Stat. Phys.* **19**, 563 (1978).
23. B. Li and D. Y. Kwok, *Phys. Rev. Lett.* **90**, 124502 (2003).
24. J. de Coninck, M. J. de Ruijter and M. Voue, *Curr. Opin. Colloid Interface Sci.* **6**, 49 (2001).
25. J. A. Diez and L. Kondic, *Phys. Rev. Lett.* **86**, 632 (2001).
26. J. Zhang, B. Li and D. Y. Kwok, *Phys. Rev. E* **69**, 032602 (2004).
27. A. E. van Giessen, D. J. Bukman and B. Widom, *J. Colloid Interface Sci.* **192**, 257 (1997).
28. J. Zhang and D. Y. Kwok, *J. Phys. Chem. B.* **106**, 12594 (2002).
29. J. Zhang and D. Y. Kwok, *J. Colloid Interface Sci.* **282**, 434 (2004).
30. J. Zhang and D. Y. Kwok, *Langmuir* **20**, 8137 (2004).
31. J. Zhang and D. Y. Kwok, *Phys. Rev. E* **70**, 056701 (2004).
32. J. Rowlinson and B. Widom, *Molecular Theory of Capillary*. Claredon, Oxford (1982).
33. J. Zhang and D. Y. Kwok, *Langmuir* **19**, 4666 (2003).

34. J. Zhang and D. Y. Kwok, *J. Adhesion* **80**, 745 (2004).
35. A. J. M. Yang, P. D. Fleming and J. H. Gibbs, *J. Chem. Phys.* **64**, 3732 (1976).
36. X. Shan and H. Chen, *Phys. Rev. E* **47**, 1815 (1993).
37. L. D. Landau and E. M. Lifshitz, *Fluid Mechanics*. Pergamon Press, New York, NY (1959).
38. D. Y. Kwok and A. W. Neumann, *Adv. Colloid Interface Sci.* **81**, 167 (1999).
39. A. W. Neumann and J. K. Spelt (Eds), *Applied Surface Thermodynamics*. Marcel Dekker, New York, NY (1996).
40. T. Young, *Philos. Trans. Roy. Soc. London* **95**, 65 (1805).
41. A. B. D. Cassie, *Discuss. Trans. Faraday Soc.* **3**, 11 (1948).
42. A. B. D. Cassie and S. Baxter, *Faraday Soc.* **40**, 54 (1944).
43. M. H. Adao, M. de Ruijter, M. Voue and J. de Coninck, *Phys. Rev. E* **59**, 746 (1999).
44. D. Urban, K. Topolski and J. de Coninck, *Phys. Rev. Lett.* **76**, 4388 (1996).
45. R. H. Dettre and R. E. Johnson Jr., *J. Phys. Chem.* **69**, 1507 (1965).
46. J. Drelich, J. L. Wilbur, J. D. Miller and G. M. Whitesides, *Langmuir* **12**, 1913 (1996).
47. M. Fabretto, J. Ralston and R. Sedev, *J. Adhesion Sci. Technol.* **18**, 29 (2004).
48. M. Gleiche, L. Chi, E. Gedig and H. Fuchs, *Chem. Phys. Chem.* **2**, 187 (2001).
49. P. E. Laibinis and G. M. Whitesides, *J. Am. Chem. Soc.* **114**, 1990 (1992).
50. P. S. Swain and R. Lipowsky, *Langmuir* **14**, 6772 (1998).
51. B. Wu, G. Mao and K. Y. S. Ng, *Colloids Surfaces A: Physicochem. Eng. Aspects* **162**, 203 (1999).
52. A. Lafuma and D. Quéré, *Nature Mater.* **2**, 457 (2003).
53. F. Domingues, D. Santos and T. Ondarcuhu, *Phys. Rev. Lett.* **75**, 2972 (1995).
54. S.-W. Lee, D. Y. Kwok and P. E. Laibinis, *Phys. Rev. E* **65**, 051602 (2002).
55. T. Ondarcuhu and M. Veyssie, *J. Phys. II (France)* **1**, 75 (1991).
56. G. C. H. Mo and D. Y. Kwok, *Colloids Surfaces A: Physicochem. Eng. Aspects* **232**, 169 (2004).
57. M. K. Chaudhury and G. M. Whitesides, *Nature* **256**, 1539 (1992).
58. S. Daniel, M. Chaudhury and J. C. Chen, *Science* **291**, 633 (2001).
59. P. G. de Gennes, *Rev. Mod. Phys.* **57**, 827 (1985).
60. S. Sciffer, *Chem. Eng. Sci.* **55**, 5933 (2000).
61. P. Seppecher, *Int. J. Eng. Sci.* **34**, 977 (1996).
62. D. Jacqmin, *J. Fluid Mech.* **402**, 57 (2000).
63. T. D. Blake, in *Wettability*, J. C. Berg (Ed.), pp. 251–309. Marcel Dekker, New York, NY (1993).
64. G. M. Fermigier and P. Jenffer, *Ann. Phys. (Paris)* **13**, 37 (1988).
65. R. L. Hoffman, *J. Colloid Interface Sci.* **50**, 228 (1975).
66. W. Rose and R. W. Heinz, *J. Colloid Interface Sci.* **17**, 39 (1962).
67. L. Tanner, *J. Phys. D* **12**, 1473 (1979).
68. R. G. Cox, *J. Fluid Mech.* **168**, 169 (1986).
69. Y. D. Shikhmurzaev, *Int. J. Multiphase Flow* **19**, 589 (1993).
70. N. G. Hadjiconstantinou, *Phys. Rev. E* **59**, 2475 (1999).
71. J. Koplik and J. R. Banavar, *Phys. Rev. Lett.* **84**, 4401 (2000).
72. J. Koplik, J. R. Banavar and J. F. Willemsen, *Phys. Rev. Lett.* **60**, 1282 (1988).
73. T. Qian, X.-P. Wang and P. Sheng, *Phys. Rev. E* **68**, 016306 (2003).
74. P. A. Thompson and M. O. Robbins, *Phys. Rev. Lett.* **63**, 766 (1989).
75. P. Sheng and M. Zhou, *Phys. Rev. A* **45**, 5694 (1992).
76. T. D. Blake and J. M. Haynes, *J. Colloid Interface Sci.* **30**, 421 (1969).
77. N. G. Hadjiconstantinou and A. T. Patera, *Int. J. Numer. Mech. Fluids* **34**, 711 (2000).
78. E. Huh and L. E. Scriven, *J. Colloid Interface Sci.* **35**, 85 (1971).
79. A. J. Briant, A. J. Wagner and J. M. Yeomans, *Phys. Rev. E* **69**, 031602 (2004).
80. G. Batchelor, *An Introduction to Fluid Dynamics*. Cambridge Univ. Press, Cambridge (1970).
81. C. Navier, *Mem. Acad. Roy. Sci. Inst. France* **1**, 414 (1823).
82. J. Maxwell, *Philos. Trans. Roy. Soc. I Appendix* (1879).
83. N. Churaev, V. Sobolev and A. Somov, *J. Colloid Interface Sci.* **97**, 574 (1984).

84. V. S. J. Craig, C. Neto and D. R. M. Williams, *Phys. Rev. Lett.* **87**, 054504 (2001).

85. R. Pit, H. Hervet and L. Leger, *Phys. Rev. Lett.* **85**, 980 (2000).

86. D. Tretheway and C. Meinhart, *Phys. Fluids* **14**, L9 (2002).

87. Y. Zhu and S. Granick, *Phys. Rev. Lett.* **88**, 106102 (2002).

88. J.-L. Barrat and L. Bocquet, *Phys. Rev. Lett.* **82**, 4671 (1999).

89. M. Cieplak, J. Koplik and J. R. Banavar, *Phys. Rev. Lett.* **86**, 803 (2001).

90. S. Gupta, H. Cochran and P. Cummings, *J. Chem. Phys.* **107**, 10316 (1997).

91. M. Sun and C. Ebner, *Phys. Rev. Lett.* **69**, 3491 (1992).

92. X. Nie, G. Doolen and S. Chen, *J. Stat. Phys.* **107**, 279 (2002).

93. C. Lim, C. Shu, D. Niu and Y. Chew, *Phys. Fluids* **14**, 2299 (2002).

94. X. He, Q. Zou, L.-S. Luo and M. Dembo, *J. Stat. Phys.* **87**, 115 (1997).

95. M. J. P. Nijmeijer, C. Bruin, A. F. Bakker and J. M. J. van Leeuwen, *Phys. Rev. A* **42**, 6052 (1990).

Contact Angle, Wettability and Adhesion, Vol. 4, pp. 29–41
Ed. K.L. Mittal
© VSP 2006

Contact line motion: Hydrodynamical or molecular process?

G. CALLEGARI,[1,2,*] A. CALVO[1] and J. P. HULIN[3]

[1] *Grupo de Medios Porosos, Facultad de Ingeniería, Universidad de Buenos Aires, Paseo Colón 850, Capital Federal, 1063, Argentina*
[2] *Textile Research Institute (TRI), 601 Prospect Avenue, Princeton, NJ 08536, USA*
[3] *Laboratoire FAST, Bâtiment 502, Campus Paris-Sud, 91405 Orsay, France*

Abstract—An experimental study of the constant velocity displacement of various water/glycerol solutions by air in poly(vinyl chloride) (PVC) capillary tubes is reported. This topic is of particular interest in relation to dewetting processes on surfaces covered by a liquid film. More specifically, variations of the dynamic contact angle with velocity and their relation to the physicochemical properties of the systems studied are investigated. These results and those of other authors are analyzed in the framework of both hydrodynamical and molecular approaches of the dynamic contact-angle problem. These comparisons indicate that either the molecular or the viscous dissipation mechanism may be dominant, depending on the system studied. These results are used to suggest explanations for apparent discrepancies between dewetting velocity measurements in different systems previously reported by the authors.

Keywords: Dynamic contact angle; wetting; dewetting; viscous dissipation; molecular dissipation.

1. INTRODUCTION

When a volume of liquid is placed in contact with a solid substrate, capillary forces induce motions of the contact line leading to the build-up of a film (complete wetting condition) or of a drop with a finite contact angle (partial wetting condition).

When a partially-wetting liquid is forced to spread on a solid, the film is unstable and tends to retract. This phenomenon, known as dewetting, has attracted much attention in the last 15 years [1–17]. On the one hand, there are countless applications for which dewetting is either necessary (drying of surfaces, detergency) or harmful (coatings). On the other hand, dewetting is closely related to the motion of a contact line with a finite contact angle, which is still an open problem, in spite of many attempts during the last four decades [18–44].

---

*To whom correspondence should be addressed, at TRI. Tel.: (1-609) 430-4816;
e-mail: gcallegari@tri.princeton.org

Dewetting processes induced by nucleation have been studied in different systems and configurations for relatively thick liquid films (from micrometers to hundreds of micrometers) [1–14]. Besides, many studies have been devoted lately to the rupture of thin and ultra-thin films (e.g. spinodal decomposition [15–17]).

Redon and co-workers were the first to study systematically the dependence of the dewetting velocity on the different parameters of the system, for macroscopic films, using different alkanes and poly(dimethylsiloxane) (PDMS) on silanized silicon wafers [2]. These authors observed that a bump built up between the receding contact line and the liquid film: the latter remained static when the dewetting velocity V_d was constant with time. For viscous liquids (viscosity μ and surface tension γ) and small static contact angles (up to 50°) the non-dimensional velocity, $Ca_d = \mu V_d/\gamma$, scales as the cube of the static contact angle θ_s, while the prefactor varies weakly with the system studied.

These results were explained [2, 3, 8, 11] by using a simple model based on the lubrication approximation and assuming a small contact angle and separate balances of the driving capillary forces and viscous forces at the two edges of the bump. The bump is assumed to have a circular profile so that the dynamic contact angle between the liquid and solid surfaces at both edges is equal. Subsequent experiments [4, 12, 14] indicate that this assumption is not always valid. However, the following variation of Ca_d predicted by this theory is in agreement with the experimental results of Refs. [2, 4].

$$Ca_d = \frac{\theta_s^3}{12\sqrt{2}\,\ln} \tag{1}$$

The factor $\ln = \ln(K/l_s)$ (ln is the natural logarithm) results from the divergence of the viscous dissipation at the contact line. Here l_s is a microscopic length of a few tens of nanometers and K is the characteristic global size of the bump.

Recent experimental results obtained by the authors [12], with an experimental setup similar to that of Redon and co-workers [1, 2] and Andrieu and co-workers [4, 8], but using liquids and solid surfaces displaying static contact angles up to 90° (much higher than in the other works) disagree with these predictions. Comparison of data from Refs. [2] and [12] can be seen in Fig. 1, where the corresponding values of Ca_d are plotted as a function of θ_s^3.

While data from Ref. [2] display a linear dependence, this is not the case for data from Ref. [12]. Note that both kinds of experiments were performed in similar conditions: planar films were extended on the solid surface and spontaneous dewetting was initiated by injecting air in the center.

The objective of the present paper is to attempt to explain the discrepancy between the variations in the behavior of the dewetting velocities of planar films displayed in Fig. 1: this discrepancy suggests that different mechanisms are operative in these different sets of experiments. The fact that the results of Redon and co-workers can be explained by the hydrodynamical model [2, 3, 8, 11], while this is not the case for those of Ref. [12], led us to investigate the influence of viscous

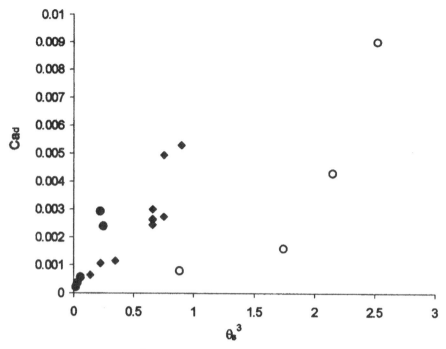

Figure 1. Variations of Ca_d vs. θ_s^3 for alkanes (●), PDMS with different molecular masses on silanized silicon wafers (♦) from Ref. [2], and (○) glycerol/water solutions on PVC surfaces from Ref. [12].

and molecular dissipations in the experimental systems of Ref. [12] (films of water/glycerol solutions dewetting on PVC surfaces). In order to estimate the relative importance of these two dissipation mechanisms, experiments in which the dynamic contact angle and the (controlled) velocity of the meniscus could be simultaneously measured were performed using the same solid surface (PVC) and liquids (water/glycerol solutions) as in the dewetting experiments of Ref. [12]. The results are compared to the predictions of both hydrodynamical and molecular-dissipation models. This comparison suggests an explanation for the discrepancy shown in Fig. 1.

2. THEORETICAL MODELS

The first approach is based on the hydrodynamical models already discussed above [19–22, 24–26, 28, 30–32]. The theory of this type which applies to the broadest range of systems has been developed by Cox [31] and introduces a general slip condition in the microscopic zone. The following relation between Ca, θ_s and θ_d is then obtained:

$$F(\theta_d, M) = F(\theta_s, M) + Ca \ln(K/l_s), \tag{2}$$

where θ_d and θ_s are, respectively, as above, the dynamic and static contact angles, M is the ratio between the viscosities of the displacing and displaced fluids, K is a characteristic macroscopic distance and l_s the microscopic slip distance. For $M = 0$ (as in the present experiments), $F(\theta_d)$ can be expressed as:

$$F(\theta_d) = \int_0^{\theta_d} \frac{((\pi - x)^2 + \sin^2 x)(x - \sin x \cos x)}{2 \sin x[(\pi - x)^2 - \sin^2 x]} \, dx. \tag{3}$$

An alternative approach entails molecular models [18, 23, 27, 33, 35]. Specifically, Blake's model [18, 33] is based on the loss or gain of energy during the adsorption or desorption of molecules onto or from the solid surface near the contact line. The motion of the latter is determined by the combination of capillary forces and thermal activation. This results in the following relation between the velocity (V) of the contact line, and the static (θ_s) and dynamic (θ_d) contact angles:

$$V(\theta_d) = 2\lambda \left(\frac{k_B T}{\mu V_L}\right) e^{(-\Delta G_s / N k_B T)} \sinh \left(\frac{\gamma(\cos \theta_d - \cos \theta_s)\lambda^2}{2 k_B T}\right), \tag{4}$$

where λ is the length of molecular jumps (or the distance between adsorption sites on the solid surface), k_B is the Boltzmann constant, μ is the viscosity of the liquid, V_L is a typical molecular volume (for simple liquids V_L is taken equal to the volume of one molecule and for polymers to the volume of a monomer), N is the Avogadro number and T is the temperature in K.

Equation (4) is obtained by assuming that the total energy has two contributions: ΔG_s and ΔG_v. ΔG_s takes into account for solid–liquid molecular interactions and ΔG_v corresponds to fluid–fluid molecular interactions (in the same phase). ΔG_v is implicitly included in equation (4) through the viscosity μ.

3. EXPERIMENTAL SETUP AND MEASUREMENTS

The solid (PVC) and liquids (aqueous solutions of glycerol) used in the experiments were the same as in Ref. [12]. The experimental setup used to measure the dynamic contact angle for a liquid slug receding at a constant velocity inside a capillary tube is displayed in Fig. 2.

The PVC capillary tube (radius 1 mm, length 400 mm) was connected to a constant flow rate pump allowing for a constant velocity motion of the liquid–air interface. Images of the receding interface of the liquid slug (air displacing the liquid) were captured by a CCD sensor and digitized. A 3-mm-wide field of view was obtained with a 55 mm lens. The capillary tube was inserted inside a glass channel with parallel walls, filled with glycerol to match refractive indexes and to obtain images with a minimal distortion. A transparent ruler (100 μm for each division) located immediately behind the capillary tube allowed to determine the location of the interface as a function of time.

The measured velocity remained constant with time: its value was such that capillary force was larger than the viscous one (Ca $= \mu V / \gamma \ll 1$), which was

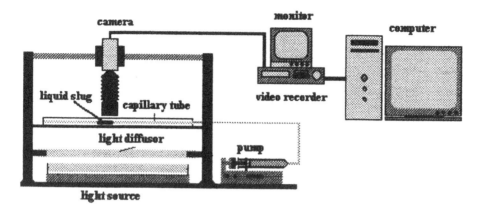

Figure 2. Experimental setup for measuring dynamic contact angles and velocity of air–liquid interfaces.

Table 1.
Physicochemical parameters of the systems studied

Liquid	ρ (g/cm^3)	μ (cP)	θ_s (degrees)	γ (mN/m)
Pure water	1	1	55	71.5
76% solution	1.22	30	74	69
88% solution	1.23	119	78	67
Pure glycerol	1.26	1050	80	64.5

ρ, liquid density; μ, viscosity; θ_s static contact angle; γ, surface tension.

larger than the inertial term (Re $= \rho LV/\mu \ll 1$ (Re is the Reynolds number, ρ is the density of the liquid and L a characteristic length, here the radius of the capillary tube).

Table 1 shows the measured values of the physicochemical parameters of the liquids used in the experiments. θ_s is the asymptotic value of θ_d in the limit of low capillary numbers Ca $< 10^{-6}$ [40, 43]. Both the fluid solutions and the solid surface material (PVC) were the same as in the dewetting experiments [12].

Images from different experiments in which the air (to the left of each image) displaces the glycerol are displayed in Fig. 3. Each digitized image corresponds to a different interface velocity, such that the dynamic contact angle θ_d is different. As expected, the dynamic contact angle (defined in the liquid phase) decreases when the velocity of the interface increases (from left to right in Fig. 3).

The dynamic contact angle is determined by fitting the interface with a circular arc of radius R, so that:

$$\cos \theta_d = \frac{D}{2R}, \tag{5}$$

in which D is the tube diameter, which is determined independently. Measurements of the contact angle were performed at 5 different times during each experiment.

All these values are within ±2°. This represents also the typical error in the measurement of the dynamic contact angle.

Experimental variations of the dynamic contact angle as a function of the capillary number are displayed in Fig. 4 for several liquids of different viscosities. In all cases, the contact angle decreases with the capillary number.

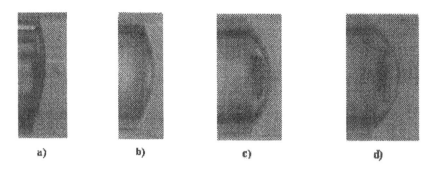

Figure 3. Air (on the left of each image) displacing glycerol. Velocity increases from left to right images: (a) $Ca = 7 \times 10^{-5}$, $\theta_d = 68°$; (b) $Ca = 2.7 \times 10^{-4}$, $\theta_d = 60°$; (c) $Ca = 2.7 \times 10^{-3}$, $\theta_d = 43°$; (d) $Ca = 5 \times 10^{-3}$, $\theta_d = 37°$.

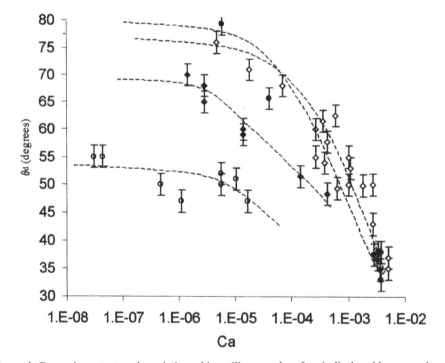

Figure 4. Dynamic contact angle variation with capillary number, for air displaced by: pure glycerol (•); 88% water/glycerol solution (◊), 76% water/glycerol solution (♦) and pure water (o). Dotted lines are drawn as an aid to the eye.

4. DISCUSSION OF EXPERIMENTAL RESULTS

These results have first been compared to the predictions of the hydrodynamical model [31] leading to equation (2). This relation has been tested on data points corresponding to air displacing pure glycerol and the 88% water/glycerol solution ($M = 0$). Both data sets are expected to display similar variations, since θ_s has almost the same value.

Figure 5 indicates a poor agreement between the experimental data and the predictions of the hydrodynamical model. ($\ln(K/l_s)$) has been taken equal to 12, which is a typical value for such systems [31], but this choice does not influence significantly the variation). For clarity, data obtained for water and 76% glycerol/water solution are not included, but they display a similar disagreement with the model. Similar results were reported by Fermigier and Jenffer [43] for a liquid displaced by another liquid and by Hayes and Ralston [44] in water/glycerol solutions over poly (ethylene terephthalate) (PET), in both advancing and receding configurations.

Next, the molecular model represented by equation (4) is applied to the same sets of data in Fig. 6 in which $\gamma(\cos\theta_d - \cos\theta_s)$ is plotted as a function of log (μV). The fit is clearly improved provided a different set of values of λ and ΔG_s is used in both the low and high velocity regimes as suggested in Refs. [18, 36]. At low velocities, molecules of the displaced (resp. displacing) fluid have enough time to

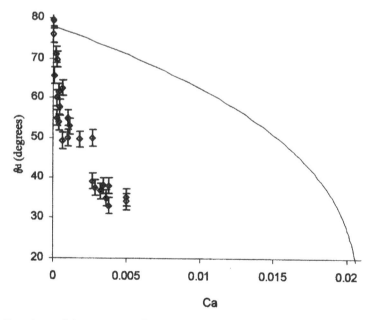

Figure 5. Experimental data corresponding to air displacing pure glycerol and 88% glycerol/water solution fitted with the hydrodynamical model (equation (2)) ($M = 0$). The data points and the corresponding symbols are the same as those displayed in Fig. 4 for the same solutions.

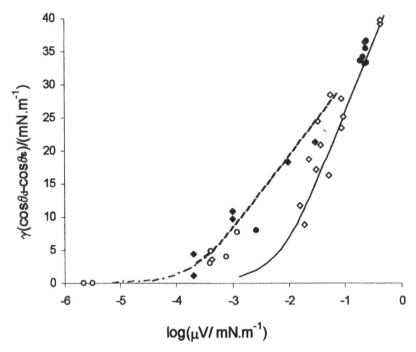

Figure 6. Experimental data fitted with the molecular model (equation (4)). The same symbols as used in Fig. 4. Dotted line: $\lambda = 12.9 \times 10^{-8}$ cm, $\Delta G_s = 31.5$ kJ mol^{-1}; full line: $\lambda = 9.5 \times 10^{-8}$ cm, $\Delta G_s = 21.5$ kJ mol^{-1}.

desorb (resp. adsorb) on the solid surface. On the contrary, for large velocities, there is not enough time for desorption and the motion of the displacing fluid takes place over a few molecular layers of the displaced fluid [33].

The values of ΔG_s and λ for low (dashed line) and high (continuous line) velocities estimated from the data of Fig. 6 are listed in Table 2 (V_L is taken equal to the molecular volume, for glycerol $V_L = 12 \times 10^{-23}$ cm^3). The values estimated from the experimental data of various authors are listed for comparison [33].

The following important features appear from the data of Table 2:

(i) The distance λ between adsorption sites varies from 0.46 to 1.4 nm and is, as expected, of the order of molecular distances.

(ii) ΔG_s varies between 5.6 and 31.5 kJ mol^{-1}. This rather broad range of values is not surprising since the adsorption–desorption process is expected to depend significantly on the fluids involved as well as on the substrate material.

(iii) At low contact line velocities, the value of ΔG_s from the present work (31.5 kJ mol^{-1}) is very close to the one given in Ref. [33] (30 kJ mol^{-1}) for water/glycerol solutions on PET surface, a very similar system. Both sets of values are larger than those obtained for glycerol/water solutions in air on glass

Table 2.
Fitting parameters of Blake's molecular model (ΔG_s and λ) for different systems obtained in this and previous works

System	ΔG_s (kJ mol^{-1})	λ (nm)	Ref.
PDMS–air–silanized glass			
$10^{-3} < \mu V < 10$	8.8	0.8	[40]
Polyester resin solutions in styrene–air–glass			
$10^{-2} < \mu V < 10^2$	10	0.69	[38]
Polystyrene/xylene solutions–air–glass			
$10^{-4} < \mu V < 10^2$	13	1	[38]
Glycerol/water solutions–air–glass			
$10^{-4} < \mu V < 10$	17	0.66	[38]
Glycerol/water solutions–air–PET			
$10^{-3} < \mu V < 10^2$	22	0.92	[41]
Glycerol/water solutions–air–PET			
$10^{-4} < \mu V < 10^{-2}$	30	1.4	[33]
$10^{-2} < \mu V < 10^2$	5.6	0.46	
Air–glycerol/water solutions–PVC			
$10^{-6} < \mu V < 10^{-2}$	31.5	1.29	Present work
$10^{-2} < \mu V < 1$	21.5	0.95	

In all cases, the first fluid is the displacing one, and the second the displaced one. Parameters for the first six systems were found in Ref. [33], and parameters for the 7th one were obtained in the present work; the original data are reported in the references listed in the last column.

surfaces (Table 2). The smallest values are obtained for the systems in which the molecular interactions are the weakest.

(iv) For larger contact line velocities, the values obtained on PET and PVC surfaces are, however, quite different: (respectively $\Delta G_s \cong 6$ kJ mol^{-1} and 21.5 kJ mol^{-1}). As pointed above, this difference may be due to the fact that in this regime a few layers of displaced fluid are expected to be present between the wall and the displacing fluid. The characteristics of the motion should then indeed be different in the experiments of Ref. [33] in which the receding fluid was air and in the present experiment in which it was the advancing fluid.

5. DEWETTING PROCESSES AND MOLECULAR MODEL

The molecular model accounts well for the experimental dynamic contact angle values reported in the present work. In the present section, we use this model to predict contact line velocities in other dewetting experiments [12] also performed with water/glycerol solutions on PVC surfaces.

For this purpose, we shall first assume that the relation between the contact line velocity and the static and dynamic contact angles represented by equation (4), and verified experimentally above, remains valid in the dewetting experiments of Ref. [12] in which the dynamic contact angle could not be measured directly.

More specifically, the values of the parameters ΔG_s and λ will be assumed in the following to be the same as those reported in the previous section for the same fluids and surfaces ($\Delta G_s = 21.5\,\text{kJ mol}^{-1}$, $\lambda = 9.5 \times 10^{-8}$ cm). Static contact angles were measured independently in the dewetting experiments of Ref. [12]: their values were very similar to those measured for the same fluids in the experiments discussed above.

In order to relate the dewetting capillary number Ca_d to the static contact angle, a relation between the angles θ_d and θ_s is needed both in the hydrodynamical and in the molecular models. Indeed, in Ref. [11] it is assumed that the bump has a circular profile and that both the dissipation and the driving capillary forces have equal values at the two ends of the bump. This leads to the relation:

$$\theta_d = \frac{\theta_s}{\sqrt{2}}. \tag{6}$$

Equation (6) was found to be valid in the experiments reported in Ref. [2] corresponding to air displacing PDMS on a silanized surface. This is, however, not the case in the dewetting experiments of [12] (air displacing water/glycerol solutions on a PVC surface) for which the bump profile is neither symmetrical nor circular.

However, in the range of capillary number values corresponding to the dewetting velocity, the dynamic contact angle θ_d is observed to be about half the static contact angle θ_s for the fluids and surfaces investigated (Table 1). So, assuming $\theta_d = \theta_s/2$ and $\cos\theta \approx 1 - \theta^2/2$ in equation (4) leads to the following relation between μV_d and the static contact angle θ_s:

$$\mu V_d(\theta_s) \approx 2\lambda \left(\frac{k_B T}{V_L}\right) e^{(-\Delta G_s/Nk_B T)} \sinh\left(\frac{3\gamma\theta_s^2\lambda^2}{16k_B T}\right). \tag{7}$$

This prediction is displayed in Fig. 7 by a continuous line for the values of ΔG_s and λ determined above. The experimental data from Ref. [12] are represented by open symbols.

The agreement obtained in this way between the experimental data and the theory is clearly much better than that obtained by applying the hydrodynamical model to the same set of data (Fig. 1).

We have then applied the same approach to the data of Ref. [2] which, in contrast, agree well with the hydrodynamical model. In this case, the molecular parameters were not reported but those obtained by Blake, by fitting Hoffman's data [40] for a similar system (PDMS–air–silanized glass) can be used as an estimate ($\Delta G_s = 8.8\,\text{kJ mol}^{-1}$ and $\lambda = 0.8$ nm). The corresponding variations of μV_d as a function of $\gamma\theta_s^2$ predicted from equation (7) are also plotted in Fig. 7 (dashed curve at left) together with the experimental data of Ref. [2] (solid circles).

In this case, the values of Ca_d predicted by the model are up to one order of magnitude larger than the experimental values. This indicates that in this case the molecular dissipation resulting from the model is only a small fraction of the total dissipation so that for these systems viscous dissipation is dominant. Thus, it is not

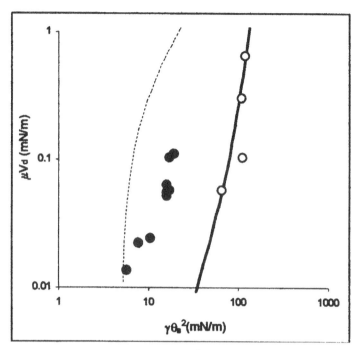

Figure 7. Variations of μV_d as a function of $\gamma\theta_s^2$. Experimental results from Ref. [12] (o) and from Ref. [2] (•). Theoretical predictions from equation (7) for these experimental systems: $\Delta G_s = 21.5$ kJ mol^{-1}, $\lambda = 9.5 \times 10^{-10}$ m (solid line) and $\Delta G_s = 8.8$ kJ mol^{-1}, $\lambda = 8 \times 10^{-10}$ m (dashed line), respectively.

surprising that the data of Ref. [2] are in good agreement with the predictions of the hydrodynamical model ($Ca_d \propto \theta_s^3$) based on the hypothesis that viscous dissipation is the only mechanism of dissipation. In the systems studied in the present work and in Ref. [12] (water/glycerol solutions on PVC), in contrast, molecular dissipation is at least of the same order as viscous dissipation.

These differences reflect essentially the different values of the parameter ΔG_s for the two systems studied. While the orders of magnitude of the fluid viscosities are similar, the parameter ΔG_s is 3-times higher ($\Delta G_s = 21.5$ kJ mol^{-1}) for the system for which molecular dissipation is important (glycerol/water on PVC) than for that for which it is negligible ($\Delta G_s = 8.8$ kJ mol^{-1} for PDMS on silanized glass). The values of the characteristic length λ are, in contrast, comparable (9.5 and 8×10^{-10} m, respectively).

6. CONCLUSIONS

The present work has allowed to characterize experimental dynamic contact-angle variations for water/glycerol solutions displaced by air on PVC surfaces. It has been demonstrated that these variations are much better accounted for by a molecular model based on adsorption and desorption processes in the vicinity of the contact line than by hydrodynamical models, and the results are in agreement with other works on similar systems [44, 45]. The latter paper, published during the review process of the present one, also concludes, with a different approach, the need to combine hydrodynamical and molecular dynamical effects to explain the dynamic contact-angle behaviour. The values of the molecular parameters determined by fitting the variations of the dynamic contact angle to a molecular model [33] also agree with those obtained by other authors in similar systems. These conclusions are, however, not valid for all systems: for instance, hydrodynamical effects related to viscous dissipation may be dominant in the experiments of Ref. [2] (PDMS on silanized surfaces). The dominance of dissipation mechanism is largely determined by the solid–liquid interaction parameter ΔG_s: molecular dissipation should be dominant when the value of this parameter is large.

The same approach has been used to study the variations of the dewetting velocity with the static contact angles and accounts for discrepancies between the results obtained for different systems. Hydrodynamical models [11] account well for dewetting velocity variations in systems for which ΔG_s is small so that molecular dissipation can be neglected compared to viscous effects [2]. In contrast, in systems with larger values of ΔG_s (water/glycerol solutions on PVC surfaces), molecular dissipation is at least of the same order of magnitude as viscous effects. Then, variations of the dewetting velocity with the contact angle are much better predicted by the molecular model.

REFERENCES

1. F. Brochard-Wyart, C. Redon and F. Rondelez, *C. R. Acad. Sci. II* **306**, 1143 (1988).
2. C. Redon, F. Brochard-Wyart and F. Rondelez, *Phys. Rev. Lett.* **66**, 715–718 (1991).
3. F. Brochard-Wyart, P. Martin and C. Redon, *Langmuir* **9**, 3682 (1993).
4. C. Andrieu, C. Sykes and F. Brochard-Wyart, *Langmuir* **10**, 2077 (1994).
5. C. Redon, J. B. Brzoska and F. Brochard-Wyart, *Macromolecules* **27**, 468 (1994).
6. P. Martin, A. Buguin and F. Brochard-Wyart, *Eur. Phys. Lett.* **28**, 421–426 (1994).
7. G. Debregeas, P. Martin and F. Brochard-Wyart, *Phys. Rev. Lett.* **75**, 3886 (1995).
8. C. Andrieu, C. Sykes and F. Brochard-Wyart, *J. Adhesion* **58**, 15–24 (1996).
9. L. Bacri, G. Debregeas and F. Brochard-Wyart, *Langmuir* **12**, 6708 (1996).
10. A. Buguin, V. Vovelle and F. Brochard-Wyart, *Phys. Rev. Lett.* **83**, 1183 (1999).
11. F. Brochard-Wyart and P. G. de Gennes, *Adv. Colloid Interface Sci.* **39**, 1 (1992).
12. G. Callegari, A. Calvo and J. P. Hulin, *Colloids Surfaces A* **206**, 167 (2002).
13. G. Callegari, A. Calvo, J. P. Hulin and F. Brochard-Wyart, *Langmuir* **18**, 4795 (2002).
14. G. Callegari, A. Calvo and J. P. Hulin, *Eur. Phys. J. E* **16**, 283 (2005).
15. G. Reiter, *Phys. Rev. Lett.* **68**, 75 (1992).
16. G. Reiter, *Langmuir* **9**, 1344 (1993).
17. S. Herminghaus, R. Seeman and K. Jacobs, *Phys. Rev. Lett.* **89**, 5 (2002).
18. T. D. Blake and M. J. Haynes, *J. Colloid Interface Sci.* **30**, 421 (1969).
19. C. Huh and L. E. Scriven, *J. Colloid Interface Sci.* **35**, 85 (1971).
20. O. V. Voinov, *J. Fluid Dynam.* **11**, 714 (1976).
21. L. M. Hocking, *J. Fluid Mech.* **79**, 209 (1977).

22. C. Huh and S. G. Mason, *J. Fluid Mech.* **81**, 401 (1977).
23. E. Ruckenstein and C. S. Dunn, *J. Colloid Interface Sci.* **59**, 135 (1977).
24. H. P. Greenspan, *J. Fluid Mech.* **84**, 125 (1978).
25. L. H. Tanner, *J. Phys. D* **12**, 1473 (1979).
26. E. B. Dussan V., *Ann. Rev. Fluid Mech.* **11**, 371 (1979).
27. T. D. Blake and K. J. Ruschak, *Nature* **282**, 489 (1979).
28. L. M. Hocking and A. D. Rivers, *J. Fluid Mech.* **121**, 425 (1982).
29. P. Neogi and C. A. Miller, *J. Colloid Interface Sci.* **92**, 338 (1983).
30. P. G. de Gennes, *Rev. Mod. Phys.* **57**, 827 (1985).
31. R. G. Cox, *J. Fluid Mech.* **168**, 169–194 (1986).
32. E. B. Dussan V., E. Ramé and S. Garoff, *J. Fluid Mech.* **230**, 97 (1991).
33. T. D. Blake, in: *Wettability*, J. G. Berg (Ed.), pp. 251–309, Marcel Dekker, New York, NY (1993).
34. Y. D. Shikhmurzaev, *Int. J. Multiphase Flow* **19**, 589 (1993).
35. E. Ruckenstein, *J. Colloid Interface Sci.* **170**, 284 (1995).
36. Y. D. Shikhmurzaev, *J. Fluid Mech.* **359**, 313 (1998).
37. T. D. Blake, M. Bracke and Y. D. Shikhmurzaev, *Phys. Fluids* **11**, 1995 (1999).
38. G. Inverarity, Ph.D. Thesis, University of Manchester, Manchester (1969).
39. G. Inverarity, *Br. Polym. J.* **1**, 254 (1969).
40. R. L. Hoffman, *J. Colloid Interface Sci.* **50**, 228 (1975).
41. R. Burley and B. S. Kennedy, *Br. Polymer J.* **8**, 140 (1976).
42. C. G. Ngan and E. B. Dussan V., *J. Fluid Mech.* **209**, 191 (1989).
43. M. Fermigier and J. Jenffer, *J. Colloid Interface Sci.* **146**, 226 (1991).
44. R. Hayes and J. Ralston, *J. Colloid Interface Sci.* **159**, 429 (1993).
45. S. Ranabothu, C. Karnezis and L. Dai, *J. Colloid Interface Sci.* **288**, 213 (2005).

Contact Angle, Wettability and Adhesion, Vol. 4, pp. 43–59
Ed. K.L. Mittal
© VSP 2006

The detailed structure of a perturbed wetting triple line on modified PTFE

CATHERINE COMBELLAS,[1] ADRIEN FUCHS,[1] FRÉDÉRIC KANOUFI,[1,*]
and MARTIN E. R. SHANAHAN[2]

[1]*Laboratoire Environnement et Chimie Analytique, Ecole Supérieure de Physique et Chimie,
Industrielles, 10 rue Vauquelin, 75231 Paris Cedex 05, France*
[2]*Centre des Matériaux de Grande Diffusion, Ecole des Mines d'Alès, 6 avenue de Clavières,
30319 Alès Cedex, France*

Abstract—The essential form of an initially straight wetting triple line perturbed by the presence of
a (higher surface free energy) "defect" on the solid surface has been recognised for a long time, and
it corresponds to a logarithmically decaying form. However, less attention has been paid to the behaviour of the triple line within the domain of the defect. This was actually studied a few years ago
from a theoretical viewpoint, leading to the prediction of an inversion of curvature. Recent experimental work has been concerned with the electrochemical treatment of PTFE, leading to small
etched areas of higher wettability with typical widths of 100–300 μm. Wetting experiments have
been carried out on such solids and the results confirm the general conclusion of inverted curvature
of the triple line in the treated zones. However, the "excess wettability" in the treated zones, as
evaluated experimentally, was found to be greater than predicted theoretically. Possible causes are
discussed.

Keywords: Defect; electrochemical treatment; fine structure; surface modification; wetting line;
PTFE.

1. INTRODUCTION

Surface treatments of polymers have many uses, such as the enhancement of aes-
thetic properties (optical or tactile feeling), in increasing wear resistance, in decreasing friction coefficient, or, on the other hand, in increasing adhesion when
bonding the polymer in question to another material. Fluoropolymers, in particular, have innately low surface free energies (or tensions), which is a useful feature
when low friction is required but is a bane for bonding.

Electrochemical techniques have been developed to modify the surface free energy and, therefore, the wettability of polymeric solid surfaces. Scanning Electro-

---

*To whom correspondence should be addressed. Tel.: (33-1) 4079-4526; Fax: (33-1) 4079-4425;
e-mail: frederic.kanoufi@espci.fr

chemical Microscopy (SECM) is one of these promising techniques [1]. Therefore, we have carried out surface treatment of poly(tetrafluoroethylene) (PTFE)
using SECM. This method is capable of producing very small areas (typical dimensions: 100 μm) of treated surface presenting higher values of surface free energy than their surrounding zones.

Classical wetting techniques, such as the measurement of contact angles of sessile drops of probe liquids, are very sensitive to variations in surface properties,
yet are only applicable to regions of typical dimensions of 1 mm or greater. As a
consequence, such an approach is precluded as a method for the characterization
of the small SECM-treated surface areas mentioned above.

However, in recent years developments in wetting theory have considered the
deformed shape of a wetting triple line when in the proximity of a "defect", or
small region presenting either a higher or a lower (higher in the present case) surface free energy and, therefore, wettability, as defined by contact angle [2–6],
than the neighboring solid region. The "defect" can have dimensions of about 100
μm, and yet produce noticeable and quantifiable perturbations to the triple line
shape. Thus, the analysis of observed distorted wetting fronts lends itself well to
the characterization of solid surfaces modified on the size scale discussed above.

In this work we have compared the expression for a deformed triple line, obtained recently, to experimental data and have extended our analysis to the case of
diffuse defects that are generally obtained by chemical or optical techniques. Indeed, chemical or optical techniques rarely produce local surface modification
with sharp boundaries, but rather generate modifications whose edges are diffuse.
Thus, as a continuation of earlier work [7], we consider here the wetting properties of PTFE surface, locally marked by higher surface free energy regions of
small dimensions, obtained by targeted chemical modification.

2. THEORETICAL EQUATIONS OF TRIPLE LINE

The triple wetting line for the liquid/homogeneous solid/vapour (V) system obeys
Young's equation at equilibrium [8]:

$$\gamma_{SV} = \gamma_{SL} + \gamma \cos \theta \tag{1}$$

where γ_{SV}, γ_{SL} and γ are, respectively, solid/vapour, solid/liquid and liq-
uid/vapour interfacial tensions, and θ is the characteristic equilibrium contact angle of the system. However, if the solid surface presents a heterogeneity also
called "defect" (and, thus, is no longer entirely homogeneous) in the immediate
vicinity of the contact line, which we shall assume to represent an increase in wettability, or a decrease in local contact angle, a deformation of the triple line will
occur: the liquid front will protrude to cover at least part of the defect and gently
return to its unperturbed level, following a logarithmic law, as distance from the
defect increases (see Fig. 1). The triple line is then "pinned" on the defect and yet

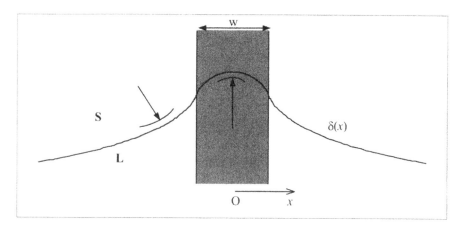

Figure 1. Schematic representation of liquid/solid/vapour triple line perturbed by solid surface energy heterogeneity of width w.

is "pulled" by the far field of the unperturbed contact line. This "elasticity" effect of the triple line has been studied both theoretically and experimentally by several workers [2, 4–6, 9]. Using the nomenclature of Fig. 1, the basic form of the triple line can be described by the protrusion distance, $\delta(x)$, as a function of distance $|x|$ from the centre of the defect [2].

$$\delta(x) \approx \frac{f}{\pi\gamma\theta^2}\ln\frac{r_0}{|x|}, \quad |x| > \frac{w}{2} \tag{2}$$

where f is the "pull force" of the heterogeneity, w is heterogeneity width and r_0 is a macroscopic cut-off distance. θ represents the equilibrium contact angle on the homogeneous solid, outside of the defect. The pull force is defined as the force due to variation in the surface free energy ε at the heterogeneity:

$$\varepsilon(x,y) = (\tilde{\gamma}_{SV}(x,y) - \tilde{\gamma}_{SL}(x,y)) - (\gamma_{SV} - \gamma_{SL}) = \gamma(\cos\tilde{\theta}(x,y) - \cos\theta) \tag{3}$$

where $\tilde{\gamma}_{SV}$, $\tilde{\gamma}_{SL}$ and $\tilde{\theta}$ refer to interfacial tensions and contact angle on the heterogeneous material, $|x| < w/2$, and γ_{SV}, γ_{SL} and θ are the same quantities for the virgin homogeneous solid, $|x| > w/2$.

The pull force, f, is then expressed as:

$$f = \int \varepsilon(x,\delta(x))dx \tag{4}$$

Equation (2) was derived using energy minimization principles and Fourier transform methods, assuming an initially straight triple line [2].

In a different, but related, approach, the meniscus shape of a slightly deformed axisymmetric drop, of radius r_0, was derived using the Fourier series analysis [6]. Briefly, in cylindrical coordinates (r is the radial coordinate and φ the azimuthal one), the local deformation of the contact line is given by:

$$\delta(\varphi) = \frac{r_0}{\gamma\theta^2}\left\{-B_0(r_0)/2 + \sum_{n=2}^{\infty}\frac{A_n(r_0)\sin n\varphi + B_n(r_0)\cos n\varphi}{n-1}\right\} \quad (5)$$

where θ is the unperturbed contact angle and the coefficients A_i and B_i are obtained from the Fourier form of ε near the drop periphery:

$$\varepsilon(r_0,\varphi) = B_0(r_0)/2 + \sum_{m=1}^{\infty}A_m(r_0)\sin m\varphi + B_m(r_0)\cos m\varphi \quad (6)$$

This analysis leads to a considerably more complicated expression for $\delta(x)$, but intrinsically also allows to account for the behaviour within the defect, $|x| < w/2$, which equation (2) cannot.

2.1. Square profile (sharp stripe defect)

Let us consider a single heterogeneity near the periphery of the drop such that ε is expressed by a square function:

$$\varepsilon(r_0,\varphi) = \begin{cases} 0 & -\pi < \varphi < -\chi \\ \varepsilon_0 & -\chi < \varphi < \chi \\ 0 & \chi < \varphi < \pi \end{cases} \quad (7)$$

where ε_0 corresponds to the gain in surface free energy introduced by the heterogeneity.

The Fourier coefficients obtained from (7) inserted in (5) lead to a complicated, yet straightforward, expression for the contact line deformation:

$$\delta(\varphi) = \frac{r_0\varepsilon_0}{\pi\gamma\theta^2}\left\{-\chi + 2\sum_{n=1}^{\infty}\frac{\sin(n\chi)}{n(n-1)}\cos(n\varphi)\right\} \quad (8)$$

It may be simplified with standard trigonometric rules by using the following identities:

$$\sum_{n>0}\frac{\cos(n\alpha)}{n} = \ln\left(\frac{1}{2\sin\frac{\alpha}{2}}\right) \quad (9)$$

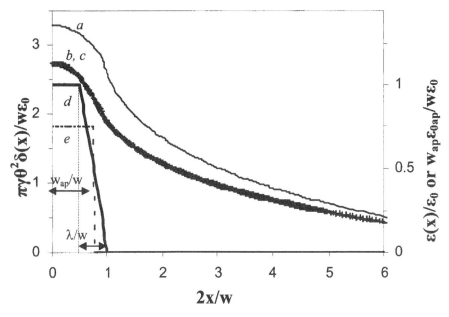

Figure 2. (a, b, c) Triple line profiles for different defect types. (a) Triple line deformation according to equation (11) for a sharp defect of width w and strength $w\varepsilon_0$. (b) Triple line deformation according to equation (15) for the diffuse defect profile (d), with $\lambda = 0.25$ w and strength $w\varepsilon_0$. (c) best fit of the front (b) by the sharp defect model according to equation (11) with the values of $w_{ap} = 0.75$ w and $w_{ap}\varepsilon_{0ap} = 0.75$ $w\varepsilon_0$. On the left axis, unit for the triple line profiles (a, b, c) obtained for $w\varepsilon_0/\gamma\theta^2 = 1$. On the right axis, unit for the defect surface free energy profiles (d, e).

$$\sum_{n>0} \frac{\sin(n\alpha)}{n} = \frac{1}{2}(\pi - \alpha) \qquad (10)$$

and allowing the radius of the curved triple line of the sessile drop, r_0, to tend to infinity [10], we obtain the following basic equation for the perturbed contact line:

$$\delta(x) = \frac{w\varepsilon_0}{2\pi\gamma\theta^2}\left\{ (x^* + 1)\ln\left(\frac{1}{x^* + 1}\right) - (x^* - 1)\ln\left(\frac{1}{|x^* - 1|}\right) \right\} + C \qquad (11)$$

where $x^* = 2|x|/w$ and C is a constant. The triple line deformation obtained from equation (11) for a defect of width w and strength $f = w\varepsilon_0$ with $w\varepsilon_0/\gamma\theta^2 = 1$ is shown schematically in Fig. 1 and as curve a in Fig. 2.

An interesting consequence of equation (11) is that the triple line is predicted to be concave, with respect to the liquid phase, outside of the zone occupied by the defect, in accordance with equation (2) and yet is convex within the defect. Equation (2) does not consider the latter aspect.

2.2. Trapezoidal profile (diffuse defect)

Using the same strategy, one may similarly derive the contact line deformation for a stripe of a diffuse defect of total width w. In this case, ε is described by a trapezoidal function:

$$\varepsilon(r_0,\varphi) = \begin{cases} 0 & -\pi < \varphi < -\chi_2 - \chi_1 \\ \varepsilon_0(\varphi + \chi_2 + \chi_1)/\chi_2 & -\chi_2 - \chi_1 < \varphi < -\chi_1 \\ \varepsilon_0 & -\chi_1 < \varphi < \chi_1 \\ \varepsilon_0(-\varphi + \chi_2 + \chi_1)/\chi_2 & \chi_1 < \varphi < \chi_2 + \chi_1 \\ 0 & \chi_2 + \chi_1 < \varphi < \pi \end{cases} \tag{12}$$

where the linear decrease (see Fig. 2, curve d) corresponds, for instance, to mixing of the low and high surface free energy materials along a diffusion length, $\lambda \approx \chi_2 r_0$. Typically, in the local chemical transformation of a low surface energy into a high surface energy material by a source of a chemical species, λ is the so-called reaction length where the chemical substance is depleted due to its reaction with the surface.

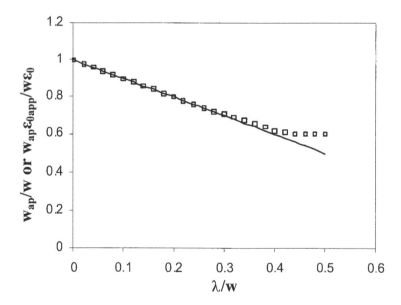

Figure 3. Values of the apparent sharp defect width (□), w_{ap}, and strength (−), $w_{ap}\varepsilon_{0ap}$, required in the sharp defect model of triple line front deformation according to equation (11) to fit the triple line front deformation by a diffuse defect of width w and strength $w\varepsilon_0$ according to equation (15).

From the Fourier series corresponding to this surface free energy profile, the theoretical triple line profile may be easily obtained as a Fourier series expression:

$$\delta(\varphi) = \frac{r_0\varepsilon_0}{\pi\gamma\theta^2}\left\{-\chi_1 - \frac{\chi_2}{2} - \frac{2}{\chi_2}\sum_{n=1}^{\infty}\frac{\cos(n(\chi_1+\chi_2))-\cos(n\chi_1)}{n^2(n-1)}\cos(n\varphi)\right\} \quad (13)$$

It is converted using trigonometric transformations (9), (10) and (14)

$$\sum_{n>0}\frac{\cos(n\alpha)}{n^2} = \frac{1}{6} + \frac{\alpha^2}{4} - \pi\frac{|\alpha|}{2} \quad (14)$$

into the following equation (when allowing r_0 to tend to infinity):

$$\delta(x) = \frac{w\varepsilon_0}{2\pi\gamma\theta^2}\left\{\frac{(1+x^*)^2}{2\lambda^*}\ln\left(\frac{1}{x^*+1}\right) + \frac{(1-x^*)^2}{2\lambda^*}\ln\left(\frac{1}{|1-x^*|}\right)\right.$$

$$\left. -\frac{(1-\lambda^*+x^*)^2}{2\lambda^*}\ln\left(\frac{1}{1-\lambda^*+x^*}\right) - \frac{(1-\lambda^*-x^*)^2}{2\lambda^*}\ln\left(\frac{1}{|1-\lambda^*-x^*|}\right)\right\} + C' \quad (15)$$

with $\lambda^* = 2\lambda/w$.

The surface free energy profile $\varepsilon(x)$ obtained by extension of $\varepsilon(r,\varphi)$ to infinity and the corresponding theoretical triple line profiles (15) are presented in Fig. 2 (curves b and d, respectively) in the case of $\lambda = 0.25$ w.

Equation (15) has been computed for a trapezoidal-shaped stripe defect (diffuse defect) of total width w, for the special case of $\lambda = 0.25$ w and is compared to equation (11) for the sharp-defect case of the same width w. As expected, a diffuse defect of width w exerts a lower pinning force on the liquid front than a sharp defect of the same width. Moreover, for the whole range of diffuse lengths, λ, the triple line front of the diffuse defect may be satisfactorily fitted (with less than 5% error) by the sharp defect for which the stripe defect has an apparent surface free energy equal to ε_{0ap} and an apparent size defect w_{ap}. This situation is illustrated in Fig. 2, where the diffuse defect ($\lambda = 0.25$ w) defined by the excess surface free energy given in curve d deforms the triple line front (curves b and c) as much as a sharp defect of apparent width $w_{ap} = 0.75$ w and apparent strength $w_{ap}\varepsilon_{0ap} = 0.75$ $w\varepsilon_0$ (see curve e). The variations of the best values of w_{ap} and $w_{ap}\varepsilon_{app}$ for different diffuse lengths λ are presented in Fig. 3. For $\lambda < 0.4w$, we find $w_{ap} \approx w-\lambda$ within 5% error and $w_{ap}\varepsilon_{0ap} = f = w<\varepsilon> = (w-\lambda)\varepsilon_0$ where $<\varepsilon>$ is the mean surface free energy of the diffuse defect over the whole defect length.

From a practical point of view, the diffusive part of a defect may not be observable. From the fitting of the triple line deformation to the sharp defect model, one obtains the apparent width of the defect $w_{ap} = w-\lambda$. This value can be compared to the observed value of the defect width, w_{exp}. If $w_{exp} < w_{ap}$, then the defect is certainly diffuse and $w_{ap}-w_{exp}$ gives an estimate of λ, the defect diffuse length. One then obtains the real defect width $w = 2w_{ap} - w_{exp}$. Moreover, the fit of the experimental triple line front with the sharp defect leads to a value of the proportionality constant $C = w_{est}\varepsilon_{0,est}/(\pi\gamma\theta^2)$. In C, the value chosen for w_{est} is w_{ap}, the apparent width $w_{ap} = (w-\lambda)$. Thus, from the preceding discussion comparing the two theoretical triple line fronts for sharp and diffuse defect profiles, the value $\varepsilon_{0,est}$ used in C is ε_0, the value of the difference in surface free energy on the sharp part of the diffuse defect. The fit of the triple line front, under such conditions, is then equivalent to the fit of the data with the diffuse defect triple line front, with the value of $w\varepsilon_0/(\pi\gamma\theta^2)$ and a defect width w.

2.3. Comparison with the literature

Outside of the zone occupied by the defect, the triple line follows a logarithmic law in accordance with the original, simpler form (2).

For comparison within the defect zone, we have compared our analysis with the recent work of Nikolayev and Beysens [11]. They examined the dynamics of triple line deformation induced by a defect on a vertical surface. From energy minimisation principles and Fourier transform methods, they proposed an analytical expression for the triple line deformation, denoted δ_{NB}, by a stripe type defect of width w:

$$\delta_{NB}(x) = \frac{\varepsilon_0}{\pi\gamma}\left(\int_0^{w/2+x} K_0(u/l_c)du + \int_0^{w/2-x} K_0(u/l_c)du \right) \tag{16}$$

where K_0 is the modified Bessel function of zeroth order. The capillary length, l_c, appears here, as gravity is the main force that counterbalances the capillary effect. We have adapted their strategy in order to depict the triple line shape on a plane tilted at an angle α from the horizontal plane. Therefore, as stated elsewhere [9, 12, 13], the pull force is divided by θ^2 and l_c is changed into $l_c^* = l_c(\sin\alpha)^{-1/2}$ and δ_{NB} becomes δ_{NB2}, for a stripe defect of width w:

$$\delta_{NB2}(x) = \frac{\varepsilon_0}{\pi\gamma\theta^2}\left(\int_0^{w/2+x} K_0(u/l_c^*)du + \int_0^{w/2-x} K_0(u/l_c^*)du \right) \tag{17}$$

We have computed expression (17) for various values of w/l_c^* (Fig. 4) and compared $\delta_{NB2}(x)$ to $\delta(x)$ obtained from equation (11). The maximum pinning is obtained for small angles (when gravity is negligible), meaning small values of α

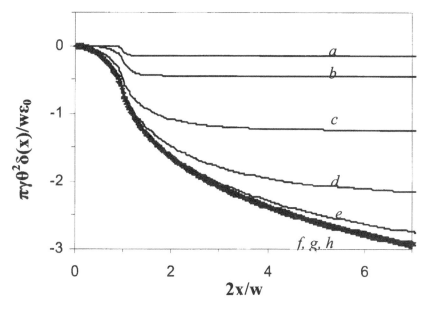

Figure 4. (a-g) Triple line front deformation by a sharp defect of width w according to equation (17) for $2l_c/w$ = (a) 0.1, (b) 0.3, (c) 1, (d) 3, (e) 10, (f) 30, (g) 100. (h) Triple line front deformation by the same defect according to equation (11).

or large values of l_c^*. Moreover, under these conditions, equation (17) is identical to equation (11).

3. EXPERIMENTAL

3.1. PTFE surface preparation

PTFE samples (Goodfellow, $2.5 \times 1.5 \times 0.1$ cm^3) were polished with emery paper (P4000, Presi, France), rinsed in distilled water, then polished on a wet cloth (DP-Nap, Struers, France) until a shiny "mirror" surface was obtained.

Following this, a narrow band (50–200 µm) of treated material was produced by electrochemical treatment using (i) disk and (ii) band ultramicroelectrodes. In a few cases, the treatment was restricted to local circular spots.

The ultramicroelectrodes used were: (i) Pt disks made from wires of 25, 50 or 100 µm diameter embedded in an insulating glass as described previously [14], and (ii) a Au band (width 50 µm, length *ca.* 3.5 mm) microelectrode assembly as recently proposed [15]. The latter assembly consisted of a plaque of Au separated from two Pt wires embedded in epoxy. After polishing, the assembly revealed a Au band of width 50 µm and length 3.5 mm surrounded by two disks of radius 25 µm. The electrolytic solution consisted of 3 ml dimethylformamide (DMF) as the solvent, 0.1 M NBu$_4$BF$_4$ as the electrolyte and typically contained 0.05 M 2,2'-

bipyridine (M_1, E^0_{M1} = - 2.10 V *vs.* saturated calomel Electrode, SCE) and 2 mM terephthalonitrile (M_2, E^0_{M2} = - 1.51 V *vs.* SCE). The scanning electrochemical microscope (SECM) consisted of a three-stage assembly of micro-step motors driven by a controller ESP300 (Newport) piloted by a PC. The current flowing through the ultramicroelectrode was controlled by a potentiostat CHI720 (CH Instruments, USA) piloted by a PC.

It was necessary to ensure the parallelism between the PTFE surface and the (i) microelectrode disk path or (ii) the plane of the band microelectrode assembly. For this reason the electrolytic solution used contained, in a more dilute concentration, a second redox mediator, M_2 (terephthalonitrile), whose E^0_{M2} = -1.51 V *vs.* SCE is not negative enough to ensure reduction of PTFE. If the electrode is held at a potential of E = -1.60 V *vs.* SCE (enough to generate M_2 radical anion but not M_1 radical anion), the PTFE surface behaves as an insulator, i.e., moving the electrode towards the PTFE surface results in a decrease in the current.

In case (i), the PTFE and electrode surface parallelism was adjusted so that any lateral displacement of the disk electrode at a constant distance from the PTFE surface would result in a constant current. Parallelism was then obtained by mechanical adjustment of the PTFE surface plane until a constant current corresponding to d = 0.64 *a*, (where *a* is the electrode radius) was obtained during the whole lateral microelectrode displacement.

In case (ii), the parallelism between the PTFE surface and the band microelectrode assembly was adjusted by changing the inclination of the PTFE surface until the current at the two external disks microelectrodes was identical. The band microelectrode assembly was then moved toward the PTFE surface until the band current corresponded to the value expected for d = 0.64 *a* [15].

The (i) disk or (ii) band electrode was then biased at a more negative potential (E = -2.15 V *vs.* SCE) to ensure M_1 reduction into $M_1^{\bullet-}$ which, in turn, reduced the PTFE.

In case (i), local transformation of PTFE along the band was achieved when the disk microelectrode was displaced at a constant distance of d = 0.64 *a*, and at a constant speed of 5 μm/s. The band obtained was as long as the displacement and its width depended on the electrode radius, *a*, the distance between the electrode and the PTFE surface, d, and on the relative speed of the electrode. It was possible to produce bands having widths ranging from *ca.* 80 to 250 μm. Typically 0.3 to 1 mm long bands were fabricated within 60 to 200 s.

In case (ii), the PTFE was instantaneously transformed along a band of length equal to the length of the Au band. Its width was *ca.* 6 times that of the Au band for a treatment time of 150 s.

3.2. Wetting experiments

Tricresyl phosphate (TCP, Aldrich, 90%, mixture of isomers) either pure or with a 3.5 mM concentration of a fluorescent dye (Rhodamine B (RB) or Curcumine

Table 1.
Wetting data on PTFE

Test liquid	γ (mN m^{-1})	θ_r (°)	θ_a (°)	$\varepsilon_{max} = \gamma\,(1 - \cos\theta_{r/a})$ (mJ/m^2)
TCP	40.9	55	66	17/24
TCP + Rhodamine B (RB)	41.9	51	62	15/22
TCP + Curcumine (Cur)	41.9	50	62	15/22
Water	72.6	98	108	83/95

γ is the liquid surface tension, θ_r and θ_a are the receding and advancing contact angles, respectively, of the liquid on virgin PTFE. ε_{max} is the maximum wettability difference as given by equation (3) for ε with $\cos\tilde{\theta} = 1$.

(Cur), Aldrich) in attempts to facilitate observation of the triple line, and ultra-pure water (Milli-Q grade) were used as the test liquids. The liquid surface tension, γ, and the virgin PTFE wetting ability by the different mixtures used were obtained from Wilhelmy plate and contact angle measurements, respectively (see Table 1). The different values obtained and used are also given in Table 1. The column, ε_{max}, corresponds to the maximal values of ε from equation (3), as defined when $\cos\tilde{\theta} = 1$ and where θ is taken either as θ_r the receding or as θ_a the advancing angle.

Treated sheets of PTFE were mounted almost horizontally (with an angle of *ca.* 5° with respect to the horizontal) in a clamp. By means of a micro-displacement control, the PTFE sample could be immersed or emersed into or from the TCP, along the direction parallel to the treated band, in a small liquid bath. Owing to the large contact angle of water on PTFE, this procedure could not be used with water as a test liquid. In this case, a large continuous film of water was deposited with a syringe onto the tilted PTFE surface. We carefully checked that the treated band was only partially covered by the film of water and that the film was straight and perpendicular to the band 2 mm away from it. For this purpose the films used typically had dimensions of 20×10 mm^2.

The system described was placed under a vertically mounted optical microscope (magnification up to $\times 32$) with a camera (magnification $\times 3$, 4 Mpx resolution image), video recorder and PC set-up attached. Observation and recording of static wetting fronts corresponding to the solid/liquid/air triple line were undertaken so that the contact line shape could be analyzed and compared with theoretical predictions.

4. RESULTS AND DISCUSSION

4.1. SECM etching of PTFE

The PTFE was locally reduced to a polymeric carbon material by SECM. Figure 5 shows the schematic principle of PTFE surface reduction by SECM using both disc and band electrodes. The stable radical anion of a redox mediator, denoted $M_1^{\bullet-}$, is generated at an ultramicroelectrode biased at a potential $E < E^0_{M1}$ in the close vicinity of the PTFE surface. Typically, if the electrode, of radius (or half-width) a, is held within a distance of $2a$ from the sample surface, the radical anion generated can react with the polymer surface provided it is a strong enough reducer [16]. This reaction transforms PTFE into a polymeric carbonaceous material (black material in Fig. 5) and regenerates M_1, according to the following global equation [17]:

$$(2+\delta)n\ M_1^{\bullet-} + \delta n\ NBu_4^+ + (-CF_2-)_n \rightarrow$$

$$\rightarrow (2 + \delta)n\ M_1 + ((-C^{\delta-}-)_n ; \delta n\ NBu_4^+) + 2n\ F^- \qquad (18)$$

where NBu_4^+ is the cation of the electrolyte and $((-C^{\delta-}-)_n ; \delta n\ NBu_4^+)$ represents the n-doped carbonaceous material resulting from reduction.

The amount of current flowing through the microelectrode is strictly correlated to the amount of charge used for the transformation of the PTFE material. For a stationary electrode (band or disk), a carbonaceous area, which is the image of the electrode geometry (band or disk), is imprinted on the PTFE surface. When a disk electrode was moved above the PTFE sample, a band was imprinted on the PTFE surface. In fact, the transformation of the material extends also to the bulk of the material, but at a rate much slower than the surface transformation rate. The surface reaction rate is dependent on the reaction time and also on the disk scan rate.

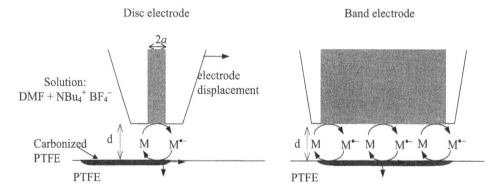

Figure 5. Schematic representation of the PTFE surface treatment by SECM with a disc (left) or a band (right) electrode.

In the time region used for the PTFE band etching, if we assumed the expansion of the band etching as diffusion-like, we may estimate the length, λ, of the transformation diffuse layers, located at each border of the band. In this diffuse area the PTFE transformation is likely only partial, and we obtain the electrode current i:

$$i \approx nFD_S \frac{C_P}{\lambda_r} rh \qquad (19)$$

where F is the Faraday constant, D_S is the diffusion rate of the reducing species for surface carbonization, C_P is the concentration of the monomer unit in the polymer ($C_P = 22$ mol·l^{-1}) and n the number of electrons exchanged for reduction of each monomer unit ($n = 2+\delta = 3$), λ_r is the length of the reducing species diffusion layer along the surface in the carbonized region, r is the surface etching propagation length ($r \approx w/2$) and h is its bulk propagation length. The bulk propagation is diffusion-like and $h = (D_b t)^{1/2}$ with $D_b \sim 10^{-11}$ cm^2·s^{-1} and t is the time required for the PTFE etching.

The propagation of the carbonization at the material surface is typically 100-times faster than that in the bulk; thus, $D_S^{1/2} \sim 100 D_b^{1/2}$, which leads to $D_S \sim 10^{-7}$ cm^2·s^{-1}. Typically, the etching of a band of width w = 80 μm was achieved with an electrode of radius a = 12.5 μm moved at v = 5 μm/s with a current of 2×10^{-7}A. The electrode might be assumed to be a discoid source of etchant whose diameter is about twice larger than that of the electrode and the band etching occurs during a time-of-flight t $\sim$ 4a/v = 10 s. A value of $\lambda_r \sim$ 2 μm then ensues. Even for the smallest carbonaceous band (w = 80 μm) the reaction layer represents less than 5% of the apparent black part of the transformed band. Therefore, to a first approximation, it is not worth using the trapezoidal-profile model and the defect may be considered as sharply defined.

It is difficult to estimate the surface free energy (dispersion and non-dispersion components) for an entire PTFE surface treated under the same conditions, due to the heterogeneity of the treatment on large surfaces. However, the values of the surface free energy of PTFE modified under strongly reducing conditions are reported to be in the range 40–50 mJ/m^2 [18–20]. This large increase in the surface free energy is mainly due to the oxidation of the carbonaceous material by air or water. It tends to oxygenate the surface and then to increase the hydrophilicity of the treated band compared to original PTFE.

With TCP as a wetting liquid, quasi-total or total wetting is expected with γ_{SV}-γ_{SL} of the order of γ, leading to values of the excess surface free energy ε_0 of about 20 mJ/m^2. If water is the wetting liquid, the wetting is partial and a receding contact angle of about 60° is expected on the etched band and, consequently, the expected excess surface free energy ε_0 is about 40 mJ/m^2.

4.2. Triple line front analysis

Figure 6 represents typical experimental triple line fronts obtained on an etched band of width 80 μm with TCP and water as test liquids. After digitization, the images were used in conjunction with regression analysis and equation (11) to obtain the best fits as shown in Fig. 6a and 6b, respectively, for the wetting by TCP and water of the same etched band. As stated in our earlier work [7] with TCP, the whole wetting profile fitted well to equations (2) and (11) outside the defect. Equation (11) also correlated successfully with the wetting profile inside the defect. Both the experimental and theoretical profiles demonstrate the transition from a concave triple line outside the "defect" to a convex triple line within the

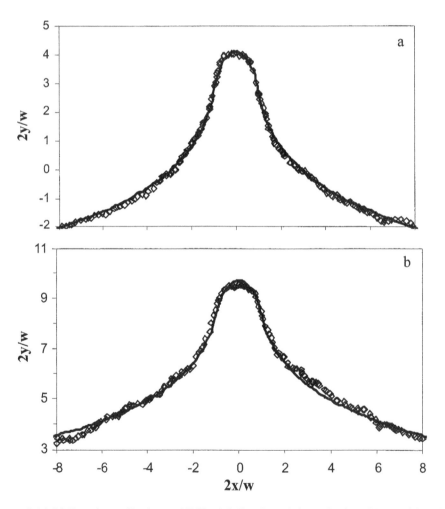

Figure 6. (a) (–) Experimentally observed TCP triple line front deformation by a 80 μm wide etched PTFE band. (◊) Best fit of the triple line front by equation (11). (b) (–) Experimentally observed water triple line front deformation by a 80 μm wide etched PTFE band. (◊) Best fit of the triple line front by equation (11).

confines of the heterogeneity. The inflexion points of the triple line correspond to the borders of the heterogeneity. The apparent width, w_{ap}, of the defect is obtained from the second derivative of the triple line front. It is then compared to the width of the black etched material, $w_{exp} = 80$ μm. We find $w_{exp} = w_{ap}$ to within 5%, which indicates that if the defect is diffuse λ is lower than 2 μm in agreement with its proposed estimate.

Thus, the regression fits allowed us to calculate the values of ε_0, the "excess wettability", from equation (11) either using w_{ap} and ε_0, or from equation (2) with $f = \varepsilon_0 \, w_{ap}$; both approaches give the same results.

A summary of the principal experimental findings from these wetting experiments on etched PTFE (w = 80 or 290 μm) is presented with those previously obtained [7] in Table 2. With TCP as test liquid, we obtained a value of $\varepsilon_0 = 47 \pm 6$ mJ/m^2 in good agreement with those already reported. Again, the actual values of ε_0 are considerably higher than the maximum theoretical value given by

$$\varepsilon_{max} = \gamma \, (1 - \cos\theta) = 17 \text{ mJ/m}^2$$ taking the highest possible value of $\cos\tilde{\theta}$, i.e. unity. The situation thus corresponds to total wetting of the etched material with a contact angle of zero. Owing to the high surface energy value of the reduced PTFE, a complete wetting is likely. The excess surface free energy of *ca.* 30 mJ/m^2 could characterize the pinning force due to the spreading of TCP on the etched band. Indeed from time to time, thin isolated droplets of TCP were observed on the dewetted parts of the etched PTFE band. The possible role of the liquid spreading on the triple line profile has not been studied so far, and we are currently investigating it from both theoretical and experimental viewpoints.

The case of water as test liquid is also critical. Indeed, the value of $\varepsilon_0 = 105 \pm 5$ mJ/m^2 deduced from equation (11) is again higher than the maximum theoretical value, $\varepsilon_{max} = 83$ mJ/m^2. However, a complete wetting of the etched band is less likely than that for reduced PTFE. Indeed, after the whole PTFE surface was reduced by dipping into a reductive solution, water droplets were shown to spread

Table 2.
Summary of principal wetting experiments performed with TCP and water on etched PTFE

Geometry of treated zone (width in μm)	Liquid	Irradiation time in visible light (h)	Number of experiments	ε_0 (mJ/m^2)
Band (80, 120, 180, 200, 220, 290)	TCP/RB/Cur	<1[a]	18	47 ± 6
Band (180)	RB	1	1	34
Band (180)	TCP/Cur	>4	5	20 ± 2
Band (80)	Water	<1[a]	3	105 / 34[b]

Variability of band width was typically ± 5 μm. The wetting triple line was observed in the static receding mode. The parameter ε_0 corresponds to "excess wettability" as defined by equation (7). Values for band widths 80 < w < 290 μm are from Ref. [7].
[a]Low intensity irradiation.
[b]Second value obtained when using $\sin^2\theta$ instead of θ^2 in (11).

more easily on the reduced surface and a receding contact angle of about 60° was observed [18, 19]. The overestimate of ε_0 might be due to the assumptions made during the calculation of the triple line front. Indeed, water is a non-wetting liquid for virgin PTFE and the receding contact angle is greater than 90°: this contradicts the assumption of small values of θ. This argument has already been mentioned from a theoretical point of view [21]. The variation of the liquid–vapor interfacial energy due to the deformation of the surface of the liquid was originally estimated for small angles [2]. It was generalized to arbitrary values of θ [21], and consists, for geometrical reasons, in substituting θ^2 by $\sin^2\theta$ in the expression for the capillary energy and, thus, leads to the same substitution in equation (2). Similarly, in the methodology we have adopted [6], we have used the same substitution, since the approximation of small angles comes mainly from the relation between the axisymmetric droplet contact radius, r_0, and its curvature radius, $R = r_0/\sin\theta \sim r_0/\theta$. However, the experimental work demonstrated the validity of the use of θ^2 when a drop of a non-wetting liquid was placed on a periodic array of square hydrophilic defects patterned on a hydrophobic surface [22].

Consideration of more general values of θ in the establishment of equation (11) also leads to the replacement of θ^2 by $\sin^2\theta$. A much lower value of $\varepsilon_0 = 34 \pm 4$ mJ/m^2 is obtained when taking into account this $\sin^2\theta$ term. A contact angle of 70 $\pm$ 5° then ensues for water on the reduced PTFE, in better agreement with the estimated value of 60°. The same consideration with TCP slightly modified the excess surface free energy (40 mJ/m^2 instead of 50) and, therefore, does not change our discussion on excess wettability.

5. CONCLUSIONS

This paper links the electrochemical modification of initially low surface free energy polymers and the wetting theory. Earlier predictions suggested that the general form of an initially straight wetting triple line perturbed by a local (high surface energy) "defect" should be logarithmically decaying outside of the "defect". This concave wetting line should present an inflexion point at the (assumed) sharp barrier between the defect and the homogeneous material on each side, and then a change to convexity should ensue within the defect. By using microelectrodes, "defect" dimensions (of the order of tens of μm) suitable for wetting studies can be produced on polymer surfaces.

It has been found that the wetting behavior is certainly qualitatively in agreement with theoretical predictions: good fits between theoretical equations and observed wetting fronts have been found. The theory assumes sharp cut-offs between the defect and surrounding material. Modifying the theory, allowing for diffuse boundaries, extends it to a more precise description of heterogeneous surfaces obtained by chemical means. A special feature of the theoretical treatment is a quantity which we term "excess wettability" and that estimates of this quantity were made and found sometimes to be too high.

By allowing for some corrections in the treatment of wetting, specifically by no longer assuming small contact angles, an improved correlation between experiment and theory has been obtained.

REFERENCES

1. D. Mandler, in: *Scanning Electrochemical Microscopy*, A. J. Bard and M.V. Mirkin (Eds.), p. 593. Marcel Dekker, New York, NY (2001).
2. J. F. Joanny and P. G. de Gennes, *J. Chem. Phys.* **81**, 552 (1984).
3. P. G. de Gennes, *Rev. Mod. Phys.* **57**, 827 (1985).
4. Y. Pomeau and J. Vannimenus, *J. Colloid Interface Sci.* **104**, 477 (1985).
5. L. W. Schwartz and S. Garoff, *J. Colloid Interface Sci.* **106**, 422 (1985).
6. M. E. R. Shanahan, *J. Phys. D: Appl. Phys.* **22**, 1128 (1989).
7. F. Kanoufi, C. Combellas and M. E. R. Shanahan, *Langmuir* **19**, 6711 (2003).
8. T. Young, *Phil. Trans. Roy. Soc. (London)* **95**, 65 (1805).
9. G. D. Nadkarni and S. Garoff, *Europhys. Lett.* **20**, 523 (1992).
10. M. E. R. Shanahan, *Colloids Surfaces A* **156**, 71 (1999).
11. V. S. Nikolayev and D. A. Beysens, *Europhys. Lett.* **64**, 763 (2003).
12. J. A. Marsh and A. M. Cazabat, *Phys. Rev. Lett.* **71**, 2433 (1993).
13. J. F. Joanny and M. O. Robbins, *J. Chem. Phys.* **92**, 3206 (1990).
14. A. J. Bard, F. R. F. Fan and M. V. Mirkin, in: *Electroanalytical Chemistry*, A. J. Bard (Ed.), Vol. 18, p. 243. Marcel Dekker, New York, NY (1994).
15. C. Combellas, A. Fuchs and F. Kanoufi, *Anal. Chem.* **76**, 3612 (2004).
16. C. Combellas, J. Ghilane, F. Kanoufi and D. Mazouzi, *J. Phys. Chem. B* **108**, 6391 (2004).
17. C. Amatore, C. Combellas, F. Kanoufi, C. Sella, A. Thiébault and L. Thouin, *Chem. Eur. J.* **6**, 820 (2000).
18. K. Brace, C. Combellas, M. Delamar, A. Fritsch, F. Kanoufi, M. E. R. Shanahan and A. Thiébault, *J. Chem. Soc. Chem. Commun.*, 403 (1996).
19. K. Brace, C. Combellas, E. Dujardin, A. Thiébault, M. Delamar, F. Kanoufi and M. E. R. Shanahan, *Polymer* **38**, 3295 (1997).
20. L. Kavan, *Chem. Rev.* **97**, 3061 (1997).
21. M. O. Robbins and J. F. Joanny, *Europhys. Lett.* **3**, 729 (1987).
22. G. Wiegand, T. Jaworek, G. Wegner and E. Sackmann, *J. Colloid Interface Sci.* **196**, 299 (1997).

Contact Angle, Wettability and Adhesion, Vol. 4, pp. 61–76
Ed. K.L. Mittal
© VSP 2006

A simple geometrical model to predict evaporative behavior of spherical sessile droplets on impermeable surfaces

LECH MUSZYŃSKI,[1,*] DIOGO BAPTISTA[2] and DOUGLAS J. GARDNER[2]

[1]*Department of Wood Science & Engineering, 119 Richardson Hall, Oregon State University, Corvallis, OR 97331-5751, USA*
[2]*Advanced Engineered Wood Composites Center, University of Maine, Orono, ME 04469-5793, USA*

Abstract—A simple model to predict the evaporative behavior of spherical sessile droplets on impermeable surfaces is presented. The model is capable of predicting changes in shape and volume of spherical droplets resulting from evaporation from the droplet's cap area. It is demonstrated that at any moment all geometrical parameters of a spherical droplet on a surface (volume, contact angle, contact radius and area, cap radius and area, droplet height) can be easily calculated from basic geometrical relations if any two of them are known (e.g., initial droplet volume and initial contact angle, or droplet cap radius and height). Droplet dynamic behavior due to evaporation is further determined using the known value of the evaporation intensity from a unit area of the droplet cap (or evaporation flux). The bulk evaporation rate from the droplet cap decreases proportionally to the shrinking cap area. More complex droplet behavior can be simulated if a receding contact angle value is known. This model was used to simulate experiments performed by the authors as well as reported by other researchers. It is demonstrated that the simple geometrical relations actually account for many features of the sessile droplet dynamic behavior reported in the literature. It is also demonstrated that the often reported bulk evaporation rate, not adjusted for changing droplet cap area, should not be used as a meaningful indicator of droplet dynamics.

Keywords: Droplet dynamics analysis; mathematical model; image analysis.

1. INTRODUCTION

The behavior of sessile droplets in non-saturated environments is a complex, multi-physics phenomenon that in most cases involves heat and mass transfer between phases, besides surface forces equilibrium. Evaporation occurs when the partial pressure of the vapor in the ambient air is below saturation. Even if initially the temperature of the droplet is in equilibrium with both the substrate and

---

*To whom correspondence should be addressed. Tel.: (1-541) 737-9479; Fax: (1-541) 737-3385; e-mail: Lech.Muszynski@oregonstate.edu

ambient air, the vaporization process will drain the heat from the droplet, disturbing the thermal equilibrium. As the temperature of the droplet near the cap surface decreases, a temperature gradient within the droplet is created and heat transfer between the droplet and the solid surface, as well as between the droplet and the ambient air is initiated to restore the equilibrium. Generally, the droplet is expected to cool with time. However, the heat transfer between the droplet and the environment, and thus the magnitude of the heat deficit, will depend on the heat capacity and conductivity of the phases. It may be reasonably assumed that the deficit may be negligible for slow evaporating droplets on substrates which are good heat conductors and have substantial heat capacity. On the other hand, the cooling may be significant for fast evaporating droplets (e.g., ethanol) on substrates of low heat conductivities (most polymers). Significant cooling may slow down the evaporation process.

A simple model to predict evaporative behavior of spherical sessile droplets on impermeable surfaces was developed as a part of larger project in which droplet dynamics analysis methods are used to predict water resistance of coated surfaces with various protective treatments [1, 2]. In this method, a single digital camera with microscopic lenses is used to acquire a series of images of water droplets deposited on a range of isotropic surfaces. Because in a general case both evaporation from the free droplet surface and penetration into the test surface are expected, impermeable reference surfaces were used to evaluate the effect of evaporation. The method proved relatively accurate, as long as the volumes and contact angles of the droplets on test and reference surfaces were similar. The results were less accurate when these contact angles were significantly different, because the intensity of evaporation from a unit surface of the droplet cap (or the evaporation flux) was affected by droplet geometry. One possible way of improving the accuracy of the method could be matching each test sample with an impermeable reference sample of similar surface energy. However, the number of reference samples, as well as the total test time, can be significantly reduced if the effect of evaporation at given test conditions could be theoretically predicted for all test surfaces.

The effect of evaporation on contact angle of water droplets on polymer surfaces was studied by Shanahan and Bourges [3] and Bourges-Monnier and Shanahan [4]. There are also studies on thermodynamics of droplet evaporation on hot metal surfaces where the reverse relation (i.e., effect of droplet shape on evaporation rate) is analyzed (e.g., Refs [5, 6]). However, in most studies where surface properties of materials at ambient temperature are investigated by means of droplet analysis the influences of diffusion and evaporation on the droplet shape are simply neglected. Panwar *et al.* [7] proposed a simple theoretical model for the evaporative behavior of droplets in non-saturated vapors based on measured evaporation rate, where an empirical correlation between the evaporation rate and droplet size was determined. Although a good correlation with data used

Table 1.
Empirical correlations between normalized average flux values and droplet contact angles and volumes

Correlation	Model function	Parameter		R^2	Discrete change of the evaporation flux	
		a	b			
$f(V)_{\text{lin}}$	$\left(\dfrac{\overline{f_V}}{f_{V_0}}\right) = a \cdot V + b$	-0.1006	1.6245	0.2391	$\Delta \overline{f_t} = \overline{f_0} a \Delta V_t$	(5)
$f(V)_{\text{log}}$	$\left(\dfrac{\overline{f_V}}{f_{V_0}}\right) = a \cdot \ln(V) + b$	-0.2855	1.5788	0.2638	$\Delta \overline{f_t} = \overline{f_0} \dfrac{a}{V_t} \Delta V_t$	(6)
$f(\theta)_{\text{lin}}$	$\left(\dfrac{\overline{f_\theta}}{f_{60°}}\right) = a \cdot \theta + b$	-0.0069	1.4067	0.3632	$\Delta \overline{f_t} = \overline{f_0} \dfrac{a}{a60° + b} \Delta \theta_t$	(7)

to derive the model was achieved, no further validation of the model was presented. Hu and Larson [8] proposed an analytical model for calculating evaporation rates from sessile droplets. The model incorporated the effect of non-uniform distribution of evaporation flux over the droplet cap; however, it was restricted to droplets of constant ("pinned") contact areas and contact angles below 90°. The simulation also anticipated that the distribution of evaporation flux (*f*) from a sessile droplet depended on the contact angle and was not uniform when the angle was smaller than 90°. Generally, the smaller the contact angle the bigger the difference between the evaporation intensities on droplet perimeter and in the center. The parameters for the flux distribution model were derived from theoretical speculation based on analogy with electrostatic potential fields and finite element modeling, rather than from empirical experiments. Good correlations of the simulated evaporation rates with experimental results by other researchers were reported. However, these authors predict that the evaporation flux from the droplet cap should decrease with decreasing contact angle, while experimental data collected (not shown) indicated an opposite trend, i.e., the average evaporation flux increased slightly as the evaporation process advanced, and the contact angle decreased. This latter observation was quantified in terms of empirical correlations, of which best fitting was obtained for logarithmic and linear correlations between the evaporation flux and residual droplet volume. A relatively weak correlation was also obtained for evaporation flux and contact angle for droplets with contact angles above the receding values. These findings are summarized in Table 1.

In this paper an alternative approach to modeling the evaporative behavior of sessile droplets from impermeable surfaces is presented. A simple geometrical model was developed, which relies principally on strict geometrical relations between various droplet measures, combined with empirical correlations between an average evaporation intensity from the droplet cap and the various geometrical measures of the droplet shape (data not shown).

2. MODEL DESCRIPTION

2.1. General assumptions

Droplet dynamic behavior due to evaporation is best characterized by the local evaporation intensity (or evaporation flux, f) from a unit area (dA_S) of the droplet cap, which for flat liquid surfaces can be assumed uniform and given for known ambient conditions (temperature and partial vapor pressure at a certain distance from the surface), i.e.,

$$f = -\frac{d}{dA_S}\left(\frac{dV}{dt}\right) \tag{1}$$

The intensity of vaporization is driven by the gradient of the partial pressure of the vapor in the ambient air near the droplet surface. The process may be characterized in terms of generalized diffusion. In a general case the gradient should be assumed variable over the droplet surface. It is also sensitive to the actual temperature (which may change as a result of heat transfer between the droplet and the environment).

In that case the evaporation rate (F), or the amount of liquid evaporating from the entire droplet surface (A_S) in a unit time would be an integral of the local evaporation intensities over the surface, and proportional to the surface area (i.e., in any given period of time more liquid would evaporate from a larger surface):

$$F = \int_{A_S} f \, dA_S \tag{2}$$

For flat liquid surfaces at stable ambient conditions the evaporation intensity is the same for the entire surface and, thus, equation (2) can be simplified to:

$$F = f \cdot A_S \tag{3}$$

Should this be the case for small sessile droplets exposed to such conditions, the evaporation rate would gradually decrease in proportion to their decreasing surface area.

In the literature, however, constant or nearly constant evaporation rates are often reported (e.g., Refs [7–10]), which indicates that in fact the average intensity of evaporation gradually increases as the droplet gets smaller. This was empirically demonstrated by the determination of correlations between the average evaporation flux (equation (4)) *vs.* droplet volume and contact angle for a range of glass and polymer surfaces characterized by a relatively wide range of surface energies (contact angles between 40° and 105°) (data not shown).

$$\bar{f} = \frac{F}{A_S} \tag{4}$$

Table 2.
Formulae used for quantitative droplet geometry description

Quantity	Formula
Contact angle	$\theta = a\cos(1 - h/R)$
Base radius	$r = R\sin\theta$
Free droplet area (cap area)	$A_s = 2\pi Rh$
Droplet base area	$A_b = \pi r^2$
Droplet volume	$V = \frac{\pi}{3}(3R - h)h^2$

Droplet radius (R) and height (h) were measured from the digital images.

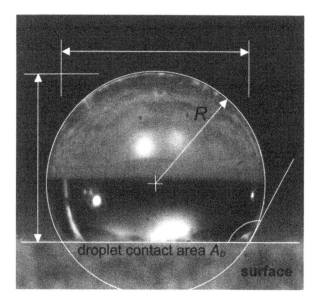

Figure 1. Quantities characterizing droplet shape: droplet radius (R); base or contact area radius (r); droplet height (h); contact angle (θ). From these values droplet free surface area (A_s), droplet contact area (A_b) and droplet volume (V) are calculated.

Limitations of these correlations will be discussed elsewhere. The empirical model formulas and correlation parameters are summarized in Table 1.

Certain simplifying assumptions can usually be made. When compared to most solids, air is considered a good heat insulator; therefore, the heat transfer between the liquid and ambient air can be assumed to be relatively insignificant. Then, the local vapor concentration gradient in the air will depend only on the droplet geometry. The temperature gradient in substrates that are considered good heat conductors (e.g., metals, glass) can also be neglected so that the temperature at the contact with the droplet can be assumed constant.

Generally, the distribution of the evaporation intensity over the droplet cap should be calculated from the heat and mass transfer on the gas–liquid boundary. For practical purposes, however, the average evaporation intensities measured on reference droplets in the same conditions can be used. In such a case the effect of the droplet geometry and boundary conditions on the average evaporation intensity should be examined and, if necessary, accounted for.

Then, if average evaporation/diffusion intensities over the droplet boundaries can be theoretically calculated or empirically determined the dynamic droplet behavior can be modeled as a purely geometrical problem.

The basic quantities characterizing the shape of the droplet are defined in Fig. 1. These include: droplet radius (R) and droplet height (h), as well as base (or contact) area radius (r), contact angle (θ), droplet cap area (A_s), droplet contact area (A_b) and volume (V). It can be demonstrated that at any moment all geometrical parameters of a spherical droplet on a surface can be easily calculated from basic geometrical relations if any two of them are known (Table 2). Although in reality it is the initial droplet volume and physical interaction between the solid surface, liquid and ambient vapor, manifested in a specific contact angle, that determine all other geometrical characteristics; in practice, other characteristics (e.g., droplet cap radius and height) are much easier to measure while the contact angle and volume are usually calculated values.

2.2. Calculations

The calculations were conducted using a simple recurrent algorithm, where the droplet parameters and changes in the average evaporation flux from the droplet cap for every time step were based on values from the previous step and known constant parameters. The simplified block diagram of the simulation algorithm is presented in Fig. 2.

The input parameters required for a simulation are: (1) initial droplet volume V_0 (µl), (2) initial contact angle θ_0 (deg), (3) receding contact angle θ_R (deg), (4) the initial evaporation rate F_0 (µl/min, or mg/min) and (5) empirical model parameters for the adjustment of the average flux as it changes with the droplet geometry (a and b).

For given ambient conditions the values of parameters (1)–(4) can either be taken from the literature or previous tests, or be easily determined empirically. Calculation of all initial ($t = 0$) droplet characteristics from the initial droplet volume (V_0) and the initial contact angle (θ_0), are based on formulas derived directly from the set of equations given in Table 2. The average evaporation flux ($\bar{f}_0$) is calculated by substituting the initial evaporation rate (F_0), which can be easily determined as the initial gradient of the residual droplet volume diagrams

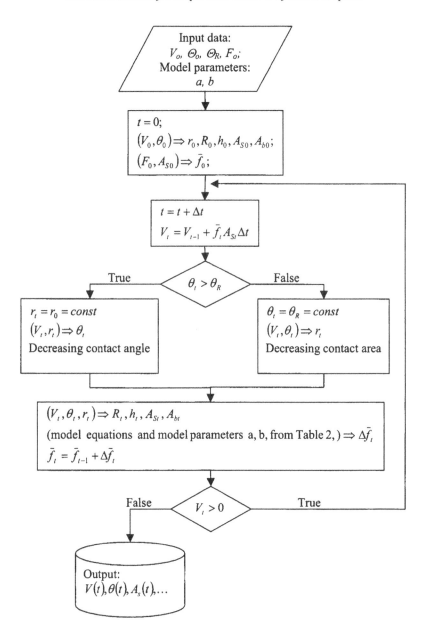

Figure 2. Schematic diagram of the algorithm used in simulations (symbols as in Fig. 1).

and the cap area A_S in equation (4). The recurrent procedure begins with determination of the droplet volume loss (ΔV_t) for a given time step (Δt). Next, if the previous contact angle value was greater than the receding angle, it is assumed that the droplet base area (A_b) and base radius r remain constant ("pinned"), while the volume change causes the angle to decrease to a new value (θ_t). If the receding contact angle is reached, the angle remains constant and a new value of the base radius (r_t) is determined. Other droplet parameters can then be computed. Finally, the step change in the average evaporation flux ($\Delta \bar{f}_t$) is determined using differentials of the empirical formulas (equations (5)–(7) in Table 1). The loop is repeated with constant time increments until the entire volume "evaporates". All droplet characteristics calculated at each time step can be sent to the output.

The important features of this algorithm are its simplicity, small number of easily accessible input parameters and that it can simulate evaporative dynamics of droplets beyond the receding point.

3. RESULTS AND DISCUSSION

In this study, the initial parameters for simulations were taken from tests with small (about 4 µl) water droplets on acrylic and acrylonitrile butadiene styrene (ABS) polymer surfaces (data not shown), as well as from the data published by Birdi *et al.* [10] and Panwar *et al.* [7], where larger water droplets (5–90 µl) were tested on glass and polycarbonate surfaces (Table 3). Formulas for the flux adjustments used in simulations (equations (5)–(7) in Table 1) were based on the empirical correlations. It is important to note that particularly the linear correlation between the evaporation flux and volume despite good fit with the experi-

Table 3.
Input data used in the simulations

Reference	Liquid/Surface	Initial droplet volume (µl)	Initial contact angle (deg)	Receding contact angle (deg)	Initial evaporation rate (µl/min)
[10]	Water/Glass	5.0	41	2	0.1175
[10]	Water/Glass	10.0	41	2	0.1627
[10]	Water/Glass	15.0	41	2	0.2028
[7]	Water/Glass	15.4	43	2	0.2460
[7]	Water/Glass	32.4	49	2	0.2880
[7]	Water/Glass	45.0	50	2	0.3600
[7]	Water/Glass	73.5	52	2	0.4560
[7]	Water/Polycarbonate	28.1	76	2	0.1140
[7]	Water/Polycarbonate	34.5	72	2	0.1320
[7]	Water/Polycarbonate	57.6	72	2	0.1860
[7]	Water/Polycarbonate	89.0	73	2	0.2280
Our experiments	Water/ABS	3.15	79	51	0.0600
Our experiments	Water/Acrylic	4.65	68	52	0.1080

mental data is physically unrealistic, and should not be extrapolated beyond the droplet volume range for which it was determined (4–8 μl). Consequently, it was not used to simulate the experiments reported in Ref. [7] which cover a much wider range of volumes (15–90 μl). The logarithmic model for flux and volume, and the linear model for flux and contact angle do not pose such difficulties.

In addition, reference calculations were performed on the assumption that the average evaporation flux remained constant during the entire evaporation process. The simulated residual droplet volume and contact angle diagrams were compared with the reported experimental data.

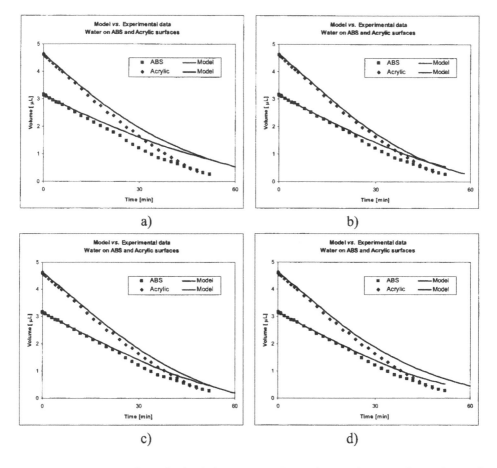

Figure 3. Comparison of sample simulation results and experimental data (not shown) for small droplets evaporating from the ABS and Acrylic polymer surfaces. Residual droplet volumes calculated assuming the evaporation flux as: (a) f = constant (reference), (b) $f(V)_{lin}$, (c) $f(V)_{log}$, (d) $f(\theta)_{lin}$ (see descriptions in Table 1).

Sample simulation results and experimental data for small droplets evaporating
from the acrylic and ABS polymer surfaces are shown in Figs 3 and 4. Generally
the simulations under-predict the evaporation rate to different degrees, which may
result from the fact that the model is unable to predict the decrease of contact angle
below the receding value later in the process (Fig. 4). The difference between the
experimental data and simulation is greatest for the reference model, where the
evaporation flux is not adjusted for changing droplet shape (Fig. 3a). It is also least
accurate in predicting the development of the contact angle for the ABS surface
(Fig. 4a). The best results were obtained when the linear empirical correlation
between flux and droplet volume is used (equation (5) in Table 1; Fig. 3b and 4b).

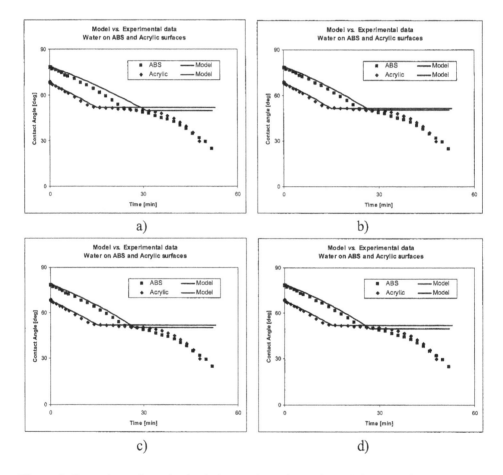

Figure 4. Comparison of sample simulation results and experimental data (not shown) for small
droplets evaporating from the ABS and Acrylic polymer surfaces. Contact angles calculated assum-
ing the evaporation flux as: (a) f = constant (reference), (b) $f(V)_{lin}$, (c) $f(V)_{log}$, (d) $f(\theta)_{lin}$ (see descriptions in Table 1).

Comparison of simulation results with graphs published by Birdi *et al.* [10] for water droplets on glass is presented in Fig. 5. Because the numerical data for all points were not published, the simulation curves are superimposed on the copies of the original graphs with careful matching of the diagram scales. Here the reference model under-predicts the evaporation rate for the two larger droplets (10 and 15 mg), while a good agreement was achieved for the smallest one (Fig. 5a). The weakest matching between the simulation and experimental results was obtained when the linear correlation between flux and droplet volume (equation (5) in Table 1) was used to predict development of the two larger droplets (Fig. 5b). This, however, was expected, since the linear correlation is not realistic for droplets larger than tested in the experimental study (data not shown) and, thus, should not be extrapolated. Relatively good agreement with experimental data was achieved for the logarithmic correlation (equation (6) in Table 1; Fig. 5c). Almost perfect

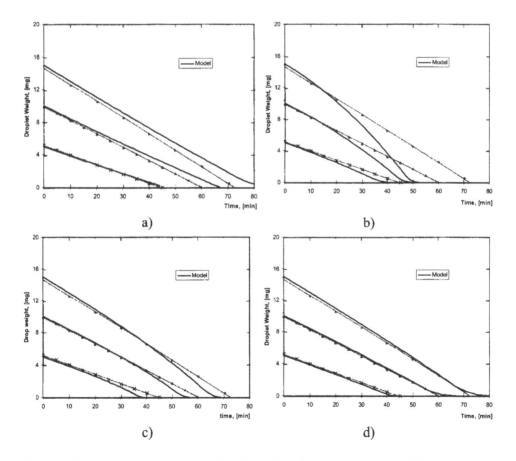

Figure 5. Comparison of simulation results with residual droplet mass graphs published by Birdi *et al.* [10]. The evaporation flux was assumed as: (a) $f =$ constant (reference), (b) $f(V)_{lin}$, (c) $f(V)_{log}$, (d) $f(\theta)_{lin}$ (see descriptions in Table 1).

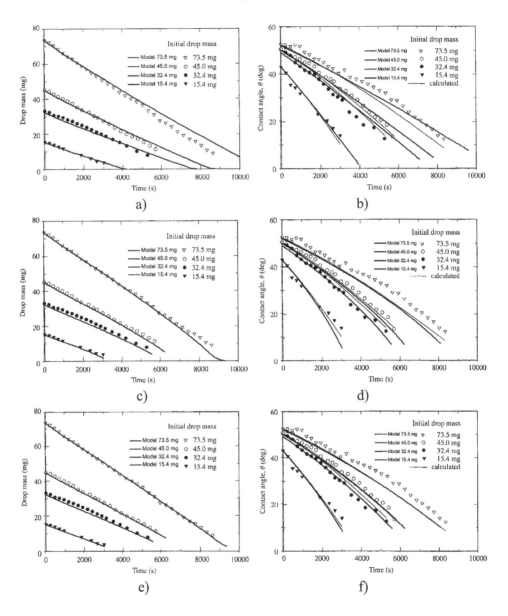

Figure 6. Comparison of simulation results with residual droplet mass and contact angle graphs for glass surfaces published by Panwar *et al.* [7]. The evaporation flux was assumed as: (a, b) f = constant (reference); (c, d) $f(V)_{\log}$; (e, f) $f(\theta)_{\lin}$ (see descriptions in Table 1).

agreement between the published data and the model for all droplets was found when the linear correlation between evaporation flux and contact angle was used (equation (7) in Table 1; Fig. 5d).

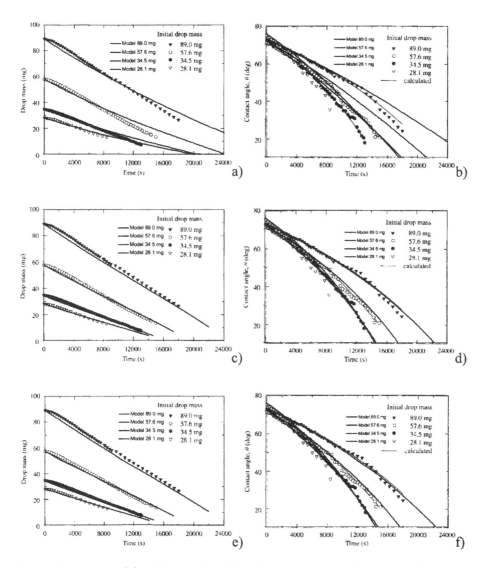

Figure 7. Comparison of simulation results with residual droplet mass and contact angle graphs for polycarbonate surfaces published by Panwar *et al.* [7]. The evaporation flux was assumed as: (a, b) f = constant (reference); (c, d) $f(V)_{log}$; (e, f) $f(\theta)_{lin}$ (see descriptions in Table 1).

In Fig. 6 the simulations results are compared with experimental data for even larger water droplets (15–73 mg) on glass, published by Panwar *et al.* [7]. Although the reference simulation for the two largest droplets (45 and 73 mg) again under-predicted the evaporation in the later stage of the process, in this case a reasonable agreement between the reference simulation and the experimental data for volumes and contact angles was achieved (Fig. 6a and 6b). However, much better agreement was achieved when the logarithmic correlation between flux and droplet volume was used (equation (6) in Table 1; Fig. 6c and 6d), and almost a perfect match for correlation between flux and contact angle (equation (7) in Table 1; Fig. 6e and 6f). Although in the latter two cases predictions of contact angles were not as close, they were almost identical with theoretical curves generated by the authors [7].

In Fig. 7 a similar comparison with experimental data reported by Panwar *et al.* [7] for water droplets on polycarbonate surfaces is presented. The experimental data show apparently slower evaporation in the initial stage of the process. The authors did not provide any theoretical explanation for this behavior and the algorithm presented here does not account for it. The reference simulation is less accurate here (Fig. 7a and 7b). The other two models, however, match very closely the experimental data for all four droplets (Fig. 7c and 7e). The prediction of the contact angle development was also excellent (Fig. 7d and 7f).

The summary of the model performance when compared to experimental data is presented in Table 4. In Table 4 each comparison is assigned one of four grades, reflecting an assessment based on the percentage of data range for which good agreements between the simulated and experimental curves were achieved.

Table 4.
Summary of the model performance assessment when compared to experimental data from the literature

Reference	Initial droplet volume (μl)	f=const		$f(V)_{lin}$		$f(V)_{log}$		$f(\theta)_{lin}$	
		$V(t)$	$\theta(t)$	$V(t)$	$\theta(t)$	$V(t)$	$\theta(t)$	$V(t)$	$\theta(t)$
[10]	5.0	***	—	*	—	**	—	***	—
[10]	10.0	*	—	X	—	**	—	***	—
[10]	15.0	X	—	X	—	**	—	***	—
[7]	15.4	***	**	—	—	***	**	***	***
[7]	32.4	***	*	—	—	***	***	***	***
[7]	45.0	**	***	—	—	***	***	***	***
[7]	73.5	**	***	—	—	**	**	***	**
[7]	28.1	**	**	—	—	***	**	***	**
[7]	34.5	**	**	—	—	***	***	***	***
[7]	57.6	*	*	—	—	***	***	***	***
[7]	89.0	*	**	—	—	***	***	***	***
Our experiments	3.15	*	X	**	**	**	**	**	**
Our experiments	4.65	*	**	**	**	**	**	**	**

Grades: ***, excellent, good agreement for the entire range of experimental data; **, good agreement for more than 75%; *, satisfactory, agreement for more than 50%; X, unacceptable, agreement for less than 50% of the data range.

Best predictions of the evaporative behavior of the droplets were achieved when a linear correlation between flux and contact angle was used. Good results were also achieved when a logarithmic correlation between the evaporation flux and droplet volume was used. In both cases the empirical formulas determined for relatively small droplets (4–8 μl) allowed an accurate prediction of evaporative behavior of droplets 2–18-times larger.

4. CONCLUSIONS

A simple geometrical model to predict evaporative behavior of spherical sessile droplets on impermeable surfaces has been developed.

The important features of the model algorithm are its simplicity, small number of easily accessible input parameters and that, to certain degree, it can simulate evaporative dynamics of droplets beyond the receding point.

Comparison of simulation results with experimental data generated by the authors as well as with data published by other researchers confirmed that the model was quite robust and capable of predicting changes in shape and volume of spherical droplets resulting from evaporation from the droplet's cap area.

Three empirical formulas determined by the authors were used to account for the effect of droplet geometry on the average evaporation flux from the droplet cap. It was confirmed that including this effect in the simulation algorithm improved the accuracy of the model.

Best predictions of the evaporative behavior of the droplets were achieved when the effect of droplet geometry on the average evaporation intensity from the surface was modeled as a linear correlation between the evaporation flux and contact angle. Good results were also achieved when a logarithmic correlation between the evaporation flux and droplet volume was used. In both cases the empirical formulas determined for relatively small droplets (4–8 μl) allowed an accurate prediction of evaporative behavior of droplets 2–18-times larger.

As expected, the applicability of the linear correlation between the evaporation flux and droplet volume was limited to droplets of volumes within the range for which this correlation was determined.

Acknowledgements

The project was supported by the National Research Initiative of the USDA Cooperative State Research, Education and Extension Service, grant number 2003-35103-12890.

REFERENCES

1. L. Muszyński, M. E. P. Wålinder, C. Pîrvu, D. J. Gardner and S. M. Shaler, in *Contact Angle, Wettability and Adhesion*, Vol. 3, K. L. Mittal (Ed.), pp. 463–478. VSP, Utrecht, The Netherlands (2003).
2. L. Muszyński, M. E. P. Wålinder, C. Pîrvu and D. J. Gardner, in *Proceedings of the 8th International IUFRO Wood Drying Conference*, Brasov (2003).
3. M. E. R. Shanahan and C. Bourges, *Int. J. Adhesion Adhesives* **14**, 201–205 (1994).
4. C. Bourges-Monnier and M. E. R. Shanahan, *Langmuir* **11**, 2820–2829 (1995).
5. M. di Marzo, P. Tartarini, Y. Liao, D. Evans and H. Baum, *Int. J. Heat Mass Transfer.* **36**, 4133–4139 (1993).
6. S. Chandra, M. di Marzo, Y. M. Qiao and P. Taitorini, *Fire Safety J.* **27**, 141–158 (1996).
7. A. K. Panwar, S. K. Barthwal and S. Ray, *J. Adhesion Sci. Technol.* **17**, 1321–1329 (2003).
8. H. Hu and R. G. Larson, *J. Phys. Chem. B.* **106**, 1334–1344 (2002).
9. S. Birdi and D. T. Vu, *J. Adhesion Sci. Technol.* **7**, 485–493 (1993).
10. K. S. Birdi, D. T. Vu and A. Winter, *J. Phys. Chem.* **93**, 3702–3703 (1989).

Part 2

Contact Angle Measurements/ Determination and Solid Surface Free Energy

Contact Angle, Wettability and Adhesion, Vol. 4, pp. 79–99
Ed. K.L. Mittal
© VSP 2006

About the possibility of experimentally measuring an equilibrium contact angle and its theoretical and practical consequences

CLAUDIO DELLA VOLPE,* MARCO BRUGNARA, DEVID MANIGLIO, STEFANO SIBONI and TENZIN WANGDU

Department of Materials Engineering and Industrial Technologies, University of Trento, Via Mesiano 77, 38050 Trento, Italy

Abstract—The measurement of contact angles and, thus, determination of solid surface tension has been considered for years as a "comedy of errors", but in recent years the introduction of more sophisticated techniques and of computer-controlled devices, along with a general better understanding of surface structures, has led to a greater precision and accuracy of measurements. However, it is common to neglect the difference between the advancing contact angle and the Young's angle or to underestimate the role and significance of receding contact angles. In previous papers an experimental procedure has been developed, called the Vibration Induced Equilibrium Contact Angle (VIECA), applied to a Wilhelmy experiment, which appeared to be able to provide a really stable and equilibrium-like value: this procedure was based on previous, rare literature attempts at providing an operational and satisfactory definition of equilibrium contact angle. The VIECA results seem to be related to the advancing and receding values through simple, but approximate, relations. Moreover, the VIECA appears to be independent of the roughness and heterogeneity of the surfaces analysed in the majority of cases. In the present paper, the VIECA method is extended to the sessile drop technique and comparison is made with the common advancing or "static" estimates of contact angle. A theoretical modelling of the physical situation induced by the application of mechanical vibrations to the meniscus or to the drop is proposed. The main consequence of these results is that the contact angles on common surfaces for common liquids are overestimated; a more subtle consequence is the effect on the evaluation of the surface free energy *via* the most common semiempirical models.

Keywords: Equilibrium contact angle; sessile drop vibration; constrained-drop oscillation.

1. INTRODUCTION

The measurement of contact angles, for years considered as a "comedy of errors" [1, 2], has been greatly improved in recent years by the introduction of more sophisticated techniques and computer-controlled devices. Nevertheless, the common measurement of the contact angle is still performed in conditions not corresponding

---

*To whom correspondence should be addressed. Tel.: (39-461) 882-409; Fax: (39-461) 881-977; e-mail: devol@devolmac.ing.unitn.it

to true stable-equilibrium states and, thus, gives metastable-equilibrium values, the so-called advancing and receding contact angles.

In the field of thermodynamics, experimental measurements not corresponding to real equilibrium situations are quite common and one can easily find many examples of similar phenomena. Consider the case of water freezing. Using very pure degassed water in very clean containers and cooling the liquid very slowly one can reach temperatures lower than 0°C, without water freezing. By this procedure water can attain the temperature of $-38°C$ in a repeatable way. This, however, does not correspond to the measurement of the water freezing/melting phase-transition temperature, because the system is not at equilibrium, but is in a metastable state, due to kinetic reasons; however, even slight perturbations can move the system into an equilibrium state. No physical chemist would conclude from such an experiment that the freezing/melting transition temperature of water is $-38°C$ or that it can vary in a certain interval $(0/-38°C)$ according to the details of the experimental procedure adopted. On the contrary, the physical chemist would search for an experimental setup able to eliminate the reasons for the metastability.

In the field of contact angle, unfortunately, the situation is different. It is customary to consider that the advancing contact angle is a good approximation of the equilibrium value or to neglect the different significance of the experimental angles (advancing and receding).

A relatively large literature concerns with the theoretical aspects of the problem and the reasons for hysteresis [3–24], but presentation of experimental methods able to overcome the metastability problem is not a very common [25–35]. However, it is possible to find about 10 papers in which different kinds of mechanical or acoustic mehods have been used to modify the state of the meniscus.

In previous papers [31–33] an experimental procedure, called the Vibration Induced Equilibrium Contact Angle (VIECA), applied to a Wilhelmy experiment, was developed in our laboratory, which provides a really stable and equilibrium-like situation.

For this purpose a well-defined frequency vibration, produced by a loudspeaker, placed under the liquid container of a Wilhelmy microbalance during a temporary stop of immersion or withdrawal, was able to relax the liquid meniscus to the same final state, independent of advancing or receding contact angle, the speed of immersion, the presence of roughness and/or heterogeneity in the sample (at least to a certain extent). The effects on the common experimental results are shown in Figs 1 and 2. It is important to note that the amplitude of the oscillation after an initial constant period is slowly reduced, allowing the meniscus to find the lowest energy state by itself. The final state appears to be stable.

Viscosity, very large roughness or the combined presence of roughness and heterogeneity make it difficult, or even impossible, to attain the same final state from both advancing and receding. When Young's contact angle can be reasonably assumed to be zero or the presence of a liquid film on the sample is detected, the contact angle in the final state of the meniscus after vibration turns out to be typically

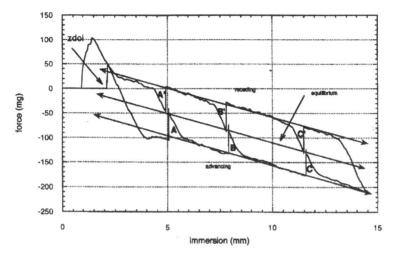

Figure 1. A typical Wilhelmy VIECA experiment. The arrow labeled with "ZDOI" indicates the zero depth of immersion of the sample. At each immersion level the movement is stopped and an equilibrium relaxation cycle is carried out. In the picture each relaxation cycle is represented by a wedge-shaped portion of the diagram, as indicated by the letters A, B, C and A', B', C'. The equilibrium points reached after relaxation are the vertices of the "wedges". The line extrapolated from advancing, receding and equilibrium points allows to calculate the contact angle from the equation $F = \rho g \cos\theta$, obtaining θ_{adv}, θ_{rec} and θ_{equ}.

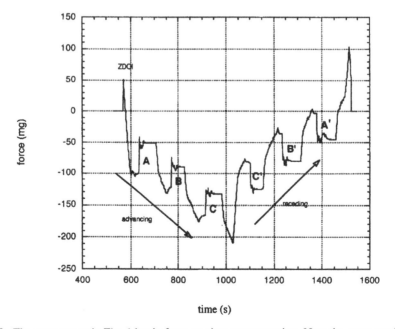

Figure 2. The same run as in Fig. 1 but in force *vs.* time representation. Note the correspondence of the different relaxation steps of the measurement, still marked by letters A, B, C (advancing) and A', B', C' (receding).

zero: a high-energy surface acts as an "attractor" for the vibrated meniscus and the VIECA generally vanishes.

Moreover, the VIECA results seemed well related to the advancing and receding values through simple but approximate relations, already proposed in the past [3–6, 15, 18, 20, 29]. For instance, the arithmetic mean of cosines of advancing and receding angles generally corresponds to the cosine of the experimental equilibrium angle, i.e.

$$\cos \theta_{\text{equ}} = (0.5 \cos \theta_{\text{adv}} + 0.5 \cos \theta_{\text{rec}}). \tag{1}$$

A final comment is that VIECA experimental angles appear also in very good agreement with the model proposed by Kamusewitz and co-workers [36, 37]; in this model the Young angle on a rough surface is obtained by extrapolating the advancing and receding angles on rough surfaces to zero hysteresis. The results of VIECA measurements are in good agreement with Kamusewitz estimates, even on rough surfaces. This double agreement will be reconsidered in the Discussion.

In the present paper the VIECA method is extended to the sessile-drop technique and some comparisons are made between the common advancing or "static" evaluations of contact angle and the VIECA results obtained from Wilhelmy experiments.

2. MATERIALS AND METHODS

Contact-angle measurements were performed in a static way using a self-developed instrument, schematically shown in Fig. 3. The core of the instrument consists of a light source, a sample holder eventually replaceable with a metallic membrane loudspeaker and an optic device for image detection, all mounted on a rigid support which can rotate around the axis of the optics. In this way it is possible to vary the inclination of the sample and to carry out measurements of advancing and receding contact angles by the tilting plate method. The sample holder can be replaced with a metallic membrane loudspeaker (8 W nominal impedance, 6 cm diameter). In this way, using a simple amplifier circuit we can induce controlled vibrations in the drop. The acoustic signal could be generated both by a function generator (MITEK MK1050) or by an NiDAQ card and a LabView program. The latter approach is better, because it allows to have a good control on the duration of the applied vibration and it is possible to decrease the vibration amplitude in a controlled and continuous manner. The parameters used to perform drop vibration were: sine waveform at frequencies of 55 Hz or 263 Hz, empirically found as the most efficient; and fading time of the oscillation, 20 s. The amplitude was evaluated through its effects on the drop and was measured by the applied voltage.

The optics were comprised of a 10× Ealing lens connected to a Nikon 995 digital camera, which is piloted *via* software using a specific program called Krinnicam (http://krinnicam.cjb.net/). The drops are manually deposited using a Gilson Pipetman P10 micropipette. It is important to underline that the drop

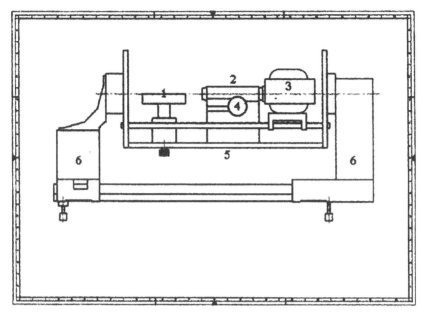

Figure 3. The scheme of the self-developed goniometer: (1) sample holder or loudspeaker, (2) lens tube, (3) digital camera, (4) screw for optical focusing, (5) rotating carriage, (6) stator. The dotted line denotes the optical axis of the instrument.

deposition is a really critical step and it could be a great source of errors. In particular, we found different values of contact angle depending on whether the drop was gently placed on the surface or it fell from the micropipette tip. All the values presented in this work were measured from gently dispensed drops keeping the micropipette perpendicular to the surface. The images obtained were analyzed using a Java plug-in, written for the powerful Open-Source program ImageJ. The plug-in can be downloaded for free from the link http://rsb.info.nih.gov/ij/plugins/contact-angle.html; it calculates the contact angle by applying the tangent approach, for drop shape to be considered as.

In order to assume the effect of gravity to be negligible, which deforms the drop equilibrium shape, thus invalidating the spherical drop-shape hypothesis, we used liquid volumes smaller than 5 μl. All the contact angle measurements were made at room temperature.

For Wilhelmy measurements a Cahn 322 microbalance and the same experimental procedure as described elsewhere [31–33] were used. The procedure is partly commented in the Discussion section. As a measuring liquid, ultrapure water (resistivity 18 MΩ·cm, produced by a Millipore MilliQ device) was employed.

The solid materials analyzed were poly(methyl methacrylate) (PMMA, Goodfellow, research grade) and commercial wax sheets for artistic application. Wax melting temperature was evaluated by DSC as 52°C, but the melting process covered a very large temperature interval, probably indicating a low purity of the material.

The commercial PMMA samples were cleaned using the procedure described in Ref. [38] for the removal of the thin protective film of static agent mentioned by the manufacturer.

The samples were first washed with trichloroethylene, then immersed in a non-ionic anionic detergent (Ausilab, 5 wt%, Antibioticos, Milan) and finally rinsed with water. Experimentally, it was observed that exposure of PMMA samples to trichloroethylene and detergent for a long time resulted in a considerable swelling of the sample. To avoid such a degree of swelling, two processes for sample cleaning were carried out in the analysis of contact angles, depending on duration of the trichloroethylene bath. We used 8 s in the first case and 12 in the second one. A longer time results in a considerable swelling of the surface, but a shorter one does not allow to clean the surface. Wax samples were used to compare surfaces of different roughnesses. Three different kinds of wax samples were analysed: (a) cut from the untreated sheet, (b) spin-coated on a glass slide and (c) obtained by pressing the sheet surface against a clean rough support, such as a metallic file.

To evaluate roughness, AFM images were acquired in the contact mode in air using an NT-MDT Solver P47 Pro AFM, equipped with silicon contact tips (conical shape, angle < 22 deg, typical curvature 10 nm, spring constant 0.03 N/m). The scan lateral size was 80 μm (corresponding to 512×512 pixels) and the determination of R_a, the roughness expressed as the absolute value average of surface point heights relative to the mean surface height, was done using the AFM built-in software.

Some statistical evaluations were made to compare the sets of results obtained by different methods (e.g., static and VIECA); the comparison was made using a standard Student's t-test. P-probability determines if there is a statistically significant difference between the two means. If this value is below a certain level (usually 0.05), the conclusion is that there is a difference between the two group means.

3. RESULTS

A sessile-drop goniometer adapted to VIECA measurements was used to determine the equilibrium contact angles of water on various surfaces of different compositions and roughnesses. More precisely, measurements were performed on commercial wax and PMMA samples produced by different techniques.

The wax samples were of three kinds:

- common sheets, with a mean roughness $R_a = 0.067$ μm and the Wenzel ratio [39] (between the actual area and the geometrical one) $r = 1.004 \pm 0.002$;

- wax spin coated onto glass slides, with a roughness coefficient $R_a = 0.250$ μm and the Wenzel ratio $r = 1.037 \pm 0.003$;

- common sheet roughened using a file. In this case the roughness was $R_a > 4$ μm and the Wenzel ratio r was not directly available, although a crude estimate provided values between 1.2 and 1.4.

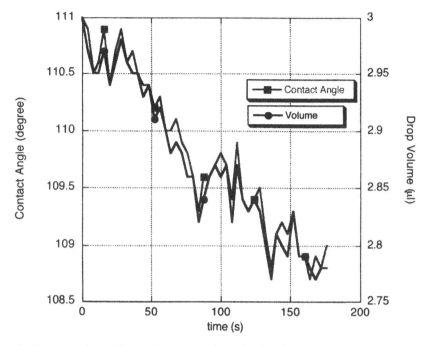

Figure 4. Contact angle and drop volume *versus* time, showing the effect of liquid evaporation.

As for the commercial PMMA (Goodfellow), it should be noted that such PMMA sheets are usually covered with a protective thin film, which must be removed prior to use by appropriate cleaning methods. The surface was cleaned with procedures which do not induce roughness and was monitored with ESCA for purity [38].

In principle, due to the relatively small size of drops, measurements are appreciably affected by liquid evaporation. As illustrated in Fig. 4, in our experimental conditions the natural relaxation of the drop and the effect of evaporation may typically cause a variation in the contact angle of 2 degrees within 3 min. In order to avoid such difficulty, the typical equilibrium measurement on the static device was performed within 40 s of the drop placement on the surface.

3.1. Wax results

The contact angle values obtained on wax samples are summarized in Tables 1–3 and plotted in Fig. 5 *versus* roughness (as R_a). Contact angles were obtained by both static measurements and the VIECA technique, the latter using vibrations at 55 or 263 Hz with a sine or square waveform, on 3 and 5 μl drops. The data obtained show a strong dependence of the static contact angle on the sample roughness, corresponding to the typical contact angle hysteresis due to surface roughness. In contrast, the VIECA angles are well reproducible, providing an almost constant estimate of the contact angle. Such an estimate is in satisfactory accord with the

Table 1.

Contact angles (CA) of water measured on cast wax samples

Type of CA	Wave frequency (Hz)	Waveform	Drop volume (μl)	Number of data points	CA (degrees)	SD (degrees)
Static			2	9	106.5	2.0
Static			3	10	106.4	1.0
Static			5	11	106.4	0.9
VIECA	55	sine	3	10	106.4	0.9

All the comparisons among the four CA means result in a P value much higher than 0.05, showing that the CA estimates are indistinguishable and are likely equal.

Table 2.

Contact angles (CA) of water measured on wax sheet samples

Type of CA	Wave frequency (Hz)	Waveform	Drop volume (μl)	Number of data points	CA (degrees)	SD (degrees)
Static			3	42	108.1	1.2
Static			5	29	107.4	1.6
VIECA	55	sine	3	13	106.2	1.4
VIECA	55	square	3	14	108.1	1.6
VIECA	263	sine	3	13	108.5	1.8
VIECA	263	square	3	5	106.9	2.2

For all the CA means comparisons the P values are larger than 0.05: CA means seem to be the same. The only exception comes from the comparison between data of row 1 (static CA, drop volume 3 μl) and row 3 (VIECA, frequency 55 Hz, sine waveform).

Table 3.

Contact angles (CA) of water measured on wax sheet samples roughened using a file

Type of CA	Wave frequency (Hz)	Waveform	Drop volume (μl)	Number of data points	CA (degrees)	SD (degrees)
Static			3	11	115.8	3.5
Static			5	21	115.4	4.4
VIECA	55	sine	5	6	108.6	5.8
VIECA	55	square	5	26	109.6	5.0
VIECA	263	sine	5	10	107.2	3.8
VIECA	263	square	5	6	106.1	3.5

For all the comparisons of the six CA estimates we calculate a P value lower than 0.05, thus evidencing that CA estimates are significantly different.

equilibrium value found by Wilhelmy microbalance measurements. It should be pointed out that, typically, the equilibrium angle is better determined using higher vibration frequencies, which turn out to be more efficient in inducing relaxation to equilibrium of the drop on the sample surface, no matter what the waveform of the vibration is, sine or square.

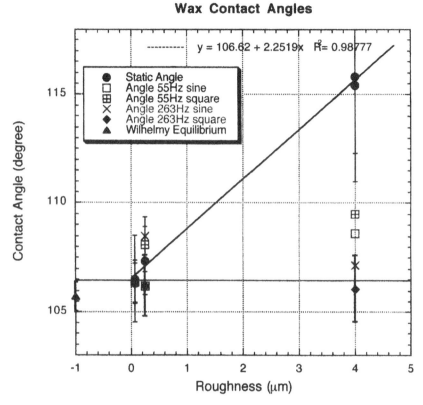

Figure 5. Contact angles of water measured on wax surfaces of various roughnesses. The smallest roughness corresponds to the cast wax samples of Table 1, while the largest roughness concerns the wax sheet samples roughened by a file (Table 3). The intermediate value of roughness is measured on the untreated wax sheet samples of Table 2.

For these samples the means of the results obtained through static and VIECA techniques are different (i.e., $P < 0.05$ in Student's t-test) only for rough surfaces. All the VIECA results are indistinguishable (except one case) among them.

3.2. PMMA results

PMMA data are plotted in Fig. 6a–6c and Tables 4–7 for 3 μl or 5 μl water drops on PMMA surfaces which were cleaned according to the two different procedures described earlier. The experimental results clearly show that the proposed cleaning prescriptions do not affect measurements in an appreciable way.

As previously mentioned in Materials and Methods, in both static and VIECA measurements contact-angle estimates were obtained in a spherical approximation, by assuming that the effect of gravity could be neglected and the surface of the drop was well approximated with a spherical cap. This approach imposes an important

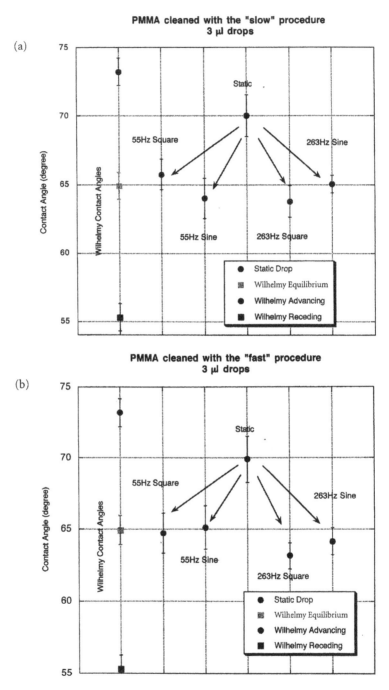

Figure 6. (a) Contact angles of water measured on PMMA. VIECA data are for 3 μl drops placed on PMMA surfaces cleaned with the slow procedure (long time) described in the text, data of Table 4. (b) Contact angles of water measured on PMMA. VIECA data are for 3 μl drops placed on PMMA surfaces cleaned with the fast procedure (short time) described in the text, data of Table 5. (c) Contact angles of water measured on PMMA. VIECA data are for 5 μl drops placed on PMMA surfaces cleaned with the fast procedure (short time) described in the text, data of Table 7.

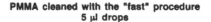

PMMA cleaned with the "fast" procedure
5 μl drops

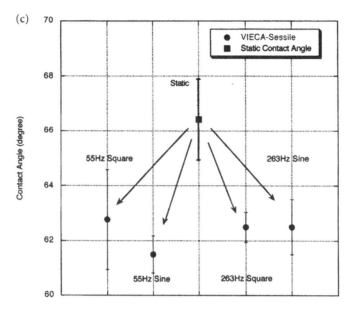

Figure 6. (Continued).

Table 4.
Contact angles (CA) of water drops on PMMA samples cleaned in a trichloroethylene bath for 12 s (drop volume is 3 μl)

Type of CA	Wave frequency (Hz)	Waveform	Drop volume (μl)	Number of data points	CA (degrees)	SD (degrees)
Static			3	43	69.9	1.5
VIECA	55	sine	3	10	64.0	1.5
VIECA	55	square	3	9	65.7	1.1
VIECA	263	sine	3	12	65.0	0.6
VIECA	263	square	3	13	63.8	1.2

The comparisons between static and VIECA estimates provide P values lower than 0.05 and thus suggest that static and VIECA results are actually different. All the calculated P values are higher than 0.05 for comparisons among VIECA data: VIECA estimates appear self-consistent.

limitation to the volume and, therefore, to the typical size of the drop used, in order to avoid contribution of gravity to the free energy of the system.

The successful application of the VIECA technique requires an effective supply of vibrational/mechanical energy to the drop/surface system. This goal can be achieved by exciting the normal modes of oscillation of the system by means of external vibrations of appropriate frequency, which induce resonant capillary waves mainly at the surface, but also in the bulk, of the drop. Example images of resonant capillary

Table 5.
Contact angles (CA) of water drops on PMMA samples cleaned in a trichloroethylene bath for 8 s (drop volume is 3 μl)

Type of CA	Wave frequency (Hz)	Waveform	Drop volume (μl)	Number of data points	CA (degrees)	SD (degrees)
Static			3	42	69.9	1.6
VIECA	55	sine	3	11	65.1	1.5
VIECA	55	square	3	11	64.7	1.4
VIECA	263	sine	3	11	64.1	0.9
VIECA	263	square	3	11	63.1	0.9

All the calculated P values are lower than 0.05 for comparisons between static and VIECA results; the calculated P values turn out to be higher than 0.05 for comparisons among VIECA data. Yet, static and VIECA angles seem significantly different, whereas VIECA results show a good self-consistency.

Table 6.
Contact angles (CA) of water drops on PMMA samples cleaned in a trichloroethylene bath for 12 s (drop volume is 5 μl)

Type of CA	Wave frequency (Hz)	Waveform	Drop volume (μl)	Number of data points	CA (degrees)	SD (degrees)
Static			5	43	67.1	1.7
VIECA	55	sine	5	11	60.7	0.9
VIECA	55	square	5	12	61.3	0.8
VIECA	263	sine	5	11	61.5	1.0
VIECA	263	square	5	11	65.6	0.6

All the calculated P values are lower than 0.05 for the comparison between static and VIECA results, with the only exception of data in row 4 (VIECA angle, frequency 263 Hz, sine waveform): VIECA and static CA appear different. The calculated P values are larger than 0.05 for all the comparisons among VIECA angles, which thus must be regarded as self-consistent.

Table 7.
Contact angles (CA) of water drops on PMMA samples cleaned in a trichloroethylene bath for 8 s (drop volume is 5 μl)

Type of CA	Wave frequency (Hz)	Waveform	Drop volume (μl)	Number of data points	CA (degrees)	SD (degrees)
Static			5	42	66.4	1.5
VIECA	55	sine	5	11	61.5	0.7
VIECA	55	square	5	10	62.8	1.8
VIECA	263	sine	5	11	62.5	1.0
VIECA	263	square	5	11	62.5	0.5

The calculated P values are smaller than 0.05 for all the comparisons between static and VIECA results, thus signalling that the two measurement procedures lead to significantly different estimates. The calculated P values are larger than 0.05 for all the comparisons among VIECA angles and must be regarded as self-consistent.

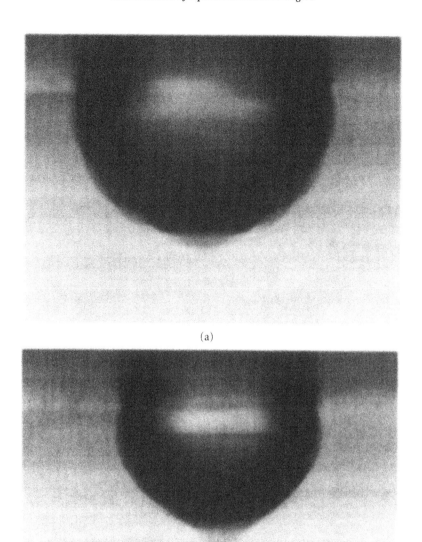

(a)

(b)

Figure 7. (a) Typical appearance of a resonant oscillating drop. The oscillations normal mode represented is axisymmetric with mode number $M = 3$, as evidenced by the number of nodal lines (the three horizontal circles on the drop surface, represented by pairs of points at the same height in this view, where the liquid remains approximately at rest during vibration). After partial evaporation of the liquid, the drop was then vibrated at 265 Hz by a square waveform. (b) The same as in Fig. 7a, but for a non-axisymmetric normal mode.

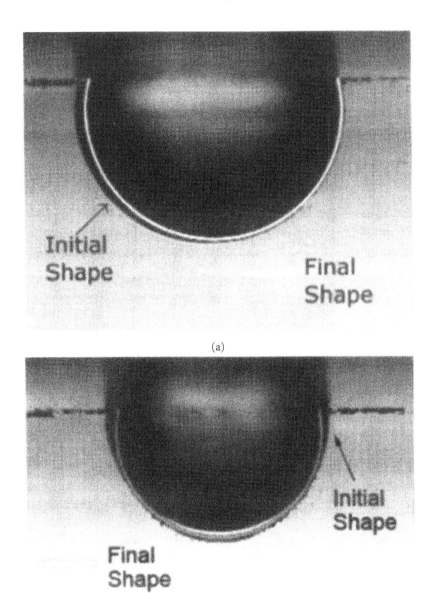

(a)

(b)

Figure 8. (a) Effect of vibration on a water drop on PMMA. The initial shape corresponds to a receding contact angle obtained by a slight evaporation and the final shape is due to relaxation. (b) Comparison of the drop profile before and after vibration for a water drop on wax. The initial shape corresponds to a receding contact angle obtained by evaporation of the liquid.

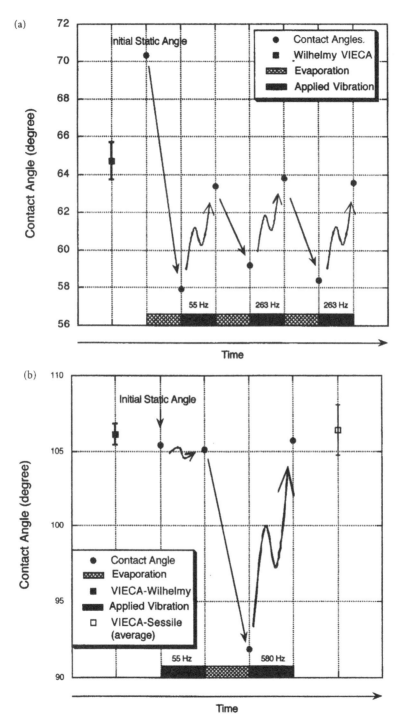

Figure 9. (a) VIECA estimates obtained starting from receding drop profiles, obtained by evaporation. Effect of repeated cycles of evaporations and vibrations on water drops on wax. (b) Water drop on wax: effect of vibration on receding drop profiles obtained by evaporation. Effect of a vibration–evaporation–vibration sequence on water drops on wax.

waves in the drop are illustrated in Fig. 7a for the axisymmetric case, a vibrational normal mode of order $M = 3$, and in Fig. 7b for the non-axisymmetric case. After partial evaporation, both drops were vibrated at 265 Hz by a square waveform. A comparison of drop profiles before and after vibration is shown in Fig. 8a and 8b.

The case of drops close to the receding state is particularly interesting; the receding state was obtained by allowing evaporation of the drop. The supply of mechanical energy induces the system to relax to a well-reproducible value of contact angle, which does not change in an appreciable way whenever subsequent evaporation/vibration cycles are performed on the system. This estimate of equilibrium contact angle is in good accord with the determination of both static and Wilhelmy VIECA, as illustrated in Fig. 9a for PMMA and in Fig. 9b for wax.

In the case of PMMA all the static results are statistically different (i.e. $P < 0.05$) from the VIECA ones. All the VIECA results are indistinguishable (except one case) among them.

4. DISCUSSION

The previous results have important implications in both the theory as well as the practice of surface free energy calculation of solids.

First of all we should remember that the differences between static and VIECA angles appear statistically significant for both rough wax as well as for PMMA, whose roughness is very low [38]. The reason for this difference is not clear; the roughnesses of cast wax or common sheet and PMMA are comparable and very low. A possible reason may be that increasing the interaction between the liquid and the solid (i.e., decreasing the contact angle, or increasing the work of adhesion) the potential barriers, which contribute to the metastability, increase. Further work is necessary to settle this argument.

From a purely numerical point of view it is evident that the significant difference between the advancing or static contact angle and the VIECA contact angle, which constitutes a closer approximation to equilibrium contact angle, appreciably affects the estimate of surface free energy of solids, in agreement with some warnings already available in the literature [10, 36]. In particular, it can be claimed that typically the use of the advancing contact angle underestimates the surface free energy of the solid, and that a more realistic estimate should be made using the VIECA contact angles.

Another important consequence, which has significant implications for practical calculations, is that Young and VIECA angles of a stable equilibrium state of the system could be considered coincident for a Wenzel ratio up to $r = 1.2$–1.4 on randomly enough, essentially homogeneous surfaces. The statament is true at least in the interval of values covered by the two materials studied (70–110°).

However, one should consider the effect of the Wenzel correction on the values of intrinsic angle and the results obtained by Kamusewitz and co-workers [36, 37].

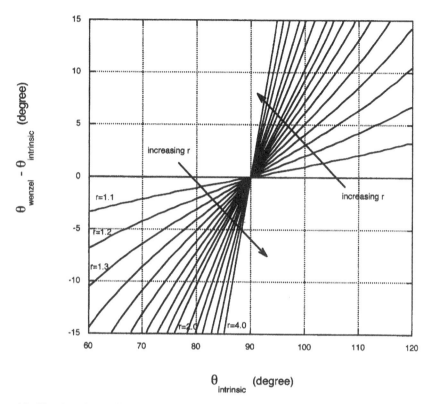

Figure 10. The plot of the difference between the Wenzel angle $\theta_{Wenzel} = \arccos(r \cdot \cos \theta_{intrinsic})$ and the intrinsic contact angle $\theta_{intrinsic}$ for values of the Wenzel ratio r from $r = 1.0$ to 4.0 and $\theta_{intrinsic}$ in the interval $60-120°$.

From Fig. 10 it is possible to accept the idea that in the indicated interval the difference between the intrinsic angle and the Wenzel angle is, however, low. Out of this interval, as repeatedly pointed out by other authors, such as Kamusewitz and co-workers [36, 37], the apparent symmetry between advancing and receding contact angles with respect to the intrinsic angle fails (in other terms, the Kamusewitz graph is asymmetric).

A possible interpretation of our findings is that the VIECA angle could be considered as an experimental evaluation of the Wenzel angle, whose value is very close to the intrinsic Young's angle only for contact angles close to 90° on moderately rough surfaces. On higher roughness surfaces (e.g., for "composite" surfaces) and/or for different values of contact angles the difference between the Wenzel and the intrinsic or Young angle should become significant. A logical consequence is that the simple equation (1) would not be valid when contact angles are not close to 90° and are measured on a slightly rough surface.

Equation (1) assumes in fact symmetry of the shape of the curve of surface free energy *vs.* cosine of contact angles (or *vs.* contact angle). Symmetry around the

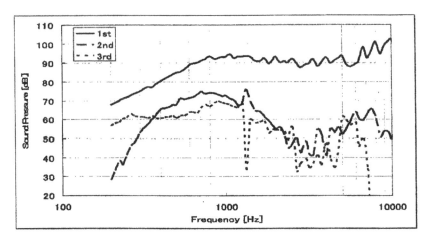

Figure 11. The power spectrum of three common commercially available loudspeakers: the efficiency of the loudspeaker is expressed by the sound pressure *vs.* the frequency.

equilibrium point simply implies:

$$\Delta G_{\text{interf}} = g_1(\cos\theta - \cos\theta_{\text{equ}}), \tag{2}$$

where g_1 is any non-negative even function.

This hypothesis appears probably too restrictive to be valid in all cases. A wider set of measurements is necessary to answer this question.

A third point to consider is the correspondence between the theory of mechanical oscillations of a constrained drop and our findings in terms of frequencies able to obtain the resonance of drop profile and the relaxation of drop meniscus.

In the case of Wilhelmy VIECA setup the size of the loudspeaker is, in fact, comparable to the size of the water phase used in the experiment; on the contrary, in the case of sessile VIECA setup the loudspeaker is about 10–100-times larger than the drop. This situation is different from other experimental setups [35, 40–44]. Moreover, the efficiency of a common loudspeaker, expressed by the sound pressure *vs.* the frequency, strongly reduces at the low frequencies useful in our case (see Fig. 11).

There are two main theories [40–42] about the bounded or constrained drop; the one by Atachi and Takaki [40] is valid for the polygonal vibrations, for gravity flattened spherical drops and is expressed by the formula

$$F = \frac{1}{2\pi}\sqrt{\frac{\gamma M^3}{\rho_1 R^3}}, \tag{3}$$

where F is the frequency, M the mode number, R the radius of the sphere, ρ_l the liquid density and γ the surface tension of the liquid.

Table 8.
Gross comparison of calculated and used frequencies

Mode number	Frequency (Hz), according to Ref. [40]	Coefficient λ, according to Ref. [42]	Frequency (Hz), according to Ref. [42]
1	21	0.08	75
2	60	0.01	180
3	110	0.005	300
4	170		
5	238		
6	313		

The second model, by Strani and Sabetta [41, 42], is valid for axisymmetric vibrations of drops in contact with a spherical bowl and is expressed by

$$F = \frac{1}{2\pi} \sqrt{\frac{\gamma}{\rho_l R^3 \lambda(\theta, \rho_v/\rho_l)}}, \tag{4}$$

where the mode number is substituted by a non-dimensional coefficient λ, which is a function of the contact angle θ and of the density ratio ρ_v/ρ_l between the vapor and liquid phases. This function cannot be written in an explicit form, but is only estimated numerically by solving a generalized nonlinear eigenvalue problem; numerical solutions are plotted in Ref. [42].

A gross comparison of the calculated and used frequencies shows that:

– in the Wilhelmy method the frequency was invariably 53–55 Hz;

– in the sessile method the frequencies ranged from 50 to about 600 Hz, especially 55 and 263 Hz, the empirically found values.

Remarkably, in Ref. [42] the values of λ are tabulated only for θ larger than 90°. Typical results for a water drop of 5 μl, between 1 and 2 mm radius, on wax are presented in Table 8, comparing different mode number oscillations, where, for the axisymmetric oscillations considered here, the mode number is simply the number of nodal lines (horizontal circles on the drop surface where the liquid remains essentially at rest during vibration).

One can see that there is a qualitative agreement among these calculated frequencies and those used experimentally, as illustrated before. However, given the low efficiency of the loudspeakers in the the low-frequency interval, there are only two choices to obtain measurements to compare with theory: to use large or very large drops (as in Ref. [35]) or to change the vibration device.

We want to comment also on the lack of statistical difference among the results obtained by different cleaning methods for PMMA. These methods were developed in a previous paper [38] and slightly adjusted here, modifying the treatment time. From a commercial material such as PMMA, using a treatment time between 8 and 12 s allows to obtain a surface chemically and topologically very close to ideality (pure material, well correlated to the formal chemical composition and with a very low roughness [38]).

5. CONCLUSIONS

The VIECA method has been applied to sessile drops and has provided an equilibrium contact angle whose value is independent of the starting state (advancing or receding) and roughness (in the chosen moderate interval of roughness). The results are in good agreement also with those obtained by the Wilhelmy VIECA setup.

These findings corroborate the idea that the "vibrated angle" is a better approximation to the equilibrium quantity for use in surface free energy calculations.

A wider set of data is, however, necessary to test the proposed simple equation (1), especially using systems for which the contact angles are outside the interval 70–100°.

A qualitative agreement is found between the frequencies calculated by the theory and those used in the experiment. This is partly due to the low efficiency of loudspeakers to transfer mechanical energy at low frequencies (under 100 Hz), particularly to objects very small compared with the loudspeaker membrane size.

Acknowledgements

We gratefully acknowledge JEPA-Limmat Foundation for the 1-year Scholarship to T. W. at the DIMTI (University of Trento, Italy).

REFERENCES

1. J. F. Padday, in *Surface and Colloid Science*, E. Matijevic (Ed.), Vol. 1, p. 101. John Wiley, New York, NY (1969).
2. J. M. Douillard, *J. Colloid Interface Sci.* **188**, 511–514 (1997).
3. A. B. D. Cassie and S. Baxter, *Trans. Faraday Soc.* **40**, 546–551 (1944).
4. A. B. D. Cassie, *Disc. Faraday Soc.* **3**, 11 (1948).
5. R. Shuttleworth and C. L. J. Bailey, *Disc. Faraday Soc.* **3**, 16–22 (1948).
6. R. E. Johnson Jr. and R. H. Dettre, *J. Phys. Chem.* **68**, 1744–1750 (1964).
7. D. O. Jordan and J. E. Lane, *Aust. J. Chem.* **17**, 7–15 (1964).
8. A. W. Neumann and R. J. Good, *J. Colloid Interface Sci.* **28**, 341–358 (1972).
9. D. Li and A. W. Neumann, in *Applied Surface Thermodynamics*, A. W. Neumann and J. K. Spelt (Eds), p. 109. Marcel Dekker, New York, NY (1996).
10. C. Huh and S. G. Mason, *J. Colloid Interface Sci.* **60**, 11 (1977).
11. J. F. Joanny and P. G. de Gennes, *J. Chem. Phys.* **81**, 552–562 (1984).
12. V. M. Starov, *Adv. Colloid Interface Sci.* **39**, 147–173 (1992).
13. E. L. Decker, B. Frank, Y. Suo and S. Garoff, *Colloids Surfaces A* **156**, 177–188 (1999).
14. U. Öpik, *J. Colloid Interface Sci.* **223**, 143–166 (2000).
15. R. E. Johnson Jr. and R. H. Dettre, *J. Phys. Chem.* **69**, 1507–1514 (1965).
16. R.-D. Schultze, W. Possart, H. Kamusewitz and C. J. Bischof, *J. Adhesion Sci. Technol.* **3**, 39–48 (1989).

17. J. M. Di Meglio, *Europhys. Lett.* **17**, 607–612 (1992).
18. R. E. Johnson Jr. and R. H. Dettre, in *Wettability*, J. C. Berg (Ed.), pp. 2–71. Marcel Dekker, New York, NY (1993).
19. L. W. Schwartz and S. Garoff, *J. Colloid Interface Sci.* **106**, 422–436 (1985).
20. R. E. Johnson Jr. and R. H. Dettre, *J. Adhesion* **2**, 3–15 (1970).
21. A. Marmur, *Colloids Surfaces A* **116**, 55–61 (1996).
22. G. Wolansky and A. Marmur, *Langmuir* **14**, 5292–5297 (1998).
23. G. Wolansky and A. Marmur, *Colloids Surfaces A* **156**, 381–388 (1999).
24. E. Wolfram and R. Faust, in *Wetting, Spreading and Adhesion*, J. F. Padday (Ed.), pp. 213–238. Academic Press, London (1978).
25. G. R. M. del Giudice, *Eng. Mining J.* **137**, 291–294 (1936).
26. F. M. Fowkes and W. D. Harkins, *J. Am. Chem. Soc.* **62**, 3377–3386 (1940).
27. W. Phillipoff, S. R. B. Cooke and D. E. Caldwell, *Mining Eng.* **4**, 283–286 (1952).
28. T. Smith and G. J. Lindberg, *J. Colloid Interface Sci.* **66**, 363 (1978).
29. C. Andrieu, C. Sykes and F. Brochard, *Langmuir* **10**, 2077 (1994).
30. E. L. Decker and S. Garoff, *Langmuir* **12**, 2100 (1996).
31. C. Della Volpe, D. Maniglio, S. Siboni and M. Morra, *Oil Gas Sci. Technol.* **56**, 9–22 (2001).
32. C. Della Volpe, D. Maniglio, M. Morra and S. Siboni, *Colloids Surfaces A* **206**, 46–67 (2002).
33. C. Della Volpe and S. Siboni, in *Contact Angle, Wettability and Adhesion*, Vol. 2, K. L. Mittal (Ed.), pp. 45–71. VSP, Utrecht (2002).
34. M. Fabretto, R. Sedev and J. Ralston, in *Contact Angle, Wettability and Adhesion*, Vol. 3, K. L. Mittal (Ed.), pp. 161–173. VSP, Utrecht (2003).
35. X. Noblin, A. Buguin and F. Brochard-Wyart, *Eur. Phys. J. E* **14**, 395 (2004).
36. H. Kamusewitz, W. Possart and D. Paul, *Colloids Surfaces A* **156**, 271–279 (1999).
37. H. Kamusewitz and W. Possart, *Appl. Phys. A* **76**, 899–902 (2003).
38. C. Della Volpe, S. Siboni, D. Maniglio, M. Morra, C. Cassinelli, M. Anderle, G. Speranza, R. Canteri, C. Pederzolli, G. Gottardi, S. Janikowska and A. Lui, *J. Adhesion Sci. Technol.* **17**, 1425–1456 (2003).
39. R. N. Wenzel, *Ind. Eng. Chem.* **28**, 988 (1936).
40. K. Adachi and R. J. Takaki, *J. Phys. Soc. Jpn.* **53**, 4184 (1984).
41. M. Strani and F. Sabetta, *J. Fluid Mech.* **141**, 233 (1984).
42. M. Strani and F. Sabetta, *J. Fluid Mech.* **189**, 397–421 (1988).
43. H. Rodot and C. Bisch, in *Proc. 5th European Symp. on Material Sciences under Microgravity*, Schloss Elmau, Germany, 5–7 November 1984, European Space Agency, Paper ESA SP-222, pp. 23–29 (1984).
44. R. W. Smithwick III and D. M. Hembree, *J. Colloid Interface Sci.* **140**, 57–65 (1990).

Contact Angle, Wettability and Adhesion, Vol. 4, pp. 101–114
Ed. K.L. Mittal
© VSP 2006

The static contact angle hysteresis and Young's equilibrium contact angle

H. KAMUSEWITZ[1,*] and W. POSSART[2]

[1]*GKSS Research Centre, Institute of Polymer Research, Kantstrasse 55, D-14513 Teltow, Germany*
[2]*Saarland University, Geb. 22-6, P.O. Box 151150, D-66041 Saarbrücken, Germany*

Abstract—Compared to an ideally flat surface, the roughness profile of a real solid results in an additional deformation of the triple line region. This can be described by Good's contortional energy parameter F. In the literature, equal F-values are assumed for both wetting and dewetting experiments on homogeneous surfaces. Here we present a new approach for the missing experimental evaluation of this assumption. A series of ethylene glycol contact angle measurements on three pure solids (paraffin wax, polypropylene and polyetherimide), each covering the whole range from smooth to rough surfaces, show that this assumption is correct for these materials. As a consequence, Wenzel's contact angle is obtained as the arithmetic mean of the cosines of advancing and receding contact angles on a given solid. The additional determination of Wenzel's roughness factor r by an independent Scanning Force Microscopy study enables determination of Young's contact angle at thermodynamic equilibrium from the advancing and receding contact angles without varying solid surface roughness in the wetting experiments.

Keywords: Contact angles; equilibrium; contact angle hysteresis; roughness.

1. INTRODUCTION

For many years, wetting experiments have been utilized for calculating the surface energy of polymer solids. Different approaches and models are employed for these calculations, which require wetting experiments with at least two test liquids with different surface tensions on the solid surface provided that the roughness does not give rise to the inclusion of gas bubbles at the solid–liquid contact, as is the case for a 'composite' surface considered by Cassie and Baxter [1]. A comparison of the various approaches and a discussion of their potentials and limitations are given by Good [2]. Generally, all of these approaches require a knowledge of Young's contact angle at thermodynamic equilibrium, θ_e, which has to be derived from the experiment on an ideally smooth, homogeneous and rigid solid in contact with the pure test liquid (no solvent for the solid, no gas bubbles

---

*To whom correspondence should be addressed. Tel.: (49-3328) 352-458;
Fax: (49-3328) 352-452; e-mail: helmut.kamusewitz@gkss.de

trapped!) surrounded by the liquid vapor. Normally this angle is not measurable because on real solids a wide spectrum of contact angles appears depending on the experimental conditions. Instead of θ_e, the common static and quasi-static measurements give two contact angles on a real solid at the triple line of the three-phase system in local thermodynamic equilibrium: the advancing angle, θ_a, after moving the triple line forward on the dry area of the solid and the smaller receding angle, θ_r, after retreating the triple line from a formerly wetted region of the surface. The difference in the two values is called the contact angle hysteresis, $\Delta\theta$, with $\Delta\theta = \theta_a - \theta_r > 0$. Hence, a way must be found for deriving θ_e from the data measured on real surfaces with their usual statistical roughness. This is a key issue in the interpretation of wetting experiments because θ_e is required for the calculation of the work of adhesion and in all other thermodynamic considerations of wetting. Although the investigation of model surfaces has improved understanding of the physics of contact angle hysteresis (e.g., Refs [3–5]), but that is beyond the scope of this paper which deals with real solids with random roughness.

It is a common practice to discuss the θ_a data themselves or the arithmetic mean of θ_a and θ_r instead of θ_e. Such attempts cause an error which increases with the observed contact angle hysteresis. As shown in our previous works [6–9], θ_e can be obtained for homogeneous low surface energy solids, like polymers, from wetting experiments by varying the sample roughness provided that no gas bubbles are trapped at the solid–liquid contact. For each roughness value, both contact angles θ_a and θ_r are measured. We showed that the linear functions

$$\theta_a = \theta_e + A^a \Delta\theta$$

$$\theta_r = \theta_e + A^r \Delta\theta$$

(1)

with A^a and A^r as the corresponding slopes fit very well to the measured contact angle data. Extrapolation to $\Delta\theta = 0$ results in θ_e. However, it is worth noting that this approach does not require any additional information about the surface roughness. However, it is not applicable in the case of heterogeneous or 'composite' surfaces.

Wenzel [10] proposed to relate the topological description of a rough surface by the roughness factor, r, as:

$$r = \frac{A_{real}}{A_{geom}}$$

(2)

with the contact angle given by the equation

$$\cos\theta^{W} = r \cdot \cos\theta_{e}. \tag{3}$$

Again, this approach refers only to contact angle data measured on non-composite surfaces. Obviously, the so-called Wenzel angle θ^{W} depends on two unknown quantities: the roughness factor r and the equilibrium contact angle θ_{e}. According to equation (2), we always have $r > 1$ for rough surfaces and $r = 1$ for smooth surfaces. It is evident that additional assumptions are necessary in order to relate θ^{W} to the actually measured angles θ_{a} or θ_{r}. There is no good reason to assume that expressions such as $\theta^{W} = \theta_{a}$ or $\theta^{W} = \theta_{r}$ are valid.

In any case, introducing equation (3) in Young's equation

$$\gamma_{sv} = \gamma_{sl} + \gamma_{lv} \cdot \cos\theta_{e} \tag{4}$$

provides the following expression

$$r \cdot (\gamma_{sv} - \gamma_{sl}) = \gamma_{lv} \cdot \cos\theta^{W} \tag{5}$$

which relates the two unknowns r and θ^{W} to the interfacial tensions in the wetting experiment. Due to equation (3), θ^{W} may possess only a single value for a given r. Hence, Wenzel's approach does not explain why two contact angles are measured.

Therefore, Adam and Jessop [11] proposed the following equations:

$$r \cdot (\gamma_{sv} - \gamma_{sl}) = \gamma_{lv} \cos\theta_{a} + F \tag{6}$$

$$r \cdot (\gamma_{sv} - \gamma_{sl}) = \gamma_{lv} \cos\theta_{r} - F \tag{7}$$

In equations (6) and (7), the new empirical parameter F allows to relate the measured contact angles θ_{a} and θ_{r} to the roughness factor r.

Later, Good applied thermodynamic arguments and explained the parameter F as the 'contortional energy', which is caused by the deformation of the liquid meniscus in the vicinity of the rough solid as compared with the meniscus shape on the corresponding smooth surface [12]. For simplicity, the same F value is used in cases of both advancing and receding contact angle measurements.

According to [13], the combination of equations (5)–(7) yields

$$\cos\theta^{W} = 0.5 \cdot (\cos\theta_{a} + \cos\theta_{r}), \tag{8}$$

and using equation (3) we obtain

$$\cos\theta_{e} = \frac{1}{2r}(\cos\theta_{a} + \cos\theta_{r}). \tag{9}$$

Obviously, θ_e could be obtained by knowing only the roughness parameter r. As shown in [13], a suitable r-value can be derived from appropriate surface profile measurements with a scanning force microscope (see Appendix for a brief discussion).

However, it is not clear if the assumption of the same F value can be justified in equations (6) and (7). The contortional energy could be different in the measurements of advancing and receding contact angles. Therefore, we analyze the validity of the assumption regarding F and discuss the consequences in this paper. In order to do so, some previously published results on wetting experiments on paraffin wax and polypropylene will be reconsidered.

2. MODEL CONSIDERATIONS

In the general case, equations (6) and (7) have to be replaced by

$$r \cdot (\gamma_{sv} - \gamma_{sl}) = \gamma_{lv} \cos\theta_a + F^a \tag{10}$$

$$r \cdot (\gamma_{sv} - \gamma_{sl}) = \gamma_{lv} \cos\theta_r - F^r \tag{11}$$

and, hence,

$$\cos\theta^W = 0.5 \cdot (\cos\theta_a + \cos\theta_r) + \frac{(F^a - F^r)}{2\gamma_{lv}} \tag{12}$$

is obtained instead of equation (8).

The term $(F^a - F^r)/2\gamma_{lv}$ represents the difference between the special case of equation (8) and the generalized model given by equations (10)–(12).

Introducing Young's equation (4) in equations (10) and (11), F^a as well as F^r can be calculated separately, provided θ_e and r have been obtained from independent measurements like wetting experiments and scanning force microscopy investigations of solids with varying roughness [13].

Moreover, subtraction of the two equations (10) and (11) yields

$$F^a + F^r = \gamma_{lv} \cdot (\cos\theta_r - \cos\theta_a). \tag{13}$$

Hence, the sum of the two F-parameters can always be calculated from the measured contact angle data.

3. MATERIALS AND METHODS

The wetting measurements were performed with the Wilhelmy balance for paraffin wax and with the sessile drop method for polypropylene and polyetherimide. Scanning force microscopy was used for the characterization of selected samples with surfaces of different roughnesses made from polypropylene and polyetherimide.

The paraffin wax surfaces were prepared on the walls of polypropylene tubes by dipping them quickly into molten paraffin wax. As a result, the tubes were filled with the paraffin wax and their outer surface was covered with a smooth paraffin wax layer. These surfaces were roughened by gentle contact with different brushes, soft textiles or tissues. The paraffin wax layer thickness did not change appreciably during this procedure (see Ref. [9] for more details of paraffin wax surface preparation and dynamic measurements). The quasi-static experiments with the Wilhelmy technique were performed at 20°C for each paraffin wax sample along with three wetting/dewetting cycles at a rate of 2 mm/min. The maximum immersion depth was 12 mm.

On flat polypropylene and polyetherimide samples, contact angle measurements were performed with large sessile drops (base diameter 3–12 mm). Additionally, contact angle data from the literature [3] are included in the discussion which were measured on drops with up to 60 mm diameter. The static contact angles θ_a and θ_r were measured after increasing and reducing the drop-base area, respectively. Each advancing or receding value is the average of more than 30 measurements at three different places on each surface. The contact angle values scattered by about 1° on the unroughened, cleaned materials and by less than 5° on the roughened surfaces. The polypropylene samples were cut from a commercial film (Goodfellow). Polyetherimide films were obtained by spin coating or casting using methylene chloride as solvent. The film roughness was varied by employing sandpapers of different grades. More details on static measurements and sample preparation are published in Refs [6, 9, 13].

The scanning force microscope (SFM) used was a NanoScope IIIa equipment (Digital Instruments). Constant force mode images were taken in air at 3 Hz scan rate for scan areas between 100×100 nm^2 and 120×120 μm^2. The cantilever (ESPW-CONT, NanoProbe) was made from silicon with a spring constant of 0.020–0.066 N/m and with *ca.* 10 nm tip radius. The images were processed by the plan-fit operation without additional image manipulation (Software version: 5.12b15; *cf.*, Appendix for details).

4. REVIEW OF PUBLISHED DATA, OUR EXPERIMENTAL RESULTS AND DISCUSSION

4.1. Paraffin wax

Figure 1 compares the contact angle data obtained with ethylene glycol on paraffin wax surfaces with varying roughness, with the sessile drop method [3] or the Wilhelmy technique [9]. Using linear regression according to equation (1), the results in both cases show a high quality of the linear fits. The arithmetic mean of θ_a and θ_r as well as the average of $\cos\theta_a$ and $\cos\theta_r$ depend on $\varDelta\theta$. Hence, such mean values cannot be equal to θ_e as it is proposed in [14–16] and many other papers. Instead, θ_e is obtained on the ordinate by extrapolating the two linear fits to $\varDelta\theta = 0$. In Fig. 1, the θ_e-values around 80° are fully reproducible both for the literature data and our own data. We note that Bartell and Shepard [3] prepared paraffin wax model surfaces consisting of regular four-sided pyramids, whereas we studied surfaces with random roughness profiles [9]. Hence, neither the slow motion of the triple line in the Wilhelmy experiment, nor the differences in the surface profiles is important for θ_e in the case of these two wetting experiments. The different values for the slopes $|A_a|$ and $|A_r|$ are considered to result from the different surface topographies given by the regular pyramids and the random profiles. The r-values have not been determined for our paraffin wax surfaces.

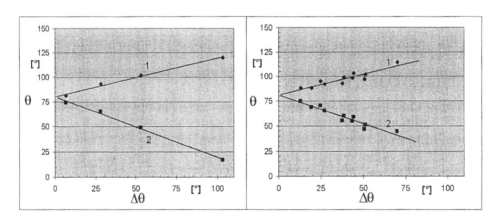

Figure 1. Advancing (1) and receding (2) contact angles measured with ethylene glycol on paraffin wax samples with varying contact angle hysteresis, $\varDelta\theta$, due to roughness variation. (Left) Data from Ref. [12] (A^a = 0.3039; A^r = -0.6061; θ_e = 80°). (Right) Our measurements [9] (A^a = 0.4231; A^r = -0.5769; θ_e = 80°).

For the model surfaces, the roughness factor is given by $r = \cos^{-1}\beta$, where β is the inclination angle of the four side walls with respect to the base of the pyramids. From the data in Ref. [3], we obtain $r = 2$ for $\beta = 60°$, $r = 1.414$ for $\beta = 45°$, $r = 1.155$ for $\beta = 30°$ and, of course, $r = 1$ for smooth paraffin wax. Now, with these r-data and with θ_e from the linear fits according to Fig. 1, equation (3) provides the data for $r \cdot \cos\theta_e = \cos\theta^W$ for the four model surfaces. A second set of $\cos\theta^W$-data is obtained with equation (8) and the measured data for θ_a and θ_r.

Figure 2 compares the two sets for $\cos\theta^W$. The data for $r \cdot \cos\theta_e$ from equation (3) for each paraffin wax surface provide the x-axis. The data points (■) obtained from equation (8) do not fit the straight line with 45° slope. This indicates that equations (3) and (8) are not equivalent for the model paraffin wax surfaces. At a first glance, this could indicate that the assumption $F^a = F^r$ could be not correct for the paraffin wax. As a check, the r-data, the deduced θ_e values and the data for γ_{lv}, θ_a and θ_r can be used to calculate the values of F^a and F^r for each rough paraffin wax surface from equations (4), (10) and (11). With these F-data, equation (12) provides a third set of $\cos\theta^W$ values, the points marked (♦) in Fig. 2. Again, they do not follow the line with 45° slope. We conclude that the r-data obtained for the pyramids do not describe the wetting behavior correctly.

Of course, usual solids with random surface profiles do not possess regular roughness elements. Therefore, in future work measurements with paraffin wax samples and random roughness are needed for determining Wenzel's r-factor from SFM and for checking the assumption $F^a = F^r$.

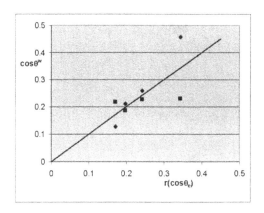

Figure 2. Comparison of $\cos\theta^W$ for the system paraffin wax–ethylene glycol [12], applying different equations. (■) $\cos\theta^W$ according to equation (8); (♦) $\cos\theta^W$ according to equation (12). Abscissa: reference data for $\cos\theta^W$ according to equation (3): $\cos\theta^W = r \cdot \cos\theta_e$.

4.2. Polypropylene

As a second homogeneous solid, polypropylene (PP) with varying random surface roughness was investigated using ethylene glycol (γ_{lv} = 48.4 mN/m) and diethyl-ene glycol (γ_{lv} = 45.1 mN/m) as the test liquids. The experimental details have been published in Ref. [13]. The thermodynamic contact angle θ_e = 68° for eth-ylene glycol on PP (Fig. 3, left graph) is lower than the value for ethylene glycol on paraffin wax due to the higher surface energy of this commercial PP. θ_e = 60° is measured with diethylene glycol (Fig. 3, right-hand side) on PP. This θ_e is lower than for ethylene glycol because the surface tension of diethylene glycol is lower. Therefore, wetting of PP is better for diethylene glycol on surfaces with similar surface roughness. As for paraffin wax, the arithmetic means of θ_a and θ_r, as well as of $\cos\theta_a$ and $\cos\theta_r$ depend on $\Delta\theta$. Therefore, θ_e cannot be ob-tained on polypropylene as such simple average values.

The roughness factor r was determined for four PP surfaces with different roughnesses using SFM (cf., Appendix for details). With these values for r and for θ_e, the first data set for $\cos\theta^W$ is calculated from equation (3) on each PP sur-face. Again, a second data set for $\cos\theta^W$ is obtained with equation (8) and the measured angles for θ_a and θ_r – symbol ($\blacksquare$) in Fig. 4. This set implies that the r-data and $F^a = F^r$ could be valid on the PP surfaces with different roughnesses for both test liquids provided that there is no accidental compensation of experi-mental errors in that wetting experiment.

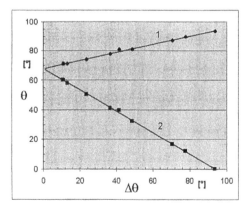
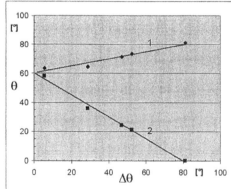

Figure 3. Advancing (1) and receding (2) contact angles measured with ethylene glycol (left graph) and diethylene glycol (right graph) on polypropylene samples with different contact angle hysteresis $\Delta\theta$ due to roughness variation. (Left) A^a = 0.2783; A^r = -0.7217; θ_e = 68°; (Right) A^a = 0.2456; A^r = -0.7544; θ_e = 60°.

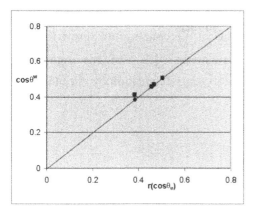 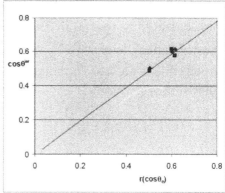

Figure 4. Comparison of $\cos\theta^{\mathrm{W}}$ for polypropylene–ethylene glycol (left graph) and polypropyl-ene–diethylene glycol (right graph) applying different equations. Abscissa: reference data for $\cos\theta^{\mathrm{W}}$ according to equation (3): $\cos\theta^{\mathrm{W}} = r \cdot \cos\theta_{\mathrm{e}}$. (■) $\cos\theta^{\mathrm{W}}$ according to equation (8); (◆) $\cos\theta^{\mathrm{W}}$ according to equation (12).

As a check, F^{a} and F^{r} are determined separately with θ_{e} and the r-data from SFM *via* equations (4), (10) and (11). Then equation (12) is applied for the calculation of $\cos\theta^{\mathrm{W}}$ providing the points (◆) in Fig. 4. These data follow the line with 45° slope also very well. This proves that the r-data have been deter-mined correctly by SFM. Now, we can also conclude from the good fit of the points (■) to the line with 45° slope that the assumption $F^{\mathrm{a}} = F^{\mathrm{r}}$ is acceptable for the PP samples and the test liquids considered.

In order to confirm these conclusions an additional solid is considered in the following.

4.3. Polyetherimide

Figure 5 shows the advancing and receding contact angles measured with ethyl-ene glycol on polyetherimide homopolymer samples with varying roughness. Young's contact angle $\theta_{\mathrm{e}} = 34°$ is lower than for PP and paraffin wax. Hence, the surface energy of polyetherimide is higher than that for the other materials. Again, Fig. 5 reveals that it is impossible to obtain θ_{e} from an arithmetic mean of θ_{a} and θ_{r} or of their cosines.

It should be noted that the part of equation (1) concerning θ_{r} is limited to the range $\theta_{\mathrm{r}} > 0$. For high roughness, the contact angle hysteresis is not defined any more in the system ethylene glycol/polyetherimide, since only θ_{a} is different from zero.

As for the other solids, Fig. 6 presents data points for the rough polyetherimide from equations (8) and (12). The results support the conclusions drawn for PP. They show that the term $(F^a - F^r)/2\gamma_{lv}$ in equation (12) is of less importance for polyetherimide, too. The simple model according to equation (8) provides acceptable accuracy again.

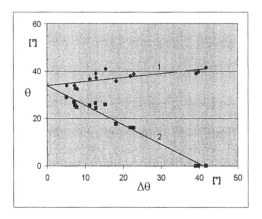

Figure 5. Advancing (1) and receding (2) contact angles measured with ethylene glycol on polyetherimide samples with varying contact angle hysteresis $\Delta\theta$ due to varying roughness. ($A^a = 0.1723$; $A^r = -0.8277$; $\theta_e = 34°$).

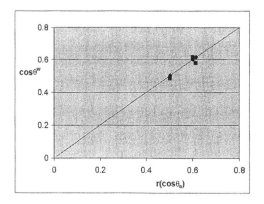

Figure 6. Comparison of $\cos\theta^W$ for polyetherimide–ethylene glycol applying different equations. (■) $\cos\theta^W$ according to equation (8); (◆) $\cos\theta^W$ according to equation (12). Abscissa: reference data for $\cos\theta^W$ according to equation (3): $\cos\theta^W = r \cdot \cos\theta_e$.

5. SUMMARY AND CONCLUSIONS

Equation (1) proved to be valid for deducing Young's equilibrium contact angle θ_e from wetting experiments on rough paraffin wax, polypropylene and polyetherimide surfaces with ethylene glycol as the test liquid, see Table 1.

The decrease in θ_e corresponds to the decrease in the slopes A^a, A^r of the straight lines for $\theta_a(\Delta\theta)$-, $\theta_r(\Delta\theta)$-curves as a result of the increasing wicking effect of surface capillarity. These results support equation (1) and our knowledge about many similar investigations on different polymers and test liquids [6, 9].

For the first time, our data show that the contortional energy parameter F is indeed almost the same for the advanced and retreated liquid menisci on the rough solids considered. However, the results for such a few organic materials and test liquids are not sufficient for the conclusion that $F^a = F^r$ will be valid in general. More wetting experiments are necessary to confirm such a conclusion.

If the assumption $F^a = F^r$ can be applied, then a second way opens to determine θ_e on surfaces like thin layers, membranes, graft-polymer-modified films, etc., where roughness cannot be modified by mechanical treatment. According to equation (9), only Wenzel's roughness factor r and the values for θ_a and θ_r are required. It was shown that r could be derived from appropriate surface profile analysis by scanning force microscopy [13]. Of course, the surface roughness can also be described by optical techniques [5]. In each case, however, additional model considerations are needed to obtain r and, hence, θ_e by equation (3) because *a priori* it is not clear if these techniques for roughness measurement provide the r-value which is relevant for wetting. Hence, it is not surprising that remarkable differences appear between such θ_e-data and the values for θ_e obtained directly according to equation (1) [5].

Finally, with the measured data for γ_{lv}, θ_a and θ_r, equation (13) provides the sum $F^a + F^r$. We plot $F^a + F^r$ as a function of the arithmetic mean of $\cos\theta_a$, $\cos\theta_r$ for the rough surfaces investigated. On ideally smooth surfaces, both

Table 1.
Results of wetting experiments on homogeneous rough solids with ethylene glycol applying equation (1)

	θ_e	A^a	A^r	$R^2(A^a)$	$R^2(A^r)$
Paraffin wax	80°	0.4231	-0.5769	0.8427	0.9088
Polypropylene	68°	0.2783	-0.7217	0.9939	0.9991
Polyetherimide	34°	0.1723	-0.8277	0.9387	0.9642

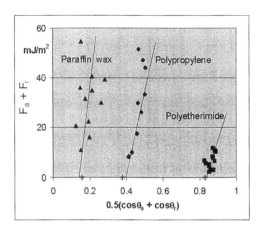

Figure 7. Sum of Good's contortional energy parameters $F^a + F^r = \gamma_{lv} \cdot (\cos\theta_r - \cos\theta_a)$ *vs.* the arithmetic mean of the corresponding cosines of advancing and receding angles. The appropriate extrapolation to $F^a + F^r = 0$ must result in $\cos\theta_e$ (in comparison equation (1) provides for these systems the intercepts marked by +: paraffin wax $\cos 80° = 0.1556$; for PP: $\cos 68° = 0.3773$; for polyetherimide: $\cos 34° = 0.8291$).

F-values become zero and the contact angles obey the equation $\theta_a = \theta_r = \theta_e$. Therefore, the extrapolation to $F^a + F^r = 0$ results in $\cos\theta_e$ for the given solid–liquid system. Theoretically, this could be another way to determine θ_e from θ_a and θ_r but we do not know what kind of function has to be chosen for the curve fit in Fig. 7 and the data points scatter considerably. Therefore, a linear fit is applied as the simplest possibility. Obviously, the extrapolation to $F^a + F^r = 0$ in Fig. 7 is less precise than the extrapolations given by Figs 1, 3 and 5. Hence, our first approach using equation (1) is the more reliable way to obtain Young's contact angle.

Recent experiments have aimed at measuring θ^W directly on vibrated surfaces with different statistically random roughness profiles [6, 7]. These measurements start with either the advancing or the receding triple line while the sample vibrates. Obviously, the triple line will reach an equilibrium state in its final position on the solid when the vibration is switched off. Due to the influence of the vibration, these equilibria and the corresponding contact angles θ_a^{vib}, θ_r^{vib} differ from the equilibria obtained after moving the triple line on a sample at rest and from the resulting contact angles θ_a, θ_r considered in this paper. It is claimed in Refs [5, 17] that $\theta_a^{vib} = \theta_r^{vib}$. Our concept (equation (1)), should be applicable to the experiments on vibrating surfaces. For physical reasons, we expect that the functions $\theta_a^{vib}(\varDelta\theta)$, $\theta_r^{vib}(\varDelta\theta)$ will be linear, too. Moreover, extrapolating these straight lines to $\varDelta\theta = 0$ should also provide θ_e.

APPENDIX

In Ref. [13] we described a way to measure the real surface area A_{real} directly by scanning force microscopy (SFM) in order to calculate Wenzel's roughness factor r from equation (2). In the constant force mode, the very sharp SFM tip (radius $r_{tip} \approx 15$ nm) glides over the surface and tracks the roughness profile line by line. Each line is a convolution between the tip shape and the local roughness profile of the sample. This image consists of 512 lines with 512 pixels per line. After triangulation of that data set employing the original software the resulting model surface fits the convolution and the sum of the triangles provides the area A_{model} which is used as an approximation for the true surface area, A_{real}. With our microscope, NanoScope IIIa, a scan size up to 120×120 μm^2 can be selected and this corresponds to the geometric surface area A_{geom}. Hence, r is calculated with A_{model} and the selected scan size A_{geom} using equation (2). A_{model} is always somewhat smaller than A_{real} due to three effects. First, the convolution is a function of the tip radius $r_{tip} > 0$. Second, fitting the image data by triangles. A third and more serious effect is connected with the pixel distance that increases with increasing scan size. As a result, r drops with rising scan size. This digitalization error is eliminated with a set of scans of varying sizes. The extrapolation of the resulting function r(scan size) to zero scan size provides the r-value for the calculations in this paper. With the small radius of the SFM tip, the remaining error in r is smaller than for mechanical profilometry (e.g., $r_{tip} \approx 2$ μm in Ref. [18]). Our error in r should be comparable to the error obtained with the optical surface measurements and the model calculations presented in Ref. [5].

Acknowledgements

The authors thank Mrs. Y. Pieper and Mrs. Heidinger for sample preparation and contact angle measurements as well as Dipl.-Ing. M. Keller for scanning force microscopy.

REFERENCES

1. A. B. D. Cassie and S. Baxter, *Trans Faraday Soc.* **40**, 546–551 (1944).
2. R. J. Good, in: *Contact Angle, Wettability and Adhesion*, K. L. Mittal (Ed.) pp. 3–36. VSP, Utrecht (1993).
3. F. E. Bartell and J. W. Shepard, *J. Phys. Chem.* **57**, 458–463 (1953).
4. R. E. Johnson Jr. and R. H. Dettre, in: *Contact Angle, Wettability and Adhesion*, Adv. Chem. Series, No. 43, pp. 112–135, Am. Chem. Soc., Washington, DC (1964).

5. T. S. Meiron, A. Marmur and I. S. Saguy, *J. Colloid Interface Sci.* **274**, 637–644 (2004).
6. R.-D. Schulze, W. Possart, H. Kamusewitz and C. Bischof, *J. Adhesion Sci. Technol.* **3**, 39–48 (1989).
7. W. Possart and H. Kamusewitz, *Intl. J. Adhesion Adhesives* **13**, 77–83 (1993).
8. H. Kamusewitz, W. Possart and D. Paul, in: *Polymer Surfaces and Interfaces*, K. L. Mittal and K.-W. Lee (Eds.), pp. 125–143. VSP, Utrecht (1997).
9. H. Kamusewitz, W. Possart and D. Paul, *Colloids Surfaces* A **156**, 271–279 (1999).
10. R. N. Wenzel, *Ind. Eng. Chem.* **28**, 988–994 (1936).
11. N. K. Adam and G. Jessop, *J. Chem. Soc. London* **127**, 1863–1868 (1925).
12. R. J. Good, *J. Amer. Chem. Soc.* **74**, 5041–5042 (1952).
13. H. Kamusewitz and W. Possart, *Appl. Phys.* A **76**, 899–902 (2003).
14. R. Ablett, *Phil. Mag.* **46**, 244–256 (1923).
15. P. A. Thiessen and E. Schoon, *Zschr. Elektrochem.* **46**, 170–181 (1940).
16. I. Sakai, K. Nakamae, K. Nonaka and T. Matsumoto, *Kobunshi Ronbunshu* **35**, 209–214 (1978).
17. C. Della Volpe, D. Maniglio, M. Morra and S. Siboni, *Colloids Surfaces* A **206**, 47–67 (2002).
18. H. J. Busscher, A. W. J. Van Pelt, P. De Boer, H. P. De Jong and J. Arends, *Colloids Surfaces* **9**, 319–331 (1984).

Contact Angle, Wettability and Adhesion, Vol. 4, pp. 115–141
Ed. K.L. Mittal
© VSP 2006

Wettability of porous materials, I: The use of Wilhelmy experiment: The cases of stone, wood and non-woven fabric

MARCO BRUGNARA,[1,*] CLAUDIO DELLA VOLPE,[1] DEVID MANIGLIO,[1] STEFANO SIBONI,[1] MARTINO NEGRI[2] and NADIA GAETI[2]

[1]*Department of Materials Engineering and Industrial Technologies, University of Trento, Via Mesiano 77, 38050 Trento, Italy*
[2]*CNR IVALSA, Via Biasi 75, 38010, S. Michele all'Adige (TN), Italy*

Abstract—From the 1980s, the computer controlled Wilhelmy experiment has been the common method to measure contact angles. The typical samples have parallelepiped shapes and are relatively flat and homogeneous, and the test liquids are not very viscous. Some results have been published for more complex and asymmetric shapes or for more complex experimental situations. However, one of the commonest difficulties in experiments using porous samples is the absorption of the liquid. In the case of porous stones, wood or nonwoven fabrics the results can be doubtful because of significant absorption of the test liquid and swelling of the sample. In the present paper a new procedure is presented for correctly using the Wilhemy experiments on porous samples. The method consists in simple modelling of the Washburn-like mechanism of absorption and it produces consistent results. Some experiments are presented and analysed for calcareous stones, different kinds of woods and non-woven fabrics. A companion paper explores the possible use of the Washburn equation in estimating the contact angles.

Keywords: Contact angle; porous material; Wilhelmy technique; absorption.

1. INTRODUCTION

For the analysis of wettability of porous materials there are two approaches documented in the literature:
- the measurement of the contact angle using the commonest techniques (e.g., sessile-drop technique, Wilhelmy technique) appropriately adapted to the specific characteristics of the porous material;
- the use of Washburn equation to calculate the contact angle directly from the liquid absorption itself.

---

*To whom correspondence should be addressed. E-mail: marco.brugnara@ing.unitn.it

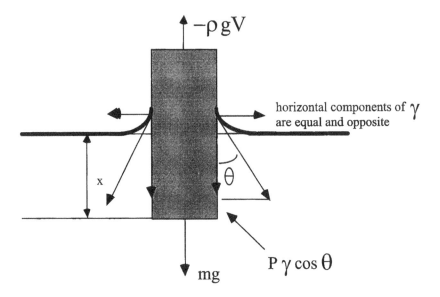

Figure 1. The force scheme of a Wilhelmy experiment.

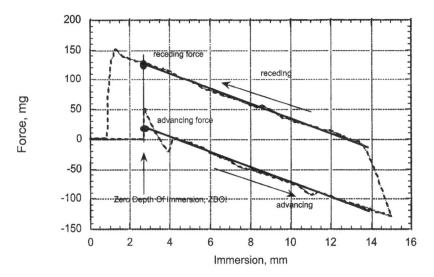

Figure 2. The description of a typical Wilhelmy experiment as force *vs.* immersion; the different portions of the run are indicated. The mg unit corresponds to mg-force, which is exactly equivalent to 0.986 dyne.

The subject of this paper is a new point of view in the application of the Wilhelmy method to porous materials, whereas a companion paper will explore instead the possibility of using the Washburn equation to calculate contact angle.

The introduction of computer controlled Wilhelmy experiments in the 80's appeared as a possible solution to the challenges faced in measuring contact angles and surface tensions, to which Padday refers as a "comedy of errors" [1]. On the contrary, the use of a modern apparatus, driven by a powerful computer and controlled by a complex software, can induce the false idea that all what one needs is simply to use clean samples and liquids and to accurately repeat as many experiments as one can. This is not true.

The technique often is not satisfactorily described in textbooks, see, e.g., Ref. [2], although more specialised texts [3] provide an appropriate description of the common calculation procedure.

We will discuss in some detail how to obtain reasonable contact angles estimates on porous samples such as porous stones, woods, and nonwoven fabrics; this topic has been discussed in recent literature with controversial conclusions [4, 5]. In doing so we will try to clarify aspects of the Wilhelmy technique which can also be helpful in the case of non-porous samples.

As it is well known, the Wilhelmy method consists of the immersion-emersion in a liquid of a sample with a well-defined shape, suspended from a microbalance. The immersion/emersion is performed along a symmetry plane or axis of the sample. Analysing this physical setup (see Fig. 1) in a reference frame where the container is at rest, one obtains the following equations for the total force acting on the sample:

$$F(x) = \left\{ mg + P\gamma \cos\theta - \rho gV \right\}$$
(1)

$$F_1(x) = F(x) - mg = \left\{ P\gamma \cos\theta - \rho gV \right\}$$
(2)

where x denotes the immersion depth, m is the sample mass, g the gravity acceleration, P the wetted perimeter of the sample, γ the liquid surface tension, θ the contact angle, ρ the liquid density and V the immersed volume of the sample. In the case of a parallelepipedic or cylindrical sample the volume can be replaced with the product xA of the immersion depth x and the sample cross-sectional area A. Owing to the constancy of the sample weight, the equation is usefully written as equation (2). The force-depth curves $F_1(x)$ in the initial portion of the immersion, at the inversion of the movement and in the final part of receding are strongly influenced by the details of the sample shape (e.g., the exact shape of the bottom face of the sample) and, thus, cannot be easily used.

By a best-fit of the central portions of such curves, the advancing and receding trends are extrapolated to ZDOI (Zero Depth of Immersion), where the buoyancy component vanishes and the two values of the force thus obtained are equal to $P\gamma\cos\theta$. It is possible then to recalculate one of the unknown parameters P, γ or, more commonly, $\cos\theta$ (see Fig. 2). The choice of the advancing and receding intervals to be fitted is not obvious. The fitting lines should be parallel (if they are straight lines) and their tangent should be comparable to the value calculated from the immersion section of the sample (i.e., $-\rho g A$). To our knowledge, no commercial software allows to perform this choice correctly. The force profile before immersion can be slightly up- or down-curved: this will depend on the presence of protruding fibers, on the electrostatic charge of the material, and on the lack of parallelism between the sample and the liquid. Finally, the constancy of the sample weight before and after the experiment is important to exclude the absorption of the test liquid.

In the past many different corrections were proposed to equations (1) and (2) to take into account the effects of liquid viscosity [6], the details of surface structure [7] and of samples with different faces [8, 9] or complex geometrical shapes [10–12]. Equations (1) and (2) refer to both parts of the experiment: advancing and receding. The equations implicitly assume that the meniscus is substantially equivalent to a "static" or equilibrium one. This nontrivial assumption can be justified only for very small values of the capillary numbers considered [13], i.e. about 10^{-6}–10^{-7}, which in water corresponds to a maximum immersion speed of 10^{-4} m·s^{-1}, taking water viscosity as 1.2×10^{-3} Pa·s and surface tension as 72 mN/m. Typical values of the immersion speed should, however, be lower. For more viscous liquids this condition requires a decrease of the immersion rate and some liquids, such as glycerol, should be used at a very low immersion speed, much less than 10^{-6} m·s^{-1}.

Some other conditions are not considered in the software packages commonly controlling the Wilhelmy microbalances.

For a sample with a constant immersion section (a parallelepiped or a cylinder) of area A the buoyancy effect is $-\rho g A x$, where x denotes the immersion depth. The term $-\rho g A$ needs to be a constant and the immersion and emersion portions of the Wilhelmy graph must be parallel straight lines. Moreover, the value of the term $-\rho g A$ should be in acceptable agreement with the measured cross-sectional area of the sample. For example, a common parallelepipedic sample of side length L and thickness ℓ has, obviously, the perimeter $P=2\cdot(L+\ell)$ and the area $A=L\cdot\ell$, so that the measurement of the sample perimeter can be used also to check the term $-\rho g A$. This simple test lacks practically in all the most common software packages controlling the microbalances.

The lack of parallelism between advancing and receding fitting lines, the difference between the calculated and experimental $-\rho g A$, or both, always indicate that an unexpected phenomenon has occurred and provide an important "warning" about the validity of the experiment and the need for removal of systematic errors. In the simplest case the problem may be due to the inclination of the sample, but it may also be due to the surface dynamics of the sample or to the absorption/adsorption of the liquid or to more complex phenomena.

One should consider that the calculation of the contact angle (or of other parameters) from a Wilhelmy experiment is made by extrapolation, one of the most risky procedures in numerical analysis. A slight change in the slope of the straight line implies a significant error in the intercept and thus in the contact angle (or in other calculated parameters).

If the sample sizes are such that the distance between the sample and the container walls is 5 mm or less, a curved meniscus is developed, with a significant effect on the measured force. Moreover, for large immersed volumes the liquid reference height may vary during the experiment. In fact, whenever the ratio V/S between the immersed volume V of the sample and the surface S of the container is large, then the reference height changes significantly during the run. A glass slide used for the surface tension measurement with a cross section of 24×0.15 mm immersed for 1 cm has an immersed volume of 0.036 cm³; in a container of 4 cm diameter this yields a variation Δ of about 28 μm in the reference level of the liquid. Consequently, the immersed volume is not $V=Ax$, as expected, but $V=Ax+\Delta$. Generally speaking, x becomes $x+(Ax/S)$ and the straight line has a tangent $A(\ell+(A/S))$. A polymer sample with a section of 10×1 mm immersed for the same length as before in the same container has a final immersed volume of 0.1 cm³ and the change in the reference level is about 80 μm. The observed difference in the straight line slope is about 1%. Higher V/S ratios increase this error further.

Porous samples are the most complex to analyse. A porous material is always rough and, thus, its wetted perimeter generally does not coincide with the geometrical one; to correct the perimeter length one can use the value measured by a low surface tension liquid which perfectly wets the solid. In this case $\cos\theta$ becomes zero in the Wilhelmy equation and, the liquid surface tension being known, the perimeter can be derived; this same value can be used also during the experiments with other liquids, although two different liquids could in principle "see" different perimeters.

The size of a porous sample may remain constant, but it may also vary during the experiment. Porous stones maintain constant size. Wood is a strongly non-isotropic material; it absorbs water and its size strongly varies in all directions, with a considerable effect on the results. Other materials, such as soft non-woven webs, may be compacted by the effect of the capillary forces and their sizes may decrease during receding. One should also consider that the open porosity of nonwoven fabrics is so high (higher than 90–95%) and their pores are so large that during immersion the liquid freely penetrates in the sample, so that the volume of the sample to be considered in equations (1) and (2) is not the sample "envelope" or geometrical volume, but the effective one (i.e., the envelope volume less the pores volume).

All these materials can absorb significant amounts of liquid; their mass is not constant. Moreover, they can easily lose a certain percentage of the absorbed liquid through evaporation; this effect also reduces the volume of the liquid and decreases its reference height. All these effects combined contribute to variations in the slopes of the advancing and receding lines and of the zero reference height.

In light of the above discussion, the analysis of the Wilhelmy experiments made on porous materials is not easy and one can legitimately pose the question if and how is it possible to apply the Wilhelmy experiment to evaluate the wettability properties of porous materials.

The literature shows some examples of analysis [4, 5, 14–19] which conclude that the method is useful, but in our opinion the analysis is not complete, especially in the case of the Wilhelmy method. For this reason we shall try to develop a simple model able to extract useful information from a Wilhelmy experiment on porous materials.

A proposed method [4] is based on the idea that varying the sample speed it is possible to obtain an acceptable compromise. In fact, a higher speed reduces absorption, but at the same time it deforms the shape of the meniscus. The authors proposed to use higher speeds, so as to be able to produce constant values of the advancing and receding contact angles; no indication is given of the effective absorption in these cases. The absorption is reduced also through end-sealing of the samples, a practice useful in some cases.

A different approach is used in the case of sessile, captive-bubble or Axisymmetric Drop Shape Analysis (ADSA) measurements [5, 20–22]; the authors claimed that the absorption was slower than their measurements and so their results could be considered as true contact angles, not significantly modified by absorption. Obviously it has to be noted that in the case of captive bubble the material absorbs the liquid during the measurement (exactly as in the case of Wilhelmy runs with a low immersion speed) and so the contact angle is evaluated on a "already" wetted surface. However, in the case of ADSA or sessile method a strong time dependence of contact angle is observed.

The case of wood, certainly the most complex among the porous materials, has been analysed by Walinder and Gardner [23], who point out at least three problems: the release of components from the wood, which may modify the surface tension of the liquid; the absorption of the liquid, which may change the size and the weight of the samples; and the surface dynamics of wood molecules. The case of other porous materials appears simpler, because the main phenomenon is the wicking of the liquid. We will focus on this aspect and thus we will not consider in the case of wood the release of components in the test liquid.

The approach we propose is different; we accept that absorption occurs during Wilhelmy runs and try to model this phenomenon in a manner consistent with the basic ideas of a Wilhelmy experiment. An obvious consequence is that our results refer not to a dry or original surface, but to an "already" wetted surface, as in the captive-bubble method, or in sessile and ADSA long-time measurements. In doing so we do not modify samples at all and use a standard speed, without affecting the meniscus shape. This proposal, together with other already used approaches for correction to the wetted perimeter, allows to obtain useful information from wettability analysis of porous materials; obviously, the contact angle in such a complex case should be regarded as an "apparent" contact angle, because the conditions are, however, far from ideality.

2. THE MODEL

Figure 3 is the result of a Wilhelmy run on a porous calcareous stone (Noto stone, see below) in heptane (continuous line); one can easily note that the sample weight is not constant. The difference between the forces at points A and B corresponds to water absorption; we can neglect at the moment the difference between points B and C, which is due to the liquid evaporation from the sample. The sample had the shape of a parallelepiped with a cross section of 10.34×4.50 mm (perimeter $P = 29.68$ mm and cross-sectional area $A = 46.53$ mm^2); heptane (ρ=0.684 g/ml) perfectly wetted the solid and the contact angle was close to zero.

The calculated buoyancy coefficient ($-\rho g A$) is -31.82 mg/mm, while the experimental buoyancy coefficients are 21.82 mg/mm in advancing and 35.3 in receding mode. The quantity $F/p\gamma$=cosθ, extrapolated to ZDOI, has the values 2.26 and 3.66, greater than the expected value 1.

The core of our model consists in the idea that the liquid penetration follows a Washburn-like law and defines a coefficient $\alpha = \dfrac{(m_B - m_A)}{\sqrt{t_B - t_A}}$, where m is the mass of the sample and t the time of the measurement, while the subscripts refer to the points A and B, corresponding to the beginning and the end of absorption, respectively. α is obviously a simple empirical coefficient, because the liquid

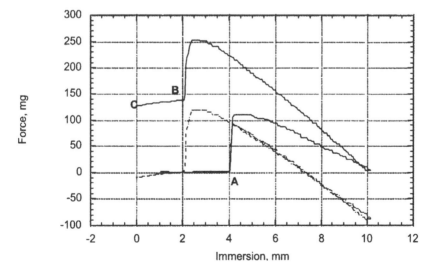

Figure 3. A Wilhelmy experiment with a porous stone sample shown in the force *vs.* immersion representation. The points A, B and C mark, respectively, the beginning and the end of the contact between the sample and the test liquid (corresponding to the beginning and the end of the liquid absorption (wicking) in the sample) and the end of the experiment. The continuous curve shows the experimental force, while the dotted one corresponds to the force obtained by correcting the sample mass for the liquid absorption. The mg unit corresponds to mg-force, which is precisely equivalent to 0.986 dyne.

penetration follows both vertical and horizontal pathways and may be different at different points; thus, it may be considered as a mean absorption coefficient. The dependence on the square root of time is related to the well-known phenomenon governed by the Washburn law.

At each point, at a time t, the total force will be corrected by subtracting the mass of the absorbed liquid:

$$F_{corr} = F_{original} - \alpha\sqrt{t - t_A}$$ (3)

The value of α can be easily calculated from the experimental data; the corrected force profile is shown in Fig. 3; the initial time is t_A=45.04 s and t_B is 208.06 s, while m_B-m_A=137 mg and α=10.73 mg·s$^{-0.5}$.

Applying this correction, the original curve force vs. immersion depth becomes the dotted one; the buoyancy coefficient is –30.86 mg/mm, in good agreement with the calculated one (–31.82 mg/mm), with a 3.0% difference only.

Now the value of overlapped advancing and receding $F/p\gamma$=$\cos\theta$ extrapolated to ZDOI becomes 1.66, again larger than 1. It can be reasonably interpreted as the ratio between the effective perimeter and the geometrical one, which roughly corresponds to the Wenzel parameter r. A roughness analysis with a roughness meter gives a lower value of r, about 1.2, at 1 µm resolution; this difference may be due to the effective ability of the liquid to penetrate the tortuosity of the perimeter. As an example, increasing the resolution of the roughness analysis on the same samples to 0.25 µm increases the r parameter to 1.3–1.4.

It is worth noting that, however, the liquid is touching, even in the advancing mode, a solid already wetted by the liquid (due to the absorption process) and this could be a valid objection to the correction validity, at least in cases where the contact angle is nonzero.

In some cases the interval between the beginning and the end of absorption can be shorter than the contact time of the sample and the liquid. In these cases one should refer the correction to this specific interval only and introduce a constant weight correction in the rest of the measurement.

The difference between points B and C is related to evaporation of the liquid from the solid sample. One can also easily model this phenomenon, assuming that the evaporation takes place at a constant rate per unit area, i.e. $\Delta m_{evap} = \int_A^B \beta \cdot S(t)dt$,

where $S(t)$ is a function expressing the evaporation area versus time and β is a mean evaporation coefficient, per unit time and area (the absorption coefficient was defined per unit time only). Using the difference between points B and C, where the evaporating surface is constant, we define $\beta = \dfrac{m_C - m_B}{t_C - t_B}$. $S(t)$ generally increases with the height of the liquid in the sample and, thus, with the square root of time, because of the absorption mechanism.

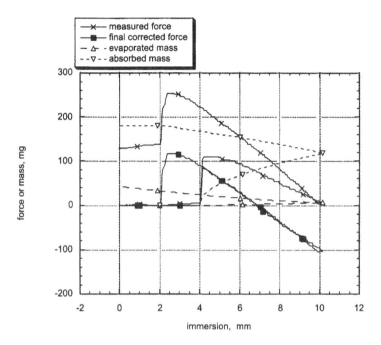

Figure 4. The same run as in Fig. 3, but showing separately the measured force, the absorbed liquid mass, the evaporated liquid mass and the final force, calculated eliminating the absorption and evaporation effect; the symbols are defined in the figure. The mg unit corresponds to mg-force, which is precisely equivalent to 0.986 dyne.

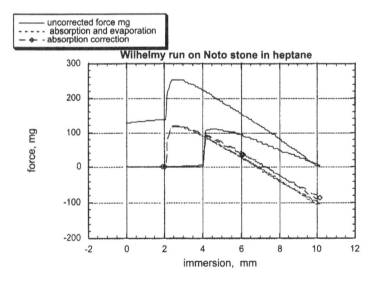

Figure 5. The same run as in Fig. 3 with the measured force, the force calculated eliminating the effect of the absorption and the force calculated eliminating the effects of absorption and evaporation. The mg unit corresponds to mg-force, which is precisely equivalent to 0.986 dyne.

An experimental check has been made on this hypothesis, simply by controlling (through a set of digital photos) that the height of the liquid in the sample follows a square root law *versus* time and subtracting the height of the liquid meniscus. A height of the liquid less than the meniscus or negative was considered as a zero contribution to evaporation. The experimental results correspond well to $S(t) = \sigma\sqrt{t - t_A}$, where σ now is an empirical constant, which equals the evaporating area after $t_A + 1$ s. The sum of the amounts evaporated per s, $\beta \cdot S(t)$, gives the total quantity of evaporated heptane, which is then summed up to the previous experimental absorbed quantity, to recalculate the new absorption coefficient, α'. By posing $\beta' = \beta\sigma$, the force is, thus, recalculated in the following way:

$$F_{corr} = F_{original} - \alpha'\sqrt{t - t_A} + \beta'(t - t_A)\sqrt{t - t_A} \tag{4}$$

where the prime indicates that the absorption coefficient has been recalculated as explained and the evaporation coefficient incorporates the variation of evaporating surface.

The final results are shown in Fig. 4. The final estimates of the coefficients are $\alpha = 10.748$ mg/s$^{0.5}$, $\beta = 0.00076$ mg/mm^2 s, $\alpha' = 14.109$ mg/s$^{0.5}$, $\beta' = 0.0207$ mg/s$^{1.5}$ and the masses exchanged by the sample were as follows: total absorbed 180 mg, total evaporated 43 mg, net absorption 137 mg.

The final Wilhelmy correction is shown in Fig. 5, where the curve corrected for absorption and evaporation is the dashed one, with -32.41 mg/mm of buoyancy coefficient (calculated value -32.82 mg/mm).

The extrapolated value of $F/P\gamma$, corresponding to the new correction of the perimeter, decreases to 1.51.

3. EXPERIMENTAL

3.1. Materials

We have analyzed the following porous materials: (A) porous stones, (B) woods and (C) non-woven fabrics.

In class A there were (1) Noto stone, an organogenic calcarenite coming from the caves of Palazzolo (Siracusa, Italy), largely employed in the local historical buildings; (2) an artificial ceramic material with a controlled porosity. It was a silica aluminate of Ca, Mg, K and Na; the content of carbonate was very low, with the presence of hematite (Fe_2O_3), quartz and unreacted feldspar. The samples, with different sizes, were obtained by cutting the material with a diamond blade saw that was cooled with deionized water. The samples surfaces were treated with P8000 abrasive pape (without any silicone additive). After abrasion,

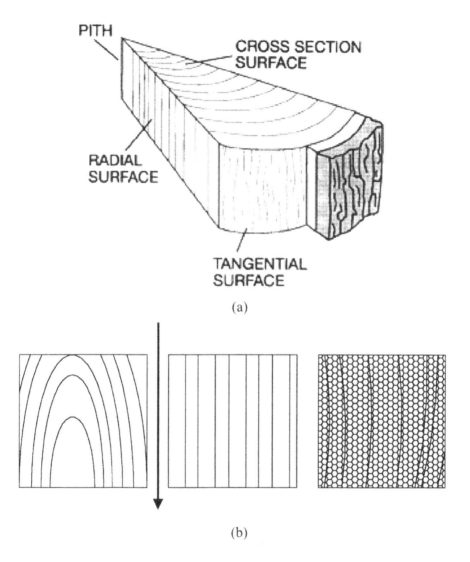

Figure 6. (a) The macroscopic structure of wood; the three principal directions are defined with respect to the annual rings; they define the cross, the tangential and the radial surfaces. (b) All the wood samples were tested according to the wood direction shown on the left of the figure (i.e., the so called tangential section of the wood). The immersion direction is represented by the black arrow. On beech, also the radial direction (in the middle) and the transverse or cross section (on the right) were tested; note that on transverse section, the annual ring influence was avoided by orienting it parallel to the immersion direction.

the samples were cleaned with MilliQ water in an ultrasonic bath for 5 min and subsequently dried in an oven at 60°C for 24 h and kept under vacuum.

In class B we considered three wood species, belonging to the hardwood group and characterised by cellular elements for sap distribution (vessels) larger than the elements with mechanical functions (fibres). The wettability tests on wood were carried out on specimens from sapwood of yellow poplar (*Liriodendron tulipifera*), from sapwood of beech (*Fagus sylvatica*) and from heartwood (the inner part, while the sapwood is the outer part of the transverse section of a tree) of eucalyptus (*Eucalyptus triticornis*).

All the wood samples were approximately 30 mm × 30 mm × 1 mm, respectively, in longitudinal, tangential and radial directions with respect to the main axis of the trunk of the tree, as described in Fig. 6a and 6b.

The beech specimens were also tested with the thickness (1 mm) in tangential direction, and the width (30 mm) in radial direction as described in Fig. 6a and 6b.

The width and thickness measurements were carried out before test and immediately after, with a 1/100 mm caliper (obtaining the mean of 5 measurements of width) and a 1/1000 mm micrometer (obtaining the mean of 10 thickness measurements).

The samples were conditioned before the measurements in an environmental chamber set at 65% relative humidity and 20°C; this was done to reduce the moisture content of the wood to 12% (i.e. the so-called normal conditions). This treatment is standard in wood analysis.

The measured density of the three species was: yellow poplar 445 kg/m^3; beech 730 kg/m^3; eucalyptus 980 kg/m^3. The computed porosity of the three species was: yellow poplar 79%; beech 56%; eucalyptus 38%.

In class C there were two different kinds of non-woven fabrics, kindly provided by Tredegar Films Products (Chieti, Italy), made of poly-propylene (PP) and PP partially oxidized through an undefined surface oxidation treatment. Here they are indicated as Web1 (density 18 g/m^2, composition 100% PP, thickness 0.25 mm and Web3 (density 15 g/m^2, composition 100% surface oxidized PP, thickness 0.24 mm).

Here the density is expressed with reference to the weight per square meter.

3.2. Roughness characterization

Different roughnessmeters were used to characterise the samples; the standard profilometer DEKTAK 3 was used to analyse the Noto stone and the ceramic material. A scan length of 2 mm was used, repeating it 5 times for each face of each sample, with 2000 sampling points.

The data were analysed by the standard software, obtaining in each case the arithmethic mean roughness value (R_a) [24]. This quantity is defined by an ANSI norm in the reference cited. It corresponds to the absolute mean value of the differences among the height of each measured point and the height of a centre line. The centre line is defined by the mean of the collected height values and is a straight line parallel to the general direction of the profile within the limits of the sampling length, such that the sums of the areas contained between it and those parts of the profile that lie on either side are equal.

The wood was evaluated by a Mitutoyo SJ301 roughnessmeter. On beech specimens, roughness measurements were carried out according to ISO11562/1997-standard.

The ratio between the effective perimeter P_{corr} and the geometric perimeter P is calculated through a simple formula:

$$P_{corr} = \sum_{i=1,n} \sqrt{(Y_{i+1} - Y_i)^2 + l^2} \text{ and} \dots r = \frac{P_{corr}}{P}, \tag{5}$$

where n is the number of steps, l is the length of each step and Y_i is the value of roughness in the step i, and r is the perimeter correction factor.

The equivalent quantity obtained from the Wilhelmy experiments is calculated from the $F/p\gamma$ at ZDOI obtained in a perfectly wetting liquid: if its value is higher than 1 it cannot be interpreted as a cosine and, assuming a perfect wettability, the number obtained is an estimate of the effective perimeter. The penetration of different liquids in the perimeter tortuosity can be different, but due to the significant absorption in our samples, we assume that all liquids "see" the same perimeter length. The two estimates of r are not necessarily correlated.

3.3. Wettability characterization

The Wilhelmy plate method was employed on wood using a KSV Sigma701 surface balance. The reference liquid used was bi-distilled water.

Experimentally, the wood sample was placed vertically above the bi-distilled water, and the liquid vessel raised on an elevator moving with a pre-set speed of 3 and 6 mm/min to immerse the wood sample.

For most of the specimens from yellow poplar and eucalyptus the immersion speed was 6 mm/min and the immersion depth was 5 mm; two specimens were tested reaching an immersion of 10 mm (those samples are designated as ET10 and TUL10).

For most of the beech specimens the immersion speed was 3 mm/min and the immersion depth was 10 mm.

Dynamic Contact Angles (DCAs) on Noto stone were measured using a TDS Wilhelmy microbalance by Gibertini (Milan, Italy) at a speed of 42 or 82 µm/s (equivalent to 2.52 and 4.92 mm/min, respectively) at room temperature. The

Table 1.
Physical properties of the liquids used

Liquid	Surface tension (mN/m)	Density (g/cm³)
MilliQ and bi-distilled water	72.8	0.997
Heptane	19.65	0.68
n-Dodecane	25.4	0.749
n-Hexadecane	27.5	0.77

liquid was contained in a vessel with a wide diameter (about 16 cm) to neglect the correction for change in the reference level due the significant volume of the immersed sample or to the water absorption. The measurements on ceramic material and non-woven fabrics were performed with a Cahn322 microbalance at a speed of 20 µm/s (equivalent to 1.2 mm/min) and at a controlled temperature of 20 ± 5°C. The liquids employed were MilliQ water (resistivity =18.2 MΩ·cm), n-heptane (Acros, purity 99%), and n-hexadecane (Merck, purity 99%). It was impossible to completely control the humidity in the measurement chamber. The surface tension and density of the liquids used are listed in Table 1. The estimate for the perimeter correction for wood was obtained as described previously using dodecane as the liquid and immersing the sample at a speed of 20 µm/s (equivalent to 1.2 mm/min) at room temperature.

4. RESULTS AND DISCUSSION

The contact angle results on different kinds of porous materials are collected in Tables 2–4.

Table 2.
Absorption coefficients and contact angles (in degrees) for stone samples in water

Material	α (mg s$^{-0.5}$)	θ_{adv} uncorr	θ_{rec} uncorr	θ_{adv} corr abs	θ_{rec} corr abs	θ_{adv} corr per	θ_{rec} corr per
Ceramic	5.113	58.7	0	65.4	17.3	68.9	34.2
Ceramic	5.366	0	0	40.6	0	48.9	27
Ceramic	5.039	45.0	0	50.4	7.2	56.5	30.8
Noto stone	7.547	0	0	46	18.5	62.6	51
Noto stone	9.149	0	0	22.9	0	52.4	47
Noto stone	12.760	0	0	27.2	0	53.9	47.7
Noto stone (82)	16.729	0	0	34.2	28.2	56.8	54.3
Noto stone (82)	12.975	0	0	21.7	0	52.0	46.7
Noto stone (82)	15.975	0	0	23.9	9.7	52.7	49.2
Noto stone	10.514	0	0	20.6	7.2	51.7	48.9
Noto stone	11.903	0	0	35.6	1.6	57.4	48.5
Noto stone	7.768	0	0	0	0	46.7	46.7

The contact angles indicated as zero correspond to extrapolated $F/p\gamma$ values higher than 1. In the Material column the numbers in parentheses correspond to the speed used in immersion if different from that indicated in the Materials section. α, is the absorption coefficient, as defined in the text; the reported contact angles are calculated from the experiment without any correction (column uncorr), with the correction for absorption (col. corr abs) and with the correction for absorption and perimeter roughness (col. corr per). The perimeter correction factor, r, used for ceramic samples was 1.155 and for Noto stone 1.51. It was obtained as described in the text from Wilhelmy experiments in heptane or hexadecane, corrected only for liquid absorption.

A general comment is that in all cases the variation in the results obtained appears very large; this is, however, common for materials of natural origin or for far from ideal surfaces. As a consequence, the large values of standard deviations should not be surprising. Referring to the literature in analogous cases it appears that our data are not far from the common results on these kinds of samples.

Table 3.
Absorption coefficients and contact angles (in degrees) for wood samples in water

Material	α (mg s$^{-0.5}$)	θ_{adv} uncorr	θ_{rec} uncorr	θ_{adv} corr abs	θ_{rec} corr abs	θ_{adv} corr size	θ_{rec} corr size	θ_{adv} corr per	θ_{rec} corr per
Fagus-rad	5.256	44.4	11.0	48.6	41.8	49.7	43.2		
Fagus-rad	4.976	33.3	0	42.4	30.1	43.8	32.3		
Fagus-rad	4.356	44.2	0	45.8	33.2	46.7	35.0		
Fagus-tan	4.332	42.0	0	45.5	31.5	50.0	38.6		
Fagus-tan	3.261	34.8	0	38.1	18.1	41.4	25.0		
Fagus-tan	3.877	30.1	0	41.8	27.2	44.1	31.1		
Fagus-trn	7.857	26.9	0	51.1	34.8	52.4	37.1		
Fagus-trn	7.127	57.2	0	57.2	29.6	58.7	33.5		
Fagus-trn	7.593	48.6	0	54.1	31.4	55.6	34.6		
Eucalyptus	1.0318	100.5	0	89	3			89.1	28.8
Eucalyptus	0.9038	92	0	87	12.7			87.4	31.2
Eucalyptus	0.8772	83	12.8	88.5	17.4			88.7	33.2
Eucalyptus	0.4184	90.7	0	84.7	4.8			85.4	29.1
Eucalyptus	0.8365	83.9	7.2	79.6	13.6			80.9	31.5
Eucalyptus	0.8007	76.5	0	74.7	0			76.6	26.0
Eucalyptus	1.2144	70.3	19.7	76.0	23.7			77.7	36.6
Eucalyptus	1.1000	77.8	25.3	78.2	29.6			79.7	40.3
Yellow Pop	16.86	24	0	35.7	35.7	36	36		
Yellow Pop	12.99	25.9	0	34.6	34.6	35	35		
Yellow Pop	13.13	35.3	23	46	50	47	51	53	56
Yellow Pop	12.67	38	0	46	46	47	47	53	53
Yellow Pop	13.21	23.4	6.9	41.4	41.4	42	42	49	49
Yellow Pop	12.99	34.3	23.2	45.7	49.9	46.4	50.5	52	56

Fagus-rad, Fagus-tan and Fagus-trn indicate, respectively, the samples of beech cut in the three directions radial, tangential and transverse, as shown in Fig. 6a and 6b. Eucalyptus and yellow pop indicate eucalyptus and yellow poplar samples. The contact angles have been calculated from the experiment without any correction (column uncorr), with the correction for absorption (col. corr abs), with the correction for liquid absorption and the correlated size variation (col. corr size); and finally for perimeter roughness (col. corr per). The perimeter correction factor, r, used for eucalyptus samples was 1.14 and for yellow poplar 1.127; for beech the correction was estimated to be negligible. The empty columns correspond to corrections estimated as negligible.

Table 4.
Absorption coefficients and contact angles (in degrees) for non-woven fabrics

Material	θ^1_{adv}	θ^1_{rec}	θ^2_{adv}	θ^2_{rec}	α (mg t$^{-1/2}$)	α/p
Web1	91±1	44±1			0.629	0.027
Web1	101±1	52±1			0.657	0.025
Web1	105±1	49±1			0.747	0.026
Web1	95±1	42±1	72	42	0.587	0.029
Web1	110±1	47±1	83	47	0.52	0.025
Web1	103±1	48±1	78	40	0.628	0.027
Average	101±7	48±4	78±6	43±4	0.63±0.08	0.027±0.002
Web3	134	91	136	87	0	0.005
Web3	106	87	110	91	0	0
Web3	151	112	151	116	0	0
Web3	137	113	143	113	0	0
Web3	150	97	150	105	0.028	0.001
Average	137±18	101±12	139±17	103±13	0.0	0.0
Web3*	115	49	90	49	0.29	0.015

The superscripts 1 and 2 for contact angles refer to the first and second immersions; mean perimeter corrections of 1.12±0.05 for Web1 and 0.99±0.05 for Web3 were calculated from Wilhelmy experiments performed in dodecane. As explained in the text to neglect the different sizes of the samples, the absorption coefficient has been "normalized" by dividing the experimental value by the geometric perimeter of each sample. α is the absorption coefficient introduced in the model, with units mg divided by the square root of time; p is the perimeter of the sample.
*This sample was compacted.

4.1. Stones

As for the stones, Table 2 and Fig. 7a, where the experimental data are plotted vs. the calculated ones, show that the buoyancy coefficient is reproduced very well, with a mean error of 1–2%.

Reasonably, for the stones used there was no effect (swelling, dissolution, etc.) able to modify the surface tension of the liquid or the size of the sample. The sample sizes used for the calculations are those of the dry samples, because they cannot be modified by the immersion process. The perimeter, however, should be adjusted according to the estimated "true" perimeter, corrected for the roughness, and expressed as the Wenzel factor r; the extrapolated $F/p\gamma$ obtained by immersion in a perfectly wetting liquid, as described previously, provides an estimate of r. To reduce the impact of the evaporation one can simply use organic liquids, such as hydrocarbons, with a relatively high molecular weight. In the case of Noto stone the correction parameter for the "true" perimeter appears to be about 1.5±0.1 using hexadecane, a hydrocarbon whose evaporation is negligible.

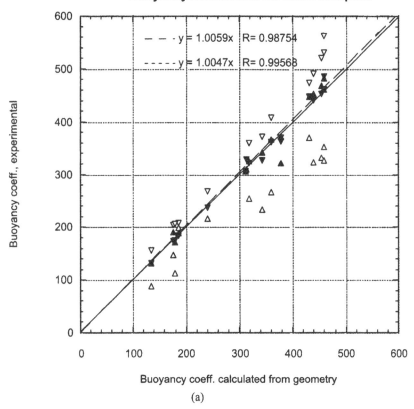

Buoyancy coefficients for stone samples.

Figure 7. (a) The comparison between the experimental and calculated buoyancy coefficients, i.e., the slope of the force *vs.* immersion curve for stone samples. The empty symbols correspond to values not corrected for the liquid absorption, while the filled symbols correspond to the values calculated from the model and thus corrected for the effects of the liquid absorption. (b) The comparison between the experimental and calculated buoyancy coefficients, i.e., the slope of the force *vs.* immersion curve for non-woven web samples. The symbols correspond to the values calculated from the model and thus corrected for the effect of the liquid absorption. The model underestimates the experimental quantity. (c) The comparison between the experimental and calculated buoyancy coefficients, i.e., the slope of the force *vs.* immersion curve for wood samples, using the initial size of the samples. The open symbols correspond to values not corrected for the liquid absorption, while the filled symbols correspond to the values calculated from the model and, thus, corrected for the effect of the liquid absorption. Triangles, beech; circles, yellow poplar; squares, eucalyptus. (d) The comparison between the experimental and calculated buoyancy coefficients, i.e., the slope of the force *vs.* immersion curve for wood samples, using the final size of the samples. The empty symbols correspond to values not corrected for the liquid absorption, while the filled symbols correspond to the values calculated from the model and, thus, corrected for the effect of the liquid absorption. Triangles, beech; circles, yellow poplar. Note that the eucalyptus samples are not present because their sizes have not been corrected.

M. Brugnara et al.

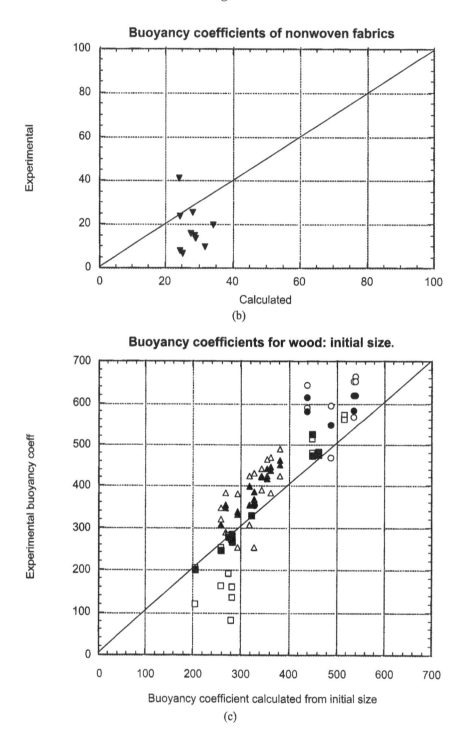

Figure 7. (Continued).

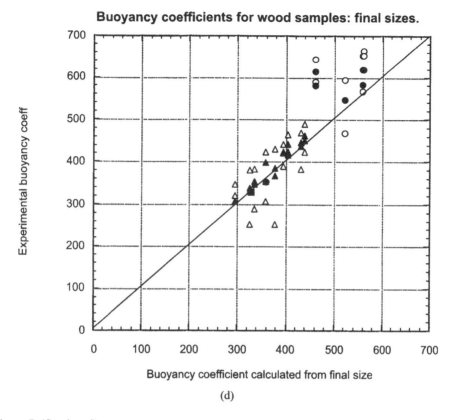

Figure 7. (Continued).

The perimeter correction evaluated by the roughness measurements (at 1 μm resolution) is substantially lower, about 1.2, but one should note that the tip of a roughness meter is certainly less able than a perfectly wetting liquid to penetrate the solid contour.

Using the value 1.5 to correct the sample perimeter, the advancing and receding contact angles of water on Noto stone change from 30°±8 and 12°±9 to 54±4° and 49±3°, respectively.

In the case of the ceramic material, whose roughness also requires to correct the geometric perimeter, measurement made with a perfectly wetting organic liquid (dodecane, which does not evaporate significantly during the measurement) allows to calculate the perimeter correction factor of about 1.155; the recalculated angles are 59±9° in advancing and 31±4° in receding modes.

For both stones it seems that the standard deviation decreases after the perimeter correction; moreover, in the majority of cases the extrapolated $F/p\gamma$ ratio (corresponding to the cosines of the contact angles) decreases below 1 after all the corrections and allows to obtain reasonable values of contact angles.

While the perimeter length was corrected as described, we did not correct the cross-sectional area of the samples. The reader can easily be convinced that if the very flat profile of a closed curve with a certain enclosed area and a certain perimeter is transformed by adding a certain roughness to the perimeter, then the perimeter length is significantly increased, but the enclosed area is generally not modified at all. The good agreement between the experimental and calculated buoyancy coefficients supports this hypothesis.

As a conclusion, the model appears to work reasonably well for stones, when there is no apparent effect of the liquid on the sample or of the sample on the liquid.

4.2. Non-woven webs

As for the non-woven webs, we compared two webs with two completely different chemical character. Both are used for hygienic applications, but in different cases and both are extremely porous materials, with an open porosity higher than 90%. The one designated Web3 is essentially hydrophobic and it does not absorb water during the time interval of the experiment; in contrast the other, denoted Web1, is able to absorb a certain amount of water. Moreover, as one can see it is initially hydrophobic, but its properties change abruptly and in the second immersion run it appears more hydrophilic, with a significant decrease of the advancing angle. This is consistent with the different contact angles we have obtained using the perimeter corrections (obtained from measurements made in dodecane) and the absorption corrections made using the proposed model.

It is very interesting that the Web3 material in its native form shows a perimeter correction factor equal to 1 (i.e., the geometrical perimeter is equal to the real perimeter, notwithstanding the porosity); if compacted it changes significantly. Its contact angle decreases, its perimeter correction increases and its ability to absorb water increases.

This can be interpreted on the basis of the structure. Web3 is made of untreated PP fibers, so that it works as a "composite" surface and is able to avoid any water absorption. If the sample is compacted, the elimination of air decreases the sample hydrophobicity.

Web1 material is made of PP, but has been treated (the exact treatment is covered by a patent) probably by a partial oxidation in air. The contact angle seems to reduce after immersion essentially in the advancing mode and at the same time the material is able to absorb a certain quantity of water.

In the cases of webs the immersion section and, thus, the buoyancy coefficients are generally lower than those calculated from the geometrical sizes of the samples (see Fig. 7b). In fact, as already noted in the Introduction, webs are very porous (>90%) and with very large diameter pores; and thus during immersion a fluid is relatively free to move in and out of the immersed sample. For this reason the effective volume of the immersed sample is actually the difference between the "envelope" or geometrical volume and the pore volume. Moreover, a

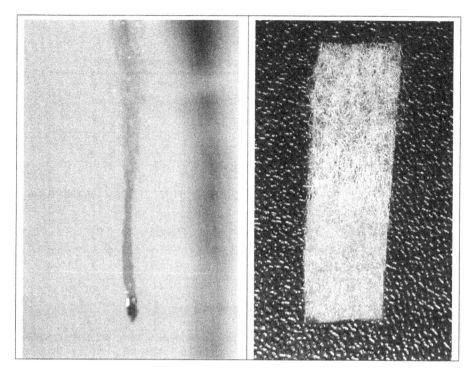

Figure 8. The dry non-woven material shown on the right; on the left the same sample after having absorbed water in its lower portion; note the different thickness of the lower wetted portion with respect to the higher dry one.

"squeezing" effect on the "envelope" volume of the samples, caused by the surface tension of the absorbed liquid in the soft structure of the material, is evidenced in Fig. 8; this effect works only when the wet sample is in the air. In Fig. 8 also the gold clip used as a weight to keep the samples flat and vertical during immersion is shown. In fact this kind of soft sample cannot easily penetrate the liquid surface; a little gold clip, easily cleaned, is attached to the sample to increase its weight and to maintain a vertical position. The gold clip volume increases the buoyancy force by a fixed quantity, which must be subtracted during the calculation. In the case of webs, due to their different thicknesses, we divided the coefficients α by the section perimeters, obtaining substantially constant parameters.

4.3. Wood

The case of wood is the most complex because of the role of many phenomena: leaching of wood components, with variation of the surface tension of the liquid; and swelling of the samples, with variation of their sizes. Our model does not deal

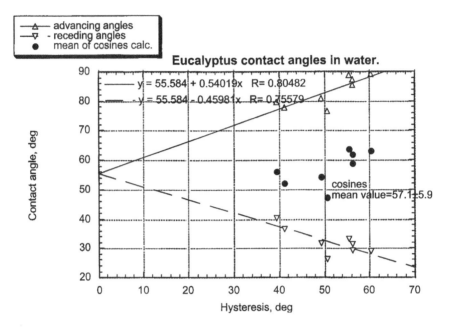

Figure 9. The final values of contact angles of eucalyptus samples in water in the representation of Kamusewitz (contact angle *vs.* hysteresis) and the equilibrium values calculated from the mean of cosines.

with the phenomenon of leaching and thus is limited to wicking and swelling phenomena.

The data are presented in Table 3. It appears that two different situations occur: some woods (as yellow poplar or beech) absorb significant quantities of water during the experiment, while eucalyptus absorbs much less water. The effect of this absorption on the sample size is very different, as it is well known. If the size of the sample changes significantly, then the analysis of the Wilhelmy experiment should be performed using the final size, not the initial one. This is true for all three directions!

The data on beech show that the buoyancy coefficient is completely different from that expected from the initial size, while using the final size the difference reduce to 2–6% (Fig. 7c and 7d).

A similar result is shown for the case of yellow poplar. For eucalyptus the situation resembles that of stone, because the agreement between the experimental values and those calculated from the initial size is acceptable (see Fig. 7c and 7d).

Sample sizes, immersion time and depth are sufficiently similar to allow a direct comparison of the values of the mean absorption coefficient (the α parameter of the model), whose numerical value is about 1 for eucalyptus, about 3–7 for beech and 12–16 for yellow poplar.

It seems quite obvious that the coefficients should be strictly comparable if and only if they refer to samples of the same final size, immersed for the same time

Table 5.
Contact angles (deg) and absorption coefficients (mg s$^{-0.5}$) for beech in water

	Rad	Tang	Transv
Adv	47±3	45±4	55±3
Rec	37±6	32±7	35±2
α	4.9±0.5	3.8±0.5	7.5±0.4

The contact angles obtained on beech samples in water in three different directions, indicated as radial (rad), tangential (tang) and transverse (transv); the definitions of the terms refer to Fig. 6a and 6b.

and to the same depth. Unfortunately, this is very difficult to obtain, because of the wood expansion, web contraction and for the different original sizes of certain samples, which cannot be reduced or modified.

However, a more useful method could be to "normalize" the data by dividing the mean absorption coefficient by the wet perimeter or the effective wet area (measured immediately after the experiment) or some other parameter. Such kinds of comparisons are not reported here (apart from the case of non-woven fabrics), because it is not clear what is the best parameter to use.

In the case of the eucalyptus the contact angle data appear sufficiently similar to that of a common polymer to be plotted using the two methods proposed by Kamusewitz and co-workers [25, 26] and by us [27, 28] to evaluate an "equilibrium" contact angle. Considering the standard deviation, the two evaluations are in satisfactory agreement, as shown in Fig. 9.

It should be considered that the size variation due to water absorption is well known in the case of wood; it affects the perimeter, as well as the section of the sample and the final immersed volume. The final effect is to change both the buoyancy coefficient and the final contact angle. In the case of the cross-sectional area the percent variation may be 20–25% or more, as one can easily check through a simple geometric consideration. (for example, increasing the two sides of a rectangle by 5 and 10%, respectively, the section area increases by 15.5%, i.e., 1.05×1.10=1.155).

A further consideration concerns the experiment made on wood samples cut along different orientations. In the case of beech we used three different orientations, which give slightly different contact angle results (see Table 5). On the other hand, the absorption properties appear significantly different and this difference comes mainly from the different orientations of the material, not from its surface properties. These results should be refined by increasing the number of samples.

In the case of yellow poplar, however, it should be noted that the experiments cannot be corrected in all cases successfully, even using the final sizes. This reflects a failure of the proposed model, probably due the fact that liquid penetration cannot be modelled through a single-coefficient model (see Fig. 7d). Three reasons could be envisaged:

(a) in some cases liquid absorption is very fast and ends before the sample de-tachment; in these cases the time corresponding to the end of the absorption should be independently detected and used instead of t_B;

(b) liquid penetration follows the pathway of the least resistance due to the wood structure and orientation; this makes the model inapplicable;

(c) a third reason may be the importance of leaching and surface dynamics, as in-dicated in Ref. [23]. We have not considered these phenomena in our ap-proach, in spite of their important role; we have focused on wicking and swelling only.

Walinder and co-workers have dedicated many papers [23] to the wood analy-sis; they considered leaching and the subsequent surface tension variation as the main event in wood Wilhelmy analysis, proposing a complete analysis of the phe-nomena. However, they do not propose an explicit function for the absorption and for its consequences; in a certain sense the present model is an explicit form of equation (8) in Ref. [23].

A final consideration should be made of the roles of roughness and the perime-ter correction. In Table 6 we report different roughness evaluations made by dif-ferent methods. We indicate the estimated values of the ratio between the effec-tive perimeter and the geometrical perimeter, because this parameter (or a very similar one), indicated as r, is referred to in the Wenzel equation. It has a great

Table 6.
Roughness results on porous materials analysed

	R_a (μm from roughness meter)	r (from roughness meter)	r (from Wilhelmy experiment using a perfect wetting liquid)
Siliceous ceramic	2.38	1.02	1.155
Noto stone	13.00	1.125	1.51
Eucalyptus triticornis (heartwood)	15.8	1.06	1.14
Fagus silvatica (beech)			
Radial	3.7	1.05	1.04*
Tangential	4.6	1.05	
Transverse	3.7	1.04	
Liriodendron tulipifera (yellow poplar)	Dry=4.8 Wet=7.0	1.12	Wet=1.127
Web1	nd	nd	1.12
Web3	nd	nd	Untreated=0.99 Compacted=1.25

The liquids used were: heptane and hexadecane for stones; dodecane for wood and non-woven fabrics.
*Mean value.

effect on the final results and cannot be neglected, in our opinion. While in the case of stone and webs it may be estimated quite directly, it may appear naive or even wrong to evaluate it for wood, where the liquid interaction with the material is particularly strong. However, it is evident in the case of eucalyptus and yellow poplar that the application of the model allows a good reproduction of the experimental parameters (particularly the buoyancy coefficient); the situation is less clear for beech.

5. SUMMARY AND CONCLUSIONS

In the present paper the physical rules governing the Wilhelmy experiments have been summarized; moreover, a simple model has been presented to treat the data obtained from Wilhelmy runs on porous materials of different origins, considering the wicking and swelling phenomena.

The application of the model, which explicitly considers the wicking and evaporation phenomena, and the need to obey the physical rules of the Wilhelmy experiment, as shown in Section 1, may strongly influence the results obtained.

We do not pretend that the proposed model is valid or is applicable in each case (e.g., the leaching of wood is not explicitly considered in the model). On the other hand, all the Wilhelmy experiments must obey the laws of Physics, as shown in Section 1.

We conclude that:

(1) Apart from the specific case of porous materials, it is convenient to consider all the results from a Wilhelmy run; particularly important is the buoyancy coefficient, whose value depends on the area section of the samples. A significant difference between the calculated and experimental values of this parameter invariably is caused by experimental problems or unexpected phenomena;

(2) Two size corrections appear important for porous materials: the recalculation of the wet perimeter from a run with a perfectly wetting liquid and the correction to the sample size due to the absorption itself, if present (swelling);

(3) A mass correction based on a Washburn-like model of the absorption (i.e., depending on the square root of the absorption time) with a single parameter can be used, eventually including the evaporation of the liquid, if significant;

(4) From the point of view of Wilhelmy results it appears useful to classify the porous materials in three groups:

(i) Rigid-like samples, such as stones. In this case the proposed correction, based on Washburn-like absorption mechanism and perimeter recalculation, allows a relatively simple and complete correction to the experimental results, whose effect on final contact angles is very significant;

(ii) Materials such as wood, whose size grows by liquid absorption; in this case one should use the final size of the sample. However, the model may fail because of the complexity due to wicking in an anisotropic material such as wood and due to the occurrence of leaching, which may modify the surface tension of the liquid;

(iii) Materials as non-woven fabrics, which (probably because of their peculiar structure) show a volume lesser than the geometrical or "envelope" one. In this peculiar case a geometric correction is more difficult, due to the softness of the material; however, in some cases it works fine, allowing the calculation of the contact angle.

The proposed corrections supplement and do not substitute the more common practice based on the increase of immersion speed or on the end-sealing of the samples. On the other hand, the model does not imply any modification of sample properties or of the experimental practice.

Acknowledgements

We are grateful to Tredegar Film Products Italia (Chieti, Italia) for permission to use the data pertaining to the non-woven materials.

REFERENCES

1. J. F. Padday, in: *Surface and Colloid Science*, E. Matijevic (Ed.), Vol. 1, p. 101. John Wiley, New York, NY (1969).
2. A. W. Adamson, *Physical Chemistry of Surfaces,* 5th edn., p. 390, Wiley, New York, NY (1990).
3. I. D. Morrison and S. Ross, *Colloidal Systems and Interfaces*, p. 120, Wiley, New York, NY (1988).
4. D. J. Gardner, N. C. Generalla, D. W. Gunnells and M. P. Wolcott, *Langmuir* 7, 2498–2502 (1991).
5. M. A. Rodriguez-Valverde, M. A. Cabrerizo-Vílchez, P. Rosales-López, A. Páez-Dueñas and R. Hidalgo-Álvarez, *Colloids Surfaces A* 206, 485–495 (2002).
6. M. Morra, E. Occhiello and F. Garbassi, *J. Adhesion Sci. Technol.* 6, 653–665 (1992).
7. C. Della Volpe, R. Di Maggio, L. Fambri, A. Pegoretti and C. Migliaresi, in: *IPCM 93 - Proceeding of the Congress on Interfacial Phenomena in Composite Materials*, Cambridge (1993).
8. C. Della Volpe, *J. Adhesion Sci. Technol.* 8, 1453–1458 (1994).
9. K. Abe, H. Takiguchi and K. Tamada, *Langmuir* 16, 2394–2397 (2000).
10. L. M. Smith, L. Bowman and J. D. Andrade, in: *Proceeding of the Durham Conference on Biomedical Polymers*, Durham, UK (1982).
11. P. Dryden, J. H. Lee, J. M. Park and J. D. Andrade, in: *Polymer Surface Dynamics*, J. D. Andrade (Ed.), pp. 9–24, Plenum Press, New York, NY (1988).
12. C. Della Volpe and S. Siboni, *J. Adhesion Sci. Technol.* 12, 197–224 (1998).
13. E. Ramè, *J. Colloid Interface Sci.* 185, 245–251 (1997).
14. K. Grundke, T. Bogumil, T. Gietzelt, H.-J. Jacobasch, D. Y. Kwok and A. W. Neumann, *Progr. Colloid Polym. Sci.* 101, 58–68 (1996).
15. M. G. Orkoula, P. G. Koutsoukos, M. Robin, O. Vizika and L. Cuiec, *Colloids Surfaces A* 157, 333–340 (1999).
16. Q. Shen, J. Nylund and J. B. Rosenholm, *Holzforschung* 52, 521–529 (1998).
17. G. I. Mantanis and R. A. Young, *Wood Sci. Technol.* 31, 339–353 (1997).
18. D. J. Gardner, *Wood Fiber Sci.* 28, 422–428 (1996).

19. E. Liptàkovà and J. Kùdela, *Holzforschung* **48**, 139–144 (1994).
20. S. Q. Shi and D. J. Gardner, *Wood Fiber Sci.* **33**, 58–68 (2001).
21. M. Kazayawoko, A. W. Neumann and J. J. Balatinecz, *Wood Sci. Technol.* **31**, 87–95 (1997).
22. M. A. Kalnins and M. T. Kaebe, *J. Adhesion Sci. Technol.* **6**, 1325–1330 (1992).
23. M. E. P. Walinder and D. J. Gardner, in: *Contact Angle, Wettability and Adhesion*, K. L. Mittal (Ed.), Vol. 2, pp. 215–238. VSP, Utrecht (2002).
24. American National Standards Institute – American Society of Mechanical Engineers, *Surface Texture ANSI/ASME B46.1*, New York, NY (1985).
25. H. Kamusewitz, W. Possart and D. Paul, *Colloids Surfaces A* **156**, 271–279 (1999).
26. R. D. Schulze, W. Possart, H. Kamusewitz and C. Bischof, *J. Adhesion Sci. Technol.* **3**, 1–78 (1989).
27. C. Della Volpe, D. Maniglio, S. Siboni and M. Morra, *Oil Gas Sci. Technol.* **56**, 9–22 (2001).
28. C. Della Volpe, S. Siboni, M. Morra and D. Maniglio, *Colloids Surfaces A* **206**, 46–67 (2002).

Contact Angle, Wettability and Adhesion, Vol. 4, pp. 143–164
Ed. K.L. Mittal
© VSP 2006

Wettability of porous materials, II: Can we obtain the contact angle from the Washburn equation?

MARCO BRUGNARA, ELIANA DEGASPERI, CLAUDIO DELLA VOLPE,[*]
DEVID MANIGLIO, AMABILE PENATI and STEFANO SIBONI

*Department of Materials Engineering and Industrial Technologies, University of Trento,
Via Mesiano 77, 38050 Trento, Italy*

Abstract—Notwithstanding the explicit limitations indicated in the original paper, the Washburn equation is often used to calculate the contact angles from wicking experiments on porous materials that do not fulfill the exact conditions of its validity. This procedure is implemented even in commercial software programs. It is based on the hypothesis that some aspects of the porosity can be expressed as "constant" characteristics of the material analyzed. However, doubts have been raised in the literature whether such a description is correct and acceptable in all cases. In the present paper some wicking measurements made on different natural and synthetic porous materials with water and different perfectly wetting ($\theta=0$) liquids are presented. It is shown that a weak correlation exists between the contact angles estimated through the Washburn equation and those directly measured. In particular we have emphasized that the value obtained depends on the initial choice of the reference liquid and on the sample size. It seems possible, however, to evaluate some aspects of the porosity from wicking experiments, although the correlation with the contact angle of the liquid on the porous material is certainly non-trivial.

Keywords: Washburn equation; equivalent radius; imbibition measurements; porosity; contact angle.

1. INTRODUCTION

In the analysis of wettability of porous materials there are two approaches documented in the literature:
 (i) The Young-equation approach, i.e., to measure the contact angle using the commonest techniques (i.e., sessile, Wilhelmy) and adapting them to the specific characteristics of porous materials;
 (ii) the Washburn-equation approach, i.e., using the Washburn equation to calculate the contact angle from the absorption phenomenon itself.
 The use of the Lucas–Washburn (LW) equation has been discussed often in the past and some modifications have been proposed [1–12], so it is useful to analyze

---

[*]To whom correspondence should be addressed. Tel.: (39-461) 882-409; Fax: (39-461) 881-977;
e-mail: devol@devolmac.ing.unitn.it

how the multiple aspects which play a significant role during liquid absorption were approached.

This equation provides a kinetic model for the rise of a liquid in a cylindrical capillary and is written in the form

$$h^2 = \frac{1}{2} r v^* t$$

(1)

with r the radius of the capillary and h the height of the meniscus at time t. The coefficient

$$v^* = \frac{\gamma_{liq} \cos\theta}{\mu}$$

(2)

measures the ability of the liquid to penetrate into the porous solid in terms of the surface tension γ_{liq}, dynamic viscosity μ and the contact angle θ of the liquid on the solid.

In the original paper of Washburn an equation for a system of porous capillaries was obtained, simply considering the case in which the porous body "can be taken as equivalent to the penetration of n cylindrical capillary tubes of radii $r_p...r_n$" [2]. However, it was clearly stated that this approach was valid unless one of the following cases occurred: (i) the pores of the body could not be taken as equivalent to a cylinder; (ii) the cross-section of the pore changed with its length; (iii) the pores contained enlargements, or were blind; (iv) pores were of molecular size.

In all these cases, in the opinion of Washburn, the applicability of the modified equation "could only be determined by experiments" [2]. These considerations would limit the validity of the Washburn law to yarns of fibers, given the regularity of their channel capillaries, but not to samples of porous rocks or ceramic materials because of their irregular shapes or, at least, we need to consider its application with suspicion. On the contrary, the analysis of porous rocks or of other natural and synthetic porous materials is commonly done using the Washburn approach, as realized even in commercial software applications, apparently without any reference to the explicit limitations pointed out by Washburn.

In 1934 Wolkowa [4] applied the Washburn equation to the case of different polar and dispersive liquids in porous silicates; all the contact angles were assumed to be zero. He found very different estimates of porous diameters for different liquids and attributed this effect to specific interactions of polar liquids. No reference to the limitations indicated in the Washburn paper was made. In 1955 Studebaker and Snow [5] proposed a procedure based on assumption that using a perfectly wetting, or a non-perfectly wetting liquid, the "mean" pore radius of the porous material did not change. As a consequence, one should be able to evaluate the contact angle of non-perfectly wetting liquids in a porous material from absorption measurements. This approach was developed in a more complete form in other papers [13–16]. It consists in an adaptation of the Washburn law to liquid absorption in porous media by introducing the average pore radius $\bar{r}$, so that the porous sample is ideally replaced with a set of equal cylindrical capillaries and the Washburn equation can be rewritten as:

$$h^2 = \frac{1}{2}\bar{r}\frac{\gamma_{liq}\cos\theta}{\mu}t$$

(3)

Moreover, by considering the density ρ of the liquid and the total cross-sectional area of pores S, it is possible to express the Washburn law in terms of the absorbed mass M of liquid instead of the height as:

$$M^2 = (\rho Sh)^2 = \frac{1}{2}(\rho S)^2\bar{r}\frac{\gamma_{liq}\cos\theta}{\mu}t$$

(4)

By introducing the mean surface porosity $\bar{\varepsilon}_S$, which coincides by definition with the volume porosity, we can also describe the area occupied by pores as a function of the geometric (or apparent) area A of the sample as:

$$\bar{\varepsilon}_S = \frac{S}{A}$$

(5)

and then the Washburn equation takes the more useful form

$$M^2 = \frac{\left(\rho\bar{\varepsilon}_S A\right)^2\bar{r}\gamma_{liq}\cos\theta}{2\mu}t$$

(6)

Some authors concluded that it was troublesome to simply use such an equation in order to calculate the intrinsic contact angle on the porous material. Although it is undoubtedly recognized that the liquid absorption by the porous medium is driven by capillarity [17, 18], it still appears unclear the role played by the many parameters involved in the phenomena (e.g., pore size distribution and their connectivity, throat length and the liquid properties). Due to the difficulty in the application of a similar idea, alternative approaches have been developed [19–23]. Moreover it has been theoretically proven [24] that the equilibrium capillary rise depends on the thickness of the sample used. Attempts to apply these laws to estimate the value of the contact angle between the invading liquid and the porous medium or comparing the pore radius values with those obtained by other techniques were not successful in many cases [5, 10, 13–16, 25]. For instance, it has been assumed [12] that the quantity $K = S^2\bar{r}$ is a constant representing the porosity geometry, that could be determined in advance by performing measurement with a liquid under condition of perfect wetting ($\cos\theta=1$). Such an approach turned out to be completely unsatisfactory, since the quantity K depends on the liquid employed.

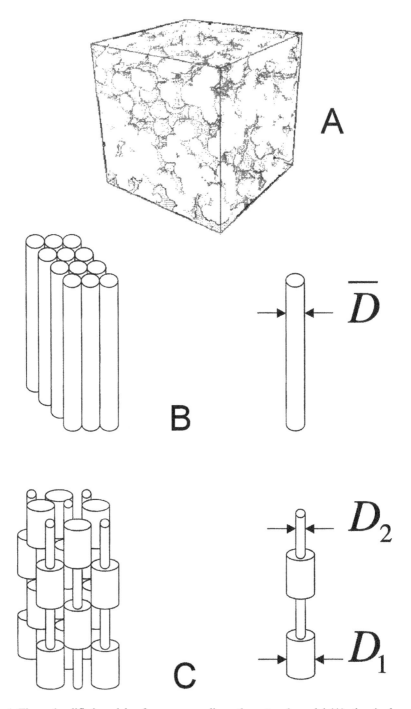

Figure 1. Three simplified models of a porous medium: the network model (A), the single capillaries tube model (B) and the Dullien model [23] (C). $\overline{D}$, D_1 and D_2 correspond to the mean, maximum and minimum diameters, respectively.

It has also been shown [25] that during the first part of the imbibition in the same porous medium each liquid dwells in pores with different sizes and the mean radius of pores determined in this way is not a constant. Moreover, this radius turns out to be one or two orders of magnitude smaller than the real radius, as measured by other techniques [18, 26].

The reasons why the Washburn approach appears to be problematic for the prediction of capillary absorption of liquids into porous media and particularly for the determination of contact angle and pore radius are many.

(1) The LW equation [1, 2] is derived by applying the Poiseuille law, based on the assumption that the motion of the fluid is stationary and slow, so that the inertial terms in the Navier–Stokes equations can be neglected with respect to the pressure and viscosity terms. Actually, the motion in a porous medium is not completely stationary, since the invading liquid always comes in contact with a new surface of the solid matrix.

(2) The effects of viscosity and compressibility of the removed fluid (typically, air) during imbibition are ignored. In principle, the rate at which the invading liquid enters the porous medium may depend on the length of the path the removed fluid has to follow to exit. As an extreme illustration of such a dependence, the speed of imbibition, from below, of a liquid in a parallelepipedic sample strongly decreases when the lateral surface of the sample is completely sealed.

(3) Filling of pores may occur due to mechanisms other than the gentle advance of a wetting front across the matrix/pore pattern of the porous medium. For instance, the wetting liquid can flow along the roughness of the walls and surround the solid grains of the matrix. At a given pressure the invading liquid that accumulates on the walls of the pore may become unstable and fill the pore. This mechanism, known as pinch-off [27] depends on the contact angle and it may lead to non-wetting liquid entrapment in adjacent pores. Instability mainly occurs within small-diameter capillaries, bypassing the large-diameter ones, so that it may happen that the non-wetting fluid is trapped inside aggregates of large-diameter capillaries surrounded by the small-diameter capillaries where the pinch-off phenomenon has taken place. This is described in the literature as bypass, or cut-off, mechanism.

(4) The co-existence of untreated high surface free energy and treated low surface free energy regions on the surface of the capillaries induces contact angle hysteresis and may increase the flow instability of the wetting liquid, giving rise to pinch-off and bypass phenomena. Chains of trapped drops of the wetting liquid, separated by bubbles of the removed fluid in the same capillary, may support large pressure differences due to hysteresis of the contact angle (the Jamin effect [28]). A similar phenomenon may occur owing to the roughness of the matrix surface.

(5) Since the description of a porous medium in terms of a unique effective capillary of suitable radius does not yield satisfactory results, it is necessary to
introduce more sophisticated models, which take into account the geometry of
the porous medium more realistically. To this end, the geometric structure of
the porous medium is usually described in terms of a material matrix partially
surrounded by empty cavities, available to the invading liquid and treated as
capillaries. Two main different classes of capillary structures are considered:
the so-called "bundle of capillary tubes model" and the "network model" [23]
(Fig. 1).

The bundle of capillary tubes provides the simplest model and consists in an effective description of the porous material in terms of an appropriate distribution
of perfectly cylindrical capillaries, with the same length but different diameters
[22, 29]. A more realistic and faithful description comes from the network model:
it consists of a three-dimensional connected network of pores and cavities. Capillary systems have various sizes and geometries and it is quite difficult to treat
them as a single, effective capillary. It has been shown in the literature, however,
that a random set of interconnected cavities in a porous matrix can be adequately
represented by a three-dimensional network of cylindrical capillaries [30]. Based
on the paper of Szekely *et al.* [20], who considered a periodic tube consisting of
alternating conical sections, Dullien and co-workers [22, 23] concluded that the
frequent disagreement between the mean radius of pores obtained by applying the
Washburn equation and by mercury porosimetry was caused by a fundamental
limitation in the Washburn model; generally the value of pore radius is 1–2 orders
of magnitude lower than that obtained from mercury porosimetry. Dullien proposed a new model by studying the liquid flow through capillaries having an alternating radius; the equivalent radius reckoned by his model is, at least qualitatively, in agreement with the radius values obtained by applying the Washburn
equation.

In the present paper, experimental absorption, contact-angle and mercury-
porosimetry measurements obtained by polar and non-polar liquids on two different calcareous and silicates porous media are reported. A physical explanation of
the equivalent radius is given while the results show the limitations of Washburn
analysis, and an unexpected effect of the size ratio of the samples on the experimental absorption results is presented. It is actually accepted that the value obtained by applying the Washburn equation is not the Young contact angle [10,
11], or an equilibrium contact angle, but it should be stressed that the parameters
involved are functions not only of the liquid employed but even of the sample geometry. Such a conclusion raises serious doubts about the possible use of this
equation for the determination of a contact angle.

2. MATERIALS AND METHODS

We analyzed two different porous materials:
(1) Noto stone, an organogenic calcarenite coming from the caves of Palazzolo (Siracusa, Italy), largely employed in local historical buildings;
(2) A ceramic material having a controlled porosity (a silica aluminate of Ca, Mg, K and Na).

The samples with different sizes were obtained by cutting the material with a diamond blade saw, that was cooled with deionized water, rinsed in MilliQ water and dried in an oven at 60°C for 24 h. As suggested in the literature [17, 18], a thin ring of silicone of height around 5 mm was applied on the lowest lateral portion of the samples in order to prevent absorption from the lateral surfaces wetted by the meniscus. The polymerization of the silicone allows good adhesion to the substratum and also prevents the formation of preferential paths for the liquid to rise under the film. The samples were finally rinsed in MilliQ water, dried and kept under vacuum.

The liquids employed were in MilliQ water (resistivity $=18.2\ M\Omega \cdot cm$), heptane (Acros, purity 99%), n-hexadecane (Merck, purity 99%), formamide (Merck, purity 99.5%), 1-bromonaphthalene (Fluka, purity 97%) and ethylene glycol (Carlo Erba, purity 99.5%). The liquids surface tension, density and viscosity are listed in Table 1.

Both mass–time and distance–time imbibition measurements were simultaneously carried out at a temperature of 22±2°C, using a Gibertini TSD Wilhelmy microbalance and a camcorder. The sample was suspended in a closed chamber, in order to equilibrate it with the test probe liquid, over a Petri dish containing the liquid in which it was dipped for 0.3 mm. The absorbed mass was measured by the instrument software, while the measurement of the height reached by the liquid front was made by analyzing the video frames with the software ImageJ [31].

For absorption measurements of different liquids and for the determination of the effective porosity and equivalent radius, we used Noto stone samples with a size of 10×3×30 mm. For the analysis of the parameters involved in the Washburn equation, ceramic samples of sizes 5×2.5×30, 10×2.5×30, 25×4.5×30 and 25×7.6×30 mm were used.

Table 1.
Physical properties of liquids used

Liquid	Surface Tension (mN/m)	Density (g/cm³)	Viscosity (mPa·s)
Heptane	19.65	0.68	0.387
n-Hexadecane	27.5	0.77	3.032
Formamide	58	1.13	3.302
1-Bromonaphthalene	44.4	1.478	4.52
Ethylene glycol	48	1.113	16.1

For the measurements performed through full immersion, we followed the method suggested by UNI 10921 [32]. The samples had dimensions of 50×25×7.5 mm and were dipped completely in the liquid in a closed polypropylene container and maintained at a temperature of 22±2°C. The samples were withdrawn and weighed at regular intervals of time. The measurement was stopped as soon as the mass variation between two consecutive weighings became less than 1%. The "effective porosity" ε_{eff} is defined as:

$$\varepsilon_{eff} = \frac{M_{abs}/\rho_{liq}}{Vol_{sample}}$$

(7)

Advancing and receding contact angles were measured with a Gibertini TSD Wilhelmy microbalance by applying a new data analysis [33].

3. RESULTS AND DISCUSSION

3.1. Porosimetry analysis

We initially investigated the porosity of Noto stone and the ceramic material by mercury porosimetry and optical microscopy. In Noto stone there are at least three

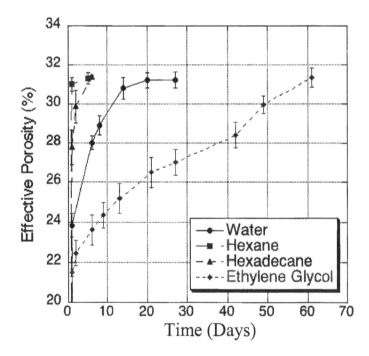

Figure 2. The effective porosity of ceramic samples measured by total immersion in different liquids; the complete imbibition time ranges from few hours to many days depending on the liquid surface tension.

different sets of capillaries: one with a mean radius of about 1 μm, a second set with a mean radius between 3 and 4 μm, which represents about 70–75% of the whole pore volume, and a third set with values between 10 and 30 μm. The ceramic material instead shows a very sharp distribution, with most pores having a radius between 0.5 and 1 μm. We also performed an optical microscopy analysis in order to investigate the size of macroporosity in Noto stone. It was not possible to obtain a quantitative evaluation of the percentage of macroporosity present in the material, but pores with sizes between 10 and 50 μm were detected on the surface.

A further measurement of the porosity of the material was performed by total immersion in a liquid. We found that by leaving the samples dipped for a sufficiently long time, the "effective porosity" ε_{eff} measured for each liquid coincides with the value of porosity as obtained by mercury porosimetry (Fig. 2).

3.2. Wicking measurements

3.2.1. Data analysis

All absorption measurements with different liquids were performed by collecting the mass of the sample and the height of the liquid front versus the imbibition time. The modified Washburn law, expressed by equations (3) and (6), can be rewritten as:

$$H_w(t) = h_1\sqrt{t} \qquad \text{with} \qquad h_1 = \sqrt{\frac{\bar{r}\gamma_{liq}\cos\theta}{2\mu}} \tag{8}$$

$$M_w(t) = m_1\sqrt{t} \qquad \text{with} \qquad m_1 = \sqrt{\frac{\left(\rho\varepsilon_{eff}A\right)^2\bar{r}\gamma_{liq}\cos\theta}{2\mu}} \tag{9}$$

In the measurement of liquid absorption by capillarity it is necessary to consider also other effects which affect and modify the time behaviour predicted by the Washburn equation:

(1) When the sample touches the liquid surface, surface tension effects determine the onset of a meniscus of mass $M_m(t)$. Within a very short time such a mass takes a value which can be assumed constant for the whole duration of the experiment.

(2) Hygroscopic liquids absorb the air moisture and their density is, therefore, variable with time. For this reason we preferred to exclude ethylene glycol, which is very hygroscopic, from the probe liquids list.

(3) During measurement liquid evaporation occurs from both the Petri dish and the lateral surfaces of the sample. The evaporation rate v_{ev} of the liquids employed is defined as the mass of liquid vaporized per unit surface and per unit of time. Water evaporation rate was measured in open air conditions by leaving a Petri dish on the pan of a precision balance and recording the mass

variation with time (Fig. 3). A linear behaviour was observed, in a time interval comparable with the typical duration of an absorption measurement. The evaporation from the Petri dish yields a variation of the reference level and a modification of the buoyancy force acting on the sample in contact with the liquid. During the experiment we, thus, had a linear time variation of the mass as detected by the Wilhelmy microbalance. The result is a linear time variation of the measured values of mass, $M_{ev}(t) = Av_{ev}t$. The evaporation of liquid from the sample surfaces is instead directly proportional to the wetted lateral surface; since the perimeter P of the sample is constant, the related variation of mass is proportional to the height $H_w(t)$, see equation (8), of the liquid front in the sample. The mass evaporated from the lateral surfaces, M_{side}, follows, therefore, a trend of the form:

$$M_{side}(t) = v_{ev}A_{side}(t) = v_{ev}PH_w(t) = v_{ev}Ph_1\sqrt{t}$$, i.e., $M_{side}(t) = v_{ev}Q_{side}(t)$ where A_{side} is the wet lateral area of the sample.

We concluded that it was impossible to experimentally distinguish the contribution of lateral evaporation from the mass absorbed according to equation (9). The experimentally determined values of v_{ev} and h_1 allow, however, to estimate Q_{side} and to verify that the latter coefficient takes a value negligible with respect to m_1.

The total mass measured by the microbalance is thus the sum of the various contributions described above:

$$M = M_w + M_{ev} + M_m \tag{10}$$

so that the equation to be used for data fitting is given by:

$$M = m_1\sqrt{t} + m_2t + m_3 \tag{11}$$

Data analysis for the height of the liquid front was performed instead using the best-fit function:

$$H = h_1\sqrt{t} + h_3 \tag{12}$$

The constant term h_3 was added to the time dependence predicted by equation (8), in order to take into account the experimental error introduced in the choice of the origin of the reference frame.

Once the parameters m_1 and h_1 have been obtained, as in Fig. 4, the effective porosity can be determined by the following relationship:

$$\varepsilon_{eff} = \frac{V_{Liq}}{V_{tot}} = \frac{\rho_{Liq}}{\rho_{Liq}}\frac{V_{Liq}}{Ah} = \frac{m_1\sqrt{t}}{\rho_{Liq}Ah_1\sqrt{t}}$$

$$\varepsilon_{eff} = \frac{m_1}{\rho_{Liq}Ah_1} \tag{13}$$

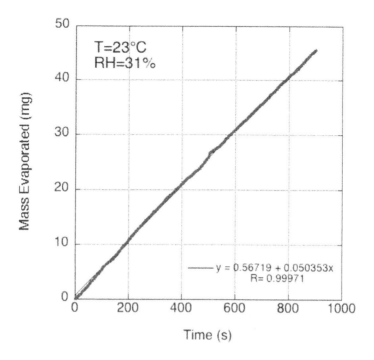

Figure 3. The evaporation rate of water from a Petri dish at 23°C and 31% RH.

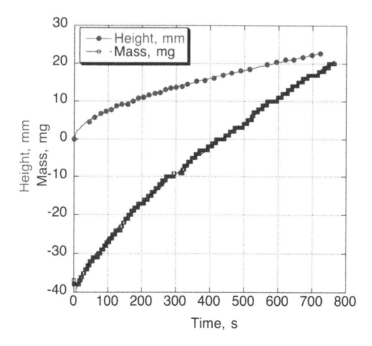

Figure 4. Typical wicking experiment with measurements of sample mass and liquid front height *versus* time.

while the equivalent capillary radius is defined by equation (8) as:

$$r_{eq} = \frac{2\mu h_1^2}{\gamma_{Liq}} \quad with \quad \cos\theta = 1$$

(14)

or using equation (9), in terms of mass and effective porosity:

$$r_{eq} = \left(\frac{m_1}{\varepsilon_{eff}A\rho}\right)^2 \frac{2\mu}{\gamma_{Liq}} \quad with \quad \cos\theta = 1$$

(15)

It is clear that equations (14) and (15) are not independent, as erroneously assumed in Ref. [25], but they simply represent the same function expressed using different variables. It is, therefore, logical that once the parameters m_1, h_1 and ε_{eff} have been determined, the same value of r_{eq} is finally found.

The contact angle θ_{nw}, for the non-wetting liquid, can be then calculated by applying the simple relationship:

$$\cos\theta_{nw} = \frac{r_{nw}}{r_{wet}}$$

(16)

where r_{wet} and r_{nw} are the values calculated with equations (14) or (15) using, respectively, a perfectly wetting and non-wetting liquid, assuming initially $\cos\theta=1$. Only in the first case the radius can be defined as "the equivalent radius".

It is important to pay attention to the units of the physical quantities involved in our calculations, in order to correctly evaluate the magnitude of the radius r_{eq}. The height H_w is expressed in mm, while time t is in seconds. The coefficient h_1 is, therefore, measured in $mm/s^{0.5}$, whereas the parameter m_1 is given in $mg/s^{0.5}$.

By using the values of the physical properties of the liquids listed in Table 1, the equivalent radius provided by equation (14) is defined as:

$$[r] = \frac{[h_1]^2[\mu]}{[\gamma]} = \frac{mm^2}{s} mPa \cdot s \frac{1}{mN/m} = \mu m$$

(17)

The results of effective porosity and equivalent radius for four different liquids on Noto stone are listed in Table 2.

It is confirmed that during capillary absorption the liquid does not use all the available porosity (column 4 of Table 2 and Fig. 5), as measured by mercury porosimetry. We assume that, because of the bypass phenomenon, the smallest capillaries are filled first, while the larger ones are invaded only partially and in a much longer time, as shown by the full-immersion absorption results (Fig. 2).

It is important to remark that not only this value of effective porosity varies according to the liquid employed, as assumed by Labajos Broncano and co-workers

Table 2.
Results of capillary absorption for four different liquids into Noto stone

Liquid employed	h_1 $(mm/s^{0.5})$	$\dfrac{m_1}{Area}$ $(mg)/(mm)^2 s^{0.5}$	ε_{eff} (%)	r_{eq} (nm)
Water	0.74 ± 0.14	0.15 ± 0.02	26.2 ± 0.6	8.0 ± 1.7
Formamide	0.26 ± 0.02	0.07 ± 0.01	25.0 ± 2	7.7 ± 1.5
1-Bromonaphthalene	0.17 ± 0.04	0.06 ± 0.01	24.0 ± 2	7.0 ± 3.0
n-Hexadecane	0.23 ± 0.05	0.04 ± 0.01	20.8 ± 0.5	19.0 ± 3.9

Parameters, h_1, m_1, and r_{eq} are defined according to equations (8), (9), (14) and (15), respectively.

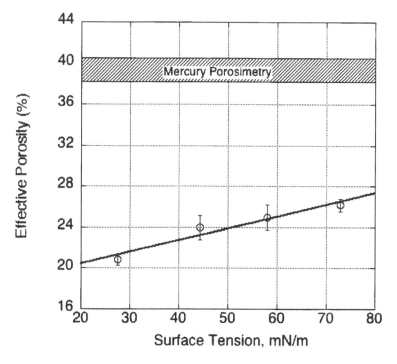

Figure 5. Comparison between mercury porosimetry results and the effective porosity obtained by imbibition analysis using test liquids with different surface tension.

[25], but that there is also a linear correlation between the effective porosity and the surface tension of the liquid.

In an analogous way, we found that the estimated value of the equivalent capillary radius changes with the liquid, as shown in Fig. 6. It is not reasonable to assume that such a value specifies a real set of capillaries because, as also observed

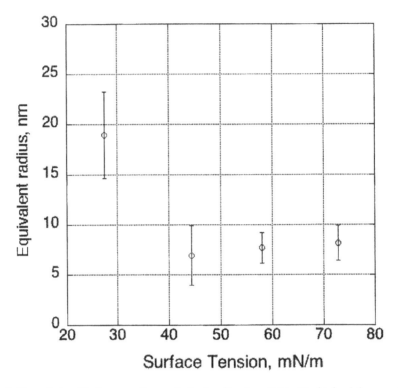

Figure 6. The relationship between the equivalent radius (see Table 2) obtained from each probe liquid and the liquid surface tension.

by some authors [18, 23, 26], the equivalent radius determined in this way turns out to be at least two orders of magnitude smaller than the average pore radius deduced from mercury porosimetry.

3.2.2. Contact angles

By using Wilhelmy technique, Noto stone gives advancing and receding contact angles equal to zero in all the organic liquids used, except water which presents instead 54±4° as advancing and 49±3° (degrees) as receding. Contact angles different than zero obtained on ionic materials with an estimated surface free energy well over that of water (more than 200 mJ/m² [34]) may depend on a roughness effect [35] or on the fact that a material with such high surface free energy may reduce it by adsorbing impurities and water from the environment [36]. This could explain why in our measurements we obtain a contact angle in water different than zero: in fact we are analyzing an "already wet" surface.

The evaluation of contact angles measured following the hypothesis of Studebaker and Snow [5] are presented in Table 3 and the results calculated using the equivalent radius are shown in Table 4. It can be seen that the contact angle for water, using the equivalent radius obtained from the n-hexadecane absorption is

Table 3.
Contact angle for the water/Noto stone system as calculated from the absorption results of other wetting liquids using the approach of Ref. [5]

Liquid employed	Contact angle of water (degree)
n-Hexadecane	73.2
Formamide	48.3
1-Bromonaphtalene	53.5

Table 4.
Contact angle for the water/Noto stone system as calculated from the absorption results of other wetting liquids using the equivalent radius approach

Liquid employed	Contact angle of water (degree)
n-Hexadecane	65.1±7
Formamide	0–23
1-Bromonaphtalene	0–27

Table 5.
Contact angle for Noto stone for different liquids as calculated from the absorption results of n-hexadecane using the equivalent radius approach

Liquid employed	Contact angle (degree)
Water	65.1±7
Formamide	65.8±8
1-Bromonaphtalene	66.6±9

not far from the value found using the Wilhelmy approach. Instead, the results calculated from formamide and 1-bromonaphtalene radii are totally different: this could mean that our reference radius refers to a non-complete wetting liquid. So it is possible to apply the same procedure in order to calculate the contact angles by equation (15) of formamide and 1-bromonaphtalene using the n-hexadecane radius and the results are presented in Table 5. The outcome is surprising and very difficult to justify: it looks like formamide and 1-bromonaphtalene show high contact angles, comparable with that of water, but our measurements confirm that the contact angles of these liquids on Noto stone are zero. This incongruity with the Wilhelmy results could be partially explained by the different behaviors of the liquids during the absorption but casts serious doubts on the possibility to use the Washburn law for the calculation of an unknown contact angle: some times it works, some times it does not, and the results depend on the initial choice of the reference liquid.

3.2.3. Physical meaning of the average pore radius

As indicated before, Dullien and co-workers [22, 23] calculated the equivalent mean square capillary radius, using the Washburn equation, in a periodically constricted tube with periodic changes in diameter, D_n.

The equivalent diameter was estimated by the relationship:

$$\left(D_{eq}\right)_{Dullien} = \frac{1}{3}\left(\sum_{k=1}^{n} a_k D_k\right)^2 \sum_{k=1}^{n}\left[a_k D_k \sum_{j=1}^{n} a_j \left(\frac{D_k}{D_j}\right)^3\right]$$

(18)

where $a_{j,k}$ denotes the length-to-diameter ratio of a capillary segment. It was found that $\left(D_{eq}\right)_{Dullien}$ was insensitive to the value of $a_{j,k}$, so $a_{j,k}=1$ was used throughout.

Moreover, for simplicity sake, only two limiting values D_1 and D_2, corresponding to the maximum and the minimum diameters of the tube respectively, were taken into account (Fig. 1). By posing $n=2$ and $a_{j,k}=1$ in equation (18), the special function obtained is strongly non-linear; its graphical representation (Fig. 7) shows that slight variations in the parameters D_1 and D_2 may significantly change the equivalent radius. Optical observations on Noto stone resulted in a meaningful, largest value of the diameter $D_1=50\mu m$, while from mercury porosimetry an average value of $D_2 = 8\mu m$ was obtained. By reckoning the equivalent radius by the above equation (18) we obtain:

$$\left(D_{eq}\right)_{Dullien} = 0.017\mu m \longrightarrow (r_{eq})_{Dullien} = 0.0085\mu m$$

(19)

The equivalent radius is comparable to that obtained from Washburn measurements (see Table 2).

A qualitative explanation for r_{eq} values being much smaller than the radius estimated by mercury porosimetry is the following: the meniscus spends the majority of time in the largest segments, where the capillary driving force is the smallest and the volume to fill is the greatest. As a consequence, r_{eq} corresponds to the radius whose Laplace pressure, driving force for capillary absorption, turns out to be comparable with the pressure present in the capillary network of the material. Such a pressure, once the value of r_{eq} is known, can be calculated as:

$$\Delta P_L = \frac{2\gamma_{liq}}{r_{eq}}$$

(20)

and the results are shown in Fig. 8.

The largest Laplace pressure corresponds to the greatest ability for filling the pores of the material. This can also explain why the liquids with different surface

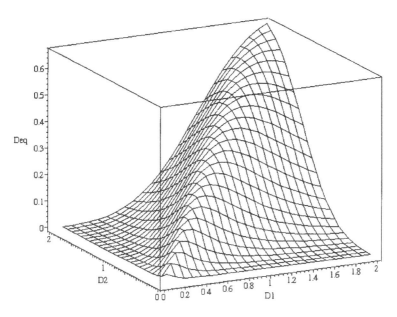

Figure 7. The Dullien equivalent diameter D_{eq} for $n=2$ as a function of the largest, D_1, and the smallest, D_2, diameters of the porous system. The function is strongly non-linear.

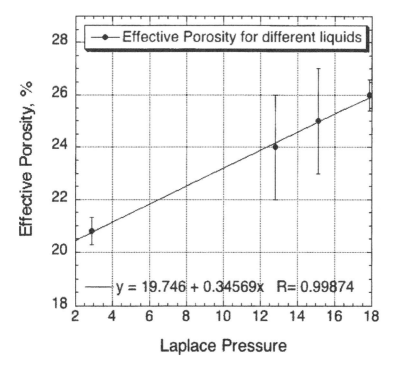

Figure 8. The effective porosity *versus* the Laplace pressure calculated using the equivalent radius obtained for different liquids by applying equations (7) and (20).

tensions correspond to different values of effective porosity ε_{eff}. The correlation between the Laplace pressure and pore filling is quite complex, and it is not sufficient by itself to understand the phenomenon of selective penetration. Indeed, the Laplace pressure is calculated as a function of r_{eq}, which, in turn, depends on the way the liquid is displaced within the capillary network. The existence of preferential paths for the liquid inside the porous matrix can be explained in terms of the rate of air displacement in the largest pores and the subsequent replacement by the invading liquid.

3.2.4. Effects of sample shape

We want to show now that the parameters obtained from the application of the Washburn model to the liquid absorption in a porous medium vary in a significant way with the size ratio of the sample. Various measurements with water and heptane were performed on samples of ceramic material with different transverse sections. By applying equations (11)–(13) we calculated the parameters h_1, m_1 and ε_{eff}.

The parameter m_1 grows linearly with the sample area (Fig. 9), as we expected theoretically from equation (9), whereas the parameter h_1 shows a maximum (Fig. 10), while a constant value was expected (see equation (8)). The factors which affect the mass absorbed with time and the rate of propagation of the liquid front as

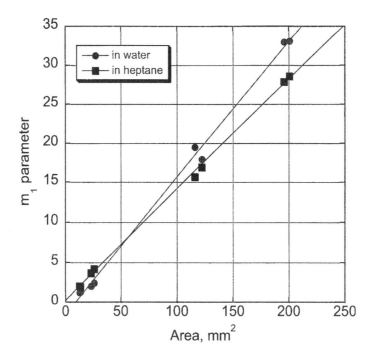

Figure 9. Correlation between the m_1 parameter and the cross-sectional area of the ceramic sample in water and heptane.

measured by the front height are, therefore, different. It is important to emphasize that the measurement of the liquid mass absorbed is performed by simply weighing the sample by the microbalance and hence it does not depend on the size or shape of the sample, while the liquid height is estimated as it appears at the external faces of the parallelepiped sample. It is quite reasonable to conjecture that the height of the liquid front, as it is visible on the lateral surface, could not correspond to the height reached by the liquid in the whole section, as assumed in the application of the Washburn law. By introducing a form or shape factor, defined as:

$$\text{Shape factor} = \frac{\text{Thickness of the sample}}{\text{Width of the sample}} \qquad (21)$$

we found a decreasing trend of the coefficient h_1 in the case of water absorption (Fig. 11). Samples with different sizes and areas, but having the same shape factor, show the same value of h_1. Moreover, the rate of liquid rise turns out to be larger in thin samples.

This result is very important because it implies that different values of effective porosity are obtained by varying the sample geometry: water and heptane show the minimum value of effective porosity at two different values of area (Fig. 12).

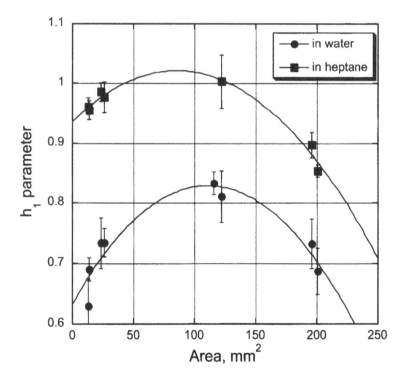

Figure 10. Correlation between the h_1 parameter and the cross sectional area of the ceramic sample in water and heptane.

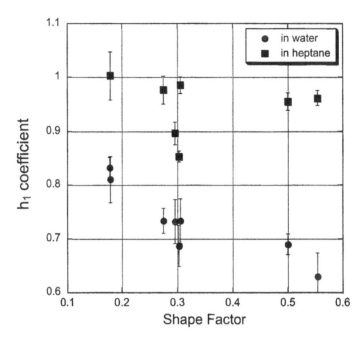

Figure 11. Correlation between the h_1 parameter and the cross section shape factor of the ceramic sample in water and heptane.

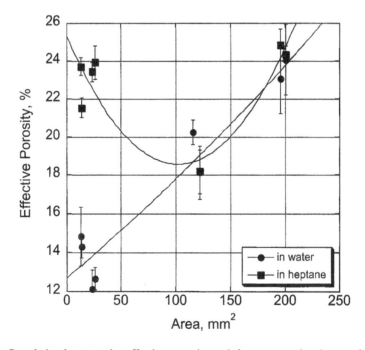

Figure 12. Correlation between the effective porosity and the cross-sectional area of the ceramic sample in water and heptane.

The subjective choice of the sample size, thus, affects the final result in a significant way, posing a further obstacle to the interpretation of absorption measurements by capillarity and making it difficult, at the present state of the art, to rigorously apply the Washburn model to porous materials.

These results are in qualitative agreement with Marmur, who found that at equilibrium the same thickness of a porous sample induces different equilibrium heights, due to the "reexposition" phenomenon [24]; it is conceivable that a similar phenomenon operates also in non-equilibrium situations.

4. SUMMARY AND CONCLUSIONS

The application of the Washburn equation for the contact angle calculation is subjected to stringent hypotheses concerning the motion of the liquid and the simplification of the capillary system network. Although the capillarity absorption is very well described from the equations of mass and height as a function of $\sqrt{t}$, the dependence of the coefficients on the physical parameters is not clarified. From the results presented in this paper we can conclude that:

(1) The value of the contact angle obtained using the Washburn equation strictly depends on the initial choice of the sample size, as well as on the reference liquid. Some liquids, although their contact angle is zero, lead to incoherent results, as different shape factors can lead to different values. There is no specific rule, which suggests *a priori* a particular geometry or a reference liquid instead of another one. In spite of these serious concerns, this procedure is implemented in many commercial software packages proposed for the contact angle measurement without pointing out its limitations!

(2) It is possible to describe a capillary system by a three-dimensional repetition of cylindrical capillaries. In this case, by applying the Dullien relationship and simplifying the network of capillaries as a periodic pattern of connected necks and bulges of constant diameter, it is possible to give a physical meaning to the value of the equivalent radius calculated by the Washburn equation.

(3) The value of the Laplace pressure, calculable from the equivalent radius, is very well correlated with the value of the measured effective porosity.

Acknowledgements

The authors are grateful to Dr. T. Poli and Prof. L. Toniolo for supplying the Noto Stone and the ceramic samples, to Ing. M. Bertoldi for the mercury porosimetry measurements and to Ing. M. Benedetti for interesting discussions.

REFERENCES

1. R. Lucas, *Kolloid Z.* **23**, 15 (1918).
2. E. W. Washburn, *Phys. Rev.* **XVII**, 3 (1921).
3. C. H. Bosanquet, *Philos. Mag.* **45**, Series **6**, 525–531 (1923).
4. Z. W. Wolkowa, *Kolloid Z.* **67**, 280–284 (1934).
5. M. L. A. Studebaker and C. W. Snow, *J. Phys. Chem.* **59**, 973–976 (1955).
6. D. H. Everett and J. M. Heynes, *Zschr. Phys. Chem. Neue Folge* **82**, 36–48 (1972).
7. R. Good and N. Lin, *J. Colloid Interface Sci.* **54**, 52 (1976).
8. D. T. Hansford, D. J. W. Grant and J. M. Newton, *Powder Technol.* **26**, 119–126 (1980).
9. Y. Yang, G. Zografi and E. Miller, *J. Colloid Interface Sci.* **122**, 24–34 (1988).
10. K. Grundke and A. Augsburg, *J. Adhesion Sci. Technol.* **14**, 765–775 (2000).
11. E. Chibowski and L. Holysz, *J. Adhesion Sci. Technol.* **11**, 1298 (1997).
12. J. K. Spelt and D. Li, in: *Applied Surface Thermodynamics*, J. K. Spelt and A. W. Neumann (Eds.), pp. 239–292, Marcel Dekker, New York, NY (1996).
13. G. Buckton, *J. Adhesion Sci. Technol.* **7**, 205–219 (1993).
14. P. M. Costanzo, R. F. Giese and C. J. van Oss, *J. Adhesion Sci. Technol.* **4**, 267 (1990).
15. C. J. van Oss, R. F. Giese, Z. Li, K. Murphy, J. Norris, M. K. Chaudhury and R. J. Good, *J. Adhesion Sci. Technol.* **6**, 413–428 (1992).
16. E. Chibowski and L. Holysz, *Langmuir* **8**, 710–716 (1992).
17. J. Schoelkopf, C. J. Ridgway, P. A. C. Gane, G. P. Matthews and D. C. Spielmann, *J. Colloid Interface Sci.* **227**, 119 (2000).
18. J. Schoelkopf, P. A. C. Gane, C. J. Ridgway and G. P. Matthews, *Colloids Surfaces A* **206**, 445–454 (2002).
19. P. M. Hertjies and W. C. Witvoet, *Powder Technol.* **3**, 339–343 (1969).
20. J. Szekely, A. W. Neumann and Y. K. Chuang, *J. Colloid Interface Sci.* **33**, 273 (1971).
21. S. Levine, P. Reed, G. Shutts and G. Neale, *Powder Technol.* **17**, 163–181 (1977).
22. F. A. L. Dullien, M. S. El-Sayed and V. K. Batra, *J. Colloid Interface Sci.* **60**, 497–506 (1977).
23. F. A. L. Dullien, *Porous Media: Fluid Transport and Pore Structure*, p. 396, Academic Press, New York, NY (1979).
24. A. Marmur, *J. Colloid Interface Sci.* **129**, 1 (1989).
25. L. Labajos Broncano, M. L. Gonzales, J. M. Bruque, C. M. Gonzales Garcia and B. Janczuk, *J. Colloid Interface Sci.* **219**, 275–281 (1999).
26. E. O. Einset, *J. Am. Ceram. Soc.* **79**, 333–338 (1996).
27. M. G. Bernadiner, *Trans. Porous Media* **30**, 251–265 (1998).
28. J. C. Jamin, *Phil. Mag.* **19**, 204–207 (1860).
29. C. M. Case, *Physical Principles of Flow in Unsaturated Porous Media*, p. 40, Oxford University Press, New York, NY (1994).
30. E. J. Garboczi, *Powder Technol.* **67**, 121 (1991).
31. W. S. Rasband, *ImageJ*. US National Institutes of Health, Bethesda, MD; http://rsb.info.nih.gov/ij/ (1997–2005).
32. UNI10921 *Protocol, Cultural heritage: Natural and artificial stones. Water repellents* (2001).
33. M. Brugnara, C. Della Volpe, D. Maniglio, S. Siboni, M. Negri and N. Gaeti, These proceedings.
34. A. Marmur, in: *Modern Approaches to Wettability: Theory and Applications*, M. E. Schrader and G. I. Loeb (Eds.), Plenum Press, New York, NY (1992).
35. R. E. Johnson Jr. and R. H. Dettre, in: *Surface and Colloid Science*, Vol. 2, E. Matijevic (Ed.), Wiley, New York, NY (1969).
36. D. S. Keller and P. Luner, *Colloids Surfaces A* **161**, 401–415 (2000).

Contact Angle, Wettability and Adhesion, Vol. 4, pp. 165–176
Ed. K.L. Mittal
© VSP 2006

Contact-angle measurements of sessile drops deformed by a DC electric field

C. ROERO*

High Voltage Laboratory, Swiss Federal Institute of Technology, CH-8092 Zürich, Switzerland

Abstract—Sessile water drop geometries on various surfaces have been studied with and without exposure to electric fields with the aim to understand the phenomena involved in the generation of the sound emission from wet high-voltage transmission lines. It has been demonstrated that noise can be reduced by the application of certain coatings to the high-voltage conductors. Various coatings and treatments have been evaluated with respect to contact angle measurements under a wide range of conditions, including discharge levels. In the experiments presented water drops of well-controlled volume were placed on horizontal stainless steel and aluminium surfaces which had undergone a variety of treatments, singly or in combination: sandblasting, glass-bead blasting, hydrophobic and hydrophilic coatings. The surfaces mentioned formed the lower electrode of a parallel plate discharge gap. Contact angles were determined by first producing an electronic image of the drop which was then assessed by means of the software developed in this study. It has been shown that the voltage at which instability, resulting in ejection of a fine jet of water of a few μm with increasing electric field, is reached depends strongly on the initial shape of the water drop and, thus, on the contact angle which, in turn, is controlled by the surface properties. This work demonstrated that the instability voltage was roughly doubled if the surface conditions changed from hydrophobic to hydrophilic.

Keywords: Contact angle; high-voltage transmission lines; acoustic emission; drop instability.

1. INTRODUCTION

After precipitation high-voltage transmission lines emit excessive levels of noise. The major component of 'wet' weather noise to be suppressed is a fixed frequency (100 Hz in Europe and 120 Hz in North America) hum called tonal noise and perceived as particularly annoying especially at night and when it can be the dominant noise in the locality. The early study of acoustic emission from high-voltage lines was done in 1969 [1]. Former investigations [2, 3] demonstrated that tonal emissions from high voltage transmission lines were directly linked to the presence of water drops on the conductors and their periodic (2f) deformation. This interpretation has recently been scrutinized in depth (see Refs [4, 5]). It has

---

*Tel.: (41-1) 632-7624; Fax: (41-1) 632-1202; e-mail: croero@eeh.ee.ethz.ch

been shown that the essential role of oscillating drops is that of injecting electrical charge carriers into the immediate surroundings of the conductor. The periodic moment of these charges will eventually lead to sound emission [5]. At high local field strengths (>20 kV/cm) the drop deformation can lead to instability accompanied by water ejection, loss of charge and sound emission. The drop elongates in the direction of the electric field and becomes unstable at a critical field strength.

Clearly the surface field strength on the drops plays a decisive role which, in turn, is largely determined by the drop shape, which is strongly influenced by the surface properties of the substrate. For instance, with a hydrophilic substrate the water drop contact angle is small and sessile drops have a very shallow contour; this implies lower surface field strengths, thus less deformation and a lower level of sound emission, as well as lower discharge activity. In these studies it was shown that a model line with a strongly hydrophilic surface had – after cessation of precipitation, only 4% of the integrated discharge current of the untreated/hydrophobic line, taken over a 20-min drying period of the lines.

The work presented here is based on characterization of the drop contact angle for different conductor surfaces with and without application of voltage. Although alternating current (AC) is the most common form of electrical energy transmission, studies with direct current (DC) were used to elucidate the process involved in the acoustic emission from overhead high voltage lines; specifically details of the deformation of water drops in the electric field in dependence on drop size, surface condition and field strength were investigated.

2. INSTABILITY OF WATER DROPS IN AN ELECTRIC FIELD

The phenomenon of water drop deformation and instability in an electric field is well documented. Lord Rayleigh [6] found a stability criterion, later refined by Sir Geoffrey Taylor for spherical drops [7], which connects the applied electric field, drop radius and surface tension. The critical electric field strength for instability of an uncharged free-floating drop [8] is given as:

$$E_{cr} = 447\sqrt{\frac{\gamma}{r}}$$

where r is the drop radius and γ the water surface tension.

Experimental investigations have shown that with the increase of the applied voltage, instability is reached, a Taylor cone is formed and the drop starts to vibrate. As a consequence of the pointed drop shape eventually attained there is a field enhancement at the apex which leads to axial extension of the drop. In this way, the apex angle decreases and this produces further field enhancement and deformation which has to be balanced by the surface tension. When the field enhancement due to deformation produces so high a force at the apex that it can no

longer be balanced by surface tension, instability is reached, i.e., the drop is rup-
tured and a fine water jet/droplet is ejected.

3. METHODS TO MEASURE CONTACT ANGLES

The contact angle is the parameter used to characterize different surfaces and to
classify the shape of a water drop resting on a horizontal surface with and without
application of an electric field. In all the experiments described here static contact
angles were measured, since we did not have the equipment to measure advancing
and receding contact angles.

The first step of the measurement was to place a water drop of defined volume
on the solid sample surface, which was always exactly horizontal. To apply reproducible uniform volume drops of deionised water, calibrated micropipettes
were used; in general, the volume of the water drop used here was in the range
20–100 µl. Drop shape was recorded with a high speed framing camera, images
were then processed by a computer and stored. In the investigations carried out to
date, the drop shape is then automatically evaluated in terms of contact angle as
represented by the angle between the substrate surface and the tangent from the
edge to the contour of the drop.

The uncertainty in the measurements depends on the light-dark contrasts of the
drop picture, in particular at the air-liquid-solid triple point, and on the method
used for the evaluation. An error of 3–4° must be assumed, which can still be accepted, since the experiments were carried out in a system where perfect surface
uniformity and cleanliness could not be guaranteed.

At this stage two methods for contact angle measurements were used depending on the degree of drop picture quality manually and using software. The problems encountered in different experimental methods are described in the following. In general, it is easy for the program to find the drop profile, but hard for it to
find the correct baseline, which may be obscured in the image; furthermore, the
specimen may be somewhat irregular. Conversely, it is easy for a person to locate
the baseline but fitting a tangent to the drop profile is more difficult.

The assumption one can use in some cases is that the drop profile is represented
by a portion of a circle of defined radius and centre. When this assumption is true
it is much easier to find the contact angle manually, because one needs only to
measure with a ruler the length of the chord and the height of the drop. Then with
a simple relation ($\theta = \arcsin (4\,s\,h\,/\,(s^2+4h^2))$, where s is the length of the chord, h
is the height of the drop and θ is the contact angle) one can find the value of the
angle.

If the assumption that the drop profile is a portion of a circle is not applicable,
one still has the choice to assess the contact angle by drawing the tangent to the
drop surface and measuring the angle formed between the solid surface and the
tangent by means of a goniometer. In this case it is easily understandable that the
measurements are strongly subjective and a larger error is expected.

Measuring contact angle using a software has the big advantage that it enables one to determine contact angles on solids rapidly and easily. Many companies in the world sell systems [9] which provide a number of features that make the procedure for contact angle measurements easy and versatile. In this study only approximate values of contact angles are required so that investment in costly systems would not be justified. Thus, for a more rapid but also a more economical method to analyze the image of a drop and, therefore, to assess contact angle, a suitable software was developed here. The program, a Matlab script, allows hands-free measurements using image acquisition and image processing techniques. First the program reduces the grey-scale image of the drop to a set of

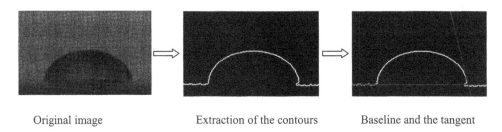

Original image   Extraction of the contours   Baseline and the tangent

Figure 1. Measurement of contact angle using image processing techniques.

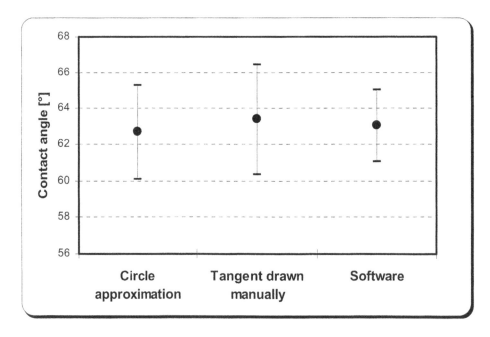

Figure 2. Comparison between different contact angle measurement methods.

equations describing the periphery, and then it obtains the intersection of the base-line with the tangent. The different steps in the processing technique are presented in Fig. 1.

A comparison between the methods described above for the case of a 20 µl water drop on sandblasted aluminium surface, where the circle approximation of the drop also applies, is presented in Fig. 2.

The results show that the measurement using the software produces a lower un-certainty (about 2°), while a larger error is encountered in measurements made manually. However, the mean value seems to show no significant difference for different methods.

4. DROP EVAPORATION ON SANDBLASTED ALUMINIUM AND STAINLESS-STEEL SURFACES

In the first test series the evaporation of drops on sandblasted aluminium and stainless steel surfaces was investigated without the influence of an electric field.

Many authors state that contact angle measurements are strongly influenced by the time t between positioning the drop on the surface and performing the meas-urement [9, 10]. Furthermore, since a drop does not behave symmetrically in the electric field, the contact angle was separately determined at the left and at right sides; the average of the left and right values is presented here. Pictures were re-corded at room temperature of 20°C and 40% relative humidity. The contact angle was measured with both methods, i.e., manually and using the program. Figure 3 illustrates the decrease of contact angle with time for three drop volumes (10, 50

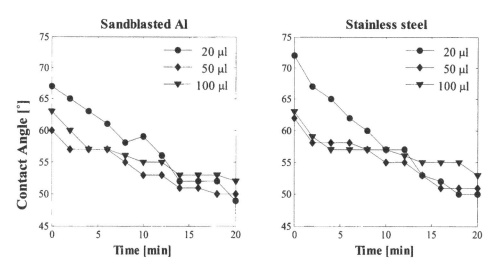

Figure 3. Contact angle *versus* time for 20, 50 and 100 µl drops on sandblasted Al and stainless-steel surfaces.

and 100 µl) for a rough/polished surface (sandblasted aluminium) and a smooth/polished surface (stainless steel). Although it has been found that the contact angle decreases with decreasing drop size, as demonstrated in Ref. [11], this experiment shows that a 100-µl water drop seems to have a smaller contact angle than a 10-µl one. This is mainly due to the non-uniformity of the analyzed surfaces, which contributes also to a large hysteresis.

Within a few minutes the loss of volume due to evaporation, manifesting itself in a decrease of contact angle, is quite evident with both surfaces and with all drop sizes. At this point it can be stated that, with both stainless steel and sandblasted aluminium surfaces, the time t elapsed until the contact angle is measured should be less than 2 min, in order not to influence the results significantly, in accordance with Ref. [6]. This is a good choice, because the experiments with electric fields are carried out within 1.5 min, including deposition of the drops and taking the normal security steps required before starting an experiment using high voltage. Moreover, in the present experiment the temperature of the electrodes on which the drops were positioned did not change during application of high voltage.

5. LABORATORY INVESTIGATIONS WITH AN ELECTRIC FIELD

In the next test series the behaviour of single drops during an increase of the applied electric field was studied.

Two parallel metal plates of defined dimensions and shapes (6.5 cm diameter, 90° Rogowski profile) were arranged with a 1-cm gap. The upper electrode was

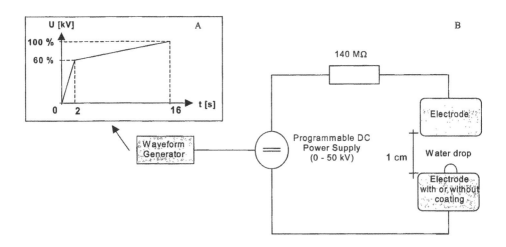

Figure 4. Experimental set-up (B). The sessile drop in a parallel-plate electrode arrangement was exposed to a ramped electric field (A), recording the voltage at the instant of instability which manifests itself as a load current on the supply DC source.

always stainless steel without any treatment, while the lower one was varied to test different surfaces and coatings available.

After each measurement the lower electrode was rubbed down with ethyl alcohol to restore uniform starting conditions. The arrangement of Fig. 4, supplemented as required by a framing camera, was used to study drop deformation and instability.

Deionised water drops of 30, 50, 80 and 100 µl were deposited on the lower electrode in the central region. The high DC voltage was produced by a stabilized HV power supply controlled by a function generator. To have sufficient time to record water drops sequences up to instability, a special function with voltage linearly increasing with time was programmed as represented in Fig. 4. With the function chosen, 60% of the maximum voltage amplitude (U) was reached in 2 s, and then the voltage was increased up to the breakdown voltage in 14 s.

To record rapid drop deformation and the development and consequences of instability, a high speed (up to 10 000 fps) digital framing camera was used. In this test series single drops were recorded with 2000 frames/s for a duration of 16 s. The water drops were always illuminated indirectly, to avoid rapid evaporation.

6. PROCEDURE AND MEASUREMENTS

The electrodes were variously prepared for each measurement: stainless steel polished, aluminium sandblasted and aluminium glass-beads blasted. Furthermore, coatings were obtained for assessment from commercial and institutional sources; for confidential reasons the names of the sources cannot be disclosed. A commercial hydrophobic coating based on silica and a TiO_2 powder was applied to all the substrates. The contact angle for different surface states (untreated, hydrophobic and hydrophilic) of the electrodes in the absence of high voltage are presented for a 50 µl drop in Table 1.

By definition, a water drop on a hydrophobic surface has a large contact angle. The deformation of such a drop of 80 µl in the electric field in the parallel plate configuration is illustrated in Fig. 5. The growing axial elongation of the drop with increasing field is quite evident (intentionally the sequence has not been taken up to instability).

The voltage at which instability occurs depends clearly on the conductor surface and on the drop size (Fig. 6). The results underscore that the critical global electric field at first instability is higher when the drop is flatter (lower contact angle) and, correspondingly, it is lower with a greater contact angle. Figure 6 also illustrates that the decrease of instability voltage with increasing drop volume in the range 30–100 µl is approximately linear with both hydrophobic and untreated substrates. The non-linearity of the hydrophilic curves is probably due to the difficulty for the program in measuring small contact angles.

Table 1.
Water drop contact angles for stainless steel, sandblasted Al, glass-beads-blasted Al with different coatings

Surface	Treatment	Contact angle (°)
Stainless steel	Untreated	72
	Hydrophobic	98
	Hydrophilic	15
Sandblasted Al	Untreated	61
	Hydrophobic	114
	Hydrophilic	approx. 0
Glass beads blasted Al	Untreated	59
	Hydrophobic	102
	Hydrophilic	approx. 0

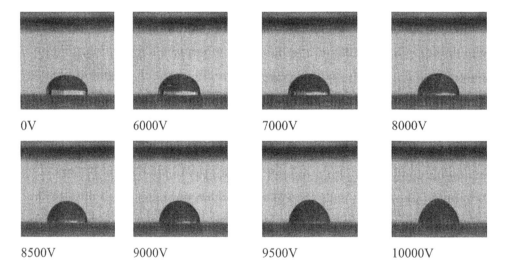

Figure 5. Deformation of a 80 µl water drop by the electric field on aluminium surface sandblasted and subjected to a proprietary hydrophobic treatment. Gap spacing was 10 mm. The sequence was terminated before reaching instability (>10 000 V).

The deformation of the drop by the electric field brings an evident change in contact angle. Figures 7 and 8 show variations of contact angle as functions of voltage and drop size for a stainless steel electrode which was untreated or made ultra-hydrophilic.

One can see that the contact angle decreases as the electric field distorts the drop. The drop has initially a higher contact angle, which then becomes somewhat smaller with increasing deformation of the drop. A comparison between the cases

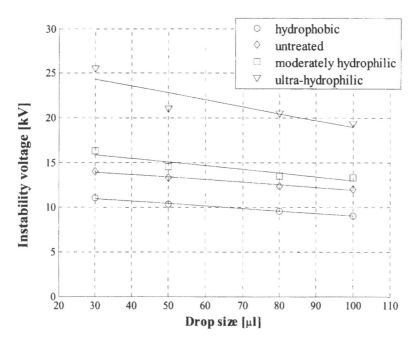

Figure 6. Instability voltage in dependence of drop size for various treatments of a sandblasted aluminium substrate.

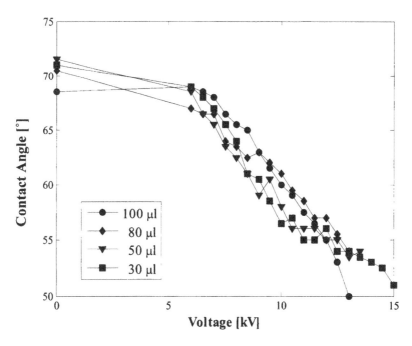

Figure 7. Variation of contact angle *versus* applied voltage for different drop sizes for an untreated stainless steel substrate.

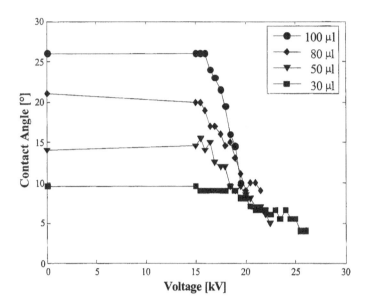

Figure 8. Variation of contact angle *versus* applied voltage and drop size for an ultra-hydrophilic treatment of a stainless steel substrate. Note that the deformation sets in at about twice the voltage for the sample in Fig. 7.

analyzed shows that variation of the contact angle has a similar tendency for both hydrophobic, as well as hydrophilic surfaces.

7. INTERPRETATION AND DISCUSSION

These investigations provided details on the instability of water drops in an electric field in dependence of drop size, surface condition and field strength. In fact, the instability controls the size of surviving drops on wet high-voltage transmission lines, which contributes to sound emission.

It was observed with all surfaces analyzed that the drop changed its shape becoming more pointed with increasing voltage but flatter at the base, which meant that the contact angle decreased considerably. This behaviour began at *ca.* 6 kV for hydrophobic, as well as for untreated surfaces, then the decrease of contact angle was nearly linear with increase in voltage; while with ultra-hydrophilic surfaces it became significant only above about 16 kV (see Figs 7 and 8).Clearly the instability voltage depends on the contact angle, as shown in Figs 6, 7 and 8. An evaluation of the entire series of contact angle measurements leads to the plot in Fig. 9, in which it is observed that with a decrease of contact angle from about 90° (hydrophobic surface) to about 5° (ultra-hydrophilic) the instability voltage roughly doubles for a 100 μl drop. For 80 and 50 μl drops the same

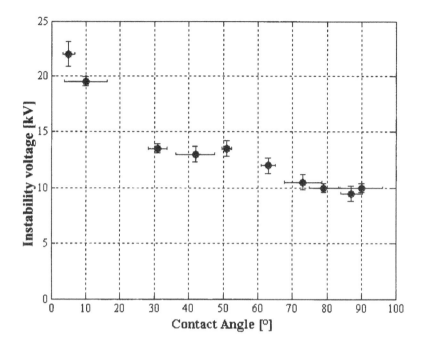

Figure 9. Instability voltage in dependence of contact angle: 100 µl water drop, 10 mm gap.

behaviour was observed, while for a 30 µl drop the instability voltage increased even by a factor of 3.

8. SUMMARY AND CONCLUSIONS

(1) A new set-up for detailed optical investigation of drop deformation was used and it proved its capability to yield significant information on the parameter dependence of deformation of water drops in an electric field.

(2) A tool written in Matlab to evaluate automatically the water drop contact angle was successfully developed, giving an uncertainty of about 2°.

(3) The instability voltage at which a water drop elongates in the direction of the electric field was demonstrated to increase strongly with a reduction in zero-field contact angle and to decrease with increasing drop volume.

Acknowledgements

Constant help and support provided by the CONOR team (T. H. Teich, M. Semmler and U. Straumann) is much appreciated, and so is the continued feedback from Prof. Klaus Fröhlich. Financial support of the work by EnBW (Germany), APG (Austria), Illwerke AG (Austria), PSEL (Switzerland) and BUWAL (Switzerland) is gratefully acknowledged.

REFERENCES

1. E. R. Taylor, V. L. Chartier and D. N. Rice, *IEEE Trans. on Power Apparatus and Systems* **8**, 666-679 (1969).
2. T. H. Teich and H.-J. Weber, *Elektrotech. Informationstech.* **119**, 103-113 (2002).
3. T. H. Teich and H.-J. Weber, *Proc. 14th Intl. Conf. on Gas Discharges and Their Applications* **1**, Liverpool, pp. 259-262 (2002).
4. C. Roero, T. H. Teich and H.-J. Weber, *Proc. 15th Intl. Conf. on Gas Discharges and Their Applications* **1**, Toulouse, pp. 335-338 (2004).
5. U. Straumann and M. Semmler, *Proc. 15th Intl. Conf. on Gas Discharges and Their Applications* **1**, Toulouse, pp. 363-366 (2004).
6. Lord Rayleigh, *Phil. Magn.* **14**, 184-186 (1882).
7. Sir Geoffrey Taylor, *Proc. Roy. Soc.* **A 280**, 383-397 (1964).
8. W. A. Macky, *Proc. Roy. Soc.* **A 133**, 565-587 (1931).
9. S. Keim and D. Koenig, *2000 IEEE Conf. on Electrical Insulation and Dielectric Phenomena*, Victoria, Canada, 792-795 (2000).
10. H. Janssen and U. Stietzel, *Proc. 10th Intl. Symp. on High Voltage Engineering* **3**, Montréal, pp. 149-152 (1997).
11. R. J. Good and M. N. Koo, *J. Colloid Interface Sci.* **71**, 283-292 (1979).

Contact Angle, Wettability and Adhesion, Vol. 4, pp. 177–182
Ed. K.L. Mittal
© VSP 2006

Dynamic contact angle of a droplet spreading on heterogeneous surfaces

J. LÉOPOLDÈS*

Department of Materials, University of Oxford, Oxford OX1 3PH, UK

Abstract—Spreading of viscous droplets on model surfaces obtained by microcontact printing is studied. Here, the special case where surface heterogeneities are in the form of micro-chemical stripes is considered. It is shown that an elongated flow occurs, which develops through the relaxation of the longitudinal curvature of the liquid–air interface. Finally, we find that the dynamic contact angle of a droplet spreading along the chemical stripes shows similar variation with spreading velocity as that for a droplet spreading on a homogeneous surface.

Keywords: Wetting; microcontact printing; dynamic contact angle.

1. INTRODUCTION

The behavior of a droplet spreading on a surface has been a topic of investigation for a long time [1]. Many questions arise from the complicated profile of the contact line and the dynamic behavior of the meniscus [2–4]. When the contact line advances at a low velocity, a critical behavior is expected and several studies, both theoretical and experimental, have been reported [5–7]. However, these studies concern a certain situation where the contact line is driven at imposed force or velocity, which allows to unravel the main physics involved, as well as proper comparison with experiments. The case of a spreading droplet (hence without applied force or velocity), though relevant to many technological applications (ink-jet, agriculture), has not been studied yet. As well, the main body of the work previously published has focused on topological defects created by different processes (chemical etching, indentation, etc.). Here we study the case of a droplet spreading on chemically patterned substrates obtained by microcontact printing. Our aim is to analyze the shape of the droplet spreading on such heterogeneous substrates.

---

*Present address: Laboratoire de Physico-Chimie des Polymères, Université de Mons-Hainaut, 2 Place duParc, Mons 7000, Belgium. E-mail: julien.leopoldes@umh.ac.be

2. CONTACT LINE DYNAMICS

On a heterogeneous substrate the contact line is expected to pin/depin on surface defects while the droplet is spreading at an average velocity v. Near the pinning threshold, the average velocity of the contact line follows:

$$v \sim (F - F_c)^{\beta}, \tag{1}$$

where F is the driving force, F_c the pinning threshold and β the critical exponent (see Ref. [7] for a review). In the case of a spreading droplet, the unbalanced driving force acting on the contact line is linked to the difference between dynamic and equilibrium contact angles (see equation (2)):

$$F = \gamma (\cos \theta_d - \cos \theta_{eq}), \tag{2}$$

where θ_d and θ_{eq} are dynamic and equilibrium contact angles, respectively. Combining equations (1) and (2) gives:

$$v \propto (\cos \theta_d - \cos \theta_{eq})^{\beta}. \tag{3}$$

A spreading droplet should behave accordingly provided that the flow occurs in a quasi-static regime and that the spreading velocity is small enough to allow a critical behavior.

3. EXPERIMENTS

The liquid used was squalene obtained from Aldrich and used as received (viscosity $\eta = 35.0$ mPa·s and surface tension $\gamma = 30.1$ mN·m^{-1} at 21°C). The surfaces were produced using microcontact printing methods on gold coated silicon wafers as described elsewhere [8, 9] to form patterns which were parallel microstripes of different wettabilites. Using mercapto-undecanoic acid and octadecanethiol as patterning chemicals, high and low surface energy microstripes of the same width w were obtained. High surface energy stripes (COOH-ended thiols, denoted as surface chemistry C_1) were completely wetted by squalene, while low surface energy ones (CH$_3$-ended thiols, surface chemistry C_2) had an equilibrium contact angle of 51°. The dispersion and polar components of the surface energy of C_1 and C_2 samples were evaluated to be ($\gamma^d_{C_1} = 27.8$, $\gamma^p_{C_1} = 26.8$) and ($\gamma^d_{C_2} = 28.0$, $\gamma^p_{C_2} = 1.7$) mJ/m^2. Several widths w are considered in this paper: 2, 5, 10, 20 and 40 μm. The droplets where deposited on the surfaces by a micro-syringe and the spreading was monitored via a video-recorder (640 $\times$ 512 pixels) linked to a computer. The acquisition rate was 25 images/s.

4. RESULTS AND DISCUSSION

In Fig. 1 a typical final conformation of a droplet deposited on a striped substrate is shown. The equilibrium shape is influenced by the substrate pattern and is

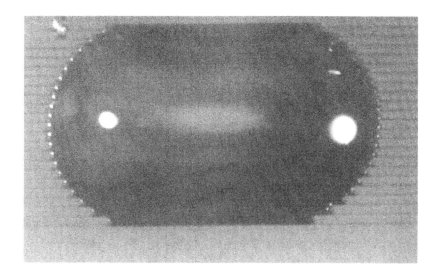

Figure 1. Squalene droplet deposited on a microstriped substrate. The width of the microstripes (w) is 80 μm.

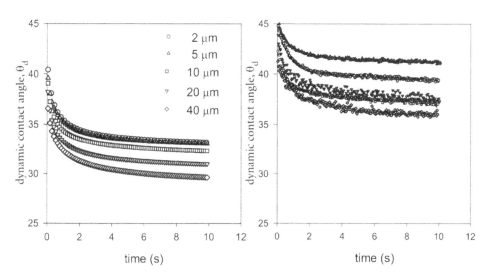

Figure 2. Evolution of dynamic contact angles as a function of time for different microstripe widths: 2, 5, 10, 20, 40 μm. (Left) Contact angle measured normal to the stripes. (Right) Contact angle measured along stripes direction. The curves are average of three different experiments.

highly elongated. All droplets spreading on striped surfaces showed identical morphologies, yet with a different aspect ratio between their final length and width. Figure 2 shows the evolution of dynamic contact angles measured along the stripes and normal to the stripes. It is important to note that normal to the stripes spreading is very limited and not accessible experimentally with the acquisition frequency

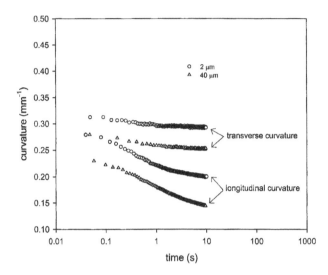

Figure 3. Curvature of the liquid–air interface for two different microstripe widths. Two examples are shown, for $w = 2$ and 40 μm.

of the camera used. The evolution of the contact angles normal to the stripes is, therefore, only related to the expansion of the droplet along the stripes. At late stages of spreading some liquid is able to flow out of the droplet on the totally wetting stripes. Therefore, capillary dragging of the liquid on individual hydrophilic stripes could contribute to the evolution of contact angles measured on the whole droplet, as soon as the spreading velocity approaches the spreading velocity on a hydrophilic stripe (approx. 1 μm/s, see Ref. [10]). A surprising feature of Fig. 2 is that the equilibrium contact angles measured normal to the stripes decrease when the width of the microstripes decreases. This could be linked to a variation in the amount of surface heterogeneity due to the number of stripe edges, although the average surface composition remains the same. Finally, note that the standard errors in the contact angles measured along the stripes are much larger than for those measured normal to the stripes (see Fig. 2), suggesting a higher contact angle hysteresis. This is in line with some results previously reported [11]. Additionaly, in Fig. 3 the curvature of the liquid–air interface seen normal and parallel to the stripes is shown. No variation in the transverse curvature is detected as soon as capillary spreading occurs. This observation could be used, for example, to model the flow observed with scaling arguments.

Another question to investigate is the evolution of dynamic contact angle with spreading velocity. To compare the patterned substrate to homogeneous ones, we grafted a homogeneous mixed monolayer of alkanethiols (0.7 mol mercaptounde-canoic acid + 0.3 mol octadecanethiol; sample immersed for 18 h at room tempera-ture) on the same type of gold coated silicon wafers as used for surface patterning, which resulted in an equilibrium contact angle of 30.1°. The spreading on such sub-

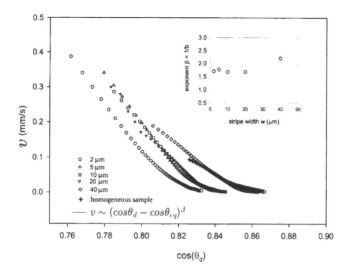

Figure 4. Spreading velocity as a function of dynamic contact angle for different stripe widths $w = 2, 5, 10, 20$ and 40 μm. Horizontal bars are the standard errors in the average values of contact angles and are shown only when larger than the symbols. The spreading occurs along the stripes. The inset represents the different values of β resulting from the fits.

strate is the same as the one occurring on the 40 μm stripes because the equilibrium contact angles are equal.

Figure 4 reports the evolution of spreading velocity as a function of the dynamic contact angle; experimental data are compared to equation (3). No significant variation of the exponent is detected and we obtain average values of $\beta \approx 1.9 \pm 0.3$ for patterned samples (spreading along the stripes) and $\beta \approx 1.9 \pm 0.1$ for the homogeneous sample (see inset in Fig. 4). We emphasize that our claim is not a comparison of the exponents obtained here to the values of critical exponent β published in the literature [7] (although a good agreement is obtained) since no measurement of $F - F_c$ is available and because the spreading occurs in a transient regime.

Rather, we show that for the system studied, there is no significant variation in the evolution rate of dynamic contact angle with spreading velocity, irrespective of the presence and size of chemical defects on the surface. Surface heterogeneities seem to modify mainly the statics of spreading through contact angle hysteresis. Whether this is a general phenomenon still needs to be clarified, mainly in situations where contact line pinning occurs.

5. CONCLUSIONS

The spreading of squalene droplet on a striped heterogeneous surface was studied. It has been shown that the orientation of surface defects can modify the orientation of wetting. The curvature of the longitudinal liquid–air interface seems to vary predominantly as compared to the lateral one. Finally, it was found that with the heterogeneities considered, the evolution of dynamic contact angle as a function of spreading velocity is identical to that on a homogeneous surface of similar equilibrium contact angle. Surfaces with other wettability contrasts and defect shapes will be studied in further work.

Acknowledgments

This study forms part of the IMAGE-IN project which is funded by the European Community through a Framework 5 grant (contract number GRD1-CT-2002-00663).

REFERENCES

1. P. G. de Gennes, *Rev. Mod. Phys.* **57**, 827–863 (1985).
2. R. L. Hoffman, *J. Colloid Interface Sci.* **50**, 228–241 (1975).
3. R. G. Cox, *J. Fluid Mech.* **168**, 169–194 (1986).
4. Q. Chen, E. Ramé and S. Garoff, *J. Fluid Mech.* **337**, 49–66 (1997).
5. P. G. de Gennes and J. F. Joanny, *J. Chem. Phys.* **81**, 552–562 (1984).
6. S. Kumar, D. H. Riech and M. O. Robbins, *Phys. Rev. E* **52**, 5776–5779 (1995).
7. E. Schäffer and P. Wong, *Phys. Rev. E* **61**, 5257–5277 (2000).
8. J. Leopoldes, A. Dupuis, D. G. Bucknall and J. M. Yeomans, *Langmuir* **19**, 9818–9822 (2003).
9. A. Kumar, H. A. Biebuyck and G. M. Whitesides, *Langmuir* **10**, 1498–1511 (1994).
10. A. A. Darhuber, S. M. Troian and W. W. Reisner, *Phys. Rev. E* **64**, 031603 (2001).
11. J. Drelich, J. L. Wilbur, J. D. Miller and G. M. Whitesides, *Langmuir* **12**, 1913–1922 (1996).

Contact Angle, Wettability and Adhesion, Vol. 4, pp. 183–202
Ed. K.L. Mittal
© VSP 2006

Wetting on chemically heterogeneous surfaces: Pseudo-spherical approximation for sessile drops

PEDRO GEA-JÓDAR,[1] MIGUEL ÁNGEL RODRÍGUEZ-VALVERDE[2] and MIGUEL ÁNGEL CABRERIZO-VÍLCHEZ[1,*]

[1]*Biocolloid and Fluid Physics Group, Department of Applied Physics, Faculty of Sciences, University of Granada, E-18071 Granada, Spain*
[2]*Polymers and Composites Laboratory, Department of Materials Engineering and Industrial Technologies, University of Trento, I-38050 Trento, Italy*

Abstract—In this work, wetting on rigid, flat and chemically heterogeneous solid surfaces is studied theoretically. For this purpose two different methods have been employed: (i) computer calculations performed with the public-domain software Surface Evolver and (ii) an approximation to describe the sessile drops on non-ideal solid surfaces, called pseudo-spherical caps in this study. The use of these two methods was complementary and allowed us to interpret the behavior of a drop on a heterogeneously patterned surface, besides predicting the values of contact parameters, such as contact angle, contact area, etc. Two different chemical heterogeneity patterns have been analyzed: a striped pattern and a chessboard pattern. The results are also compared with the values expected from the Cassie equation. The good agreement obtained using these two methods allows us to extend the pseudo-spherical approximation to more complex surfaces.

Keywords: Contact angle; sessile drop; heterogeneity; wetting; Cassie equation.

1. INTRODUCTION

Wetting phenomena are of fundamental interest in many natural and technological processes such as cell adhesion [1], self-cleaning mechanism in plants [2, 3], paintings, etc. Hence, a complete understanding of these phenomena appears essential for a correct design and control of their mechanisms. Significant efforts have been made towards understanding of wetting phenomena, both from theoretical and experimental points of view, also involving novel techniques [1–17]. However, due to its higher simplicity, much attention has been paid to the study of wetting phenomena on ideal surfaces, whereas the understanding of real surfaces remains somehow unclear.

---

*To whom correspondence should be addressed. Tel.: (34-958) 243-211; Fax: (34-958) 243-214; e-mail: mcabre@ugr.es

Drop-shape analysis is one of the simplest and most efficient and popular techniques in the study of wetting phenomena. Contact angles provide important information regarding a solid surface. Thus nowadays measurement of the contact angle is among the most widely used techniques for characterization of solid surfaces.

The simplest model of a solid surface that can be considered is a rigid, planar and chemically homogeneous surface, i.e., an ideal solid surface. Sessile drops (or captive bubbles) on such surfaces exhibit axial symmetry and a singular contact angle. The contact angle on an ideal surface is predicted by the well-known Young equation:

$$\gamma_{LV} \cos \theta_Y = \gamma_{SV} - \gamma_{SL},$$ \hfill (1)

where γ_{LV} is the liquid–vapor interfacial energy, γ_{SV} the solid–vapor interfacial energy, γ_{SL} the solid–liquid interfacial energy and θ_Y the Young contact angle. The difference $\gamma_{SL} - \gamma_{SV}$ represents the energy to wet the solid (energy to replace solid–vapor interface by solid–liquid interface) and it is denoted solid surface energy difference in this paper.

The Young equation presents an explicit relation between the solid surface energy difference ($\gamma_{SL} - \gamma_{SV}$) and the contact angle for a given liquid. This equivalence allows obtaining the experimentally inaccessible solid surface energy difference from contact angle measurements. In this context, the intrinsic contact angle is defined as the contact angle which a liquid would have at a given point of the three-phase contact line due to the solid surface energy difference. On ideal surfaces this intrinsic contact angle coincides with the Young contact angle θ_Y.

From an experimental point of view, axial symmetry is usually assumed. Thus, a drop profile is often analyzed by directly obtaining the contact angle with a goniometric procedure or indirectly with a numerical algorithm such as the Axisymmetric Drop Shape Analysis (ADSA) [4]. In the last years, however, there has been an increasing interest in the study of drops on real surfaces, on which the drop axial symmetry is not applicable [5–14].

Chemical heterogeneity and roughness are very common on surfaces of technological interest. This complexity causes two phenomena that are usually closely related: (i) the lack of drop axial symmetry involving a multiplicity of local contact angles along the contact line and (ii) the presence of hysteresis phenomena due to the existence of metastable states. These hysteresis phenomena can originate from many different causes [2, 3]. However, it must be noted that in this work we will restrict to the study of hysteresis effects due only to chemical heterogeneity and roughness.

As has been stated above, ideal solid surfaces can be adequately described by their intrinsic contact angles. Therefore, it would be very desirable to be able to obtain an effective intrinsic contact angle that would similarly describe non-ideal surfaces. Nevertheless, such a description is not a straightforward task. To this end, one has to take into account that on chemically heterogeneous surfaces, the intrinsic contact angle varies at different points on the surface. Moreover, on rough surfaces, drops exhibit an apparent contact angle which is accessible experimentally and is different from the expected intrinsic contact angle. Such an effective intrinsic contact angle was proposed by Cassie (equation (2)) for chemically heterogeneous surfaces [18] and by Wenzel (equation (3)) for rough surfaces [19]:

$$\cos \theta_{\text{Cassie}} = \langle \cos \theta_i \rangle_{\text{Surface}} \tag{2}$$

$$\cos \theta_{\text{Wenzel}} = r_{\text{W}} \cos \theta_i = \frac{A_{\text{Surface}}}{A_{\text{Projection}}} \cos \theta_i , \tag{3}$$

where θ_{Cassie} and θ_{Wenzel} represent the Cassie and Wenzel contact angles, respectively, θ_i is the intrinsic contact angle and r_{W} the Wenzel factor defined by the ratio of true contact area (A_{Surface}) and projected contact area ($A_{\text{Projection}}$).

Besides the difficulty of defining an effective intrinsic contact angle that characterizes a non-ideal surface, sessile drops on such surfaces would not necessarily be axisymmetric. The analysis of non-axisymmetric sessile drops adds some new difficulties that do not appear in the analysis of axisymmetric drops. The main difficulty is the choice of the most suitable contact angle to represent a non-axisymmetric drop, since it presents a multiplicity of local contact angles (see Section 3.1). Consequently, some key questions arise: Can this contact angle be obtained from the drop shape? Is there a direct relation between this contact angle and the solid surface characteristics? How is this contact angle related to the effective intrinsic contact angles proposed by Cassie or by Wenzel?

A theoretical analysis of the non-axisymmetric sessile drops would be an excellent way to answer these questions for a wide spectrum of possible non-ideal surfaces. Theoretical studies are usually supported by computer calculations [6–9] or numerical approximations because of the mathematical complexity of the system constituted by the Young–Laplace equation (4) and the generalized Young equation (5) in the non-axisymmetric case [14]:

$$P_{\text{L}} - P_{\text{V}} = \gamma_{\text{LV}} \left(\kappa_1 + \kappa_2 \right) = 2\gamma_{\text{LV}} H \tag{4}$$

$$\gamma_{\text{LV}} \cos \theta = \gamma_{\text{SV}} - \gamma_{\text{SL}} + \sigma_{\text{SLV}} \kappa_{\text{g}} - \nabla \sigma_{\text{SLV}} \cdot \mathbf{m} \tag{5}$$

where $P_{\text{L}} - P_{\text{V}}$ is the pressure difference across the liquid–vapor interface, σ_{SLV} the line energy associated with the triple phase contact line (also known as line tension) [5, 14], κ_1 and κ_2 the principal curvatures, H the mean curvature of the interface, κ_{g} the geodesic curvature of the contact line, θ the contact angle and $\mathbf{m}$ a unit vector orthogonal to the normal vector to the solid surface and to the tangent vector to the contact line.

Hence, the main objective of this research work is to perform a theoretical analysis of non-axisymmetric drops on chemically heterogeneous surfaces. For this purpose, two different methods are employed in the characterization of these systems. The first one consists in computer calculations achieved using the public-domain software package Surface Evolver [21, 22], which has also been used by other researchers [5–9]. Even though Surface Evolver calculations are a very powerful method, these require suitable computer capacities for acceptable running times. For this reason, an alternative approach has been introduced, which is an approximation to solve the sessile drop problem and is easier and faster than the Surface Evolver calculations.

2. NUMERICAL PROCEDURES

2.1. Sessile drop problem

A sessile drop on a solid surface is a three-phase system. It is composed of a solid and a liquid in equilibrium with its vapor. When a drop of the liquid (of a given volume) is deposited on the solid surface, it acquires a particular shape due to the gravity and the properties of the three-phase system. To find the final shape of a drop of a given liquid on a non-ideal solid surface is termed here the sessile drop problem.

The shape of a drop in thermodynamic equilibrium is governed by the Young–Laplace equation (4) with the boundary condition established by the Young equation (5). The verification of these two equations is equivalent to minimization of the total energy (E) of the system:

$$E = \int_{\text{Volume}} (\rho_L - \rho_V) \, g \, z \, dV + \int_{\text{Interface}} \gamma_{LV} \, dA + \tag{6}$$
$$+ \int_{\text{Contact area}} (\gamma_{SL} - \gamma_{SV}) \, dA + \int_{\text{Contact line}} \sigma_{SLV} \, dL$$

maintaining a constant drop volume (V):

$$V = \int_{\text{Volume}} dV = \text{const.} \tag{7}$$

where $\rho_L - \rho_V$ is the density difference between the liquid and vapor and g stands for the gravity acceleration.

2.2. Absence of gravitational and line energies

The total energy of a liquid drop on a surface is given by the sum of gravitational, interfacial and line energies (see equation (6)). Each of these terms has different importance depending on the characteristic lengths of the system under study.

Hence, the gravitational term is especially important for large drops, i.e., when the capillary length of the liquid–vapor interface is greater than the curvature radius at the drop apex. Conversely, the line energy term plays a dominant role in the case of very small drops.

Therefore, a very interesting range of drop sizes is that in which neither the gravity effects nor the line energy effects are significant. Thus, in this range of sizes it is usual to neglect both the gravitational and line terms. In this work, the sessile drops considered are in this range. In this case the first and the last terms of equation (6) vanish, and the problem is reduced to minimizing the interfacial energies, i.e.:

$$E = \int_{\text{Interface}} \gamma_{\text{LV}} \, dA + \int_{\text{Contact area}} (\gamma_{\text{SL}} - \gamma_{\text{SV}}) \, dA \tag{8}$$

At this point, energy minimization can be achieved by two different methods: Surface Evolver calculations and pseudo-spherical approximations. Let us analyze both of these independently in detail.

2.3. Surface Evolver calculations

Surface Evolver is an interactive program developed for the study of liquid surfaces shaped by various forces and constraints. The program evolves a user-defined initial surface towards the minimal energy by means of a gradient descent method. The numerical algorithm based on Lagrange's principle of virtual work, proposing that a mechanical system is in a state of equilibrium, leads to a minimum in the total energy of the system. It is generally used for solving problems involving minimization of surfaces, but it has also been used in more practical applications [21].

The Surface Evolver treats a sessile drop (the liquid–vapor interface) as a mathematical surface that corresponds with the liquid–vapor interface (see Fig. 1). Accordingly, the energies involved have to been referred to this interface and the contact line, i.e., the mathematical surface and its frontier. This was achieved using the divergence and curl theorems. Further details on this process can be found elsewhere [21].

The Surface Evolver uses a triangular tessellation to represent surfaces, which can be progressively refined to achieve the desired level of accuracy. Generally, however, the triangles are homogeneously distributed over the whole surface. Since the most interesting zone is the one close to the contact line, it would be desirable to obtain further details about this region. As stated before, the accuracy could be improved by means of an increased number of homogeneously distributed triangles. This procedure, however, implies a large increase in computing time. Nevertheless, in this work a different procedure was used. An algorithm was added to control both the size and the shape of the triangles. Hence, a triangle whose edge length, normalized by its height, is greater than a certain value was divided into smaller triangles. Additionally, a triangle whose normalized area is

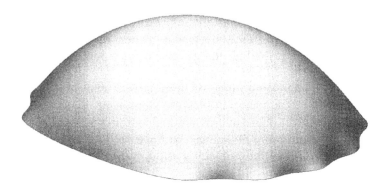

Figure 1. Picture of a drop on a rigid, flat and chemically heterogeneous surface. This shape has been computed using the public-domain software Surface Evolver.

lower than another threshold value was deleted. These limiting values were determined according to certain function of the height designed to increase the number of triangles in the proximity to the contact line. In this manner, the computing of a large number of unnecessary triangles is avoided.

Sessile drops on patterned surfaces present symmetries depending on the surface pattern (see Sections 3.2 and 3.3). It should be noted that symmetry constraints were not imposed and the drop shape symmetries are the results of the Surface Evolver calculations.

2.4. Pseudo-spherical approximation

In the case of ideal surfaces (axial symmetry situation), the two simplifications considered in our present study (absence of gravity and line energy) lead us to a spherical interface (spherical cap) with a contact angle obtained by the Young equation (1).

Our proposal (second method) appears at this point as a modification to this solution, extending the results to non-axisymmetric drops. First, a two-parameter contact line is assumed. Second, the liquid–vapor interface is described with circular arcs joining the drop apex with the contact line.

In the axisymmetric solution, the contact line has a circular shape. Nevertheless, the contact line does not have a known shape in the general non-axisymmetric case, which is determined by minimization of the total energy (8). Thus, the contact line shape is fixed in order to obtain an approximate solution to the general case. The family of curves used to describe the contact lines should be chosen according to surface symmetries and with the minimum number of parameters.

At this point, the real liquid–vapor interface would be a mathematical surface of constant mean curvature constrained by a certain family of contact lines. In order to simplify obtaining of this interface shape, another assumption is made. The

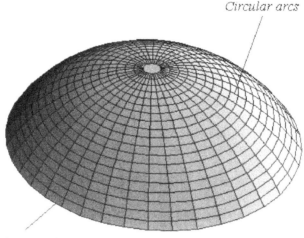

Figure 2. Pseudo-spherical cap used as an approximation for a non-axisymmetric sessile drop. The frontier of the mathematical surface (the contact line) is an ellipse and the curves joining the apex with the frontier are circular arcs.

constant mean curvature surface is replaced by another surface, similar to the real one, which obviously would not satisfy the Young–Laplace equation.

This new family of surfaces, named pseudo-spherical caps, is described by circular arcs joining the drop apex with the contact line (see Fig. 2). These curves exhibit a varying radius of curvature (r_m) given by:

$$r_m(\varphi) = \frac{\rho_C^2(\varphi) + h^2}{2h} \tag{9}$$

where ρ_C is the contact radius, φ the polar angle and h the height of the drop.

For many heterogeneous patterns an elliptical contact line can be assumed, even though more complex contact lines can be used, such as superellipses or perturbed circumferences.

In the present case, the contact radius ρ_C as a function of the polar angle φ is given by:

$$\rho_C(\varphi) = \frac{ab}{\sqrt{a^2 \sin^2(\varphi) + b^2 \cos^2(\varphi)}} \tag{10}$$

where a and b are the semi-axes of the ellipse.

The pseudo-spherical cap with an elliptical contact line is completely described by three parameters: the elliptical contact semi-axes (a, b) and its height (h). The volume of a pseudo-spherical cap can then be expressed as a function of these three parameters as:

$$V = \frac{\pi}{6}h\left(3ab + h^2\right) \tag{11}$$

Minimization procedure by this second method is easier. By considering equation (11), a pseudo-spherical cap with a given volume has only two degrees of freedom (parameters a and b). Thus, the total energy of the system (8) can be calculated and minimized in terms of a and b, as:

$$E_{V\,const.} = E\left(a, b, h_{V\,const.}\left(a, b\right)\right) = E_{V\,const.}\left(a, b\right) \tag{12}$$

As we mentioned above, a generic pseudo-spherical cap does not obey the Young–Laplace equation. Only in the axisymmetric situation the pseudo-spherical cap coincides with a spherical one (that satisfies the Young–Laplace equation). However, the predicted results are in good agreement with simulated results.

3. RESULTS AND DISCUSSION

In this section and taking into account the theoretical considerations mentioned above, the Surface Evolver calculations and the pseudo-spherical approximation have been compared using the results obtained for rigid, planar and chemically heterogeneous surfaces. First, possible definitions of a singular contact angle are briefly discussed. Second, two different heterogeneity patterns are studied. Finally, the proposed approximation is applied to the analysis of metastable states and hysteresis phenomenon.

3.1. Drop contact angle

On non-axisymmetric drops a multiplicity of contact angles along their contact lines are observed. Thus, different definitions of a singular drop contact angle can be proposed. Therefore, we present here a brief discussion of these different possibilities to define a drop contact angle.

On the one hand, the drop contact angle can be defined easily as the average over the local contact angles along the contact line. The average can also be calculated in terms of cosines:

$$\overline{\cos\theta} = \frac{1}{L_C}\int_{Contact\ line} \cos\theta\, dL\,, \tag{13}$$

where L_C is the contact line length.

On the other hand, from the energetic interpretation of the Young equation, a drop contact angle (effective contact angle) can be defined as the average of the intrinsic contact angle along the contact surface (14) or even an average along the contact line as proposed by other researchers [14] (15).

$$\cos\theta_{\text{eff-}A_C} = \left\langle\cos\theta_i\right\rangle_{\text{Contact area}} = \frac{1}{A_C}\int_{\text{Contact area}}\frac{\gamma_{\text{SV}} - \gamma_{\text{SL}}}{\gamma_{\text{LV}}}\,\mathrm{d}A \tag{14}$$

$$\cos\theta_{\text{eff-}L_C} = \left\langle\cos\theta_i\right\rangle_{\text{Contact line}} = \frac{1}{L_C}\int_{\text{Contact line}}\frac{\gamma_{\text{SV}} - \gamma_{\text{SL}}}{\gamma_{\text{LV}}}\,\mathrm{d}L \tag{15}$$

where A_C and L_C are, respectively, the contact area and the contact line length.

Neither one is easy to evaluate experimentally. Therefore, a usual choice for the drop contact angle is the contact angle of a hypothetical axisymmetric drop with the same contact area and volume (16). Similar definitions of an equivalent contact angle can be written, e.g., an equivalent contact angle obtained from a hypothetical drop with the same contact line length and volume (17).

$$\cos\theta_{\text{eq-}A_C} = \cos\theta(A_C, V) \tag{16}$$

$$\cos\theta_{\text{eq-}L_C} = \cos\theta(L_C, V) \tag{17}$$

Among these candidates for the drop contact angle, taking into account its simplicity and easy experimental access, one of them qualifies as the most convenient from our point of view, namely the equivalent contact angle obtained from the contact area (14). For this reason, from now on, contact areas are analyzed and compared as an indirect measure of the drop contact angle.

3.2. Striped pattern

One of the most studied patterns, because of its simplicity and its lack of axial symmetry, is the striped one. In this work a heterogeneous striped pattern was investigated. To this end, a sinusoidal variation of the solid surface energy difference was chosen (see equation (18)). Remember here the definition of solid surface energy difference described in Section 1 (see equation (1)).

A solid surface can be described mathematically by three distributions, which represent, respectively, the roughness (height distribution h), the chemical heterogeneity (solid surface energy difference $\gamma_{\text{SL}} - \gamma_{\text{SV}}$ and the line energy σ_{SLV} distributions):

$$h(x, y) = 0$$

$$[\gamma_{\text{SL}} - \gamma_{\text{SV}}](x, y) = T_{\text{m}} + T_0\cos\left(2\pi\frac{x}{\lambda}\right) \tag{18}$$

$$\sigma_{\text{SLV}}(x, y) = 0$$

where T_m, T_0 and λ are, respectively, the mean solid surface energy difference, the amplitude of the heterogeneity, and the characteristic length of the heterogeneity.

Sessile drops on such surfaces reach an equilibrium state that depends on the surface characteristics and occasionally on the initial states from which the drop has evolved. For an in-depth examination of this hysteresis phenomenon see Section 3.4.

In Fig. 3 normalized contact areas for drops with different volumes are shown. This normalization allows us to consider drops of different volumes (V) on the same solid surface and also to consider only a single drop of fixed volume on striped surfaces with different characteristic lengths (λ). Open circles represent the results obtained with Surface Evolver calculations and solid lines are calculated with the pseudo-spherical cap approximation.

Since the drop total energy is an analytical function (see equation (8)) minima of energy can be easily found. For drops on striped surfaces two different curves are found (curves (a) and (b) in Fig. 3), i.e., there are at least two energy minima

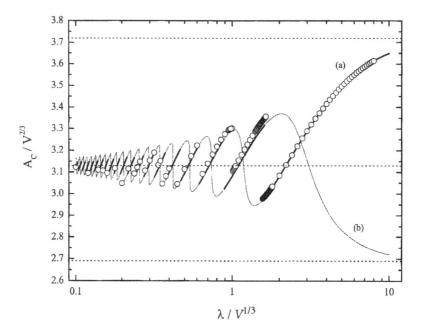

Figure 3. Normalized contact area on a heterogeneous striped surface as a function of the ratio of heterogeneity length and drop size ($\lambda/V^{1/3}$). Points represent results obtained by the Surface Evolver calculations. Solid lines are obtained with the pseudo-spherical cap approximation and show different contact areas corresponding to different equilibrium states at the same value of the ratio $\lambda/V^{1/3}$. Bold segments of these lines show the contact areas of the lowest energy states. Dashed lines show the contact area of a hypothetical drop on a homogeneous surface with, from top to bottom, the highest, the mean and the lowest surface energy difference of the heterogeneity. The results are calculated for a striped surface with $T_m/\gamma_{LV} = -0.50$ and $T_0/\gamma_{LV} = 0.15$.

for each drop (of a given value of $\lambda/V^{1/3}$). One of these minima corresponds to the absolute minimum energy (see bold segments of curves (a) and (b) in Fig. 3).

When the approximated solution (bold segments in Fig. 3) is compared with the points obtained with the Surface Evolver calculations, a good agreement is observed. The discontinuous and oscillatory features of this solution cannot be easily appreciated by considering only the points obtained with the Surface Evolver calculations. So, not only the approximated solution describes well the results obtained with Surface Evolver, but it can also be used to understand the Surface Evolver results. Moreover, the calculation of a single point with Surface Evolver is much more time consuming than the calculation of the whole figure with the pseudo-spherical cap approximation.

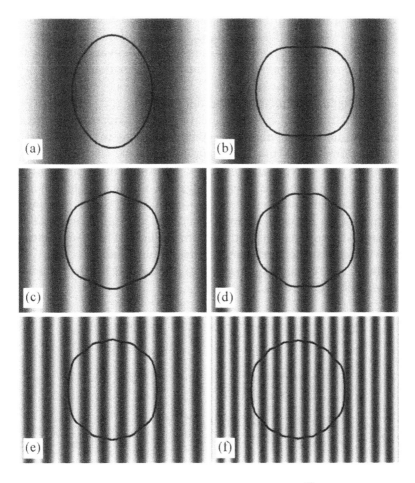

Figure 4. Contact lines on a striped pattern at different values of $\lambda/V^{1/3}$. Drops are wetting, respectively, (a) one stripe (hydrophilic), (b) three stripes (two hydrophilic and one hydrophobic), (c) five stripes, and so on. The contact line centers are at two different positions: in the middle of a hydrophilic stripe (a, c and e) or in the middle of a hydrophobic stripe (b, d and f).

Besides a good correspondence between Surface Evolver calculations and the pseudo-spherical cap approximation, some aspects in Fig. 3 deserve special attention. The first aspect to be considered is the oscillatory and discontinuous behavior. The discontinuities are a consequence of the discrete nature of a striped surface, i.e., a sessile drop wets a certain number of stripes. From the right to the left, the lines correspond to drops that wet one stripe (hydrophilic), three stripes (two hydrophilic and a hydrophobic), five stripes and so on (Fig. 4a–4f).

Other aspects to be considered are the two different tendencies that appear at the extreme values of $\lambda/V^{1/3}$. On the one hand, when the drop size ($V^{1/3}$) is greater than the characteristic length (λ), a quasi-homogeneous situation is reached. The drop wets numerous stripes and the effective solid surface energy difference in this limit tends to the average along the surface (T_m), i.e., the surface can be considered as a homogeneous surface with the mean solid surface energy difference T_m (see equation (18)). On the other hand, when the drop size is less than the characteristic length, a different quasi-homogeneous situation is reached. The drop wets only a single stripe and the effective solid surface energy difference tends to the most hydrophilic value ($T_m - T_0$, see equation (18)), i.e., the drop prefers to wet a hydrophilic stripe than a hydrophobic one.

Drops on such striped surfaces should have certain symmetry. This feature has allowed us to restrict our pseudo-spherical approximation to those stripes with an elliptical contact line centered in the middle of the stripe. Drops on a striped surface present two axes of symmetry: one is transversal to the stripes and the other is longitudinal to the stripes and goes through the middle of a hydrophilic (Figs. 4a, 4c or 4e) or a hydrophobic stripe (Figs. 4b, 4d or 4f). These two possibilities are represented with the two curves (solid lines in Fig. 3) obtained with the pseudo-spherical approximation. Remember here that the bold segments of these curves show the states which are energetically more favorable.

Finally, the Cassie contact angle, considering it as an effective value that characterizes the surface (2), is calculated as:

$$\cos\theta_{\text{Cassie}} = \left\langle\cos\theta_i\right\rangle_{\text{Surface}} = \left\langle\frac{\gamma_{\text{SV}} - \gamma_{\text{SL}}}{\gamma_{\text{LV}}}\right\rangle_{\text{Surface}} =$$

$$= \frac{1}{\gamma_{\text{LV}}}\frac{1}{A_{\text{Surface}}}\int_{\text{Surface}}(\gamma_{\text{SV}} - \gamma_{\text{SL}})\,dA =$$

$$= -\frac{1}{\gamma_{\text{LV}}}\left(T_m + T_0\frac{\int_{\text{Surface}}\cos(2\pi x/\lambda)\,dA}{A_{\text{Surface}}}\right)\xrightarrow{A_{\text{Surface}} \gg \lambda^2} -\frac{T_m}{\gamma_{\text{LV}}} \tag{19}$$

The Cassie contact angle considered here corresponds to a surface large enough to neglect the effect of the number of hydrophilic and hydrophobic stripes taken into the calculation. Thus, the Cassie contact angle tends to the intrinsic contact angle on a homogeneous surface characterized by the mean solid surface energy difference of the heterogeneity T_m (see equations (1) and (18)). In Fig. 3 the normalized contact area of a drop on a hypothetical homogeneous surface with an intrinsic contact angle equal to the Cassie contact angle has been shown (central dashed line).

3.3. Chessboard patterned surfaces

Another surface pattern was also investigated. Now, the surface is a chessboard patterned surface, where the heterogeneities also vary in a sinusoidal way. This can be seen in the following equations:

$$h(x, y) = 0$$

$$[\gamma_{SL} - \gamma_{SV}](x, y) = T_m + T_0 \cos\left(2\pi\frac{x}{\lambda}\right)\cos\left(2\pi\frac{y}{\lambda}\right) \qquad (20)$$

$$\sigma_{SLV}(x, y) = 0$$

As in the previous pattern, T_m, T_0 and λ are, respectively, the mean solid surface energy difference, the amplitude of the heterogeneity, and the characteristic length of the heterogeneity.

In Fig. 5 normalized contact areas for drops with different volumes are shown. This figure is similar to Fig. 3. Thus, much of the comments made there are applicable here. As in Fig. 3, open circles represent the Surface Evolver calculations, and solid lines are calculated with the pseudo-spherical approximation.

A difference between the striped and the chessboard patterns is that three different curves are found (see curves (a), (b) and (c) in Fig. 5). Each of these corresponds to three different symmetry situations (see below) for a drop (of a given value of $\lambda/V^{1/3}$). As in Fig. 3, the absolute minima are represented with bold segments of the curves. A good agreement between Surface Evolver calculations and pseudo-spherical calculations is also found here. In the same way, the approximated solution (bold segments of curves in Fig. 5) could be helpful to understand the Surface Evolver results.

Besides a good correspondence between Surface Evolver calculations and the pseudo-spherical cap approximation, some aspects in Fig. 5 deserve special attention. As on the striped pattern, drops on the chessboard patterned surface present two orthogonal axes of symmetry (see Fig. 6b and 6e). The chessboard pattern has a higher symmetry than the previous striped pattern; consequently, the shape of the contact line presents a higher axial symmetry. However, it should not be thought that in this case, all approximated drops degenerate into a spherical cap, i.e., the elliptical contact lines degenerate into circumferences. The three different positions of the apex projection are correlated with the three curves obtained with the pseudo-spherical cap approximation.

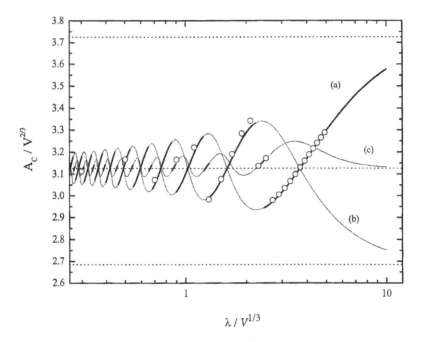

Figure 5. Normalized contact area on a heterogeneous chessboard patterned surface as a function of the ratio of heterogeneity length and drop size ($\lambda/V^{1/3}$). Points represent results obtained by the Surface Evolver calculations. Solid lines are obtained with the pseudo-spherical cap approximation and show different contact areas corresponding to different equilibrium states at the same value of the ratio $\lambda/V^{1/3}$. Bold segments of these lines show the contact areas of the lowest energy states. Dashed lines show the contact area of a hypothetical drop on a homogeneous surface with, from top to bottom, the highest, the mean and the lowest surface energy difference of the heterogeneity. The results are calculated for a chessboard patterned surface with $T_m/\gamma_{LV} = -0.50$ and $T_0/\gamma_{LV} = 0.15$.

The symmetry allows us to restrict our pseudo-spherical approximation to those drops whose contact line center is in the middle of a hydrophilic square (Fig. 6a or 6d), in the middle of a hydrophobic square (Fig. 6c or 6f), or at a vertex of a square (Fig. 6b or 6e).

As explained above (see Section 3.2) for the striped pattern, two different tendencies appear at extreme values of $\lambda/V^{1/3}$ and quasi-homogeneous situations are reached. On the one hand, when the drop size ($V^{1/3}$) is greater than the characteristic length (λ) the drop wets numerous squares and the effective solid surface energy difference in this limit tends to the mean solid surface energy difference (T_m, see equation (20)). On the other hand, when the drop size is less than the characteristic length the drop wets only one square and the effective solid surface energy difference tends to the most hydrophilic value ($T_m - T_0$, see equation (20)), i.e., the drop prefers to wet a hydrophilic square than a hydrophobic one.

Finally, the Cassie contact angle, considering it as an effective value that characterizes the surface (2), is calculated as:

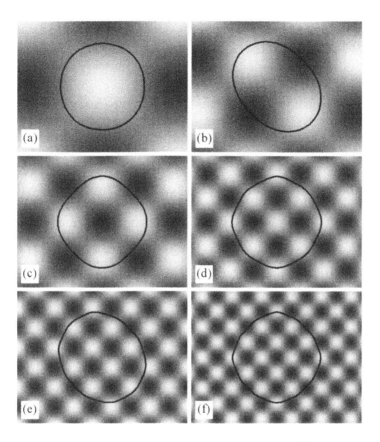

Figure 6. Contact lines on a chessboard pattern at different values of $\lambda/V^{1/3}$. The contact line centers are at three different positions: in the middle of a hydrophilic square (a and d), in the middle of a hydrophobic square (c and f) and at the vertex of a square (b and e).

$$\cos\theta_{\text{Cassie}} = \left\langle \cos\theta_i \right\rangle_{\text{Surface}} = \left\langle \frac{\gamma_{\text{SV}} - \gamma_{\text{SL}}}{\gamma_{\text{LV}}} \right\rangle_{\text{Surface}} =$$

$$= \frac{1}{\gamma_{\text{LV}}} \frac{1}{A_{\text{Surface}}} \int_{\text{Surface}} (\gamma_{\text{SV}} - \gamma_{\text{SL}}) \, dA = \tag{21}$$

$$= -\frac{1}{\gamma_{\text{LV}}} \left(T_{\text{m}} + T_0 \frac{\int_{\text{Surface}} \cos(2\pi x/\lambda)\cos(2\pi y/\lambda) \, dA}{A_{\text{Surface}}} \right) \xrightarrow{A_{\text{Surface}} \gg \lambda^2} -\frac{T_{\text{m}}}{\gamma_{\text{LV}}}$$

The Cassie contact angle (see comments on equation (19)) tends to the intrinsic contact angle on a homogeneous surface characterized by the mean solid surface energy difference of the heterogeneity T_{m} (see equations (1) and (20)). In Fig. 5 the normalized contact area of a drop on a hypothetical homogeneous surface with an intrinsic contact angle equal to the Cassie contact angle has been shown (central dashed line).

3.4. Hysteresis and metastable states

As was mentioned before, heterogeneity causes drop contact angle hysteresis. Two different closely related hysteresis consequences can be observed: (i) the lack of drop axial symmetry involving a multiplicity of local contact angles along the contact line and (ii) the presence of hysteresis phenomena due to the existence of metastable states.

The multiplicity of local contact angles has already been discussed in Section 3.1. In that section, the drop contact angle was considered to be the equivalent contact angle of a hypothetical axisymmetric drop with the same volume and

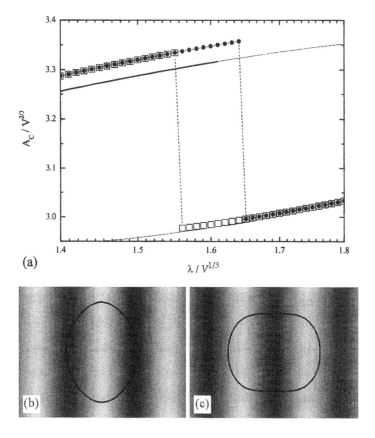

(a)

(b) (c)

Figure 7. (a) Detail of Fig. 3. Normalized contact area on a heterogeneous striped surface as a function of the ratio of heterogeneity length and drop size ($\lambda/V^{1/3}$). Points representing Surface Evolver calculations are here clearly distinguished. Contact areas for advancing and receding drops are represented by the symbols □ and ●, respectively. (b, c) Different contact line shapes for two drops with equal volumes on the same surface (identical values of $\lambda/V^{1/3}$): the advancing drop (b) and the receding drop (c).

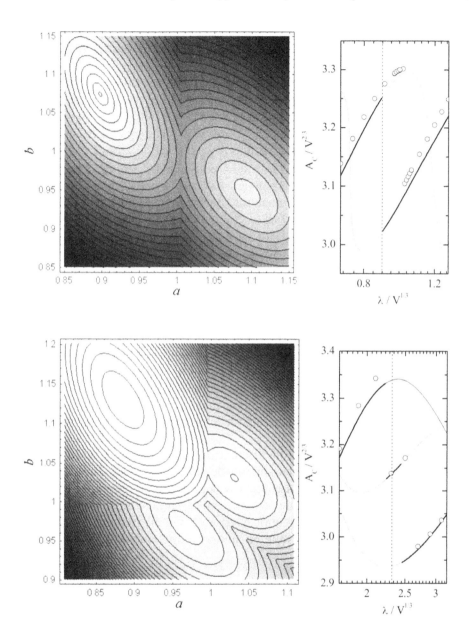

Figure 8. (Panels on the left) Energy representations of a pseudo-spherical cap as a function of its parameters *a* and *b* (semi-axes of the elliptical contact line) for a given value of the ratio of heterogeneity length and drop size ($\lambda/V^{1/3}$). (Panels on the right) Details of Figs 3 and 5 (see the corresponding figure legends). Vertical dotted lines show the $\lambda/V^{1/3}$ values used in Figs 3 and 5 on the left side. Intersections of the vertical dotted line with the solid lines correspond to the minimal energy states shown in the figures on the left side. Two chemically heterogeneous surfaces are shown: striped (top) and chessboard patterned (bottom).

contact area. In this section, however, the existence of different equilibrium states is analyzed. The pseudo-spherical approximation should be especially useful for this purpose.

First, the Surface Evolver calculations are carefully examined. In some cases, different final states were reached when the Surface Evolver was started from different initial surfaces. These states correspond to drops with the same volume on the same surface (identical value for $\lambda/V^{1/3}$), but with different contact areas.

In Fig. 7, hysteresis phenomenon for the striped pattern is shown. In Fig. 7a, final contact areas obtained from two different initial states are distinguished: advancing contact area and receding contact area. Advancing contact areas (open squares) were obtained starting from an initial drop that did not wet the solid surface. Conversely, receding contact areas (full circles) were obtained starting from a different initial drop that did wet the surface. As can be appreciated in Fig. 7, the advancing and receding values do not coincide for a determined range. Moreover, the corresponding contact line shapes of these two final states are quite different (see Fig. 7b and 7c).

The Surface Evolver calculations, however, are not able to determine all the possible metastable states. For this purpose, the pseudo-spherical approximation is used. This is an analytical approximation that describes (in the elliptical case) a drop by three parameters (a, b and h). These parameters coincide with the semi-axes of the elliptical contact line and the drop height. As we explained in Section 2.4, the analytical expression for the total energy of a drop with a given volume (V) depends only on two parameters (see equation (12)). Thus, the metastable states can be easily found by minimizing the analytical expression for the energy (for a given volume).

Figure 8 shows the existence of different metastable states (minimal energy states) for drops on striped surfaces and on chessboard patterned surfaces. On the left side, the variation of the total energy for a drop (in the pseudo-spherical approximation) with a given volume is represented. These energy representations are functions of the contact line semi-axes a and b. On the right side, the normalized contact areas for the energy minima on the left are indicated (see dotted line in Fig. 8).

4. SUMMARY AND CONCLUSIONS

In this work, the wetting on rigid, planar and chemically heterogeneous patterned solid surfaces has been studied by analyzing sessile drops. To this end, two different methods are used, namely, the Surface Evolver calculations and the pseudo-spherical approximation. In this regard, the Surface Evolver calculations can be considered a powerful method. However, this method showed to be certainly time-consuming. In order to overcome this difficulty, a new family of surfaces, denoted here pseudo-spherical caps, is introduced. In this manner, the number of surfaces used in the description of the drop shape diminishes with respect to that used by the Surface Evolver. Hence, the pseudo-spherical approximation considerably simplifies the calculations and reduces the computational time.

The results obtained with the pseudo-spherical approximation substantially corroborate those obtained by the Surface Evolver calculations. Moreover, the pseudo-spherical approximation is very useful for the interpretation of the Surface Evolver results.

With the aim of applying these methods, two different chemical heterogeneity patterns have been investigated: a striped and a chessboard. For this purpose, sessile drops placed on both surfaces are analyzed. In particular, the contact area is used as a measurement of the drop contact angle. Several conclusions arise from this analysis.

First, the contact area is found to vary in an oscillatory manner rather than monotonically with increase in drop volume. This oscillatory trend is found to be similar to a stick-slip behavior. Specifically, the tendency of the drop to wet new hydrophobic zones and the resistance of the drop to adopt this new configuration explain this behavior. Furthermore, the hysteresis phenomena are analyzed in detail, showing the relation of the different metastable states with the symmetries of the patterns.

Finally, the Surface Evolver calculations and the pseudo-spherical cap approximation appear to be complementary methods applicable to complete studies of the wetting phenomena on rigid, planar and chemically heterogeneous patterned surfaces. The excellent results and the flexibility of these procedures might make these suitable to other more complex surfaces.

Acknowledgements

Financial support from the Spanish "Ministerio de Ciencia y Tecnología, Plan Nacional de Investigación Científica, Desarrollo e Innovación Tecnológica (I+D+I), MAT2001-2843-C02-01" is gratefully acknowledged.

REFERENCES

1. A. Marmur, *Langmuir* **20**, 1317–1320 (2004).
2. A. Marmur, *Langmuir* **20**, 3517–3519 (2004).
3. A. Lafuma and D. Quéré, *Nature Mater.* **2**, 457–460 (2003).
4. O. I. del Río and A. W. Neumann, *J. Colloid Interface Sci.* **196**, 136–147 (1996).
5. J. Buehrle, S. Herminghaus and F. Mugele, *Langmuir* **18**, 9771–9777 (2002).
6. A. Marmur, *J. Colloid Interface Sci.* **168**, 40–46 (1994).
7. S. Brandon and A. Marmur, *J. Colloid Interface Sci.* **183**, 351–355 (1996).
8. S. Brandon, A. Wachs and A. Marmur, *J. Colloid Interface Sci.* **191**, 110–116 (1997).
9. S. Brandon, N. Haimovich, E. Yeger and A. Marmur, *J. Colloid Interface Sci.* **263**, 237–243 (2003).
10. S. D. Iliev and N. Ch. Pesheva, *Langmuir* **19**, 9923–9931 (2003).
11. A. Marmur, *Langmuir* **19**, 8343–8348 (2003).
12. G. Wolansky and A. Marmur, *Colloids Surfaces* A **156**, 381–388 (1999).
13. G. McHale and M. I. Newton, *Colloids Surfaces* A **206**, 193–201 (2002).

14. P. S. Swain and R. Lipowsky, *Langmuir* **14**, 6772–6780 (1998).
15. C. W. Extrand and Y. Kumagai, *J. Colloid Interface Sci.* **191**, 378–383 (1997).
16. J. T. Woodward, H. Gwin and D. K. Schwartz, *Langmuir* **16**, 2957–2961 (2000).
17. G. McHale, H. Y. Erbil, M. I. Newton and S. Natterer, *Langmuir* **17**, 6995–6998 (2001).
18. A. B. D. Cassie, *Discuss. Faraday Soc.* **3**, 11–16 (1948).
19. R. N. Wenzel, *Ind. Eng. Chem.* **28**, 988–944 (1936).
20. P. Roura and J. Fort, *J. Colloid Interface Sci.* **272**, 420–429 (2004).
21. K. A. Brakke, *Surface Evolver Manual. Version 2.20*. Susquehanna University, Selinsgrove, PA (2003).
22. K. A. Brakke, *Exp. Math.* **1**, 141–165 (1992).

Contact Angle, Wettability and Adhesion, Vol. 4, pp. 203–214
Ed. K.L. Mittal
© VSP 2006

Can contact angle measurements be used to predict soiling and cleaning of plastic flooring materials?

E. PESONEN-LEINONEN*, R. KUISMA, I. REDSVEN, A.-M. SJÖBERG and M. HAUTALA

Department of Agrotechnology, University of Helsinki, P.O. Box 28, FI-00014 University of Helsinki, Finland

Abstract—The purpose of this study was to test the feasibility of predicting the cleanability of plastic flooring materials by contact angle measurements. Advancing contact angle was used to characterize surface properties such as wetting and surface free energy. The relevance of these surface properties to cleanability of flooring materials was evaluated using a cleaning index and soil residue. A total of 10 different commercial flooring materials were soiled and cleaned using pilot scale equipment. The results indicated the potential of advancing contact angle measurements to predict the cleanability of plastic flooring materials.

Keywords: Contact angle; soiling; cleaning; plastic flooring material; colorimetric method; wetting; adhesion.

1. INTRODUCTION

Plastic flooring materials are used widely in public, office and residential buildings. The adhesion of soil to indoor surfaces is an undesirable characteristic in terms of cleaning costs. The possibility to predict the cleanability of materials as early as in the design phase of a building would make it possible to choose the most appropriate material. A first step in order to achieve this aim would be to identify the surface properties of flooring materials that are most prominent in preventing soiling or in making them easy to clean. Contact angle measurement offers an interesting possibility to evaluate the relevant surface properties.

Contact angle measurements have been widely used in surface studies [1, 2]. Numerous methods are currently available for measuring contact angles, including capillary rise, du Noüy ring, sessile drop, captive bubble, Wilhelmy balance, laser goniometry, and others which are divided into two classes: static and dynamic contact angle measurements [3]. Several authors have discussed critically

---

*To whom correspondence should be addressed. Tel.: (358-9) 1915-8497;
Fax: (358-9) 1915-8491; e-mail: eija.pesonen-leinonen@helsinki.fi

contact angle analysis [4–9] and contact angle hysteresis related to roughness [10–14] and to chemical heterogeneity [12].

Contact angle measurements have been employed to characterize various surface materials, such as wood and stone surfaces [15, 16], plastic materials, steels and glass [17] and steels with polymeric coatings [18, 19].

Cleanability studies of flooring materials have demonstrated that cleanliness depends on the type of flooring material used [20] and on the profile, roughness or texture of floor tiles [3, 21, 22]. To our knowledge, only a very few studies have concentrated on the surface properties and soil adhesion or cleanability of plastic flooring materials in indoor environments [20, 23].

It is evident that contact angle measurements are troublesome and that surface energy components obtained from these measurements may be very questionable. In the case of commercial flooring materials, errors may arise due to surface roughness or chemical heterogeneity, which prevents accurate measurements (and, thus, surface free energy calculations) [24, 25]. Despite the fact that some of the sources of errors are well known, contact angle methods are still favoured for determining the surface free energy of flooring materials.

The aim of this study was to test the feasibility of contact angle measurements in evaluating the cleanability of commercial plastic flooring materials. In addition the relationship between cleanability and surface free energies was evaluated.

2. MATERIALS AND METHODS

2.1. Materials

The 10 commercially-available solid surfaces used in this study were light-coloured, new flooring materials. They were designed to sustain heavy wear, for example, in public buildings. The PVC poly(vinyl chloride) materials with PUR (polyurethane) treatment and without PUR were provided from flooring manufacturens, cleaned according to the initial cleaning instructions of the manufacturers and coded. Detailed descriptions of the materials are presented in Table 1. Five replicates of each material (8×23 cm) per experiment were used. The samples were stored in a conditioned room at $23 \pm 2°C$ and $50 \pm 5\%$ RH for at least 24 h before use.

Black soils used were particle soil and oil soil (Table 2). Particle soil is a relevant simulation of soil that has been found on floorings in buildings, and oil soil is also a relevant model soil.

The cleaning agents used in cleaning tests were water, a surfactant solution from BASF, Germany and an alkaline model detergent from Farmos Noiro Oy, Finland (Table 3). Polyester microfibre textile (fleece) (Freudenberg Household Products Oy, Finland) was used for cleaning.

Table 1.
Description of the flooring materials as provided by the manufacturers

Code	Flooring type	Composition	Top film[a]	Thickness (mm)
S11	Vinyl composition tile (homogeneous)	Binder: Poly(vinyl chloride), PVC, 24 wt% Plasticizers, stabilizers, fillers, pigments, others	Polyurethane (PUR) surface treatment <1 wt%	2.0
S12	Vinyl sheet (homogeneous)	Binder: Poly(vinyl chloride), PVC, 27 wt% Plasticizers, stabilizers, fillers, pigments	None (bulk material)	2.0
S13	Vinyl sheet (homogeneous)	Binder: Poly(vinyl chloride), PVC[b] Plasticizers, stabilizers, fillers, pigments	Polyurethane (PUR) coating	2.0
S14	Plastic composition tile, Enomer[TM] (heterogeneous)	Binder: Thermoplastic polymer, 20 wt%, fillers, pigments	Ionomer impregnated	2.0
S15	Vinyl sheet (homogeneous)	Binder: Poly(vinyl chloride), PVC, 47 wt% Plasticizers, stabilizers, fillers, pigments	Polyurethane (PUR) coating <1 wt%	2.0
S16	Vinyl tile, static dissipative (homogeneous)	Binder: Poly(vinyl chloride), PVC[b] Plasticizers, stabilizers, fillers, pigments	None (bulk material)	2.0
S17	Vinyl sheet (heterogeneous)	Binder: Poly(vinyl chloride), PVC, 44 wt% Plasticizers, stabilizers, fillers, pigments	Polyurethane (PUR) coating <2 wt%	2.0
S18	Vinyl sheet (heterogeneous)	Binder: Poly(vinyl chloride), PVC, 45 wt% Plasticizers, stabilizers, fillers, pigments	Polyurethane (PUR) coating <1 wt%	2.0
S19	Vinyl sheet (homogeneous)	Binder: Poly(vinyl chloride), PVC, 43 wt% Plasticizers, stabilizers, fillers, pigments	Polyurethane (PUR) coating with car-borundum and quartz crystals <2 wt%	2.0
S20	Vinyl sheet (homogeneous)	Binder: Poly(vinyl chloride), PVC, 33 wt% Plasticizers, stabilizers, fillers, pigments	Polyurethane (PUR) coating <1 wt%	2.0

All flooring materials had an embossed surface. [a]Treatment/coating applied by the flooring manufacturer. [b]No information available.

Table 2.
Compositions of soils

Model soil	Composition (wt%)
Particle soil[a]	Quartz silica, 88.30 Kaolin, 9.35 Yellow iron oxide, 0.20 Black iron oxide, 0.60 Paraffin oil, 1.55
Oil soil	Paraffin oil, (Ph.Eur.) 77 Carbon Black E 153, 23 (carbon–oil mixture)

[a]Standard soil according to EN 1269 and EN 14565.

Table 3.
Composition, dilution, pH and surface tension of the cleaning agents

Model cleaning agent	Composition and concentration of active component (wt%)	Dilution	pH	Surface tension (mN/m)
Surfactant	Nonionic surfactant, C_{13}-oxoalcohol ethoxylate	1%	6.4	27.2
Weakly alkaline model detergent	Triethanol amine soap of fatty acids, 1.75% Nonionic surfactant, C_{13}-oxoalcohol ethoxylate, 9% Tetrapotassium pyrophosphate, 5%	5%	9.3	26.7
Water	3° dH	–	5.7	71.3

The probe liquids used in contact angle measurements and in the calculations of solid surface free energy were ethylene glycol, formamide and diiodomethane of high purity grade ($\geq$99%, $\geq$99% and >98%, respectively, from Merck KgaA) and distilled water. The surface tensions for the liquids were obtained from the literature [26].

2.2. Cleanliness determination

Particle soil was applied on test surfaces using the pilot equipment (Soiling Drum Tester) equipped with a hexapod (1 kg) to produce mechanical impact. The drum rotated 1000 cycles at a speed of 50 ± 2 rpm, changing the direction of rotation after 250 cycles [23]. Oil soil was applied with the Erichsen Washability and Scrubbing Resistance Tester (model 494), using two backward-and-forward movements. The soiled materials were cleaned using the Erichsen Washability Tester [23, 27]. Three backward-and-forward cleaning movements were employed using a microfibre cloth with a moisture regain of 150% and a pressure of approximately 1.4 kPa. During soiling and cleaning phases the laboratory conditions were at 27 ± 4°C and 50 ± 12% RH.

After soiling and cleaning procedures the changes in soil amounts were determined using a Chroma Meter CR-210 colorimeter (Minolta), equipped with Standard Illuminant D65 [22, 23, 27]. The color of the surfaces was measured before and after soiling and after cleaning. The use of $L*$-differences was valid for the surfaces studied since they were light and the soils were black. Averaged $L*$-values of five measurement points for each replicate were used to calculate the changes in the amount of soil in terms of color differences of the flooring materials. Cleanability of flooring materials was expressed as cleaning index (1) [22, 28], soil residue (2), and total soiling (3).

$$\text{Cleaning index} = (L*_{\text{CLEANED}} - L*_{\text{SOILED}}) / (L*_{\text{UNSOILED}} - L*_{\text{SOILED}}) \tag{1}$$

$$\text{Soil residue} = \Delta L*_{\text{RESIDUE}} = L*_{\text{UNSOILED}} - L*_{\text{CLEANED}} \tag{2}$$

$$\text{Total soiling} = \Delta L*_{\text{TOTAL}} = L*_{\text{UNSOILED}} - L*_{\text{SOILED}} \tag{3}$$

2.3. Contact angle measurements and surface energy determination

Static and dynamic contact angle measurements were performed using the CAM 100 Optical Contact Angle Meter (KSV Instruments, Finland), equipped with a video camera which collected one image per second. Image analyses were performed with CAM 100 Software using curve fitting based on the Young–Laplace equation, yielding contact angles on both sides of the droplet and their mean value.

Before contact angle measurements, the surfaces of the materials were cleaned with a cleaning agent and rinsed first with tap water and then with distilled water. For advancing contact angle measurements, the droplet was enlarged on the surface by dispensing liquid using a Hamilton syringe with a threaded plunger from under the material through the hole. The change in drop volume forces the three-phase contact line to advance. The volume of droplets was 6–8 µl. The experimental advancing contact angle θ_A was the mean value of three to eight drops. An average of 26 angle measurements (images) were taken per drop. The experimental static contact angle θ_S was the mean value of 12 drops at the time instant 15 s from deposition. The contact angle measurements were carried out in open air at a relative humidity of 40±2% and at a room temperature of 24±1°C unless otherwise stated.

The solid surface energy can be determined from the experimental contact angles θ_A and surface tensions of the probe liquids [29]. Owens and Wendt extended the ideas of Fowkes to cases in which both dispersion and polar forces operated. The approach of Owens and Wendt [30] was coupled with Young's equation to result in:

$$\gamma_{\text{LV}} (1+\cos\theta) / 2 (\gamma_L{}^D)^{\frac{1}{2}} = (\gamma_S{}^D)^{\frac{1}{2}} + (\gamma_S{}^P \gamma_L{}^P / \gamma_L{}^D)^{\frac{1}{2}} \tag{4}$$

where θ is the contact angle of the liquid on the solid surface, γ_L^D and γ_L^P, respectively, are the dispersion and polar components of the surface free energy of the liquid, and γ_{LV} is the surface free energy of the liquid [31–33]. This means that if γ_{LV} $(1+\cos\theta)$ / $(2\ (\gamma_L^D)^{\frac{1}{2}})$ is plotted against $(\gamma_L^P/ \gamma_L^D)^{\frac{1}{2}}$ (a Fowkes plot), the graph should be linear with intercept $(\gamma_S^D)^{\frac{1}{2}}$ and slope $(\gamma_S^P)^{\frac{1}{2}}$, thus allowing determination of dispersion γ_S^D and polar γ_S^P components of the surface free energy of flooring materials. Surface free energy (SFE) is the sum of the polar and dispersion components of the surface energy.

In addition, the hydrophobicity of flooring materials was estimated by the apolarity index $x^D = \gamma_S^D / \gamma_S$ [34], because it has been found that the extent of soil adhesion greatly depends on solid surface hydrophobicity, estimated by the apolar surface free energy parameter γ_S^D. The apolarity index represents solid apolarity and hydrophobicity and is a complementary parameter to surface free energy [35].

3. RESULTS

The equilibrium contact angle could not be measured on plastic flooring materials when contact angle was measured using the static sessile drop technique. The value of the contact angle decreased over time. This illustrated the challenge in using static contact angle and, hence, the necessity to decide the moment when the meaningful value should be taken. On the other hand, when the contact angle was easured using the dynamic sessile drop technique (in which liquid is injected through a hole in the material), the contact angle fluctuated around the mean value. Figure 1 shows examples of the drop behavior for flooring materials S16, S17 and S20 using both measurement techniques. In order to obtain surface properties of flooring materials, advancing contact angle θ_A was used.

The experimental static and advancing contact angles of water, ethylene glycol, formamide and diiodomethane on flooring materials are presented in Table 4. The static angle was measured 15 s after deposition of the drop. The correlation between the static contact angles and the advancing contact angles was significant ($R^2 = 0.83$, $P<0.001$) (Fig. 2).

In order to determine the relationship between contact angle measurements and cleanability of commercial plastic flooring materials, the relationships between advancing contact angle and cleaning index, soil residue and total soiling were investigated. Water contact angles θ_A were related to cleanability (cleaning index and soil residue) of the particle-soiled surfaces, regardless of which cleaning solution was employed (e.g., for surfaces cleaned with weakly alkaline detergent ($R^2 = 0.752$, $P<0.001$). The lower the contact angle, the less soil was left (i.e., higher cleaning index) on the surfaces (Fig. 3). On the other hand, for oil-soiled surfaces the relationship between the contact angle θ_A and both cleaning index and soil residue was significant only when weakly alkaline detergent was used ($R^2 = 0.437$ and $R^2 = 0.566$, $P<0.05$, respectively).

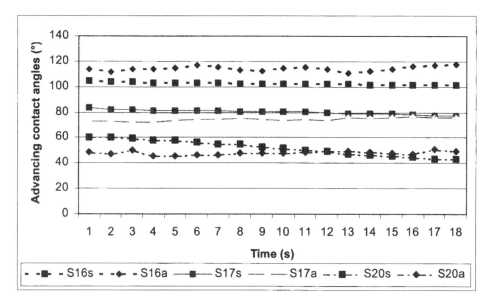

Figure 1. Advancing and static contact angles over 20 s for plastic flooring materials S16, S17 and S20 (*a* = advancing contact angle, *s* = static contact angle). Detailed descriptions of S16, S17 and S20 can be found in Table 1.

Table 4.
Experimental static (at 15 s) and advancing water contact angles (θ_A), and the advancing contact angles for ethylene glycol, formamide and diiodomethane on the surfaces of flooring materials. Detailed descriptions of the flooring materials are given in Table 1

Flooring material	Water, static (°)	Water, advancing (°)	Ethylene glycol, advancing (°)	Formamide, advancing (°)	Diidomethane, advancing (°)
S11	66.8± 0.4	67.1 ± 1.3	47.5 ± 1.0	47.3 ± 1.3	45 ± 5
S12	92.7 ± 0.4	87.1 ± 0.6	64.7 ± 0.7	68.3 ± 1.1	13.6 ± 1.2
S13	86.6 ± 1.0	72 ± 2	59.7 ± 0.8	61.2 ± 0.8	25 ± 2
S14	69.9 ± 0.7	73.0 ± 0.7	57.9 ± 2	56.6 ± 1.6	25 ± 6
S15	76.9 ± 0.6	66 ± 3	59.8 ± 1.3	47.1 ± 1.0	21 ± 4
S16	101.7 ± 0.5	115.3 ± 1.4	95.7 ± 1.1	96 ± 4	63.6 ± 0.7
S17	78.5 ± 0.6	74.9 ± 0.3	40.6 ± 1.0	55.3 ± 0.5	14.0 ± 1.2
S18	76.8 ± 0.6	76.0 ± 1.5	54.6 ± 0.9	59.5 ± 1.2	30.1 ± 1.2
S19	78.2 ± 0.5	71.8 ± 1.2	51.1 ± 2.1	56.0 ± 0.7	23 ± 3
S20	45.5 ± 0.7	48.9 ± 1.5	44.0 ± 0.9	41 ± 2	15.5 ± 0.9

Similarly, the surface free energy (SFE) of flooring materials was related to the cleanability (cleaning index) of particle soiled surfaces, e.g., for surfaces cleaned with weakly alkaline detergent ($R^2 = 0.577$, $P<0.05$). The higher the surface free

E. Pesonen-Leinonen et al.

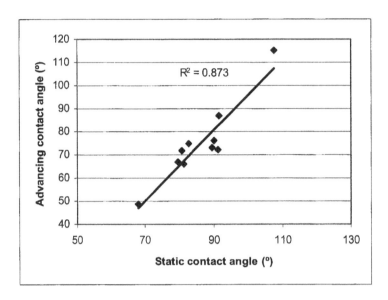

Figure 2. Correlation between static and advancing contact angles of distilled water on plastic flooring materials.

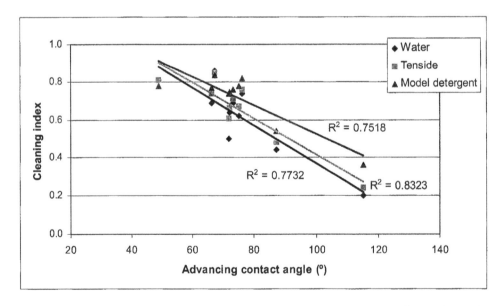

Figure 3. Effect of advancing contact angle on the cleaning index of particle-soiled surfaces. The
higher the advancing contact angle, the more the particle soil was left on the surface (i.e., the lower
the cleaning index). Compositions of the cleaning solutions, water, surfactant and model detergent,
can be found in Table 3.

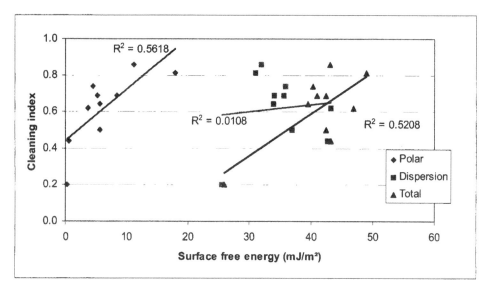

Figure 4. Relationship between the cleaning index and surface free energy and its polar and dispersion components for flooring materials soiled with particle soil.

energy, the better was the cleaning result. Considering the components of surface free energy, there appeared to be a weak relationship between the polar component and cleaning index for the surfaces soiled with both soils (Fig. 4). In addition, the apolarity index, the complementary parameter of surface free energy, was found to correlate negatively with the cleaning index of both particle- and oil-soiled surfaces.

A non-significant relationship was found between total soiling with both particle and oil soils and both contact angles θ_A and calculated surface free energy. The more hydrophobic flooring materials tended to soil more easily.

4. DISCUSSIONS

The advancing contact angle measurement was employed successfully to study the effect of surface properties on cleanability of plastic flooring materials. The advancing drops provide a good estimate of the differences between the surface properties of different materials, especially their hydrophobicity [17]. Contact angles are known to be sensitive to both surface roughness and surface chemical heterogeneity [12]. On commercial flooring materials, the contact angle of the drop applied on the surfaces continued to decrease over time, without achieving equilibrium. The advancing contact angle has been found to give results relevant to cleanability studies compared to static contact angle and receding contact angle [20].

The surface characteristics of flooring materials obtained from contact angle measurements were related to the cleanability. The cleanability was represented in terms of cleaning index, soil residue and total soiling. In contrast to earlier research [20], this study found that the lower the contact angle, the better the cleaning. This is in accordance with Young's equation: if the value of the contact angle decreases the wettability of the surface increases. In this study, the relationship between contact angle θ_A and the cleaning index of particle-soiled surfaces was significant, but only indicative for oil-soiled surfaces with the exception of oil-soiled surfaces cleaned with the weakly alkaline model detergent. Thus, both soil type and cleaning method influenced the cleaning results. Our results showed no difference between cleaning solutions if particle soil was removed. However, when oil soil was removed, the model detergent containing surfactants and alkaline salts improved oil release [3, 37, 38]. On the other hand, total soiling was not related to contact angle θ_A or surface free energy. This result is inconsistent with the past studies [20, 23].

The measurement and interpretation of contact angles in terms of surface energetics is actively debated due to the fact that some or all of the assumptions made in all energetic approaches give inconsistent values [27, 29, 39–42]. The Owens and Wendt model has been recognized as being efficient in practice for comparison of different surfaces [33]. In conclusion, the calculations of surface free energies are tentative but still valid for comparison of commercial flooring materials in experimental studies.

Only the polar component of SFE was related to the cleaning index in cases of both particle- and oil-soiled surfaces. Increase of the polar component improved the cleaning result, which agreed with the findings reported on flooring materials such as vinyl, linoleum and rubber [20] and PUR-coated PVC and linoleum [23]. Similarly, polymeric coatings with a high polar component of SFE were easily cleaned after oil soiling [18], as were the surfaces with acid–base character [43]. Consistent with these results, as the apolarity index increased the cleaning index decreased, which may indicate that hydrophobic surfaces are more difficult to clean. On the other hand, in this study the worst cleaned flooring materials were PVC materials with the lowest amount of polar component and the highest apolarity index compared to PUR-coated PVC flooring materials. These findings indicate that factors other than the polar component of surface free energy, e.g., roughness, should also be taken into consideration in evaluating the cleanability of flooring materials.

This study aimed to provide a reference for future studies. Our general aim is to develop new materials which are easier to clean and to tailor their surface topography in order to understand the reasons for the observed correlations. The cleaning properties of these materials will be compared with those of the commercial materials investigated in this study.

5. CONCLUSIONS

The results of this work indicate the potential of advancing contact angle measurements to compare and to evaluate the cleanability of real surfaces. Prediction of the cleanability of commercial flooring materials is not straightforward due to the topographical and chemical heterogeneities of the surfaces. In conclusion, the results provide a tentative relationship between the surface characteristics, such as wettability and surface free energy, and the cleanability of flooring materials.

Acknowledgements

Financial support from the Technology Agency of Finland (TEKES) is gratefully acknowledged. The authors thank the manufacturers of flooring materials, cleaning agents and textiles. This study is part of the Clean Surfaces 2002–2006-Technology Programme.

REFERENCES

1. K. L. Mittal (Ed.), *Contact Angle, Wettability and Adhesion*, Vol. 2. VSP, Utrecht (2002).
2. K. L. Mittal (Ed.), *Contact Angle, Wettability and Adhesion*, Vol. 3, VSP, Utrecht (2003).
3. A. Adamson and A. Gast, *Physical Chemistry of Surfaces*, 6th edn. Wiley, New York, NY (1997).
4. K. Kato, E. Uchida, E. T. Kang, U. Yoshikimi and I. Yoshito, *Prog. Polym. Sci.* **28**, 209–259 (2003).
5. O. I. del Río, D. Y. Kwok, R. Wu, J. M. Alvarez and A. W. Neumann, *Colloids Surfaces* A **143**, 197–210 (1998).
6. P. Dalet, E. Papon and J. J. Villenave, *J. Adhesion Sci. Technol.* **13**, 857–870 (1999).
7. E. L. Decker, B. Frank, Y. Suo and S. Garoff, *Colloids Surfaces* A **156**, 177–189 (1999).
8. S. M. M. Ramos, E. Charlaix and A. Benyagoub, *Surface Sci.* **540**, 355–362 (2003).
9. A. Amirfazli, A. Keshavarz, L. Zhang and A. W. Neumann, *J. Colloid Interface Sci.* **265**, 152–160 (2003).
10. N. Richter, J. Peggau, H. Kuhn and F. Müller, *Tenside Surface Deterg.* **40**, 202–207 (2003).
11. K. Grundke, T. Bogumil, C. Werner, A. Janke, K. Pöschel and H. J. Jacobasch, *Colloids Surfaces* A **116**, 79–91 (1996).
12. H. Kamusewitz, W. Possart and D. Paul, *Colloids Surfaces* A **156**, 271–279 (1999).
13. C. Della Volpe, A. Penati, S. Siboni, R. Peruzzi, L. Toniolo and C. Colombo, *J. Adhesion Sci. Technol.* **14**, 273–299 (2000).
14. D. Quéré, *Physica* A **313**, 32–46 (2002).
15. M. Gindl, G. Sinn, W. Gindl, A. Reiterer and S. Tschegg, *Colloids Surfaces* A **181**, 279–287 (2001).
16. M. A. Rodríguez-Valverde, M. A. Cabrerizo-Vílchez, P. Rosales-López, A. Páez-Dueñas and R. Hidalgo-Álvarez, *Colloids Surfaces* A **206**, 485–495 (2002).
17. M. C. Michalski, S. Desobry, V. Babak and J. Hardy, *Colloids Surfaces* A **149**, 107–121 (1999).
18. L. Boulangé-Petermann, C. Debacq, P. Poiret and B. Cromières, in: *Contact Angle, Wettability and Adhesion*, K. L. Mittal (Ed.),Vol. 3, pp. 501–519, VSP, Utrecht (2003).
19. L. Boulangé-Petermann, E. Robine, S. Ritoux and B. Cromières, *J. Adhesion Sci. Technol.* **18**, 213–225 (2004).
20. H. Garvens and H. Kruessmann, in: *Proceedings of Cleaning in Tomorrow's World*, pp.106–113. Finnish Association of Cleaning Technology, Monila Oy, Helsinki (1997).

21. E. Mettler and B. Carpentier, *Food Bioprod. Process.* **77**, 90–96 (1999).
22. E. Pesonen-Leinonen, I. Redsven, R. Kuisma, M. Hautala and A. M. Sjöberg, *Tenside Surface Deterg.* **40**, 80–86 (2003).
23. I. R. Redsven, R. Kuisma, L. Laitala, E. Pesonen-Leinonen, R. Mahlberg, H.-R. Kymäläinen, M. Hautala and A.-M. Sjöberg, *Tenside Surface Deterg.* **40**, 346–352 (2003).
24. A. Marmur, *J. Colloid Interface Sci.* **168**, 40–46 (1994).
25. G. Wolansky and A. Marmur, *Colloids Surfaces* A **156**, 381–388 (1999).
26. R. Mahlberg, H. E.-M. Niemi, F. Denes and R. M. Rowell, *Int. J. Adhesion Adhesives* **18**, 283–297 (1998).
27. R. Kuisma, E. Pesonen-Leinonen, I. Redsven, I. Ylä-Outinen, M. Hautala and A. M. Sjöberg, *Tenside Surface Deterg.* **40**, 25–30 (2003).
28. D. Martens, in: *Proceedings of the 40ᵗʰ International Detergency Conference*, pp. 143–150. Forschungsinstitut für Reinigungstechnologie, Strasbourg (2000).
29. F. M. Fowkes, *Ind. Eng. Chem.* **56** (12), 40–52 (1964).
30. D. K. Owens and R. D. Wendt, *J. Appl. Polym. Sci.* **13**, 1741–1747 (1969).
31. D. Y. Kwok and A. W. Neumann, *Adv. Colloid Interface Sci.* **81**, 167–249 (1999).
32. M. Rankl, S. Laib and S. Seeger, *Colloids Surfaces* B **30**, 177–186 (2003).
33. J. Comyn, L. Mascia, G. Xiao and B. M. Parker, *Int. J. Adhesion Adhesives* **16**, 97–104 (1996).
34. M. C. Michalski, J. Hardy and B. J. V. Saramago, *J. Colloid Interface Sci.* **208**, 319–328 (1998).
35. M. C. Michalski, D. M. Mousavi and J. Hardy, *J. Food Eng.* **37**, 271–291 (1998).
36. G. Schoenmakers and D. Martens, in: *Proceedings of the 39ᵗʰ International Detergency Conference*, pp. 337–340. Forschungsinstitut für Reinigungstechnologie, Luxemburg (1999).
37. D. Shaw, *Introduction to Colloid and Surface Chemistry*. Butterworth-Heinemann, Oxford (1994).
38. D. Martens, G. Schoenmakers and T. Bastein, in: *Proceedings of the 39ᵗʰ International Detergency Conference*, pp. 289–294. Forschungsinstitut für Reinigungstechnologie, Luxemburg (1999).
39. N. T. Correia, J. J. Moura Ramos, B. J. V. Saramago and J. C. G. Calado, *J. Colloid Interface Sci.* **189**, 361–369 (1997).
40. D. Y. Kwok, *Colloids Surfaces* A **156**, 191–200 (1999).
41. D. Y. Kwok and A. W. Neumann, *Colloids Surfaces* A **161**, 49–62 (2000).
42. S. Shalel-Levanon and A. Marmur, *J. Colloid Interface Sci.* **262**, 489–499 (2003).
43. L. Boulangé-Petermann, E. Robine, S. Ritoux and B. Cromières, *J. Adhesion Sci. Technol.* **18**, 213–225 (2004).

Contact Angle, Wettability and Adhesion, Vol. 4, pp. 215–236
Ed. K.L. Mittal
© VSP 2006

Surface free energy of solids: A comparison of models

FRANK M. ETZLER[*]

*Boehringer-Ingelheim Pharmaceuticals, Inc., 900 Ridgebury Road, P.O. Box 368,
Ridgefield, CT 06877-0368, USA*

Abstract—An understanding of the surface free energy and surface chemistry of solids is needed for investigation into the nature of processes involving adhesion, wetting and liquid penetration. Frequently the contact angles of several probe liquids on a given solid are used for calculation of solid surface free energy. Models by Fowkes, Kwok and Neumann, van Oss, Chaudhury and Good, as well as by Chang and Chen have been used for such calculations. Each of the above models has been championed in the literature. It has been noted by the present author and others that the use of different models may lead to different qualitative interpretations of the nature of a solid surface. A disinterested comparison of the various available models has not been made. In the present paper, a comparison of the calculations is undertaken in order to better understand the limitations of each model. Particular attention to the assumptions required for contact-angle data to be used for surface free energy calculations is given. The effect of the degree to which the experimental contact-angle data meet the required assumptions have on the calculated surface free energy is addressed in this work. When data meeting the theoretical assumptions common to the various published models are used, all of the published models fit the data, to a good approximation, equally well. A poor fit of the experimental data is an indicator that at least one liquid does not fully meet the assumptions required by the chosen model. Differences in the acid–base character of the solid surface appear to result from the acid–base scale used by the model. The paper is intended to raise the awareness of the difficulties in assigning surface free energy and predicting wetting behavior.

Keywords: Contact angle; surface free energy; surface tension; acid–base models; wetting; adhesion; surface chemistry.

1. INTRODUCTION

Both the surface chemical and surface energetic nature of materials, used in the formulation of commercial products or used in the manufacture of these products, are often important to the final quality of the product. Despite the importance of surface chemistry to the ultimate performance of the product, not as much recognition as is deserved has been given to the characterization of surface chemistry and the effects of its variation on product performance. Difficulties associated with both theoretical descriptions of interfaces and the measurement of surface

---

[*]Tel.: (1-203) 798-5445; e-mail: fetzler@rdg.boehringer-ingelheim.com

chemical characteristics make the incorporation of material surface chemical specifications for the manufacture of many products challenging. The application of the various models used to determine surface free energies from contact angle data is explored in this paper.

The literature regarding the determination of the surface free energy of solids has been reviewed elsewhere [1]. The reader is directed to this review for a more extensive discussion of the subject. Here the relative performance of the theoretical models employed in the literature is explored.

This paper does not criticaly compare the theoreticial justifications for the various modes, rather their application to an illustrative set of experimental data is explored. The author hopes the following discussion will assist others with choosing appropriate models for data interpretation and identifying appropriate liquids for study of surfaces of interest.

2. ESTIMATION OF SURFACE FREE ENERGY FROM CONTACT ANGLE DATA

Several approaches for estimating the surface free energy of solids have been used over the years. Approaches relevant to the present work are discussed below.

2.1. Zisman critical surface tension

Zisman and co-workers [2] made pioneering investigations of the thermodynamics of wetting and adhesion.

The basic thermodynamic equations describing the relation between surface free energy and wetting are given below.

$$\gamma_{SV^0} - \gamma_{SL} = \gamma_{LV^0} \cos(\theta) \tag{1}$$

$$W_a = \gamma_{S^0} + \gamma_{LV^0} - \gamma_{SL} \tag{2}$$

where the subscript LV^0 refers to the liquid–vapor interface, SV^0 is the solid saturated with vapor interface and S^0 is the solid interface in absence of liquid vapor.

From equations (1) and (2) it follows that:

$$W_a = \left(\gamma_{S^0} - \gamma_{SV^0}\right) + \gamma_{LV^0} \cos(\theta) \tag{3}$$

where

$$\gamma_{S^0} - \gamma_{SV^0} = RT \int_0^{p^0} \Gamma \, d\ln p = \pi_e \tag{4}$$

Here, Γ is the surface excess (amount adsorbed) of the liquid vapor and P^0 the saturation vapor pressure of the liquid. The vapor is assumed to be ideal. For low surface free energy solids $\pi_e \approx 0$.

Zisman and co-workers also noted that plots of $\cos(\theta)$ vs. γ_{LV^0} were nearly linear particularly when homologous series of liquids were used. It, thus, follows that:

$$\cos(\theta) = 1 + \beta\left(\gamma_c - \gamma_{LV^0}\right) \tag{5}$$

where γ_c is referred to as the Zisman critical surface tension. A plot of $\cos(\theta)$ vs. γ_{LV^0} will intersect $\cos(\theta) = 1$ when $\gamma_{LV^0} = \gamma_c$. Zisman's relation is empirical and cannot be fully justified from a theoretical basis.

From equations (1) and (5) it can be seen that:

$$\gamma_{S^0} - \gamma_{LV^0} \approx \gamma_{SV^0} - \gamma_{LV^0} = \gamma_c \tag{6}$$

Thus, the Zisman critical surface tension is not equal to the surface free energy of the solid. The Zisman model for estimating surface free energies, however, provides a number of insights into the relation between surface chemistry and contact angle. Zisman's model, furthermore, can be of importance to practical applications including, for instance, printing. The Zisman critical surface tension often closely corresponds to surface free energy calculated by the more modern models described below.

2.2. Statistical thermodynamic models for estimating surface free energy

A more detailed understanding of the relation between wetting and surface free energy and surface chemistry is offered by more recent models than that discussed by Zisman.

Several investigators including Fowkes [3–7], Good [8, 9], van Oss [10] and Chang [11] have constructed statistical thermodynamic models for wetting and adhesion. These models may be considered as statistical thermodynamic models in the sense that they offer molecular interpretations for origins of wetting and adhesion. In this section, the common features of these statistical thermodynamic theories are explored.

Intermolecular forces between molecules result from interaction between their corresponding electron orbitals. The principal non-bonding interactions result from induced dipole-induced dipole (London), dipole-induced dipole (Debye) and dipole–dipole (Keesom) interactions. The intermolecular potential energy function for each of these three types of interactions is of the same form:

$$U = \frac{-\beta_{12}}{r^6} \tag{7}$$

If London dispersion forces are considered equation (7) can be expressed as follows:

$$\beta_{12}^d = \frac{2\sqrt{I_1 I_2}}{I_1 + I_2}\left(\beta_{11}^d \beta_{22}^d\right)^{1/2} = \frac{2\beta_{11}^d \beta_{22}^d}{\beta_{11}^d\left(\dfrac{\alpha_2}{\alpha_1}\right) + \beta_{22}^d\left(\dfrac{\alpha_1}{\alpha_2}\right)} \tag{8}$$

Here the subscripts 11, 22 and 12 refer to interactions between like molecules (11, 22) and dissimilar molecules (12). β is the coefficient in equation (7). I is the ionization potential and α the polarizability. If $I_1 \approx I_2$ then

$$\beta_{12}^d \cong \left(\beta_{11}^d \beta_{22}^d\right)^{1/2} \tag{9}$$

Equation (9) forms the basis of the Berthelot principle [12] (also see, for instance, Chang and Qin [13]) which states that dispersion interactions between dissimilar molecules can be estimated as the geometric mean of the interactions between like molecules.

Alternatively, if $\alpha_1 \approx \alpha_2$, then

$$\beta_{12}^d = \frac{2\beta_{11}^d \beta_{22}^d}{\beta_{11}^d + \beta_{22}^d} \tag{10}$$

The harmonic mean estimation expressed in equation (10) has been used less often and is frequently numerically similar to the geometric mean approximation.

The interaction potentials between molecules have been used to determine the interactions between macroscopic bodies. For a column of material 1 interacting with a column of material 2, the free energy, ΔG_{12}^d, to move the planes of material from distance, d, to infinity is:

$$\Delta G_{12}^d = -W_{12}^d = \int_{d_{12}}^{-\infty} \frac{-\pi \beta_{12}^d N_2}{6x^3} N_1\ dx = -\frac{A_{12}}{12\pi d_{12}^2} \tag{11}$$

Here A_{12} is the Hamaker constant (see, for instance, Refs [13, 14]) and

$$A_{12} = \pi^2 \beta_{12} N_1 N_2 \tag{12}$$

N_1 and N_2 are the numbers of molecules of type 1 and 2, respectively.
Recalling Berthelot's principle [12]

$$A_{12} \cong \left(A_{11} A_{22}\right)^{1/2} \tag{13}$$

and further assuming that the intermolecular distance, d_{12}, can be approximated as the geometric mean of the intermolecular distances found in the pure components (d_{11} and d_{12})

$$d_{12} = (d_{11} d_{22})^{1/2} \tag{14}$$

Thus the work of adhesion, W_{12}^d, resulting from London dispersion forces is

$$W_{12}^d \cong \frac{(A_{11}A_{22})^{1/2}}{12\pi d_{11}d_{22}} = (W_{11}^d W_{22}^d)^{1/2} = 2(\gamma_1^d \gamma_2^d)^{1/2} \tag{15}$$

Fowkes [6, 7] suggested that the surface free energy of materials could be considered to be a sum of components resulting from each class of intermolecular interaction; thus,

$$\gamma = \sum_i \gamma_i \tag{16}$$

Equation (16) might be expanded as

$$\gamma = \gamma^d + \gamma^p + \gamma^{AB} + \dots \tag{17}$$

Here the superscript d refers to dispersion forces, p refers to dipole–dipole (Keesom) and dipole-induced dipole (Debye) interactions and AB refers to Lewis acid–base interactions. London, Kessom and Debye interactions are non-bonding orbital interactions and Lewis acid–base interactions which by definition involve electron acceptance and donation.

Van Oss, Chaudhury and Good chose to express surface free energy in terms of two principal components, Lifshitz–van der Waals (LW) and Lewis acid–base (AB) components.

The Lifshitz–van der Waals term is composed of the interactions covered by equation (7). The work of adhesion due to Lifshitz–van der Waals interactions is estimated using the geometric mean rule discussed above. Thus,

$$W_a^{LW} = 2(\gamma_1^{LW} \gamma_2^{LW})^{1/2} \tag{18}$$

The use of the geometric mean approximation with regard to Lifshitz–van der Waals interactions is not unique to the van Oss, Chaudhury and Good approach and is used in the models to be discussed later by Chang and Chen and by Fowkes. The harmonic mean approximation (equation (10)) has also sometimes been used although the results of the two calculations are often nearly identical. The use of the geometric mean approximation is not a subject of current controversy. The relative merits of the geometric and harmonic mean approximations have been discussed in the literature [5, 15–19].

2.2.1. Van Oss, Chaudhury and Good model
According to the van Oss, Chaudhury and Good model [8, 10], the Lewis acid–base component is modeled as follows

$$\gamma_i^{AB} = 2(\gamma_i^+ \gamma_i^-)^{1/2} \tag{19}$$

where γ^+ is the Lewis acid parameter and γ^- the Lewis base parameter. Van Oss, Chaudhury and Good further chose

$$\gamma_i^+ = \gamma_i^- \equiv 0 \tag{20}$$

for alkanes, methylene iodide and α-bromonaphthalene, which presumably interact only through Lifshitz–van der Waals interactions. For water

$$\gamma_{H_2O}^+ = \gamma_{H_2O}^- \equiv 25.5 \text{ mJ} / \text{m}^2 \tag{21}$$

Based on these above numerical choices, γ^+ and γ^- have been experimentally determined for a variety of liquids. Van Oss [10] has compiled and reviewed the determination of these values (also see Ref. [1]). Della Volpe and co-workers [20–22] have argued that the choice of γ^+ and γ^- for water is inappropriate and van Oss [23] has argued that this numerical choice is not scientifically significant. Hence, Della Volpe's acid–base scales are not considered further in this paper. It should also be pointed out that Kwok [24], for instance, has criticized the use of surface free energy components altogether.

Recalling equation (1) together with the relation

$$\gamma = \gamma^{LW} + \gamma^{AB}, \tag{22}$$

it follows that

$$W_a = \gamma_l[1 + \cos(\theta)] = 2(\gamma_l^{LW}\gamma_s^{LW})^{1/2} + 2(\gamma_l^+\gamma_s^-)^{1/2} + 2(\gamma_l^-\gamma_s^+)^{1/2} \tag{23}$$

If the Van Oss, Chaudhury and Good parameters are known for at least three liquids and the contact angles of these liquids on a solid are measured, then equation (23) can be used to determine the Van Oss, Chaudhury and Good parameters for the surface free energy of the solid. Van Oss [10] has reviewed the numerous publications which have reported the determination of the Van Oss, Chaudhury and Good parameters for various liquids.

The simultaneous solution of equation (23) for a set of several liquids would seem straightforward, at first glance. Gardner *et al.* [25], however, have pointed out that different calculation approaches may give numerically different results and that a particular author's choice may not be readily apparent to the reader.

2.2.2. Chang–Chen model

The Chang–Chen model [11, 13] for interfacial free energy is largely based on the same principles which govern the van Oss, Chaudhury and Good model. Both models treat Lifshitz–van der Waals interactions in the same way. Calculation of the surface free energy components requires the use of the same experimental data. The two models, however, differ in the way that Lewis acid–base interactions are handled.

Recall that

$$W_a = W_a^{LW} + W_a^{AB} \tag{24}$$

and

$$\gamma = \gamma^{LW} + \gamma^{AB} \tag{25}$$

The Chang–Chen model uses the same geometric mean approximation for W_a^{LW} as does the van Oss, Chaudhury and Good model. Thus,

$$W_a^{LW} = W_a^L = 2\left(\gamma_1^{LW}\gamma_2^{LW}\right)^{1/2} = P_1^L P_2^L \tag{26}$$

where

$$P_i^L = \left(2\,\gamma_i^L\right)^{1/2} \tag{27}$$

P_i^L is the dispersion component. The superscript L is equivalent to LW.

Like the van Oss, Chaudhury and Good model, acid–base interaction is modeled using two parameters. These parameters, P_i^a and P_i^b, are referred to as principal values (*i* refers to material *i*, or that of interest). The acid–base work of adhesion can be represented using the following relation:

$$W_a^{AB} = -\Delta G_a^{AB} = -(P_1^a P_2^b + P_1^b P_2^a) \tag{28}$$

The surface free energy of the material thus, is,

$$\gamma = \gamma^{LW} + \gamma^{AB} = \frac{1}{2}\left(P^L\right)^2 - P^a P^b \tag{29}$$

Tabulated P_i^a and P_i^b values [1] are substituted into equation (28) such that the work of adhesion is maximized and the free energy of adhesion is minimized.

The acid–base character of a material is characterized by the sign of P_i^a and P_i^b. If $P_i^a = P_i^b = 0$, then the material is neutral (or non-polar). If P_i^a and P_i^b are both positive, then the material is monopolar acidic and if both negative then the material is monopolar basic. If P_i^a and P_i^b are of opposite sign then the material is amphoteric.

Despite some similarities to the van Oss, Chaudhury and Good model, the Chang and Chen model differs from the former model in a number of ways. The Chang and Chen model applies the geometric mean rule only to Lifshitz–van der Waals interactions. In determining values for P_i^a and P_i^b, only n-alkanes are assumed to exclusively interact by Lifshitz–van der Waals interactions. The van Oss, Chaudhury and Good model, for instance, assumes that both methylene iodide (diiodomethane) and α-bromonaphthalene also interact with materials exclusively by Lifshitz–van der Waals interactions. A major difference is that the

Chang–Chen model allows for both attractive and repulsive interactions. In other words, $-\infty \le W_a^{AB} \le \infty$, whereas in the van Oss, Chaudhury and Good model $W_a^{AB} \ge 0$. The Lewis acid–base concept is general enough to include traditional ion–ion and dipole–dipole repulsions and, thus, it may not be unreasonable to suggest the existence of repulsive interactions [13]. Furthermore, entropic effects may contribute to the overall repulsion.

2.2.3. The Fowkes approach

As discussed above, Fowkes [6, 7] first suggested that surface free energy could be considered as a sum of components resulting from different classes of intermolecular interactions. Both, the van Oss, Chaudhury and Good, and Chang and Chen models draw upon the idea of Fowkes and as such all use the geometric mean approximation to model Lifshitz–van der Waals interactions.

Fowkes [3, 6, 19], however, has suggested a different approach to evaluating the acid–base character of surfaces (also see Ref. [26]). Fowkes has criticized the use of contact angles for determination of interfacial properties [5]. His approach is, for experimental reasons, more applicable to powdered samples. As stated previously,

$$W_a = W_a^{LW} + W_a^{AB} \tag{30}$$

W^{AB} is then, according to Fowkes, expressed by the following relation

$$W_a^{AB} = -f \cdot N \cdot \Delta H_a^{AB} \tag{31}$$

where N is the number of sites per unit area and

$$f = \left[1 - \frac{\partial \ln W_a^{AB}}{\partial \ln T} \right]^{-1} \tag{32}$$

and

$$f \approx 0.2 \ldots 1.0 \tag{33}$$

When using the Fowkes approach some authors have taken f as unity although this does not seem to be a good approximation [26]. Because f and N are generally not known, direct calculations of the work of adhesion are often not made. Determination of ΔH^{AB} for multiple probe liquids on a given solid together with models by Drago [27] or Gutmann [28, 29] can be used to assess the acid–base nature of the surface (also see Ref. [30]).

The Fowkes approach although interesting and useful will not be discussed in this work as it is not often used in conjunction with contact angle data.

2.3. Neumann's approach

Neumann and co-workers [31–41] have discussed the surface tension of solids from a purely thermodynamic point of view. Their view contrasts with the statistical thermodynamic approaches used by van Oss, Chaudhury and Good, Chang and Chen, and Fowkes (see earlier sections for references). Because the Neumann's approach does not consider the molecular origins of surface tension no statistical mechanical insight is gained.

Kwok and Neumann [31] correctly remind us that contact angle measurements can be difficult. Measured contact angles can deviate from the true Young's contact angle. Both surface topography and surface chemical heterogeneity can affect the measured value for contact angle [42–45]. Grundke *et al.* [46] have recently discussed wetting on rough surfaces.

All of the thermodynamic approaches to calculate surface free energy which are explored in this paper assume that the following conditions apply:
1. Young's equation is valid.
2. Probe liquids are pure compounds. Mixtures are likely to exhibit selective surface adsorption. See Adamson [14], for instance.
3. The surface tensions of materials are independent of the test conditions at fixed T and P (i.e., the surface tension of a solid is not modified by the probe liquid).
4. $\gamma_l > \gamma_s$ according to Neumann or more generally $\cos(\theta) > 0$.
5. γ_{sv} is not influenced by liquids.

According to Kwok and Neumann [31], contact angle can be expressed as a function of γ_{LV} and γ_{SV} only. Thus,

$$\gamma_{LV} \cos(\theta) = f(\gamma_{LV}, \gamma_{SV}) \tag{34}$$

$$\gamma_{SV} - \gamma_{SL} = f(\gamma_{LV}, \gamma_{SV}) \tag{35}$$

$$\gamma_{SL} = \gamma_{SV} - f(\gamma_{LV}, \gamma_{SV}) = F(\gamma_{LV}, \gamma_{SV}) \tag{36}$$

where f and F are suitable functions. Kwok and Neumann have observed smooth monotonic dependence of $\gamma_{LV} \cos(\theta)$ with γ_{LV} consistent with equation (34) when liquid–solid pairs conform closely to the assumptions listed above. For arbitrary solid–liquid pairs such a plot may show considerable scatter because the measured contact angles deviate significantly from the true Young's contact angle. Stick/slip behavior in advancing contact angles, time-dependent contact angles and liquid surface tension changes during the course of the experiments result from physico-chemical interactions not consistent with the use of Young's equation for determining surface free energies.

The function, F, in equation (36) has been historically modeled in several ways. Antonow [47] stated that:

$$\gamma_{SL} = |\gamma_{LV} - \gamma_{SV}| \tag{37}$$

or combining with Young's equation,

$$\cos(\theta) = -1 + 2\frac{\gamma_{SV}}{\gamma_{LV}} \tag{38}$$

Alternatively, Berthelot's rule [12] (recall equation (18), for instance), has been used such that

$$\gamma_{SL} = \gamma_{LV} + \gamma_{SV} - 2(\gamma_{LV}\gamma_{SV})^{1/2} = \left[(\gamma_{LV})^{1/2} - (\gamma_{SV})^{1/2}\right]^2 \tag{39}$$

and again, combining with Young's equation,

$$\cos(\theta) = -1 + 2\left(\frac{\gamma_{SV}}{\gamma_{LV}}\right) \tag{40}$$

Equation (39) would be identical to that used in the van Oss, Chaudhury and Good model if the solid and liquid had only Lifshitz–van der Waals interactions such that $\gamma_{LV}^{LW} = \gamma_{LV}$. If equations (37) and (38) were adequate descriptors then calculated values for γ_{SV} would be independent of the choice of probe liquid used. This is unfortunately not the case. Li and Neumann [41] have considered a modified Berthelot equation such that

$$\gamma_{SL} = \gamma_{LV} + \gamma_{SV} - 2(\gamma_{SV}\gamma_{LV})^{1/2}\, e^{-\beta(\gamma_{LV} - \gamma_{SV})^2} \tag{41}$$

and

$$\cos(\theta) = -1 + 2\left(\frac{\gamma_{SV}}{\gamma_{LV}}\right)^{1/2} e^{-\beta(\gamma_{LV} - \gamma_{SV})^2} \tag{42}$$

An alternate form of equation (42) has also been proposed by Kwok and Neumann [48] where

$$\cos(\theta) = -1 + 2\left(\frac{\gamma_{SV}}{\gamma_{LV}}\right)^{1/2} \left(1 - \beta(\gamma_{LV} - \gamma_{SV})^2\right) \tag{43}$$

Empirically it has been shown that

$$\beta \approx 0.0001247 \tag{44}$$

for equation (42) and that the measured solid surface free energy using this choice for β is nearly independent of the choice of liquid. Equation (42) can be used in two ways. First, β can taken from equation (44) and a suitable contact angle can be used to determine γ_{SV}. Second, β and γ_{SV} can be treated as adjustable parameters. Least squares analysis using contact angles measured for several liquids is then used to determine the best fit values for β and γ_{SV}. The second approach would seem to be preferable. The alternate combining rule, equation (43) as suggested recently by Kwok and Neumann [48] appears to be equally useful.

Kwok and Neumann [31] have criticized the van Oss, Chaudhury and Good approach for several reasons, including

1. The Neumann model uses fewer adjustable parameters, two *versus* six

2. Surface free energies measured using the van Oss, Chaudhury and Good model may appear to depend on the choice of liquids

3. The assigned values for the van Oss, Chaudhury and Good parameters are assigned in ways not fully satisfactory to Neumann and co-workers.

It is the present author's opinion that Kwok and Neumann have properly identified experimental factors that result in significant deviations of measured contact angles from the true Young's contact angle. Application of any of the above models requires that Young's equation be valid. Undoubtedly, contact angles reported in the literature sometimes deviate significantly from the Young's angle for a variety of reasons.

The apparent variation of surface free energy with the choice of probe liquid sets undoubtedly, in part, results from physico-chemical interactions between solid and liquid that cause Young's equation to be invalid for the present purposes. On the other hand, Kwok and Neumann have not always reported results from liquid sets containing diverse liquids. The mathematics of least squares fitting requires the use of a diverse set of liquids to achieve reliable values for the fitted parameters. Kwok, Li and Neumann [24, 31, 49] have also chosen a fitting procedure that allows $\sqrt{\gamma_i^j}$ to be negative ($j = +, -$; i is the material of interest) which appears contrary to van Oss' statements.

3. THE PRESENT PROBLEM

Despite numerous publications in the literature concerning the determination of surface free energy, a consensus as to which of the available models is the best has not been achieved. The lack of consensus is, in part, due to the relatively few papers written by disinterested parties. In this work, data from two polymer surfaces [50] are fitted to the van Oss, Chaudhury and Good model, the Chang and Chen model, as well as the Kwok and Neumann model. The impact of the quality of the data on the estimate of surface free energy and its components is also investigated. This paper is not indended to be a survey of a broad range of polymer surfaces but rather an illustrative example. It is intended to serve as guide to some of the difficulties associated with the determination of surface free energy and surface free energy components.

Kwok *et al.* [50] have collected contact angles on polymer surfaces using both axisymetric drop shape analysis and with a conventional goniometer. Their data are used in this analysis. These data offer a rare opportunity to determine surface free energy and surface free energy components under the condition where adherence to the required assumptions is at least partially known.

Contact angles may be calculated from drop dimensions. The drop shape is determined by the combination of the forces of surface tension and gravity. Several methods are available, including the axisymmetric drop shape analysis (ADSA) techniques developed by Neumann and co-workers [51]. These methods are based on the Young–Laplace equation.

Drop profiles can be determined automatically through digital images and computer analysis. Surface tension, contact angle, as well as drop characteristics such as volume and surface area are calculated from a nonlinear regression of the measured drop profile to the Young–Laplace equation. ADSA measurements produce higher accuracy and less subjectivity compared to direct measurements using a traditional goniometer. As with measurement using a goniometer, the contact angle is determined at the point where the three-phase line intersects the largest drop diameter and may not reflect any variations along the three-phase line due to heterogeneity or surface roughness.

Axisymmetric drop-shape analysis offers the advantage of being able to measure liquid surface tension and contact angle with time, as liquid is advancing over the solid surfaces.

Table 1 shows contact angle data collected on two polymers as determined by Kwok *et al.* [50]. The full description of the experimental details are described in their paper.

Figure 1 shows liquid surface tension, contact angle and drop radius, *r*, as a function of time during drop spreading. Figure 1 shows, for selected liquids, that contact angle and surface tension may not be constant over the course of the measurement. If the contact angle and liquid surface tension are not constant, then the necessary assumptions enumerated above by Kwok and Neumann are not fully met. Such data are less reliable for the purpose of determining surface free energy as the applicability of Young's equation even in approximation is in doubt. If only goniometer data, for instance, are avialable then it may be difficult to acertain whether the assumptions enumerated by Kowk and Neumann have been met.

The data in Table 1 fall into three classes for the purposes of the present discussion, as indicated in the footnotes. When a range of contact angles is shown the measured contact angle varied during the measurement indicating that the required assumptions for application of Young's equation have not been fully met. The two polymers listed in Table 1 are sometimes referred to as "hexyl" or "propyl" for convenience. *cis*-Decalin is assumed to have no acid–base component in the Chang and Chen model for the purposes of this discussion.

Table 1.
Contact angles (degrees) of various liquids on two polymer surfaces

Liquid	Poly(propene-alt-N-(n-propyl) maleamide)		Poly(propene-alt-N-(n-hexyl) maleamide)	
	ADSA-P	Goniometer	ADSA-P	Goniometer
cis-Decalin			28.81[a]	28.0[a]
2,5-Dichlorotoluene			29–37[b]	60.0[b]
Ethyl cinnamate			66–56[b]	67.0[b]
Dibenzyl amine			39–34[b]	41.0[b]
DMSO			40–65[b]	58.0[b]
1-Bromonaphthalene	30.75	33.0	40–60[c]	75.0[c]
Diethylene glycol	45–49	51.0	61.04[b]	59.5[b]
Ethylene glycol	54–57	58.5	67–70[c]	69.0[c]
Diiodomethane	45–80 ss	95.0	88–96 ss[c]	98.0[c]
Thiodiethanol	54.04	55.5	68–90 ss[b]	91.0[b]
Formamide	62–54	67.0	Dissolves	Dissolves
Glycerol	70.67	70.0	82.83[a]	80.5[a]
Water	77.51	75.0	92.26[a]	92.0[a]

ss, stick–slip.
[a]The required assumptions discussed by Kwok and Neumann appear to have been met.
[b]The van Oss, Chaudhury and Good parameters, as well as the Chang–Chen parameters are unknown.
[c]The acid–base parameters from the van Oss, Chaudhury and Good, as well as from the Chang–Chen model are unknown.

Figure 2 shows the cos (θ) *versus* liquid surface tension for poly(propene-alt-N(n-propyl) maleamide) and poly(propene-alt-N(n-hexyl)maleamide (these polymers may be referred to as "propyl" or "hexyl," respectively, for brevity). Figure 2 indicates that the gonimeter data have more scatter than the data taken using ADSA. Methylene iodide (diiodomethane) ($\gamma_{LV} \approx 50 \; mN / m$), in particular, shows the greatest deviation from expectation. The two curves drawn in Fig. 2 were fitted according to Kwok and Neumann's model using contact angle data for liquids that exhibited nearly constant contact angle during ADSA measurement (propyl polymer, water, glycerol, thioethanol and 1-bromonaphthalene; hexyl polymer, cis-decalin, ethylene glycol, glycerol and water). The fits are those calculated by Kwok and Neumann using equation (43). For the hexyl polymer, $\beta = 0.000109$ and $\gamma_{sv} = 27.8$, while for the propyl polymer $\beta = 0.000123$ and $\gamma_{sv} = 36.9$.

Equation (23) forms the basis for fitting data to the van Oss, Chaudhury and Good model. Equations (24), (26) and (28), when combined form a similar rela-

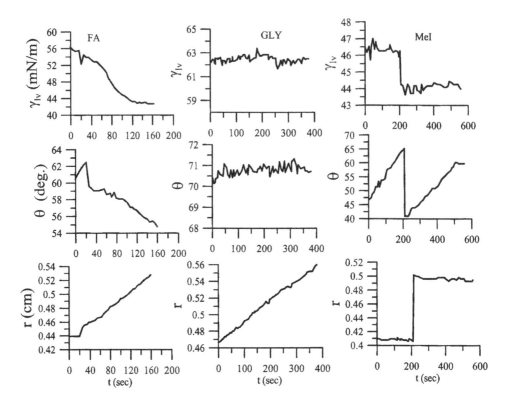

Figure 1. ADSA data for selected liquids [50]. First column formamide (FA), second glycerol (GLY) and third methylene iodide (aka diiodomethane, MeI). Bottom row shows drop radius (cm) for spreading liquid *versus* time. The first row shows calculated surface tension (mN/m) and the second row contact angle (degrees). Only glycerol shows wetting characteristics consistent with Kowk and Neumann's requirements. MeI shows stick–slip behavior, while formamide shows decreasing contact angle with time of exposure. The graphs show representative behavior of selected liquids listed in Table 1. The problems shown here might go unnoticed if only goniometer data were collected.

tion (to equation (23)), for the Chang and Chen model. These relations are, respectively, listed below.

$$W_a = \gamma_l \left[1 + \cos(\theta)\right] = 2(\gamma_l^{LW}\gamma_s^{LW})^{1/2} + 2(\gamma_l^+\gamma_s^-)^{1/2} + 2(\gamma_l^-\gamma_s^+)^{1/2} \qquad (23)$$

$$W_a = P_S^L P_L^L - (P_S^a P_L^b + P_S^b P_L^a) \qquad (45)$$

Surface free energy components were calculated using the Levenberg-Marquardt algorithm to find the best solution to n instances of either equations (23) or (45) (n = number of liquids). The Mathcad 2001i program (Mathsoft, Cambridge, MA, USA; www. mathsoft.com) was used to perform calculations.

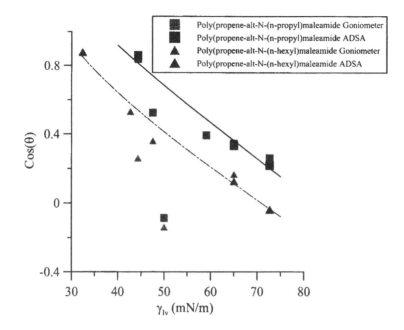

Figure 2. Cos (θ) *versus* liquid surface tension for data listed in Table 1. Gray points are goniometer data and black points are ADSA data. Lines show fit of the data meeting the Kwok and Neumann assumptions in Table 1 to the Kwok–Neumann model (equation (43)). The fitted constants are listed in the text. Solid line is for the propyl polymer. The dashed line is for the hexyl polymer. ADSA data show a better fit to the Kwok–Neumann equation than do goniometer data due to inclusion of data not fully meeting the assumptions required for application of Young's equation. The data point near 50 nN/m is for methylene iodide (diiodomethane) which exhibits stick–slip spreading.

This procedure minimizes the difference between W_a ($\gamma_{LV}[1 + \cos(\theta)]$) and W_a (calculated from the right side of either equation (23) or equation (45)). Because W_a is calculated directly from the cos(θ), calculated values of cos(θ) can be chosen to measure the quality of fit.

Figures 3–6 show calculated contact angles *versus* the corresponding experimental contact angles. The results show that all three models fitted to the ADSA data provide essentially identical estimates of calculated contact angle. Note that the Kwok and Neumann data were fitted using 4 liquids (the liquids are specified above), but only 3 are shown on the graphs as the van Oss, Chaudhury and Good, as well as the Chang–Chen parameters are unknown for the fourth liquid. The author recognizes that it would be perferable to have data from more liquids which fully meet the required model assumptions. The goniometer data, in general, show that the Chang and Chen model fits the data somewhat better than the van Oss, Chaudhury and Good model (see Fig. 7). The van Oss, Chaudhury and Good model is more sensitive to data which do not comply with assumptions required for application of Young's equation. Goodness of fit, thus, may be one measure of

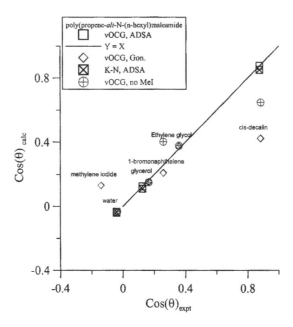

Figure 3. Cosine of the calculated contact angle *versus* cosine of the experimental contact angle for hexyl polymer. Line represents *y=x*. Fitting error is reduced when methylene iodide which exhibits stick–slip spreading, is excluded from the fitted data set. vOCG, van Oss, Chaudhury and Good; K–N, Kwok and Neumann; C–C, Chang and Chen.

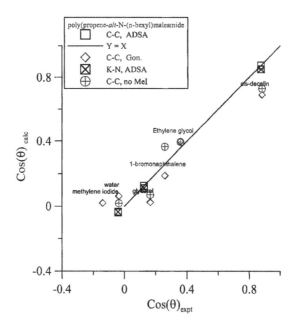

Figure 4. Cosine of the calculated contact angle *versus* cosine of the experimental contact angle for hexyl polymer. Line represents *y=x*. Fitting error is reduced when methylene iodide, which exhibits stick–slip spreading, is excluded from the fitted data set.

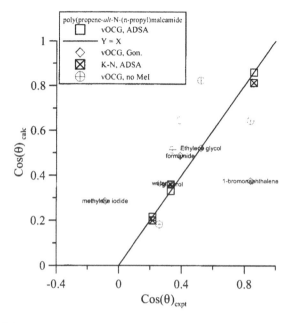

Figure 5. Cosine of the calculated contact angle *versus* cosine of the experimental contact angle for propyl polymer. Line represents *y=x*. Fitting error is reduced when methylene iodide, which exhibits stick–slip spreading, is excluded from the fitted data set.

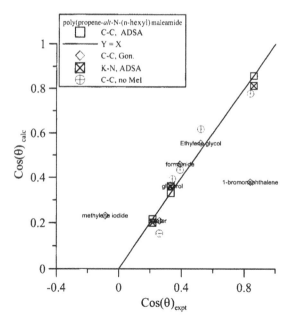

Figure 6. Cosine of the calculated contact angle *versus* cosine of the experimental contact angle for hexyl polymer. Line represents *y=x*. Fitting error is reduced when methylene iodide, which exhibits stick–slip spreading, is excluded from the fitted data set.

F. M. Etzler

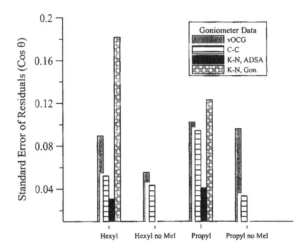

Figure 7. Standard error of the residuals (cos θ_{calc} – cos θ_{expt}) for fits of goniometer data to either the van Oss, Chaudhury and Good model, or the Chang and Chen model. Data are shown in Figs 3–6. Removal of methylene iodide from the data set improves fits for both models. The Chang–Chen model has generally smaller residuals when compared to the van Oss, Chaudhury and Good model. The data from the Kwok and Neumann model show comparable errors. The ADSA data were used by Kwok and Neumann for the purposes of fitting their model. The goniometer data used are for all liquids that Kwok and Neumann studied except for those which showed stick–slip behavior (e.g., MeI).

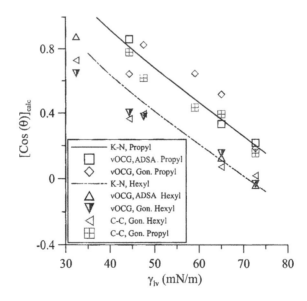

Figure 8. Cosine of calculated contact angle *versus* liquid surface tension for both propyl and hexyl polymers. Methylene iodide data have been excluded from the above plot. The predicted contact angles calculated from each of the three models discussed in this paper are comparable but the contact angles calculated from the van Oss, Chaudhury and Good model deviate the most when compared to the other models.

Table 2.
Calculated values of the surface free energy and surface free energy components (mJ/m^2) for both propyl and hexyl polymers

	Hexyl					Propyl			
	γ^{LW}	γ^+	γ^-	γ^T		γ^{LW}	γ^+	γ^-	γ^T
ADSA	28.34	0.03	3.33	28.98	ADSA	38.38	0.01	8.75	38.85
Gon.	16.28	1.54	3.97	21.22	Gon.	21.01	1.046	14.64	28.84
Gon. no MeI	21.86	0.60	3.50	24.75	Gon. no MeI	35.53	0	12.30	35.52
K–N				27.8	K–N				36.9
	P^L	P^a	P^b	γ^T		P^L	P^a	P^b	γ^T
ADSA	7.524	0.316	-2.607	29.13	ADSA	6.061	2.252	-4.633	28.80
Gon.	6.779	-0.858	-5.664	18.12	Gon.	5.876	2.791	-4.157	28.87
Gon. no MeI	6.932	-0.173	-4.312	23.28	Gon. no MeI	7.402	2.223	-2.721	33.44

data quality. It is clear that when data from methylene iodide, which exhibits stick–slip spreading, is excluded the quality of the fit is imporved particularly when the van Oss, Chaudhury and Good model is used.

Figure 8 shows the calculated cosine of contact angle *versus* liquid surface tension. Contact angle estimates resulting from application of each of the models discussed above are approximately of equal quality. The proper choice of for a given soild is important for the determination of surface free energy and its components.

Table 2 shows the estimates of the solid surface free energy and surface free energy parameters. The estimates of the total surface free energy calculated from each model are approximately equal provided data from methylene iodide which spreads in a stick–slip fashion are excluded. The spread of the estimates is only a few surface free energy units. The goodness of fit, as discussed previously, is also affected when methylene iodide data are included. As the parameters calculated from each model are unknown by independent means, the accuracy of the models cannot be assessed.

The estimates of the acid–base parameters appear to be less affected by inclusion of data not fully meeting the requirements necessary for application of Young's equation when the van Oss, Chaudhury and Good model is used. Both the van Oss, Chaudhury and Good model, and the Chang and Chen model indicate that the polymers are primarily basic. The Chang–Chen model when applied to the propyl polymer suggsts a more amphoteric nature for this material, however.

4. SUMMARY AND CONCLUSIONS

In this work, contact-angle data collected for various liquids on two polymer surfaces have been used to compare estimates of solid surface free energy and its components using models that have been frequently employed in the literature for this purpose. The ability of the models to be predictive of observed contact angles is also considered.

When the chosen liquids conform to the set of assumptions advanced by Kwok and Neumann (and which are required for application of Young's equation) the various models yield similar estimates of surface free energy within a few surface free energy units. The Chang and Chen model shows more deviation when applied to the propyl polymer.

When appropriate liquids are chosen, all models yield similar estimates of the contact angle. The inclusion of data from a liquid that exhibits stick–slip behavior, for instance, may significantly affect the estimates of surface free energy and its components. If only goniometer data are used, then the stick–slip behavior of a chosen liquid may go unnoticed.

A poor fit of the experimental data to the model can indicate that one or more of the chosen liquids do not conform sufficiently to the required assumptions. In many instances, previously reported in the literature, only three liquids have been used to estimate surface free energy. In such an instance, it may be possible to obtain inaccurate (with respect to the chosen model) estimates of surface free energy and its components. The use of a larger number of liquids, as suggested earlier by Dalal [52], would allow for better estimates of goodness of fit. ADSA analysis, when possible, could additionally be used to assess the quality of the contact angle data. A significant risk exists for obtaining incorrect values of surface free energy and its components when the nature of the solid–liquid interaction cannot be determined. The experimental data suggest that finding enough liquids conforming to the assumptions listed by Kwok and Neumann may be a difficult task in at least some instances.

The acid–base parameter estimates when using the van Oss, Chaudhury and Good model appear to be somewhat more robust than when using the Chang and Chen model. It is also noted that the Chang and Chen model fits each data set better that the van Oss, Chaudhury and Good model. It is unclear if the parameter estimate robustness of the van Oss, Chaudhury and Good is a desirable feature of the model or not. The robustness could be due to the inflexibility of this model to adjust to the data. The Chang and Chen model may describe the nature of the acid–base interactions more accurately.

At this time, it appears the each of the models discussed in this work can be used for the determination of solid surface free energies. Each of the models yields similar estimates of surface free energy.

The assumptions required for application of Young's equation and discussed extensively by Kwok and Neumann suggest that liquids chosen for study be selected carefully. Goodness of fit should be reported along with the parameter estimates. Furthermore, it may be best to use more than three probe liquids as suggested earlier by Dalal. If alternate sets of liquids result in significantly different parameter estimates then specific interactions of some of the chosen liquids may be suspected. Liquids should be chosen so that at least one non-polar liquid is selected, although it is probably better to select more than one. The other liquids should be as diverse as possible.

Further work is, also, required to determine the relative merits of the different proposed acid–base scales.

REFERENCES

1. F. M. Etzler, in: *Contact Angle, Wettability and Adhesion*, Vol. 3, K. L. Mittal (Ed.), p. 219. VSP, Utrecht (2003).
2. W. A. Zisman, in: *Contact Angle, Wettability and Adhesion*, Adv. Chem. Ser. No. 43, p. 1. American Chemical Society, Washington, DC (1964).
3. F. M. Fowkes and M. A. Mostafa, *Ind. Eng. Chem. Prod. Res. Dev.* **17**, 3 (1978).
4. F. M. Fowkes, K. L. Jones, G. Li and T. B. Lloyd, *Energ. Fuels* **3**, 97 (1989).
5. F. M. Fowkes, *J. Adhesion Sci. Technol.* **1**, 7 (1987).
6. F. M. Fowkes, *J. Colloid Interface Sci.* **28**, 493 (1968).
7. F. M. Fowkes, *J. Phys. Chem.* **66**, 382 (1962).
8. R. J. Good, in: *Contact Angle, Wettability and Adhesion*, K. L. Mittal (Ed.), p. 3. VSP, Utrecht (1993).
9. L. A. Girifalco and R. J. Good, *J. Phys. Chem.* **61**, 904 (1957).
10. C. J. van Oss, *Interfacial Forces in Aqueous Media*. Marcel Dekker, New York, NY (1994).
11. F. Chen and W. V. Chang, *Langmuir* **7**, 2401 (1991).
12. D. Berthelot, *Compt. Rend.* **126**, 1857 (1898).
13. W. V. Chang and X. Qin, in: *Acid-Base Interactions: Relavence to Adhesion Science and Technology*, Vol. 2, K. L. Mittal (Ed.), p. 3. VSP, Utrecht (2000).
14. A. W. Adamson, *The Physical Chemistry of Surfaces*. Wiley, New York, NY (1990).
15. J. Panzer, *J. Colloid Interface Sci.* **44**, 142 (1973).
16. D. K. Owens and R. C. Wendt, *J. Appl. Polym. Sci.* **13**, 1741 (1969).
17. S. Wu, *J. Polym. Sci., Part C* **34**, 265 (1971).
18. S. Wu, *Polymer Interface and Adhesion*. Marcel Deker, New York, NY (1982).
19. F. M. Fowkes, *J. Adhesion* **4**, 155 (1972).
20. C. Della-Volpe and S. Siboni, in: *Acid-Base Interactions: Relevance to Adhesion Science and Technology*, Vol. 2, K. L. Mittal (Ed.), p. 55. VSP, Utrecht (2000).
21. C. Della-Volpe, *J. Colloid Interface Sci.* **195**, 121 (1997).
22. C. Della-Volpe, A. Deimichei and T. Ricco, *J. Adhesion Sci. Technol.* **12**, 1141 (1998).
23. C. J. van Oss, in: *Acid–Base Interactions: Relevance to Adhesion Science and Technology*, Vol. 2, K. L. Mittal (Ed.), p. 173. VSP, Utrecht (2000).
24. D. Y. Kwok, *Colloids Surfaces A* **156**, 191 (1999).
25. D. J. Gardner, S. Q. Shi and W. T. Tze, in: *Acid-Base Interactions: Relevance to Adhesion Science and Technology*, Vol. 2 K. L. Mittal (Ed.), p. 363. VSP, Utrecht (2000).
26. M. D. Vrbanac and J. C. Berg, in: *Acid-Base Interactions: Relevance to Adhesion Science and Technology*, K. L. Mittal and H. R. Anderson Jr. (Eds.), p. 67, VSP, Utrecht (1991).
27. R. S. Drago, G. C. Vogel and T. E. Needham, *J. Amer. Chem. Soc.* **93**, 6014 (1971).
28. V. Gutmann, A. Steininger and E. Wychera, *Montaschr. Chem.* **97**, 460 (1966).
29. V. Gutmann, *The Donor-Acceptor Approach to Molecular Interaction*. Plenum, New York, NY (1978).
30. F. L. Riddle and F. M. Fowkes, *J. Amer. Chem. Soc.* **112**, 3259 (1990).

31. D. Y. Kwok and A. W. Neumann, in: *Acid-Base Interactions: Relevance to Adhesion Science and Technology,* Vol. 2, K. L. Mittal (Ed.), p. 91. VSP, Utrecht (2000).
32. D. Li and A. W. Neumann, *J. Colloid Interface Sci.* **148**, 190 (1992).
33. D. Li, M. Xe and A. W. Neumann, *Colloid Polym. Sci.* **271**, 573 (1993).
34. D. Y. Kwok and A. W. Neumann, *Colloids Surfaces A* **89**, 181 (1994).
35. C. A. Ward and A. W. Neumann, *J. Colloid Interface Sci.* **49**, 286 (1974).
36. D. Li, J. Gaydos and A. W. Neumann, *Langmuir* **5**, 293 (1989).
37. D. Li and A. W. Neumann, *Adv. Colloid Interface Sci.* **49**, 147 (1994).
38. D. Li, E. Moy and A. W. Neumann, *Langmuir* **6**, 885 (1990).
39. J. Gaydos and A. W. Neumann, *Langmuir* **9**, 3327 (1993).
40. D. Li and A. W. Neumann, *Langmuir* **9**, 3728 (1993).
41. D. Li and A. W. Neumann, *J. Colloid Interface Sci.* **137**, 304 (1990).
42. R. N. Wenzel, *Ind. Eng. Chem.* **28**, 988 (1936).
43. A. B. D. Cassie, *Trans. Faraday Soc.* **40**, 546 (1944).
44. S. Baxter and A. B. D. Cassie, *J. Textile Inst.* **36**, 67 (1945).
45. A. B. D. Cassie, *Disc. Faraday Soc.* **3**, 11 (1948).
46. K. Grundke, T. Bogumil, T. Gietzelt, H.-J. Jacobasch, D. Y. Kwok and A. W. Neumann, *Progr. Colloid Polym. Sci.* **101**, 58 (1996).
47. G. Antonow, *J. Chim. Phys.* **5**, 372 (1907).
48. D. Y. Kwok and A. W. Neumann, *J. Phys. Chem. B.* **104**, 741 (2000).
49. D. Y. Kwok, D. Li and A. W. Neumann, *Langmuir* **10**, 1323 (1994).
50. D. Y. Kwok, T. Gietzelt, K. Grundke, H.-J. Jacobasch and A. W. Neumann, *Langmuir* **13**, 2880 (1997).
51. P. Chen, D. Y. Kwok, R. M. Prokop, O. I. del Rio, S. S. Susnar and A. W. Neumann, in: *Drops and Bubbles in Interfacial Science,* D. M. R. Miller (Ed.). Elsevier, New York, NY (1998).
52. E. N. Dalal, *Langmuir* **3**, 1009 (1987).

Contact Angle, Wettability and Adhesion, Vol. 4, pp. 237–264
Ed. K.L. Mittal
© VSP 2006

Determination of solid surface tension at the nano-scale using atomic force microscopy

JAROSLAW DRELICH,[1,*] GARTH W. TORMOEN[1] and ELVIN R. BEACH[2]

[1]*Department of Materials Science and Engineering, Michigan Technological University, Houghton, MI 49931, USA*
[2]*The Dow Chemical Company, Analytical Sciences, 1897 Bldg., Midland, MI 48640, USA*

Abstract—Engineered surfaces, such as thin inorganic or organic films, self-assembled organic monolayers and chemically-modified polymeric surfaces, cannot be melted, dissolved, or fractured; therefore, their surface/interfacial tension (γ) cannot be determined using conventional surface tension measurement techniques. New surface tension characterization methods need to be developed. Atomic force microscopy (AFM) is capable of solid surface characterization at the microscopic and sub-microscopic scales. As demonstrated in several laboratories in recent years, and reviewed in this paper, it can also be used for the determination of surface tension of solids from pull-off force measurements. Although a majority of the literature γ results were obtained using either Johnson–Kendall–Roberts (JKR) or Derjaguin–Muller–Toporov (DMT) models, a re-analysis of the published experimental data presented in this paper indicates that these models are often misused and/or should be replaced with the Maugis–Dugdale (MD) model. Additionally, surface imperfections in terms of roughness and heterogeneity that influence the pull-off force are analyzed based on contact mechanics models. Simple correlations are proposed that could guide in the selection and preparation of AFM probes and substrates for γ determination. Finally, the possibility of AFM measurements of solid surface tension using real-world materials is discussed.

Keywords: Atomic force microscopy; chemical force microscopy; adhesion; surface energy; surface tension.

1. INTRODUCTION

Surface tension (γ) is a fundamental parameter of solids, the value of which depends strongly on the fluid in which the solid is immersed. Surface tension is the mathematical equivalent of surface free energy and the former term is used in this paper due to its common appearance in the literature. γ is one of the most important parameters affecting adhesion and friction between materials, colloidal stability of suspensions and performance of "smart" materials. For example, fine, particles contact a variety of surfaces and interfaces during the course of their storage,

---

*To whom correspondence should be addressed. Tel.: (1-906) 487-2932; Fax: (1-906) 487-2934; e-mail: jwdrelic@mtu.edu

transport/flow, manufacture, processing and use in both naturally occurring, as well as industrial operations. Adhesion (and, thus, surface tension as will be shown later) influences particle deposition, separation, processing, transport and delivery. Examples of the significance of particle–surface adhesion are often discussed in relation to cleaning of electronic materials, oil recovery with surfactant solutions, detergency, soil remediation with water, flotation of minerals, deinking of recycled paper, transportation of slurries and contaminated liquids (due to deposition of fines at the walls of pumping and storage equipment) and many others. In recent years, simulation and control of organized particle layer(s) and adsorption of colloidal particles on selected substrates have been intensely studied to fabricate nano- and micro-structured materials. Both fabrication of nanostructures and their stability depend on the interfacial properties of materials. Progress in manufacturing of micro- and nano-devices cannot continue without understanding and control of solid surfaces at their scale of manufacture. At the scale of micro- and nano-devices, the material behavior is influenced more by surfaces than by bulk: a consequence of the large surface-to-volume ratio. Furthermore, operations such as coating and lubrication are influenced by γ. Specifically, the spreading of liquids on solids is the result of competition between the liquid surface tension, the solid surface tension and the solid–liquid interfacial tension.

Surface–interfacial tension measurements for solids (mainly of macroscopic dimensions) have been the subject of research for many years. Direct and indirect γ measurement methods, such as the zero-creep method [1, 2], crystal cleavage [3, 4], change in lattice constant [5, 6], solubility studies [7, 8], heat of immersion measurements [9, 10], tensiometric techniques for molten solids [11, 12] and others, have been used on a number of different materials. The theoretical background and experimental techniques for γ determination have been reviewed previously [13, 14] and will not be repeated here.

We live in a world of materials the surfaces of which are engineered to have specific wetting, electronic, optical and other properties. Material modification is often accomplished by tailoring the material surface region at a depth of a single monolayer to several micrometers. This can be achieved by adsorption of organic modifiers, deposition of thin inorganic or organic films, surface etching, initiation of chemical reaction at the surface, etc. Many engineered surfaces cannot be melted, dissolved, or fractured; therefore, their surface/interfacial tension cannot be determined through any of the "conventional" (the above-mentioned) techniques. Additionally, these conventional techniques are usually applicable to macroscopic solids, whereas the current trend in miniaturization of products and devices poses the need for examination of nano-surfaces or surfaces with nano-heterogeneity. New surface tension determination methods need to be developed to meet these challenges. Two techniques based on either (i) contact angles or (ii) adhesion force measurements by atomic force microscopy (AFM) have been under intensive development by several research groups in recent years [14].

The backgrounds for the contact angle technique and direct pull-off force measurements using AFM are briefly reviewed in the following sections. The limitations of the contact angle technique for the purpose of γ determination are also emphasized and the need for development of the AFM technique is presented. We will then briefly discuss the basics of the AFM technique for the determination of the solid surface tension components, a research direction that might have a significant impact on the validity and accuracy of the contact angle measurement approaches.

In this paper, we also emphasize the common technical problems faced by the researchers during normalization and interpretation of the AFM pull-off force measurements. AFM pull-off force measurements have not been widely accepted as a reproducible technique for several reasons including varying surface roughness and heterogeneity characteristics and loading/unloading conditions. Without strict control of experimental conditions and quality of interacting surfaces, pull-off force measurements can vary widely for similar systems. This paper reviews progress in application of the AFM technique as an analytical tool in measurements of surface tension of engineered surfaces, and several new aspects of these measurements are presented.

2. DETERMINATION OF SOLID SURFACE TENSION FROM CONTACT ANGLES

The Lewis acid–base interaction (semi-empirical) theory proposed by van Oss, Chaudhury and Good [15–17], which considers the surface tension as the sum of apolar Lifshitz–van der Waals component (γ^{LW}) and the polar Lewis acid–base component (γ^{AB}), has received significant support over the last two decades [18]. According to this theory, the surface tension for a solid (γ_S) is expressed as follows:

$$\gamma_S = \gamma_S^{LW} + \gamma_S^{AB} \tag{1}$$

and

$$\gamma_S^{AB} = 2\sqrt{\gamma_S^+ \gamma_S^-} \tag{2}$$

where γ^+ and γ^- are the electron-acceptor (acidic) and electron-donor (basic) parameters, respectively.

To determine the surface tension of a solid, the contact angles are measured with at least three probing liquids (different liquids for which the surface tension components are defined and at least two of these liquids are polar). Then, the following equation is used to calculate the γ components of a particular solid [15–17]:

$$W_A = (1 + \cos\theta)\gamma_L^{TOT} = 2\left(\sqrt{\gamma_S^{LW}\gamma_L^{LW}} + \sqrt{\gamma_S^+ \gamma_L^-} + \sqrt{\gamma_S^- \gamma_L^+}\right) \tag{3}$$

where W_A is the work of adhesion, θ is the measured (advancing) contact angle, and subscripts L and S refer to the liquid and solid, respectively.

The γ^{LW} and γ^{AB} components of surface tension (including γ^+ and γ^- parameters) are known for a number of liquids [15–17]. These values, however, are relative and were determined based on the assumption that $\gamma^+ = \gamma^- = 25.5$ mJ/m^2 for water [19]. This assumption affects the relative values of the electron-donor and electron-acceptor surface/interfacial tension parameters for the examined surface, though it does not influence the resulting value of the total surface/interfacial tension. A detailed critical analysis of the van Oss–Chaudhury–Good theoretical model is provided by Della Volpe et al. [20]. The γ values determined for a number of different solids have been reviewed in two different monographs [16, 17].

The experimental approach to determine γ of solids based on contact angle measurements is relatively simple for flat, smooth, homogeneous and inert substrates. Problems arise with substrates that (i) are reactive with, or sensitive to, the environment (common case for solids in contact with many liquids or solutions) and/or (ii) have small dimensions (e.g., small or powdered samples, patterned heterogeneous substrates with microscopic domains). In other words, the technique is limited to materials that are stable in the environment of the probing liquids. Any ionization process of functional groups, chemical instability of the solid, or dissolution of the solid by the probing liquid invalidates this method for γ determination. Also, the interfacial tension between a solid and a solution cannot be deduced from contact angle measurement results.

The contact angle technique probes "average" surface properties of heterogeneous surfaces and thus is unable to distinguish the discrete character of such surfaces at the microscopic and sub-microscopic levels. Nanotechnology often employs molecular and nanoscopic structures; the surface properties of individual surface domains are of particular importance for the control of surface-related operations and material performance. The direct method of pull-off force measurements using AFM is a very attractive technique for surface characterization of such engineered materials. This technique can also help to validate the γ results determined from contact angle measurements and Lifshitz–van der Waals/Lewis acid–base interaction theory.

3. BASICS OF ATOMIC FORCE MICROSCOPY

The principles of the AFM have been well described in the literature [21–23]; here, only the major features of surface force measurements are reviewed. The AFM measures the deflection of a cantilever spring with a sharp cantilever tip (tip radius is usually 10 to 100 nm) or an attached particle (a particle with a diameter ranging from about 2 to 20 μm or larger is glued to the end of the cantilever) as a function of displacement from a horizontal position (Fig. 1). The deflection, monitored by a laser-photodiode system, is related to the forces acting between the probing tip and the substrate.

Figure 1 shows an idealized deflection vs. distance curve obtained by AFM. The measurement starts at a large tip–surface separation in the so-called non-touching

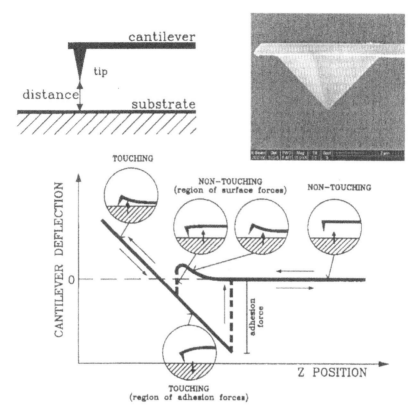

Figure 1. Schematic of a typical cantilever deflection *vs.* vertical position of cantilever curve in the AFM measurements of surface forces. Photo shows a probing pyramidal tip of triangular cantilever.

regime. Then, with or without a slight cantilever deflection, depending on the long-distance tip–surface interactions, the sample (surface) approaches the tip following the horizontal line moving right to left in the deflection *vs.* distance curve in Fig. 1 (distance between the sample and the tip is controlled by the piezoelectric scanner). At a certain point, the transition from non-touching to touching occurs and the tip jumps onto the sample surface. This transition, shown in Fig. 1 in a simplified drawing, can be complex when the measurement is carried out in a liquid. From this part of the curve, the long-distance surface forces can be calculated [23].

Moving the sample still further causes deflection of the cantilever the same distance that the sample is moved. This is called the touching regime or constant compliance region represented by the diagonal line in the left part of the deflection *vs.* distance curve. Upon retracting the surface from the tip, i.e.; going towards the right in the deflection *vs.* distance curve, the cantilever again moves with the surface. The cantilever deflects towards the surface due to adhesion force before the tip breaks contact with the surface, going through the lowest point in the deflection *vs.* distance curve. At this "jump-off" point, the tip completely loses contact with the surface and the cycle is complete. The force required to pull the tip off the substrate surface is called pull-off or adhesion force (F) and is calculated from Hooke's law:

$$F = k\Delta x \qquad (4)$$

The accuracy of the F measurement, which is directly related to the solid surface tension (as will be shown later), depends on the precision with which both the spring constant of cantilever (k) and its deflection (Δx) during pull-off force measurements are determined. The laser beam-photodiode detector systems of the AFM instruments are usually capable of recording the deflection of cantilever to sub-Ångstrom precision and the major error in determination of pull-off force is associated with the k value determination (errors in responses of piezoelectric scanner resulting from piezoelectric scanner hysteresis and creep may also cause inaccuracies in the F value determination).

4. CANTILEVER ELASTIC CONSTANT

Two basic types of cantilevers, triangular (V-frame) and single-beam cantilevers (rectangular or trapezoidal) [21, 22], are commonly used for imaging of substrate topography and are also used for force measurements. V-frame cantilevers are made of polycrystalline silicon or silicon nitride. They are available with a variety of elastic constants, from a fraction of N/m to several N/m. Single-beam silicon cantilevers are stiffer with spring constant varying from several N/m to tens of N/m. The expected strength of adhesion between the tip of the cantilever and the substrate dictates the choice of cantilever (i.e., strong adhesion forces require cantilevers with large spring constants).

Estimates for the cantilever spring (elastic) constant for every type of cantilevers is provided by the manufacturer. The values specified by producers are, however, average spring constants, usually determined for a batch of cantilevers. Small variations in the dimensions of the cantilever during manufacture result in significant changes in k value; therefore, the spring constant for a cantilever may deviate from the specified average value by as much as 10–20% for V-frame cantilevers (which are very regular in shape) and up to 100% for single beam cantilevers (which are more irregular in shape). In quantitative force studies, such as those discussed in this paper, it is necessary to determine the cantilever spring constant more accurately.

Among the several proposed approaches, three methods are commonly used for spring constant determination (see Ref. [24] for review). The first method is based on conventional static mechanics, and the spring constant is calculated from the cantilever dimensions and Young's modulus of materials used for cantilever fabrication [24]. This method is more suitable for cantilevers that are regular in shape and requires that the material properties are known and uniform over the entire cantilever. The presence of sub-microscopic defects is detrimental to the mechanical properties of microscopic cantilevers and can deviate from what is expected for bulk material. The thickness of the coating (gold, for example), if applied, must also be known.

The second method relies on measurement of the resonant frequency of the cantilever before and after an end-mass is attached [25]. A spherical particle, typically tungsten, is attached to the cantilever through capillary forces. The drawback of this non-destructive method is that particles often do not stay attached to the cantilever without glue. Also, precise weights of particles are required for accurate determination of k.

The third method is called the thermal noise method [26]. Although this method is restricted to cantilevers with $k < 1$ N/m, it has become very popular among service laboratories due to its simplicity. The method relies on measurements of mean square displacement of the unloaded cantilever due to thermal motion. The details of this method as well as two other above-mentioned methods have been reviewed by Sader [24].

These days, pre-calibrated cantilevers can be purchased from several companies, so calibration is usually no longer necessary for researchers. It should be recognized, however, that the experimental spring constants are always determined with limited precision and errors of typically 5–10% are encountered, though a 20% error is not unusual.

5. AFM PROBES/TIPS AND THEIR SIZE

Characterization of the probe size and its shape, including determination of surface defects and nanoroughness at the tip apex, is an important task for analysis of the results from AFM pull-off force measurements. A detailed characterization of the system needs to be done in order to predict the contact area between the tip and substrate, a parameter that is essential for normalization of the experimental data.

Conventional cantilevers have tips made of only two types of materials: silicon or silicon nitride; therefore, tips are often modified or replaced with particles to broaden the spectrum of materials that can be studied with the AFM technique. The tip can be coated with a thin film of metal; however, due to high surface energy of metals the control of their surface chemistry is difficult. For example, oxide films grow very fast on a majority of deposited metals. Organic contaminants adsorb very quickly on both metals and oxides when exposed to a normal laboratory environment. It is also possible to deposit a polymer on the cantilever tip, although researchers have not explored this option.

The surface of silicon or silicon nitride can be directly modified with alkylsilanes, but the quality of the adsorbed molecular layers is often questionable. Instead, tips are often coated with a thin film of gold, which is next modified with self-assembled monolayers (SAMs) of thiols with a desirable functionality. SAMs of thiols (X–$(CH_2)_n$–SH) form a very regular, close-packed structure by reacting the –SH end-group with the gold surface whereas the other end-group (X–) is exposed to the environment. Force measurements made with SAM-modified tips are categorized

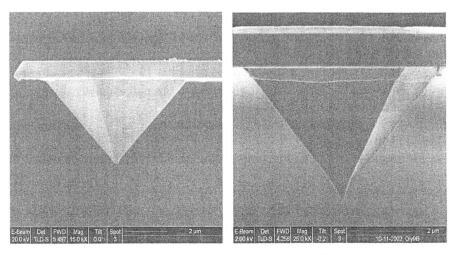

Figure 2. Field emission scanning electron micrographs of two different cantilever tips used in chemical force microscopy studies.

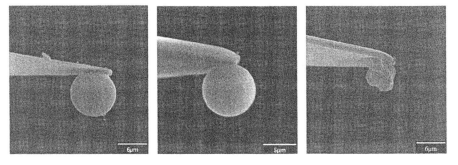

Figure 3. SEM images of colloidal particles attached to tipless AFM cantilevers. The microspheres shown are borosilicate glass (left), polystyrene (middle) and a peptide particle (right).

under chemical force microscopy (CFM) and have become an area of intensive fundamental research (see review by Noy *et al.* [27]). Two examples of cantilever tips used in CFM studies are shown in Fig. 2.

A colloidal particle with a diameter of 2–20 μm (or larger) can also be attached to the cantilever, typically by gluing or, in some cases, melting the particle to the cantilever (both tipless cantilvers and cantilevers with regular tips are used for this purpose). This allows nearly limitless combinations of materials to be used in force measurements. Figure 3 shows examples of colloidal probes.

A problem with this practice is that not many materials are commercially offered in the shape of spherical colloidal particles, and often new methodology for colloidal particle precipitation must be developed for this purpose (as, for example, in Ref. [28]). Making spherical particles is not always possible for some materials;

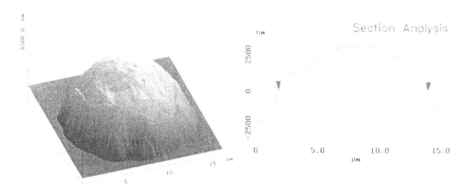

Figure 4. Three-dimensional image of the apex of a 12.1-μm-diameter polyethylene microsphere attached to an AFM cantilever and the corresponding cross-sectional view of this particle.

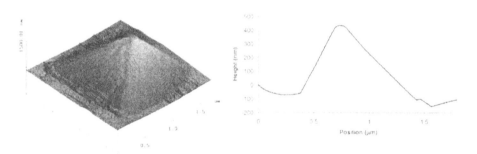

Figure 5. AFM images of the cantilever tips obtained by blind tip reconstruction method. (Left) Image of pyramidal tip obtained over the spikes. (Right) Cross-section of the left image.

therefore, force measurements are sometimes carried out with irregular particles, such as in the right image in Fig. 3, which makes quantification and interpretation of the experimental data difficult or even impossible. Even "spherical" particles usually show rough surfaces at the nanoscale that can affect the contact area between the probe and the substrate during the AFM pull-off force measurements.

The radius of curvature for AFM cantilever tips is less than 100 nm and standard thermionic emission scanning electron microscopes (SEMs) cannot resolve the edges of such small features without reducing the signal-to-noise ratio to the point that the object is indistinguishable from the background. Field emission-SEM (FE-SEM) techniques have much better resolution than thermionic emission SEM, and are often used to characterize the AFM tip shape and topography. Two alternative methods, both utilizing the AFM instrument, are used in our laboratory.

In the first method, a colloidal probe can be scanned directly and then imaged with another (much sharper) AFM cantilever. Figure 4 shows an image of a polyethylene microsphere obtained by this technique. A problem with this approach is that positioning of the scanning cantilever tip over the top of the colloidal probe is difficult using the optical microscope attached to the AFM instrument. Also it is difficult to mount the cantilever with the scanned probe such that it is stable during imaging.

The second method, called inverse imaging or blind tip reconstruction (BTR), developed by Villarubia [29], is more versatile and is used frequently in our laboratory. In BTR, the AFM tip is scanned in contact mode over an ultrasharp silicon grating (especially fabricated arrays of sharp silicon spikes or asperities such as in TGT or TGG gratings, respectively, offered by NT-MDT, Moscow, Russia). The grating features must have a higher aspect ratio than the shape of the AFM tip in order to produce an inverted image of the tip rather than the image of the spike/asperity. Also, radii of the grating feature tip or edge should be small (they are less than 10 nm for both TGT and TGG).

Usin Deconvo software (NT-MDT, Moscow, Russia) based on mathematical morphology operations, the radius of curvature of the tip is determined from images generated by the BTR method. Figure 5 shows an example of AFM image and corresponding cross section of this image produced by scanning a gold-coated AFM tip over a single spike of the TGT grating.

Noise artifacts in the image have a large influence on the results of BTR experiments. Low scan rates should be used and the AFM instrument should be placed on an anti-vibration support; cement plate suspended by bungee cords (offered by Digital Instruments) is probably the best choice in eliminating vibrations. Even after such precautions, sharp changes in height of the grating features can produce artifacts in the image, seen as streaks in Fig. 5. An additional complication in the BTR method is the possibility of scratching the gold-coated tip (or colloidal particle) with the harder silicon spike/asperity. To avoid scratching, probe–contact forces during imaging should be minimized.

6. CORRELATION BETWEEN PULL-OFF FORCE AND SOLID SURFACE TENSION

The equilibrium work of adhesion (W_A) is defined as the negative of the Gibbs free energy change per unit area (ΔG) of the interacting surfaces, and is expressed by the Dupré equation [16]:

$$W_A = -\Delta G = \gamma_{13} + \gamma_{23} - \gamma_{12}, \tag{5}$$

where subscripts 1, 2 and 3 describe three different phases of the system: solid 1, solid 2 and fluid.

If we replace one surface with a spherical particle (or probing tip), such as systems used in AFM studies, the relation between the work of adhesion, radius of particle (R) and adhesion (pull-off) force (F) can be described by the Derjaguin approximation:

$$F = 2\pi R W_A, \tag{6}$$

which is applicable to perfectly rigid particles in contact. The Derjaguin approximation is valid for systems where the separation distance between the sphere and the substrate is much shorter than the radius of the sphere (probe). In measurements of adhesion forces this assumption is always valid but this approximation can fail in the examination of (long-range) non-contact forces. However, in reality, particles and/or substrates deform elastically and/or plastically under applied loads during adhesion and particle–substrate analysis requires more accurate contact mechanics models that include a physical deformation component.

Typical AFM measurements utilize either spherical particles glued to cantilevers or commercial cantilever tips with rounded ends (as discussed in the previous section), and such systems are modeled as a sphere in contact with a flat surface. Many practical systems deviate from this idealized geometry, often causing problems in interpretation of measured AFM pull-off forces. As a result, significant effort is always concentrated on preparation of as close-to-perfect sphere-flat systems as possible in order to simplify the analysis of experimental pull-off force results. However, this has not always been accomplished and many literature reports lack complete characterization of probes and substrates (i.e., probe geometry and dimension, probe and substrate surface roughness, probe and substrate surface heterogeneity, etc.).

Two contact mechanics models, developed by Johnson *et al.* [30] and Derjaguin *et al.* [31] and named JKR and DMT model, respectively, are frequently used by researchers to interpret the pull-off forces measured by the AFM technique. These models have been reviewed in detail by many authors [23, 32–36]. In general, both JKR and DMT models apply to particle–substrate systems where the following assumptions are met: (i) deformations of materials are purely elastic, described by classical continuum elasticity theory, (ii) materials are elastically isotropic, (iii) both Young's modulus and Poisson's ratio of materials remain constant during deformation, (iv) the contact diameter between the particle and the substrate is small compared to the diameter of the particle, (v) a paraboloid describes the curvature of the particle in the particle–substrate contact area, (vi) no chemical bonds are formed during adhesion and (vii) contact area significantly exceeds molecular/atomic dimensions.

The difference between JKR and DMT models lies in the assumption of the nature of forces acting between the particle and the substrate. Johnson *et al.* [30] assumed in their model that attractive forces acted only inside the particle–substrate contact area, whereas Derjaguin *et al.* [31] included long-range surface forces operating outside the particle–substrate contact area. Both JKR and DMT models describe the correlation between the pull-off force (F) and the work of adhesion (W_A) through a simple analytical equation of the following form:

$$F = c\pi R W_A, \tag{7}$$

which leads to equation (8) for a symmetrical system (when interacting materials and their surfaces are the same):

$$\gamma = \frac{F}{2c\pi R},$$
(8)

where R is the radius of particle (probing tip) and c is a constant; $c = 2$ in the DMT model and $c = 1.5$ in the JKR model. Thus, knowing which continuum mechanics model applies to a particular system under study, and setting the operating conditions during the AFM measurements that satisfy the particular model, the γ value can be determined. Note that (i) both models, JKR and DMT, were developed based on the Hertz theory [37] and (ii) an analytical solution to the DMT model was provided by Maugis [36].

Which of the contact mechanics models should be selected for interpretation of the AFM pull-off forces is not always straightforward. In general, the DMT model is more appropriate for systems with hard materials having low surface energy and small probes. The JKR model applies better to softer materials with higher surface energy and larger probes. This generalization, however, does not bring the researcher any closer to the selection of the appropriate model and mistakes are often made (including our own [38–40]).

Maugis analyzed both the JKR and DMT models and suggested that the transition between these models could be predicted from the dimensionless parameter λ_M defined as [36]:

$$\lambda_M = \frac{2.06}{z_0} \sqrt[3]{\frac{R W_A^2}{\pi K^2}},$$
(9)

where z_0 is the equilibrium separation distance between the probe and substrate, R is the radius of the probe, W_A is the work of adhesion and K is the reduced elastic modulus for the particle–substrate system where:

$$\frac{1}{K} = \frac{3}{4}\left(\frac{1 - v_p^2}{E_p} + \frac{1 - v_s^2}{E_s}\right),$$
(10)

where v is the Poisson ratio, E is the Young's modulus, and p and s stand for probe (particle) and substrate, respectively.

For $\lambda_M \rightarrow \infty$ ($\lambda_M \geqslant 5$) the JKR model applies, whereas the DMT model is more appropriate for systems with $\lambda_M \rightarrow 0$ ($\lambda_M \leqslant 0.1$). The transition between these two models is described by the Maugis–Dugdale (MD) theory [36], and often for AFM pull-off force measurements, the MD model seems to be more appropriate than either the JKR or DMT model as is discussed in Section 9.

In the MD model, two parametric equations must be solved to describe the transition region between JKR and DMT models. These equations are [36]:

$$\overline{P} = \overline{A}^3 - \lambda_M \overline{A}^2\left(\sqrt{m^2 - 1} + m^2 \cos^{-1}\left(\frac{1}{m}\right)\right)$$
(11)

$$
\frac{\lambda_M \overline{A}^2}{2}\left[\sqrt{m^2 - 1} + (m^2 - 2)\cos^{-1}\left(\frac{1}{m}\right)\right]
$$
$$
+ \frac{4\lambda_M^2 \overline{A}}{3}\left[-m + 1 + \sqrt{m^2 - 1}\cos^{-1}\left(\frac{1}{m}\right)\right] = 1, \tag{12}
$$

where $m = b/a$ is the ratio between an outer radius at which the adhesion force no longer acts in the gap between the surfaces and the contact radius; $\overline{P}$ and $\overline{A}$ are the normalized load and normalized radius of the contact area, respectively:

$$
\overline{P} = \frac{P}{\pi R W_A} \tag{13}
$$

$$
\overline{A} = \frac{a}{\left(\frac{\pi R^2 W_A}{K^2}\right)^{\frac{1}{3}}} \tag{14}
$$

A problem with this model is that there is no single expression between a and P and both equations (11) and (12) must be solved simultaneously. Importantly, there is also no simple relation for pull-off force and iteration is needed to calculate the c value ($1.5 < c < 2$) in equation (7). Examples of calculated c values for different λ_M parameters are presented in Ref. [41].

7. PLASTIC DEFORMATION OF THE PROBE

Permanent deformation of the probe is a concern in pull-off force measurements. Loads generated during cantilever probe–substrate contact may result in plastic deformation of the material. An analysis performed by Maugis and Pollock [42] indicated that the applied load (P_p) for inducing full plasticity (irreversible deformation) of the particle with radius R is given by the following equation:

$$
P_p = 10\,800\pi \frac{R^2 Y^3}{E^2} \tag{15}
$$

and the radius of the circle of contact between the particle and the substrate during full plasticity could be predicted from:

$$
a_p = 60 \frac{R Y}{E}, \tag{16}
$$

where E is the Young's modulus and Y is the yield strength for the material.

Knowing the probe's radius of curvature (R) and mechanical properties of material (Y, E), the maximum load that can be applied in pull-off force experiments can be predicted from equation (15). For example, consider a sharp cantilever tip coated with a thick film of gold ($E = 77.5$ GPa, $Y = 0.2$ GPa, $\nu = 0.42$). For $R = 50$ nm, the full plasticity of gold will be reached at loads of $P_p = 113$ nN ($a_p = 7.7$ nm) or larger. For the commonly used V-frame cantilever with a spring constant of 0.58 N/m, the cantilever's mechanical bending of 0.2 μm in the constant compliance region of AFM force mode operation is sufficient to reach this level of load. This clearly indicates that deformation of the sharp probes, or at least of irregularities present on the probe apex [42, 43], should be a common phenomenon during the pull-off force measurements. The solution to this problem has often been the use of larger diameter probes. Two drawbacks must be considered, however, before switching to larger probes. First, larger probes are more likely to have rough surfaces and, if brittle, the nanoroughness will reduce the probe–substrate contact area and, therefore, complicate the interpretation of the measured pull-off forces. With soft materials, the nanoroughness of the probe may not be a problem as surface asperities can deform during pull-off force measurements, as discussed in the next part of this paper. Second, the pull-off forces increase with the size of the probe (see equation (7)) and the larger probes need to be attached to cantilevers with larger spring constants. It is important to select new cantilevers with a spring constant that scales approximately with the value of R, or at least less than R^2. Otherwise, if k scales with R^2 (or larger), the same (or worse) problems of probe deformation will be experienced.

Adhesion forces alone can cause plastic deformation of material (probe or substrate). Using the model presented in Ref. [42], the radius of the probe below which fully plastic deformation occurs can be calculated as:

$$R \approx 1.5 \cdot 10^{-4} \frac{W_A E^2}{(1 - \nu^2)^2 Y^3} \tag{17}$$

Consider a colloidal probe made of polystyrene ($\nu = 0.33$, $E = 2.55$ GPa, $Y = 10.8$ MPa) that is pressed against mica ($W_A = 120$ mJ/m^2) as studied in Ref. [43]. According to equation (17), the polystyrene probe will deform plastically if $R \leqslant 117$ nm. Such small colloidal probes made of polymers are not used in AFM pull-off force measurements. Nevertheless, these calculations indicate that any topographic irregularities on the probe surface that have a radius of curvature smaller than 117 nm can be deformed during polystyrene-mica adhesion contact. The experimental results presented in the literature support this possibility [38, 43].

8. SIZE LIMIT IN SURFACE IMPERFECTION AND/OR CONTAMINATION: PREDICTIONS

Surface defects such as roughness and heterogeneity (inherent or introduced by adsorbed/deposited species) are serious problems in the determination of the solid surface tension by the AFM technique. Due to the small dimensions of the probes, both nano-roughness and nano-heterogeneity influence the measured pull-off forces.

8.1. Roughness

It is well recognized that substrate roughness, for which asperities' dimensions are comparable to or smaller than the dimensions of the probe, as well as any

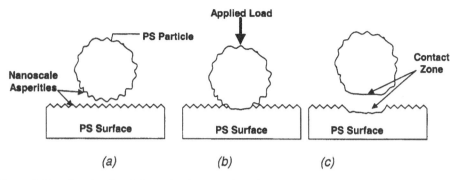

Figure 6. Drawing showing deformation in a particle–substrate system. The particle and substrate are shown with nanoscale roughness (not drawn to scale) before contact (a). After the particle has been pressed into the surface with some applied load (b), the nanoscale asperities in the contact area are flattened (c). The original shape of the particle (and/or substrate) is restored if the deformation was in response to viscoelastic properties of material, although sufficient time must be allowed before the next adhesion contact between particle and substrate is made. If the deformation is plastic, it remains permanent.

surface roughness at the probe apex, can affect the sphere-flat contact area to a degree that is not taken into account by simple contact mechanics models discussed thus far. If this happens, interpretation of the pull-off forces is difficult or even impossible. Both microscopic and sub-microscopic roughness effects can be eliminated in many experiments by appropriate selection and preparation of the substrates. Nano-roughness of colloidal probes and/or substrates often cannot be avoided. Nevertheless, as discussed in the literature [38, 43, 44], nanoscale asperities can be squeezed and flattened out during the pull-off force measurements (Fig. 6) to such a degree that the probe–substrate contact area is the same as predicted by one of the continuum contact mechanics models (usually JKR or MD model). Such experimental conditions are achievable with stiff cantilevers with which higher loads can be applied during the AFM pull-off force measurements.

We expect that the asperity may be flattened out by applying a sufficient load to initiate its plastic deformation. Therefore, if we adopt the Maugis–Pollock analysis [42], equation (15) predicts that at a specified load P, asperities with radius of curvature of R_a and smaller will flatten out:

$$R_a \leqslant \sqrt{\frac{PE^2}{10\,800\pi Y^3}} \qquad (18)$$

It is also possible that asperities of a certain size and shape may deform elastically/viscoelastically to such an extent that the probe establishes its full contact area with the substrate. This will happen if elastic deformation δ is comparable to the height of asperity h ($\delta \geqslant h$). This means that only asperities with a radius of curvature much larger than the height of the asperity can experience such deformation. If this possibility is analyzed in view of the JKR model [30], we conclude that:

$$h \leq \frac{a^2}{R_a} - \frac{2}{3}\sqrt{\frac{6\pi a W_A}{K}} \tag{19}$$

where

$$a = \sqrt[3]{\frac{R_a}{K}\left[P + 3\pi R_a W_A + \sqrt{6\pi R_a W_A P + (3\pi R_a W_A)^2}\right]} \tag{20}$$

Consider a polymeric probe interacting with a rigid material. Asperities on the probe surface are assumed to have $R_a = 100$ nm. For: $K = 65$ GPa, $W_A = 120$ mJ/m^2 and $P = 10$ μN, we calculate that asperities with a height of $h \leq 6.2$ nm can be completely deformed elastically/viscoelastically to allow for the probe–substrate contact area that is predicted by the JKR model. Alternatively, the load needed to produce desirable deformation of the surface topographic irregularities could be estimated by equatiom (19) if the dimensions of irregularities (R_a and h) are determined by AFM or another technique.

8.2. Heterogeneity

Using continuum contact mechanics models we can also estimate the maximum size of an impurity/surface heterogeneity that has a negligible effect on the measured pull-off force. In most of the systems (as we speculate), this will occur if the impurity and/or heterogeneity is smaller than 10% of the probe–substrate contact area: $\pi d^2/4 < (0.1\pi a_0^2)$; therefore, $d < 0.63 a_0$, where d is the diameter of the heterogeneity and a_0 is the probe–substrate contact radius at zero load. Assuming also that the system meets the terms of the DMT model, the size (diameter $= d$) of the heterogeneity/impurity that can be accepted in the pull-off force measurements can be estimated from the following equation:

$$d < 0.63\sqrt[3]{\frac{2\pi R^2 W_A}{K}} \tag{21}$$

For example, for $R = 50$ nm, $W_A = 100$ mJ/m^2 and $K = 63$ GPa (which are characteristic values determined for sharp tips coated with a gold film and modified with SAMs of thiols [39]), the heterogeneity with a diameter as small as 1.8 nm will influence the measured pull-off force. This means that even molecular defects in the structure of SAMs, which are common for thiol monolayers aligned into domains on gold surfaces [45], most likely affect the magnitude and variation of measured pull-off forces, and a single (reproducible) pull-off force value is practically impossible to obtain for such systems. Replacement of the sharp probe with a 10-μm-diameter colloidal probe, which more likely will operate in the regime described by the JKR model (therefore, factor 2 must be replaced with 6 under the square root in equation (21)), will increase this limit to about 90 nm. A heterogeneity size of 90 nm is more realistic in routine pull-off force measurements and the use of larger probes seems to be more appropriate in γ determination, from the perspective of surface heterogeneity. Unfortunately, colloidal probes of perfect sphericity with a smooth surface are difficult to manufacture. Cantilevers with sharp tips were more often used in the past, as they are more likely free of nano-irregularities on the tip apex. Also the effect of substrate surface roughness on measured pull-off force is reduced for sharp tips due to smaller probe–substrate contact area.

9. SOLID SURFACE TENSION VALUES DETERMINED BY THE AFM TECHNIQUE: CRITICAL REVIEW

Table 1 summarizes a number of data reported in the literature for adhesion force measurements made with atomic force microscopy. Table 1 includes data from pull-off force measurements for AFM cantilever tips, both unmodified and coated tips, as well as for some colloidal probes, performed either in liquid or gaseous environments. As for substrates, gold films modified with organic functionality, polymers, and inorganic materials were used in the AFM experiments. The spring constant is shown in Table 1 to predict possible loading conditions experienced during force measurement, a parameter often left unreported. In addition, the contact mechanics model used by the authors to calculate the work of adhesion (W_A) along with the reported W_A is included. This is contrasted with what is predicted for W_A from surface tension components determined from contact angle measurements for the tested materials. The final four columns depict calculations performed for the tested systems in order to determine what constant (c) is appropriate for determining W_A from force measurements, according to the MD model [36]. In this contribution, we used the approach proposed by Carpick *et al.* [41] for determination of the c parameter. The corresponding recalculated W_A values are shown in the last column of the table.

Table 2 shows the material properties used in calculations of the reduced elastic modulus (K) for the probe–substrate systems presented in Table 1. Finally, Table 3 shows the calculated K values. Note that for self-assembled monolayers, the material properties of the underlying substrate were used to determine K and not the properties of the monolayer films themselves due to the small thicknesses and expected pliancy of the monolayers (1–1.5 nm).

As shown in Table 1, both continuum contact mechanics models, the JKR model and DMT model, were used in analysis of the measured pull-off forces, although selection of a particular model was rarely justified. Also, based on the literature reports, there is no clear distinction between analysis of the data generated with sharp tips or colloidal probes, in spite of the fact that both the size of the probe and spring constant of cantilever differ significantly in the two cases. Additionally, although we did not analyze all details of previous experimental work, we noted that many papers also lacked in reporting the mechanical properties of materials used in experimentation and the loads that were used during pressing the tip or colloidal probe on the substrate surface. This information is important for a detailed analysis of the pull-off forces and behavior of the tip–substrate system.

J. Drelich et al.

Table 1.
Measured and calculated values of the work of adhesion for different systems

Reference	Probe–substrate functionality	Medium	R (nm or μm)	k (N/m)	W_A (mJ/m^2) Predicted	W_A (mJ/m^2) Determined	Model used	K (GPa)	λ_M (z_0 = 0.2–0.4 nm)	c	W_A (mJ/m^2)
	System		Probe		Lit. results			This study			
			Sharp tips (R in (nm))								
Noy et al. [46]	COOH–COOH	Ethanol	54	0.12	ND	8.9±3.1	JKR, c = 1.5	64.4	0.04–0.07	2.0	6.7±2.3
	CH$_3$–CH$_3$				5	3.9±1.5		64.4	0.03–0.06	2.0	2.9±1.1
Van der Vegte and Hadziioannou [47]	CH$_3$–CH$_3$	Ethanol	35	0.2	5	5	JKR, c = 1.5	64.4	0.03–0.05	2.0	3.8
	CH$_3$–CH$_3$	Water			104	103		64.4	0.2–0.4	1.8	86
	OH–OH	Ethanol			ND	6		64.4	0.02–0.04	2.0	2.3
	NH$_2$–NH$_2$	Ethanol			ND	5.6		64.4	0.01–0.03	2.0	2.1
	COOH–COOH	Ethanol			ND	9		64.4	0.03–0.05	2.0	3.4
	CONH$_2$–CONH$_2$	Ethanol			ND	10.6		64.4	0.03–0.06	2.0	4.0
Tsukruk and Bliznyuk [48]	SiOH–SiOH	Water	40–500	0.23–0.25	ND	4	JKR, c = 1.5	ND	ND	ND	ND
	Si$_3$N$_4$–Si$_3$N$_4$					8		220.7	0.01–0.06	2.0	6
	NH$_2$–NH$_2$					4.5		76.9	0.02–0.08	2.0	3.4
	SO$_3$H–SO$_3$H					1.5		76.9	0.01–0.04	2.0	1.1
	CH$_3$–CH$_3$					0.5		76.9	<0.02	2.0	0.4
Clear and Nealy [49]	CH$_3$–CH$_3$	Hexadecane	approx. 60	0.44	1.6	4.1	JKR, c = 1.5	70.2	0.02–0.03	2.0	3.1
		Ethanol			5.5	3.3		70.2	0.03–0.07	2.0	2.5
		1,2-propanediol			27.1	20.7		70.2	0.1–0.2	1.9	16.3
		1,3-propanediol			43.7	41.5		70.2	0.13–0.27	1.9	32.8
		Water			102.8	102.9		70.2	0.24–0.47	1.8	85.8
	COOH-ox-OYTS	Hexadecane			ND	19.6		70.2	0.1–0.2	1.9	15.5
		Ethanol			ND	10.1		70.2	0.04–0.08	2.0	7.6
		1,2-propanediol			ND	16		70.2	0.05–0.1	2.0	12
		1,3-propanediol			ND	6		70.2	0.03–0.07	2.0	4.5
		Water			ND	1.7		70.2	0.01–0.02	2.0	1.3

Table 1.
(Continued)

Reference	System		Probe		Lit. results			This study			
	Probe–substrate functionality	Medium	R (nm or μm)	k (N/m)	W_A (mJ/m²) Predicted	W_A (mJ/m²) Determined	Model used	K (GPa)	λ_M ($z_0 = 0.2$–0.4 nm)	c	W_A (mJ/m²)
	COOH–COOH	1,2-propanediol			ND	14		70.2	0.05–0.09	2.0	10.5
		1,3-propanediol			ND	14.6		70.2	0.05–0.1	2.0	11
	CH$_3$-ox-OYTS	Hexadecane			ND	4.6		70.2	0.02–0.04	2.0	3.5
		Ethanol			ND	2.1		70.2	0.01–0.03	2.0	1.6
		1,2-propanediol			ND	4.9		70.2	0.02–0.04	2.0	3.7
		1,3-propanediol			ND	6		70.2	0.03–0.07	2.0	4.5
		Water			ND	0.9		70.2	0.01–0.02	2.0	0.7
El Ghzaoui [50]	SiO$_2$–PTFE	Dry N$_2$	70	NA	95	88	DMT, $c = 2$	0.7	1.7–3.4	1.6	110
	SiO$_2$–PP				110	114		2	1.0–2.0	1.6	142
	SiO$_2$–PU				145	150		~(1–2)	~(1–2)	1.6	187
Skulason and Friesbie [51]	CH$_3$–CH$_3$	Water	15–130	0.23–0.28	103	110±10	JKR, $c = 1.5$	64.4	0.16–0.64	1.8	92±8
Jaquot and Takadoum [52]	Si$_3$N$_4$–Silica	Water	200	0.3	27.9	30	DMT, $c = 2$	99.9	0.1–0.2	1.9	32
	Si$_3$N$_4$–TiN				20	18.3		ND	ND	ND	ND
	Si$_3$N$_4$–DLC				13.5	12		337.7	0.03–0.06	2.0	12
	Si$_3$N$_4$–Si(111)				27.5	42		114.1	0.1–0.2	1.9	44
	Si$_3$N$_4$–Si(100)				19.9	24		114.1	0.09–0.2	1.9	25
Leite *et al.* [53]	Si–mica	Water	23	0.13	110	83	DMT, $c = 2$	36.6	0.3–0.6	1.8	92
	Si–mica	air			215	173		36.6	0.4–0.9	1.7	204

Table 1.
(Continued)

Reference	System Probe–substrate functionality	Medium	Probe R (nm or μm)	k (N/m)	Lit. results W_A (mJ/m²) Predicted	Determined	Model used	This study K (GPa)	λ_M (z_0 = 0.2–0.4 nm)	c	W_A (mJ/m²)
	Colloidal probes (R in (μm))										
Burnham et al.	W–mica	Dry N_2	2.5±0.5	50±10	369	21	DMT,	44.4	2.4–4.8	1.5	28
[54]	W–graphite				210	14	$c = 2$	ND	ND	ND	ND
	W–Al_2O_3		2.0±0.5		128	8		286	0.3–0.8	1.7	9.4
	W–CH_3				98	3		286	0.3–0.6	1.8	3.3
	W–CF_3		2.5±0.5		96	1		286	0.3–0.7	1.8	1.1
	W–PTFE		2.0±0.5		92	0		0.7	15–35	1.5	ND
Biggs and Spinks [43]	PS–mica	Dry N_2	5	27±1	102.4	148.5/ 122.5	JKR/ MP	4	7–14	1.5	148.5
Nalaskowski et al. [44]	PE–SiO_2	Water	5–9	27–30	21	4.4	JKR,	2	2–7	1.5	4.4
	PE–SiO_2 (heat)				35	21	$c = 1.5$		4–8	1.5	21
	PE–CH_3				66	64			6–12	1.5	64

R is the radius of the AFM probe; k is the spring constant of the AFM cantilever; W_A is the work of adhesion between the probe and substrate; K is the reduced elastic modulus for the probe–substrate system; λ_M is the Maugis parameter; z_0 is the equilibrium separation distance between the probe and substrate; c is the constant in equation (7) calculated from equation derived by Carpick et al. [41]: $c = \frac{7}{4} - \frac{1}{4}\left(\frac{4.04\lambda_M^{1.4}-1}{4.04\lambda_M^{1.4}+1}\right)$.

The calculated values of the Maugis parameter (λ_M) for the literature systems (Table 1) indicate that a common characteristic in a majority of publications is the use of an incorrect contact mechanics model for analysis of pull-off forces and subsequent calculation of the work of adhesion, or surface energy. The JKR model was used most frequently in analysis of pull-off forces obtained with either sharp tips or colloidal probes being in contact with various substrates. Although the selection of the JKR model for analysis of the results from colloidal probe microscopy studies is often appropriate, this selection should not be blindly made without analysis of the λ_M parameter; examples of the systems where the MD model is more appropriate are shown in Table 1. Next, the JKR model was also the

Table 2.
Elastic modulus and Poisson's ratio for materials analyzed in this study

Material	Young's modulus (GPa)	Poisson's ratio	Source
Au	78	0.44	[55]
W	411	0.28	[55]
PS	2.8–3.5	0.38	[56]
PP	1–1.6	0.4 (assumed)	[56]
PTFE	0.41	0.46	[56]
Mica	34.5	0.205	[57]
Si	107	0.27	[58]
Si_3N_4 (hot pressed)	300–330	0.22	[58]
SiO_2	94	0.17	[58]
Diamond	1035	0.1–0.29	[59]
Al_2O_3	390	0.2–0.25	[58]

Table 3.
Reduced elastic modulus (K) for AFM systems reported in the literature

System	$1/K$	K (GPA)
Au–Au	1.55077E-11	64.4
Si–Si	1.29967E-11	76.9
Si_3N_4–Si_3N_4	4.53143E-12	220.7
W–mica	2.25073E-11	44.4
W–PTFE	1.53827E-09	0.7
PS–mica	2.50004E-10	4.0
Si–mica	2.73239E-11	36.6
Si_3N_4–Si	8.76408E-12	114.1
Au–Si	1.42522E-11	70.2
W–Al_2O_3	3.5031E-12	285.5
Si_3N_4–SiO_2	1.00139E-11	99.9
Si_3N_4–DLC	2.96137E-12	337.7
SiO_2–PTFE	1.44994E-09	0.7
SiO_2–PP	4.92364E-10	2.0

most popular model used in analysis of pull-off forces measured with the sharp tips (regular and coated), in spite of the fact that the small dimensions of the tips comply better with the assumption of the DMT model. As shown in Table 1, the results of chemical force microscopy studies (sharp tips modified with organic functionality) can usually be analyzed by the DMT model. This again cannot be assumed *a priori*, and systems with softer substrates such as polymers require analysis with either the MD model or JKR model.

10. CAN SOLID SURFACE TENSION BE DETERMINED FOR ROUGH SURFACES?

The factor that complicates interpretation of the measured AFM pull-off forces the most is surface roughness. Quantitative calculations of the adhesion force between a particle and a rough surface are difficult for many reasons. The size, shape, homogeneity, mechanical properties and distribution of the asperities (deviations from an ideal planar surface) all influence the actual area of contact and, therefore, directly affect the pull-off force [39, 60]. The particle can also have an irregular geometry leading to more difficulties in quantitative analysis of pull-off force data [40] but this case is not discussed here.

Surface roughness causes the actual area of contact to vary significantly from the ideal spherical particle on flat surface contact. Roughness can increase the adhesion of particles to a surface or decrease it depending on the scale of the roughness, location on the surface where contact is made, and the size and geometry of the particle. The actual area of contact between a particle and a rough surface depends on the size and distribution of the asperities on the surface [40]. The probe–substrate contact area often increases for probes whose size is smaller than the distance between asperities of microscopically rough surface. The increased

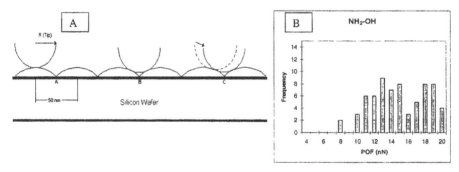

Figure 7. (A) Schematic showing single and double contact points between AFM probe and rough substrate during an adhesion force measurement. (B) Bimodal distribution of pull-off forces (POF) measured between a sharp tip ($R = 32$ nm) and rough substrate. Both the probe and the substrate were coated with a thin film of gold and modified with self-assembled monolayer of thiol of NH_2-end functionality (probe) or OH-end functionality (substrate) [39].

adhesion between the probe and rough surface can result from multiple probe–substrate contacts, if the probe penetrates only partially the space between asperities and interacts with the walls of two or more asperities [39]. Figure 7 shows a schematic representation of how multiple contact points (mainly two but also three contacts are possible) could be made between the probe and asperities of the rough substrates. If such multiple probe–substrate contact points are frequently repeated during the pull-off force measurements, the pull-off force distribution graph has a bi-modal (or even a tri-modal) character, such as shown in Fig. 7. For such results, the surface tension of the solid or the work of adhesion can still be calculated by using the data from the lower range of pull-off forces that correspond to single contact point situation.

It should be recognized, however, that the appearance of multi-modal distribution of pull-off forces does not necessarily indicate that the multiple contact points have been experienced during the pull-off force measurements. Surface heterogeneity is another cause for generation of similar graphs and, therefore, results similar to those shown in Fig. 7 should be interpreted with caution.

When the size of the probe is larger than the distance between asperities, the probe cannot penetrate the inter-asperity space and the contact area between the probe and the substrate is reduced. As a result, the pull-off forces are weaker than expected for the sphere-flat geometry. Two different scenarios are possible in such a sphere-rough surface system. First, as discussed earlier, if the loads applied during probe–substrate contact are increased to high values, plastic deformation of asperities can be induced. For applied loads that are properly managed, the probe–substrate contact area corresponds to that predicted by contact mechanics of sphere-flat geometry system.

In many systems, however, the plastic deformation of asperities cannot be initiated, due to low stiffness of available/used cantilevers and high hardness of materials used in experiments. For these systems, the work of adhesion between the probe and sub-microscopically rough substrate can still be estimated from the measured pull-off forces. This is possible with the theoretical model recently introduced by Rabinovich *et al.* [61]. Assumptions and details of the mathematics of the Rabinovich model will not be repeated here and only final equation is presented:

$$W_A = \left(\frac{F}{R} - B\right)\frac{58 R rms_2}{c\pi \lambda_2^2}$$

(22)

where

$$B = \frac{A}{6z_0^2}\frac{1}{\left(1 + \frac{58 R rms_1}{\lambda_1^2}\right)\left(1 + \frac{1.82 rms_2}{z_0}\right)^2}$$

(23)

$$rms = \sqrt{\frac{32 \int_0^r y^2 r_1 dr_1}{\lambda^2}k_p}$$

(24)

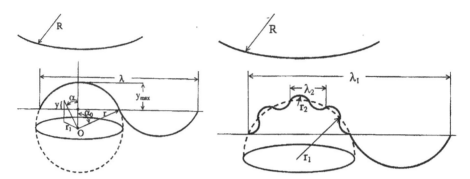

Figure 8. Geometry for substrate surface that has one (left drawing) or two scales (right drawing) of roughness according to the Rabinovich model.

c is constant equal to $c = 1.5$ if JKR contact mechanics apply and $c = 2$ if DMT contact mechanics apply (note that in the original paper by Rabinovich *et al.* [61] $c = 1.5$); λ is the peak-to-peak distance between asperities ($\lambda = 4r$); k_p is the surface packing density for close-packed spheres ($k_p = 0.907$); A is the Hamaker constant; z_0 is the distance of closest approach between the two surfaces, r and R are the radii of the asperity and particle, respectively; k_1 is a coefficient relating the rms roughness and the maximum peak height, which is equal to 1.817 in assumed geometry. The model geometry is shown in Fig. 8. In this model, surfaces which exhibit two scales of roughness, both smaller than the size of spherical probe, are considered. The first type of roughness, called rms$_1$, is associated with a longer peak-to-peak distance, λ_1. A second smaller roughness, called rms$_2$, is associated with a smaller peak-to-peak distance, λ_2.

Because for many systems $F/R \gg B$, equation (22) can be reduced to:

$$W_A = \frac{58\,F\,\text{rms}_2}{c\pi\,\lambda_2^2} \tag{25}$$

Again, if both probe and substrate are made of the same material:

$$\gamma_S = \frac{29\,F\,\text{rms}_2}{c\pi\,\lambda_2^2} \tag{26}$$

To test the Rabinovich model we analyzed the results of pull-off forces measured for silanized glass probes in contact with rough polypropylene substrates and reported in Ref. [40]. Table 4 lists the substrate characteristics, measured (average) pull-off forces, and calculated values of the work of adhesion. The work of adhesion was calculated using equation (7) assuming $c = 1.6$ (as $\lambda_M = 1.5$–2, the system complies closely with the JKR contact mechanics model). The three polypropylene substrates of very random roughness regarding the size of asperities and their distribution, which as reported earlier [40] did not fit the Rabinovich model, are not included in the analysis presented.

Table 4.

Surface roughness of polypropylene samples, average pull-off forces between polypropylene and silanized glass probe [40] and calculated polypropylene–silanized glass work of adhesion

Sample	Surface roughness characteristics				Pull-off force	Work of adhesion (mJ/m^2)	
	rms$_1$ (nm)	λ_1 (μm)	rms$_2$ (nm)	λ_2 (nm)	(nN)	Exp.	Theor.
1	25	3.7	6.5	366	76	42	
2	42	4.5	15.6	598	136	68	
3	386	5.8	39.6	964	100	49	48–54*
4	194	7.5	32.6	685	41	33	
5	31	3.1	13.6	548	74	38	
6	37	3.9	16.0	345	63	98	
Average:						55 (49)**	approx. 50

*Estimated using $W_A = 2(\gamma_1\gamma_2)^{1/2}$, where $\gamma_1 = 25.7$ mJ/m^2 is the surface tension of polypropylene [16] and $\gamma_2 = 22$–28 mJ/m^2 is the surface tension of silanized glass (our data).
**In parentheses: average value calculated for four samples, without results for the samples 4 and 6.

As shown in Table 4, an average value for the work of adhesion between polypropylene and silanized glass as calculated from the Rabinovich model relates well to the theoretical value estimated from the surface tensions of interacting surfaces.

In our previous paper [40], we also reported the pull-off forces for peptide, lactose and polystyrene probes in contact with the same polypropylene substrates. The work of adhesion for peptide and lactose particles on polypropylene samples could not be calculated with the Rabinovich model because of the irregular shape of these particles. On the other hand, we calculated the polystyrene–polypropylene work of adhesion based on measured pull-off forces reported in Ref. [40], and found these values to be 2–3-times larger than theoretical estimates. The soft character of polystyrene probes used in the experiments is probably responsible for this large discrepancy. We expect that asperities of rough substrates, which are harder than the material of the probe, penetrate into the probe *via* nanoindentations. If this happens, simple contact mechanics models including the Rabinovich model cannot describe these systems.

11. APPLICATION OF LEWIS ACID–BASE INTERACTION THEORY TO AFM STUDIES: UNTESTED HYPOTHETICAL APPROACH

The discussion presented in the previous sections indicates that either surface tension of solid or solid–solid work of adhesion can be determined from the AFM pull-off force measurements. The solid surface tension can be calculated from equation (8) if measurements are carried out for a probe made of the same material as the substrate. In many cases, preparation of a spherical probe of a specific material can be time-consuming and difficult, or simply impossible. In such situations, probing the solid surface adhesion properties can be done using probes of other materials. If the surface tension of the material used for the probe (γ_1) is known and if both substrate and probe have predominantly apolar character, the surface tension of the substrate (γ_2) can be estimated from the work of adhesion determined according to the following equation [62]:

$$W_A \approx 2\sqrt{\gamma_1 \gamma_2} \tag{27}$$

This approach, however, is not recommended if one of the materials, either probe or substrate or both, has a strong polar character in addition to its apolar one. Instead, we propose that for such a system, an approach similar to that used in contact angle measurements and the contact angle data interpretation based on the Lifshitz–van der Waals/Lewis acid–base interaction theory (see Section 2) could be applied to the AFM pull-off rce studies. Specifically, the pull-off forces between the substrate of interest and at least three AFM probes of different surface energy characteristics, one probe with non-polar functionality and two probes with a partial polar character, could be measured. Three AFM cantilever tips should have different but known γ^{LW}, γ^-, γ^+ values, where one probe should have predominantly acidic character and one probe should have predominantly basic character. Next, the surface tension components of the substrate could be calculated based on measurements of pull-off forces for at least three model AFM probes and using the following three equations:

$$W_{A1} = 2\left[\sqrt{\gamma_S^{LW}\gamma_{P1}^{LW}} + \sqrt{\gamma_S^+\gamma_{P1}^-} + \sqrt{\gamma_S^-\gamma_{P1}^+}\right] \tag{28}$$

$$W_{A2} = 2\left[\sqrt{\gamma_S^{LW}\gamma_{P2}^{LW}} + \sqrt{\gamma_S^+\gamma_{P2}^-} + \sqrt{\gamma_S^-\gamma_{P2}^+}\right] \tag{29}$$

$$W_{A3} = 2\left[\sqrt{\gamma_S^{LW}\gamma_{P3}^{LW}} + \sqrt{\gamma_S^+\gamma_{P3}^-} + \sqrt{\gamma_S^-\gamma_{P3}^+}\right], \tag{30}$$

where the subscript Pi refers to one of the three AFM probes.

A problem with this approach is that, in contrast to liquids, there are no known solids which could have large values of γ^- and γ^+. In fact, most low surface energy solids are predominantly basic [16]. Low values of γ^- and γ^+ will have profound effect on precision and reliability of the determined solid surface tension. In spite of this problem, this new approach with three probes of different surface energy characteristics may be attractive for characterization of many engineered materials, where high precision of surface properties is not always required.

Acknowledgements

Financial support provided by the Petroleum Research Fund and administrated by the American Chemical Society is gratefully acknowledged.

REFERENCES

1. J. C. Chapman and H. L. Porter, *Proc. Roy. Soc. London A* **83**, 65–68 (1910).
2. C. Herring, *J. Appl. Phys.* **21**, 437–445 (1950).
3. J. W. Obreimoff, *Proc. Roy. Soc. London A* **127**, 290–297 (1930).
4. S. M. Wiederhorn, A. M. Shorb and R. L. Moses, *J. Appl. Phys.* **39**, 1569–1572 (1968).
5. F. W. C. Boswell, *Proc. Phys. Soc. London A* **64**, 465–476 (1951).
6. T. De Planta, R. Ghez and F. Piuz, *Helvet. Phys. Acta* **37**, 74–76 (1964).
7. G. A. Hulett, *Z. Phys. Chem.* **37**, 385–406 (1901).
8. M. L. Dundon, *J. Am. Chem. Soc.* **45**, 2658–2666 (1923).
9. G. Jura and C. W. Garland, *J. Phys. Chem.* **74**, 6033–6034 (1952).
10. A. C. Zettlemoyer, *Ind. Eng. Chem.* **57**, 27–36 (1965).
11. J. J. Rasmussen and A. S. Nelson, *J. Am. Ceram. Soc.* **54**, 398–401 (1971).
12. J. D. Pandey, B. R. Chaturvedi and R. P. Pandey, *J. Phys. Chem.* **85**, 1750–1752 (1981).
13. A. I. Rusanov and V. A. Prokhorov, *Interfacial Tensiometry*. Elsevier, Amsterdam (1996).
14. H.-J. Butt and R. Raiteri, in *Surface Characterization Methods: Principles, Techniques, and Applications*, A. J. Milling (Ed.), pp. 1–36. Marcel Dekker, New York, NY (1999).
15. C. J. van Oss, R. J. Good and M. K. Chaudhury, *J. Colloid Interface Sci.* **111**, 378–390 (1986).
16. C. J. van Oss, *Interfacial Forces in Aqueous Media*. Marcel Dekker, New York, NY (1994).
17. R. F. Giese and C. J. van Oss, *Colloid and Surface Properties of Clays and Related Minerals*. Marcel Dekker, New York, NY (2002).
18. K. L. Mittal (Ed.), *Contact Angle, Wettability and Adhesion*. Vol. 3, VSP, Utrecht (2003).
19. C. Della Volpe and S. Siboni, in *Encyclopedia of Surface and Colloid Science*, A. T. Hubbard (Ed.), pp. 17–36. Marcel Dekker, New York, NY (2002).
20. C. Della Volpe, D. Maniglio, S. Saboni and M. Morra, *J. Adhesion Sci. Technol.* **17**, 1477–1505 (2003).
21. D. Sarid, *Scanning Force Microscopy*. Oxford University Press, New York, NY (1991).
22. R. Wiesendanger, *Scanning Probe Microscopy and Spectroscopy: Methods and Applications*. Cambridge University Press, Cambridge (1994).
23. B. Cappella and G. Dietler, *Surface Sci. Rep.* **34**, 1–104 (1999).
24. J. E. Sader, in *Encyclopedia of Surface and Colloid Science*, A. T. Hubbard (Ed.), pp. 846–856. Marcel Dekker, New York, NY (2002).
25. J. P. Cleveland, S. Manne, D. Bocek and P. K. Hansma, *Rev. Sci. Instrum.* **64**, 403–405 (1993).
26. J. L. Hutler and J. Bechhoefer, *Rev. Sci. Instrum.* **64**, 1868–1873 (1993).
27. A. Noy, D. V. Vezenov and C. M. Lieber, *Annu. Rev. Mater. Sci.* **27**, 381–421 (1997).
28. J. Nalaskowski, J. Drelich, J. Hupka and J. D. Miller, *J. Adhesion Sci. Technol.* **13**, 1–17 (1999).
29. J. S. Villarubia, *J. Res. (Natl. Inst. Stand. Technol.)* **102**, 425–454 (1997).
30. K. L. Johnson, K. Kendall and A. D. Roberts, *Proc. Roy. Soc. London* **A324**, 301–313 (1971).
31. B. V. Derjaguin, V. M. Muller and Yu. P. Toporov, *J. Colloid Interface Sci.* **53**, 314–326 (1975).
32. D. Maugis, *Contact, Adhesion and Rupture of Elastic Solids*. Springer, Berlin (2000).
33. N. A. Burnham and A. J. Kulik, in *Handbook of Micro/Nano Tribology*, 2nd edn., B. Bhushan (Ed.), pp. 247–271. CRC Press, Boca Raton, FL (1999).
34. O. Marti, in *Modern Tribology Handbook*, Vol. I, B. Bhushan (Ed.), pp. 617–639, CRC Press, Boca Raton, FL (2001).
35. U. D. Schwarz, *J. Colloid Interface Sci.* **261**, 99–106 (2003).
36. D. Maugis, *J. Colloid Interface Sci.* **150**, 243–269 (1992).
37. H. Hertz, *Miscellaneous Papers*. Macmillan, London (1896).
38. E. R. Beach and J. Drelich, in *Functional Fillers and Nanoscale Minerals*, J. J. Kellar, M. A. Herpfer and B. M. Moudgil (Eds), pp. 177–193. Society for Mining, Metallurgy, and Exploration, Littleton, CO (2003).
39. G. W. Tormoen, J. Drelich and E. R. Beach, *J. Adhesion Sci. Technol.* **18**, 1–18 (2004).

40. E. Beach, G. Tormoen, J. Drelich and R. Han, *J. Adhesion Sci. Technol.* **16**, 845–868 (2002).
41. R. W. Carpick, D. F. Ogletree and M. Salmeron, *J. Colloid Interface Sci.* **211**, 395–400 (1999).
42. D. Maugis and H. M. Pollock, *Acta Metall.* **9**, 1323–1334 (1984).
43. S. Biggs and G. Spinks, *J. Adhesion Sci. Technol.* **12**, 461–478 (1998).
44. J. Nalaskowski, J. Drelich, J. Hupka and J. D. Miller, *Langmuir* **19**, 5311–5317 (2003).
45. C. Schonenberger, J. Jorritsma, J. A. M. Sondag-Huethorst and L. G. J. Fokkink, *J. Phys. Chem.* **99**, 3259–3271 (1995).
46. A. Noy, C. D. Friesbie, L. F. Rozsnyai, M. S. Wrighton and C. M. Lieber, *J. Am. Chem. Soc.* **117**, 7943–7951 (1995).
47. E. W. Van der Vegte and G. Hadziioannou, *Langmuir* **13**, 4357–4366 (1997).
48. V. V. Tsukruk and V. N. Bliznyuk, *Langmuir* **14**, 446–455 (1998).
49. S. C. Clear and P. F. Nealy, *J. Colloid Interface Sci.* **213**, 238–250 (1999).
50. A. El Ghzaoui, *J. Appl. Phys.* **85**, 1231–1233 (1999).
51. H. Skulason and C. D. Friesbie, *Langmuir* **16**, 6294–6297 (2000).
52. C. Jacquot and J. Takadoum, *J. Adhesion Sci. Technol.* **15**, 681–687 (2001).
53. F. L. Leite, A. Riul Jr. and P. S. P. Herrmann, *J. Adhesion Sci. Technol.* **17**, 2141–2156 (2003).
54. N. A. Burnham, D. D. Dominguez, R. L. Mowery and R. J. Colton, *Phys. Rev. Lett.* **64**, 1931–1934 (1990).
55. Web site: www.webelements.com
56. J. R. Fried, *Polymer Science and Technology*. Prentice-Hall PTR, Englewood Cliffs, NJ (1995).
57. H. Matsuoka and T. Kato, *Proc. International Tribology Conference*. Yokohama (1995).
58. M. Barsoum, *Fundamentals of Ceramics*, McGraw Hill, New York, NY (1997).
59. W. D. Callister Jr., *Materials Science and Engineering: An Introduction*, 4th edn. Wiley, New York, NY (1997).
60. L. Sirghi, N. Hakagiri, K. Sugisaki, H. Sugimura and O. Takai, *Langmuir* **16**, 7796–7800 (2000).
61. Y. I. Rabinovich, J. J. Adler, A. Ata, R. K. Singh and B. M. Moudgil, *J. Colloid Interface Sci.* **232**, 17–24 (2000).
62. J. Israelachvili, *Intermolecular and Surface Forces*. 2nd edn., Academic Press, London (1992).

Part 3

Wettability and Its Modification

Contact Angle, Wettability and Adhesion, Vol. 4, pp. 267–280
Ed. K.L. Mittal
© VSP 2006

Study of surface charge properties of minerals and surface-modified substrates by wettability measurements

ALAIN CARRÉ* and VALÉRIE LACARRIÈRE

Corning, Fontainebleau Research Center, 7bis, avenue de Valvins, 77210 Avon, France

Abstract—The experimental and theoretical approaches for the determination of the point of zero charge (pzc) on solid inorganic substrates by contact-angle measurements are first reviewed in this paper. The experimental method consists of measuring the water contact angle in air, or in presence of another immiscible liquid, as a function of the pH of the water drop. At the pzc, the solid surface is uncharged and the water contact angle has a maximum. The classical thermodynamic analysis to interpret the variation of the water contact angle as a function of pH is presented. We have determined the pzc of bare and functionalized glass. The glass considered has a pzc around 3, whereas glass coated with 3-aminopropyltriethoxysilane (APTS) has a pzc around 7. This value is well corroborated by the data obtained from electrophoretic mobility and pH-potentiometric titration performed on aminated silica particles. Although amine groups of APTS are positively charged below pH $\approx$ 11, the APTS functionalized glass substrate is positively charged below pH 7 and negatively charged above pH 7. This particular behavior is attributed to the contribution of free SiOH groups on the functionalized glass. For the APTS-functionalized glass substrate, it was demonstrated that the electrical double layer at the solid/water interface behaved as a constant capacitor in wetting. This model was extended to uncoated metal oxide substrates such as chromium and silver oxides. An example of uncharged substrate is also presented with glass functionalized with 3-glycidoxypropyltrimethoxysilane, an epoxy silane.

Keywords: Point of zero charge; contact angle; water; pH; wetting; wettability; oxides; glass; functionalized glass; aminosilane; epoxysilane.

1. INTRODUCTION

The surface-charge properties are closely related to acid–base properties, which are important in numerous surface and interface phenomena such as wetting, adhesion, nucleic acids and proteins adsorption, catalysis, etc.

The point of zero charge (pzc) is one parameter determining the surface charge of mineral surfaces in an aqueous medium. In this paper, the experimental and

---

*To whom correspondence should be addressed. Tel.: (33-1) 6469-7371;
Fax: (33-1) 6469-7455; e-mail: carrea@corning.com

theoretical approaches for the determination of the point of zero charge by contact angle measurements on mineral and surface modified substrates is reviewed.

The basic theory is evaluated with contact angle measurements on minerals and glass substrates. Then, the surface charge behavior of functionalized glass substrates is discussed in detail. The functionalization consists of chemically bonding organosilanes to the glass surface. Two different functionalizations are considered. The first one is based on the chemical grafting of an aminosilane onto the glass surface. The functionalized substrate is used for the ionic immobilization of DNA molecules [1]. The second one utilizes an epoxysilane. The epoxy functionalized glass is used for the covalent immobilization of functionalized oligonucleotides. As shown in this paper, the two functionalized substrates have drastically different surface charge behavior.

2. THEORETICAL

The surface charge of the oxide of metal, M, results from protonation or deprotonation of hydroxyl surface functions, MOH, according to:

$$MOH \Leftrightarrow MO^- + H^+ \qquad\qquad K_{a1} = \frac{\left|MO^-\right|\left|H^+\right|}{\left|MOH\right|} \qquad (1)$$

$$MOH + H^+ \Leftrightarrow MOH_2^+ \qquad\qquad K_{a2} = \frac{\left|MOH\right|\left|H^+\right|}{\left|MOH_2^+\right|} \qquad (2)$$

where K_{a1} and K_{a2} are the acidic dissociation constants of hydroxyl and protonated hydroxyl groups, respectively. The equilibrium between positive and negative species on the solid substrate may be written as

$$MOH_2^+ \Leftrightarrow MO^- + 2H^+ \qquad\qquad (3)$$

with an equilibrium constant:

$$K = \frac{\left|MO^-\right|\left|H^+\right|^2}{\left|MOH_2^+\right|} = K_{a1} \times K_{a2} \qquad (4)$$

At the pzc, $\left|MO^-\right| = \left|MOH_2^+\right|$, so that:

$$pzc = 1/2pK = 1/2(pK_{a1} + pK_{a2}) \qquad (5)$$

The pzc may be measured by different techniques: potentiometric and electrokinetic methods (powders, fibers), colored indicators, contact angle measurements, or contact angle titration on flat substrates. The first aim of this work was to study the surface charge of flat functionalized glass substrates, and the method of measuring contact angles on flat surfaces was chosen.

2.1. Variation of the water contact angle with pH

The equilibrium contact angle, θ, of a sessile water drop satisfies Young's equation [2]:

$$\gamma_{SI} = \gamma_{SW} + \gamma \cos\theta \qquad (6)$$

where γ_{SI} is the interface free energy between the solid and phase I (water vapor or alkane), γ_{SW} is the interface free energy between water and the solid and γ is the water surface free energy (surface tension) or the water/alkane interface free energy (Fig. 1).

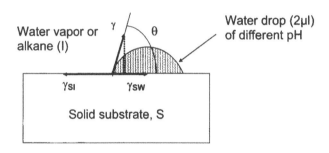

Figure 1. Experimental method for measuring contact angles of water of different pH values.

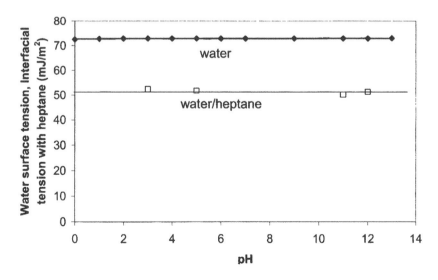

Figure 2. Water surface tension and water/heptane interfacial tension as a function of water pH.

A. Carré and V. Lacarrière

When an ionizable solid surface is in contact with an aqueous solution, it acquires a certain charge because of adsorption or desorption of a certain amount of protons [3–5]. The water surface tension [6], as well as the interface free energy between water and an alkane [4, 5], and the solid surface free energy, can be considered as constant when, for example, hydrochloric acid or sodium hydroxide is added to modify the pH of water. As an example, the invariant water/heptane interface free energy (interface tension) and water surface free energy (surface tension) as a function of pH are presented in Fig. 2. As a good approximation, we can take $d\gamma/dpH$ and $d\gamma_{SI}/dpH \approx 0$ [3–5]. Therefore, when the pH of water is changed, the wetting angle, θ, varies accordingly:

$$d\gamma_{SW} = -\gamma d(\cos\theta) \tag{7}$$

The surface charge, σ, may be simply considered as resulting from the adsorption of protons (H^+) if the solid surface is positively charged and from the adsorption of OH^- if the solid surface is negatively charged, when no other ions are adsorbed on the surface. Using the Gibbs adsorption equation it can be deduced that

$$d\gamma_{SW} = -\Gamma_{H^+} RTd\ln|H^+| = 2.303\Gamma_{H^+} RTd(pH) \tag{8}$$

where Γ_{H+} is the surface excess concentration of protons when the oxide surface becomes positively charged. When the surface charge results from the adsorption of negatively charged hydroxyl groups (OH^-), the sign of the last member of equation (8) is the opposite. In the case of protons adsorption, the surface charge density, σ, of the water/solid interface is related to Γ_{H+} as follows:

$$\sigma = \Gamma_{H^+} F \tag{9}$$

where F is the Faraday constant (96 500 C·mol^{-1}). Hence, equations (7)–(9) yield

$$\frac{d(\cos\theta)}{d(pH)} = -\frac{2.303RT\sigma}{F\gamma} \tag{10}$$

indicating that the change in cos θ and θ is controlled by σ. Note that the same equation (equation (10), but with the right-hand side of opposite sign, is obtained for a negatively charged surface, due to the adsorption of OH^- groups. At the pzc, $\sigma = 0$, so that:

$$\frac{d(\cos\theta)}{d(pH)} = 0 \tag{11}$$

Therefore, in the absence of specifically adsorbed ions, a maximum in θ or a minimum in cos θ will occur at pzc as the pH of the water drop is scanned [3–5]. A minimum of water/solid surface interaction energy, corresponding to a

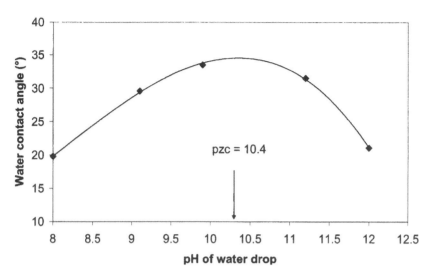

Figure 3. Water contact angle on silver oxide as a function of pH [3].

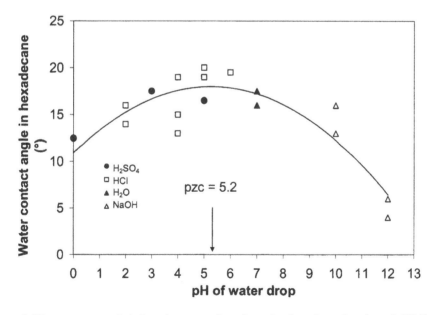

Figure 4. Water contact angle in hexadecane on chromium-plated steel as a function of pH [4].

maximum in θ, is observed in absence of ionic interactions between water and the solid surface.

On clean and high-surface-energy solids, water usually wets fully the substrate, leading to a contact angle of $0°$. In that case, the water contact angle may be measured in a second immiscible liquid phase, such as an alkane. Under this condition the water contact angle may be finite, the solid/alkane interface free energy being always lower than the surface free energy of the solid alone. To measure the wettability as a function of pH for bare glass, the substrate may be immersed in hexadecane or octane (Fig. 1).

Chau and Porter [3] reported water contact angle measurements on silver films prepared by evaporation on silicon wafers. A plot of the water contact angle θ versus pH is shown in Fig. 3. The water contact angle reaches a maximum at pH 10.4. This value was determined from a third-order polynomial fitting of the variation of θ as a function of pH.

The point of zero charge of chromium oxide was determined by McCafferty and Wightman [4] from the water contact angles at the hexadecane–aqueous solution interface measured as a function of the pH of the aqueous phase. Figure 4 shows contact angles for argon plasma treated chromium (chromium-plated steel) at the hexadecane–aqueous solution interface as a function of the water pH. The contact angle clearly exhibits a maximum at a pH of 5.2 (value obtained by the authors [4] by fitting a second-order polynomial to the data).

2.2. Water contact angle in octane

The contact angle of water in (non-polar) octane is only a function of specific, non-dispersion, interactions between water and the substrate. This arises from water and octane having the same dispersion or London contribution to their surface free energy (21.6 mJ·m^{-2} for water, 21.3 mJ·m^{-2} for octane [1, 7]).

The non-dispersion interaction between water and the substrate, I_{SW}^{nd}, may be deduced from the Young–Dupré equations:

$$\gamma_{SO} = \gamma_{SW} + \gamma_{WO} \cos \theta \tag{12}$$

$$\gamma_{Si} = \gamma_S + \gamma_i - I_{Si}^d - I_{Si}^{nd} \tag{13}$$

where subscript S represents the solid, and the subscript i the liquid phase (water, W or octane, O), θ the water contact angle under octane. I_{Si}^d is the solid/liquid dispersion interaction (the sum $I_{Si}^d + I_{SW}^{nd}$ corresponds to the solid/liquid work of adhesion, W, as expressed in the Dupre equation: $W = \gamma_S + \gamma_i - \gamma_{Si}$). Since $I_{SW}^d \approx I_{SO}^d$ and $I_{SO}^{nd} = 0$, I_{SW}^{nd} satisfies:

$$I_{SW}^{nd} \approx \gamma_W - \gamma_O + \gamma_{WO} \cos \theta \tag{14}$$

where γ_W (72.8 mJ·m^{-2}), γ_O (21.3 mJ·m^{-2}), γ_{WO} (51 mJ·m^{-2}) the water and octane surface tensions and the water/octane interface tension, respectively.

Between water and the glass surface, we will admit that the non-dispersion interactions result mainly from hydrogen bonds when the substrate in not charged. From the non-dispersion (hydrogen bond) energy of interactions, at pzc, expressed per unit of interface area, I_{SW}^{nd}, and from the molar energy of hydrogen bonds, E_H, in the order of 24 kJ/mol [8], we can estimate the number, n, of hydrogen bonds per unit interface area

$$n \approx \frac{I_{SW}^{nd}}{E_H} \times 6.02 \times 10^{23} \tag{15}$$

which corresponds to the surface density of hydroxyl groups on the solid surface.

3. EXPERIMENTAL

3.1. Sample preparation

The glass substrate was aluminoborosilicate glass composed of about 70% SiO_2, 10% Al_2O_3, 10% B_2O_3 (mol%). The glass slides had a thickness of 1.1 mm in the form of microscope slides (1×3 inch). The bare glass slides were cleaned by pyrolysis in an oven at 500°C for 4 h to burn organic contamination. The pyrolyzed samples were held in a glass rack and placed in a glass container covered with an aluminum foil and closed with a glass lid. Following opening of the container, the clean glass samples were used within 5 min. The cleanliness of the glass was controlled by checking that a water drop fully wetted the glass slides.

The glass substrate was treated with 3-aminopropyltriethoxysilane (APTS, Sigma-Aldrich) or 3-glycidoxypropyltrimethoxysilane (GPTS, Sigma-Aldrich) using a proprietary process leading to a uniform coating with thickness of the order of one or a few monolayers.

3.2. Wettability measurements

The surface charge of bare glass and APTS- and GPTS-treated glass was probed by measuring the water contact angle as a function of pH.

Contact angle measurements were made using a Ramé–Hart contact angle goniometer at a controlled temperature of 20 ± 1°C. The average water contact angle as a function of pH was obtained from measurements on 5 different glass slides of both silane coated and uncoated glass. Two measurements were made on each sessile drop for each pH value and on each slide (10 contact angle measurements per pH value, standard deviation of the order of 2°). The water drops had a volume of 2 µl. Water was purified by ionic exchange followed by organic removal leading to a resistivity of 18 MΩ·cm (Elgastat, UHP).

To measure the wettability as a function of pH for bare glass, the substrate was first immersed in hexadecane or octane (99% grade, Sigma-Aldrich). Water-wetting measurements were also made on APTS- and GPTS-coated glass in air as a function of pH. Adding HCl or NaOH to water varied the pH of water drops. The pH values were controlled with a pH meter (Basic 20, Crison).

4. RESULTS

4.1. Pzc of bare glass

Figures 5 and 6 show the variation of the water contact angle as a function of the pH for bare glass, in hexadecane and octane, respectively. We observe that the wetting angle is clearly higher at pH 3 for both alkanes, indicating a pzc at approximately pH 3. This pzc for a bare glass surface is close to that for a silica surface [9–13].

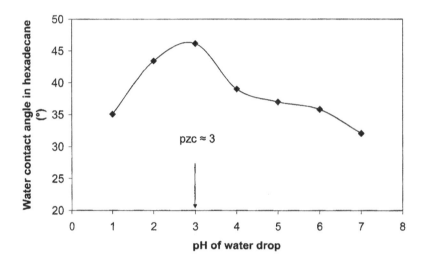

Figure 5. Water contact angle in hexadecane on glass as a function of pH.

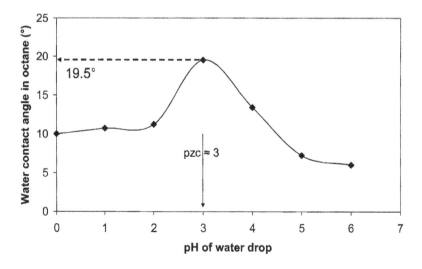

Figure 6. Water contact angle in octane on aluminoborosilicate glass around pzc.

The pK_{a1} of the acidic dissociation of silanol groups (equation (1)) is of the order of 6 [11, 12], meaning that 50% of the silanol groups are negatively charged at pH 6. From the pzc value of 3 and equation (5), we deduce that pK_{a2} (protonation of hydroxyl groups (equation (2)) is about 0 for silica, meaning that only 50% of silanol groups are protonated at pH 0. At the pzc value (pH 3), the silanol groups are practically not dissociated

$$(\frac{\left|SiO^-\right|}{\left|SiOH\right|} = \frac{\left|SiOH_2^+\right|}{\left|SiOH\right|} = 10^{-3}).$$

The water contact angle in octane (Fig. 6) can be used to estimate the density of silanol groups at the glass surface. The number of charged SiOH groups is very limited at the pzc ($\approx$ 0.2%) and the contact angle in octane is 19.5°. We may consider that non-dispersion interactions between water and the un-charged glass surface, I_{SW}^{nd}, at pH 3 are primarily generated by hydrogen bonding. From the non-dispersion (hydrogen bonds) energy of interactions expressed per unit of interface area, I_{SW}^{nd}, obtained from equation (14) and from the molar energy of hydrogen bonds, E_H (24 kJ/mol), the number, n, of silanol groups per unit interface area is obtained from equation (15). This gives about 2.5 silanol groups/nm^2 on the glass surface. This value is comparable to the density of silanol groups measured on silica that had been heat treated to 500°C [14].

4.2. Pzc of functionalized glass and constant capacitor model

The variation of the water contact angle in air as a function pH on APTS-treated glass is presented in Fig. 7. The maximum of the contact angle on APTS slides is obtained at pH $\approx$ 7, implying that the pzc of the APTS-treated glass is around pH 7. This result is in perfect agreement with the study of Golub *et al.* [15], who found a pzc of 7.2 on a silica, Aerosil A-300 modified with APTS by pH-potentiometric titration.

The pK_a of the aminopropyl group ($-R-NH_2$) must be close to the pK_a of N-propylamine which is equal to 10.7 [6]. As the pzc of the APTS-treated glass is around 7, we cannot consider that the surface of the APTS-treated glass surface is just composed of amine functions, since amine functions must be positively charged up to pH $\approx$ 11 (50% of amine functions of N-propylamine are still protonated at pH 10.7). Neutrality of the APTS treated glass surface at pH 7 implies the presence of negatively-charged species to compensate the positively charged amine functions. The presence of free SiOH groups on the functionalized glass may lead to negatively charged SiO$^-$ groups.

At the pH corresponding to the pzc, the density of positive charges is equal to the density of negative charges, implying

$$\left|SiO^-\right| = \left|SiOH_2^+\right| + \left|R-NH_3^+\right| \tag{16}$$

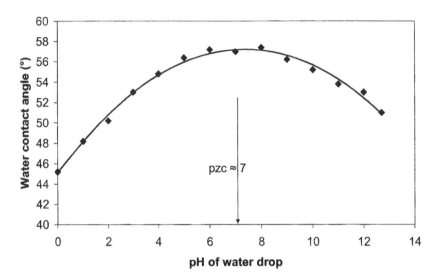

Figure 7. Water contact angle on APTS functionalized glass as a function of pH. The pzc is the pH value where the water contact angle has a maximum value.

At pH 7, the fraction of positively charged silanol groups is negligible. Therefore, equation (16) becomes simply:

$$\left|SiO^-\right| = \left|R - NH_3^+\right| \tag{17}$$

According to studies performed on high surface area silica treated with APTS, the amine density is of the order of 1 to 1.2 group per nm^2 [16]. At pH 7, about 90% of silanol groups are negatively charged ($\dfrac{\left|SiO^-\right|}{\left|SiOH\right|} \approx 10$) and 100% of amine functions are positively charged ($\dfrac{\left|-R-NH_3^+\right|}{\left|-R-NH_2\right|} = 10^{3.7}$). As a consequence, equation (17) indicates that the treated glass surface is composed of about 50% of amine groups and 50% of silanol functions.

In Fig. 7, giving the variation of the water contact angle as a function of pH, it appears qualitatively that the curve has a parabolic shape and is symmetrical about the pzc. In the following, we will demonstrate that the curve of Fig. 7 is compatible with a constant capacitor model formed by the electrical double layer.

Considering that the electrical double layer forms a capacitor of capacitance C (per unit area), the surface charge satisfies

$$\sigma = CV \tag{18}$$

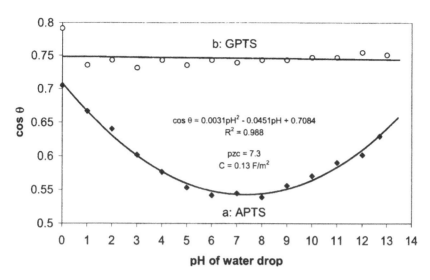

Figure 8. Cosine of the water contact angle on APTS (a) and GPTS (b) functionalized glass as a function of pH. The parabolic variation with pH of the cosine of the wetting angle of water verifies equation (20) (capacitor behavior of the water/solid interface) with the APTS (a) substrate. The constant value of the cosine of the water contact angle on the GPTS (b) indicates that this substrate is uncharged.

where V is the surface potential. The energy of the capacitor, E, contributes to the water/solid interface free energy, so that the variation of the interface free energy may also be written as:

$$dE = -d\gamma_{sw} = d(\frac{1}{2}CV^2) = \sigma dV \tag{19}$$

This is equivalent to the Lippman equation in electrowetting. Considering that σ and $V = 0$ at pzc, combining equations (7), (8) and (19) leads to the variation of the wetting angle cosine with the pH of water

$$\cos\theta = \cos\theta_{pzc} + \frac{CV^2}{2\gamma} = \cos\theta_{pzc} + \frac{C}{2\gamma}(2.303\frac{RT}{F})^2(pzc - pH)^2 \tag{20}$$

which is a parabola having the form: $\cos\theta = A\,pH^2 - BpH + C$. Equation (20) justifies the parabolic variation with pH of the cosine of the wetting angle of water on the APTS treated glass. The solid line a: APTS in Fig. 8 represents the parabolic fit of the experimental points. R^2, the correlation coefficient, is very close to 1 (0.988). From the coefficients of the parabolic fit, it is possible to determine precisely the pzc of the solid surface at pH 7.3 (pzc = $B/2\,A$).

Hence, the surface charge of the APTS-treated glass leads to the formation of an electrical double layer, comparable to a capacitor, when placed in contact with

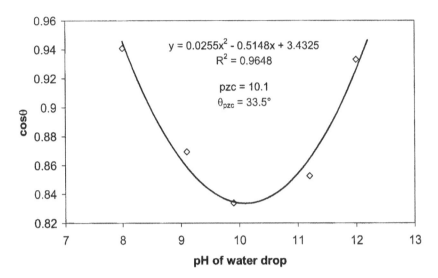

Figure 9. Cosine of the water contact angle on silver oxide as a function of pH. The parabolic variation with pH of the cosine of the wetting angle of water verifies equation (20) (capacitor behavior of
the water/solid interface) (data from Ref. [3]).

water. The value of the constant capacitance C of equation (20) is found to be
equal to 0.13 F/m². Considering a value of 80 for the water dielectric constant (ε_r)
leads to a distance d between the inner and outer planes of the electrical double
layer of 5.5 nm ($d = \dfrac{\varepsilon_0 \varepsilon_r}{C}$, $\varepsilon_0 = 8.854 \times 10^{-12} \, F/m$)).

Glass functionalized with GPTS (epoxysilane) has a different charge behavior.
As shown in Fig. 8 (line b: GPTS), the water contact angle and, therefore, its cosine is insensitive to the pH of water. With the GPTS coating, the water contact
angle is constant with a value of 42 ± 0.5° from pH 1 to 13. However, in very
acidic condition at pH 0, the surface becomes more hydrophilic ($\theta = 37°$). The
small increase of hydrophilicity at pH 0 may be due to the beginning of the hydrolysis reaction of epoxy functions, one epoxy ring leading to two alcohol functions. As a result, glass coated with GPTS is a neutral or an uncharged substrate.

The constant capacitance theory of the electrical double layer was tested with
the data published by Chau and Porter on silver oxide [3] and by McCafferty and
Wightman on oxidized chromium (Cr_2O_3)-plated steel [4]. The results are presented in Figs 9 and 10 where the cosine of the water contact angle is plotted as a
function of the pH of the water drops (in Fig. 10 the average wetting angle was
considered for each pH value). We obtain an excellent agreement for the silver
oxide substrate, and a reasonable agreement for the chromium oxide surface, with
the parabolic fitting corresponding to the constant capacitor model (solid lines
in Figs 9 and 10). The pzc value deduced from the parabolic fitting of cos θ as a
function of pH is 10.1 for silver oxide and 5.4 for chromium oxide (*versus* 10.4

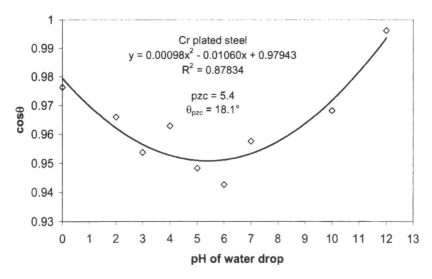

Figure 10. Cosine of the water contact angle on chromium-plated steel as a function of pH. The parabolic variation with pH of the cosine of the wetting angle of water verifies equation (20) (capacitor behavior of the water/solid interface) (data from Ref. [4]).

and 5.2 from our polynomial fitting of θ [4]). Therefore, for single oxide substrates, the model of the capacitor formed by the electrical double layer can be applied. In the case of multi-component oxide solids such as glass, the parabolic variation of the cosine of the water contact angle is only verified near the pzc point. It may be supposed that the complex shape of the curves in Figs 5 and 6 is related to the contribution of the different oxides constituting the surface of glass.

5. CONCLUSIONS

In wettability studies, water pH is a variable which impacts the water contact angle. The method of contact angle measurements at different pH values provides an easy approach for determining the point of zero charge (pzc) of solid surfaces. As in electrokinetic and potentiometric methods, it does not require powdered samples. At pzc, the solid surface is uncharged and the water contact angle has a maximum value. This method has been successfully applied to oxides, glass and functionalized glass.

The pzc of bare glass was found to be about 3, which is very close to the value usually attributed to silica (about 2). For glass coated with an aminosilane the pzc has a value of 7.3, meaning that the substrate is not positively charged at neutral pH as it could be expected from the presence of amine functions. Although amine groups of APTS are positively charged below pH $\approx$ 11, the APTS coated glass substrate is positively charged below pH 7 and negatively charged above pH 7, probably due to the contribution of free SiOH groups. The origin of these free silanol groups may be the glass surface or hydrolyzed alkoxy groups of the APTS silane. These results are in good agreement with the conclusions of former studies on the adsorption of APTS on silica. For this functionalized glass substrate, it was demonstrated that the electrical double layer formed at the solid/water interface behaved as a constant capacitor in wetting. The model was extended to the wetting of metal oxide substrates (silver and chromium oxides) by water, with a good agreement with the experimental results, especially for silver oxide.

Glass functionalized with an epoxysilane (GPTS) is uncharged. The water contact angle does not depend on the water pH, from pH 1 to 13. Therefore, the method of measuring the water contact angle as a function of pH, in air or in another immiscible liquid, allows determination of the surface charge behavior of mineral surfaces and of functionalized substrates.

REFERENCES

1. A. Carré, V. Lacarrière and W. Birch, *J. Colloid Interface Sci.* **260**, 49 (2003).
2. T. Young, *Philos. Trans. Roy. Soc. London* **95**, 65 (1805).
3. L.-K. Chau and M. D. Porter, *J. Colloid Interface Sci.* **145**, 283 (1991).
4. E. McCafferty and J. P. Wightman, *J. Colloid Interface Sci.* **194**, 344 (1997).
5. C. Vittoz, M. Mantel and J. C. Joud, *J. Adhesion* **67**, 347 (1998).
6. R. C. Weast (Ed.), *CRC Handbook of Chemistry and Physics*, 58th edition, CRC Press, Cleveland, OH (1977).
7. W. Birch, S. Mechken and A. Carré, in: *Surface Contamination and Cleaning*, Vol. 1, K. L. Mittal (Ed.), p. 85. VSP, Utrecht (2003).
8. J. N. Murrell and A. D. Jenkins, *Properties of Liquids and Solutions*. Wiley, New York, NY (1994).
9. G. A. Parks, *Chem. Rev.* **65**, 177 (1965).
10. A. Carré, F. Roger and C. Varinot, *J. Colloid Interface Sci.* **154**, 174 (1992).
11. M. L. Hair and W. Hertl, *J. Phys. Chem.* **74**, 92 (1970).
12. R. A. Van Wagenen, J. D. Andrade and J. B. Hibbs, *J. Electrochem Soc.* **123**, 1438 (1976).
13. M. Bezanilla, S. Manne, D. E. Laney, Y. Lyubchenko and H. G. Hansma, *Langmuir* **11**, 655 (1995).
14. R. K. Iler, *The Chemistry of Silica*, p. 634. Wiley, New York, NY (1979).
15. A. A. Golub, A. I. Zubenko and B. V. Zhmud, *J. Colloid Interface Sci.* **179**, 482 (1996).
16. H. Y. Huang and R. T. Yang, *Ind. Eng. Chem. Res.* **47**, 2427 (2003).

Contact Angle, Wettability and Adhesion, Vol. 4, pp. 281–294
Ed. K.L. Mittal
© VSP 2006

Study of electrically-induced wetting on silicon single-crystal substrates

DAIKI KAMIYA* and MIKIO HORIE

Precision and Intelligence Laboratory, Tokyo Institute of Technology, 4259-R2-14 Nagatsuta, Midori-ku, Yokohama 226-8503, Japan

Abstract—Electrically-induced wetting on a silicon substrate shows several stages of droplet formation during the voltage application: (1) intermittent and limited spreading, (2) continuous and unlimited spreading, (3) droplet bursting, (4) generation of microdroplets around the source droplet and (5) further movement of the microdroplets. These stages appeared to be dependent on both the potential and the rate of potential sweep in our experiments. The relationships between these stages of electrically-induced wetting behavior on the silicon substrates and the electrical conditions were experimentally investigated. The positive- and negative-type [111] silicon substrates and a solution of sulfuric acid (0.1 mol/l) for the droplet were used. Experiments were performed using the three-electrode method. When the potential of the substrate was swept from the open circuit potential in the anodic direction (the open circuit potential of the substrates with respect to the reference electrode was about −0.7 to −0.9 V), electrically-induced wetting occurred beyond a threshold of about 0 V. The spreading continued intermittently with the increase in potential, but stopped at 3 V. At this time, the wetted area enlarged about two times of the initial wetted area. Above 3–4 V, droplet spreading re-started, and appeared to be continuous. When the sweep rate was 0.05 V/s or less, the droplet was observed to burst at a potential of 3–4 V. Furthermore, just before the burst of the droplet, numerous microdroplets were generated around the original droplet. Subsequently, the microdroplets combined and moved radially away from the original droplet, and the original droplet burst. These wetting behaviors categorized by applied potential were also observed when the substrate was kept at a constant potential.

Keywords: Electrically-induced wetting; silicon single-crystal substrate; electrolytic solution; direct voltage; current-potential characteristics; stages of wetting behavior.

1. INTRODUCTION

Research and development on electrowetting *via* control of the wetting between a droplet and a hydrophobic insulator film by application of voltage between an electrode immersed in the droplet and an electrode attached on the back side of the insulator film has been brisk in this decade [1–3]. The history of electrowet-

*To whom correspondence should be addressed. Tel.: (81-45) 924-5012;
Fax: (81-45) 924-5012; e-mail: dkamiya@pi.titech.ac.jp

ting itself is not new; Lippmann had foreseen the phenomenon in the nineteenth century [4, 5]. Research on electrowetting in recent years has been furthered by development of thin polymer films with hydrophobicity and high dielectric strength [6, 7], and the recognition that electrowetting has a high potential for application in micro-electromechanical systems (MEMS) [8]. Electrowetting is also used in handling of droplets in biomedical engineering [9], digital microfluidic circuits [10], and display devices [11]. Moreover, in addition to application development based on electrowetting, improvement of the theory [12, 13] and discovery of a new phenomenon related to electrowetting [14] have also been reported.

Although an insulator and mercury have been used for materials in contact with a liquid in the previous research, we have studied electrically-induced wetting on semiconductor materials, including silicon, because of the expected similarity in performance with insulator materials [15]. From these experiments, it was observed that an aqueous droplet spread with the application of voltage between a silicon substrate and an electrode immersed in the droplet as shown in Fig. 1. With the exception of hydrofluoric acid, the aqueous droplet spread regardless of kind of ion present in the solution when the silicon substrate was anodically polarized. Moreover, it is generally known that an anodic oxide film is formed on the silicon surface during anodic polarization of the silicon substrate in a solution [16, 17]. For hydrofluoric acid the droplet did not spread, but the formation of the anodic oxide film and its dissolution repeated on the interface between the droplet

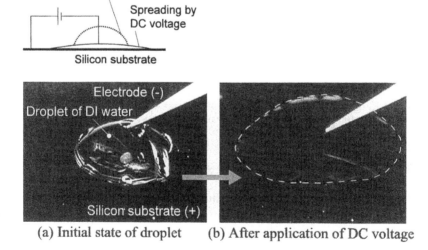

(a) Initial state of droplet   (b) After application of DC voltage

Figure 1. Electrically-induced spreading of the droplet of deionized water on an n-type [111] silicon substrate. A DC voltage of 3 V was applied between the silicon substrate and the electrode immersed in the droplet. (a) Before application of voltage and (b) after application of voltage and spreading of droplet.

and the silicon substrate during anodic polarization and resulted in formation of porous structures [18, 19]. From the above two points, it was hypothesized that the spreading of the droplet on the silicon substrate during anodic polarization would be related to formation of the anodic oxide film. We hypothesize that electrically-induced wetting occurs when an anodic oxide film works as an insulator as shown in Fig. 2. A droplet on a silicon substrate initially takes a nearly hemispherical form because of the surface tension of water and the hydrophobic nature of the silicon surface. In this state, an anodic oxide film is formed along the interface between the substrate and the droplet by anodic polarization of the substrate. Since the oxide film is naturally an insulator to some extent, electric charge is accumulated across the oxide film due to the applied voltage. With the accumulation of electric charge, electrowetting starts and the droplet spreads. An anodic oxide film is also formed on the new interface created by droplet spreading. In short, the droplet spreading occurs as a result of alternating between the anodic oxidation on

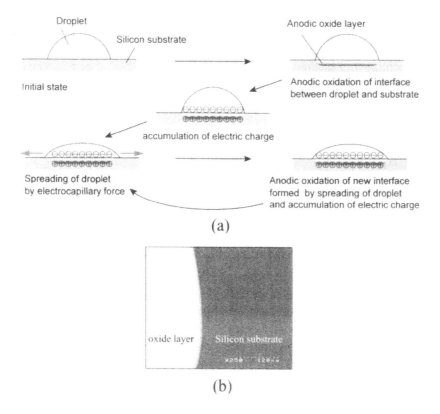

(a)

(b)

Figure 2. (a) Proposed mechanism for electrically-induced spreading of the droplet which is accompanied by anodic oxidation of the silicon surface and (b) SEM micrograph of the anodic oxide film along the wetted area.

the silicon surface and the electrowetting effects with accumulation of electric charge across the oxide film. The oxide film was actually observed along the wetted area with a scanning electron microscope after the experiment as shown in Fig. 2. Electrically-induced spreading was also observed using a cathodic potential with a solution containing sodium or potassium ions [15]. It was believed that adsorption of sodium and potassium ions on the silicon surface with cathodic polarization should affect spreading of the droplet. However, we do not discuss electrically-induced wetting by cathodic potential in this report because we have not definitely investigated the wetting behavior by cathodic polarization yet.

Electrically-induced wetting on a silicon substrate shows several stages of droplet formation during anodic polarization of the substrates: (1) intermittent and limited spreading, (2) continuous and unlimited spreading, (3) droplet bursting, (4) generation of microdroplets around the source droplet and (5) further movement of the microdroplets. These stages appeared to be dependent on both the potential and the rate of potential sweep in our experiments. In order to pursue the mechanism for spreading of the droplet in detail, it is necessary to clarify the relationship among the applied potential, the current and the spreading behavior. In this study, experiments were carried out using the potential-sweep method, the single-potential-step method and the potential-pulse method. The relationship between the electrical conditions and the spreading behavior was clarified by using the positive- and negative-type [111] silicon substrates and a solution of sulfuric acid for the droplet.

2. EXPERIMENTAL

2.1. Materials

Two types of silicon single-crystal substrates were used. Doping of these substrates was different for each type, i.e., the doping was positive or negative. Crystal orientation of the substrates was [111]. Electrical conductivity was 0.1–1.0 Ω·cm for the positive-type substrates and 5–15 Ω·cm for the negative-type substrates. The silicon substrates were 10×10-mm^2 and 0.38-mm-thick squares. The silicon substrates were treated with acid and alkali solutions for removal of contaminants from the substrate surfaces, i.e., the RCA cleaning was employed [20]. After the treatment, an indium film with a thickness of 1 μm was deposited on the back surface of the substrates via evaporation technique to form an ohmic contact as shown in Fig. 3, and a copper wire was connected to the indium film using a conductive epoxy. These electrically connected parts were covered with epoxy for protection of the indium film against the solution of hydrofluoric acid in the following treatment. Each substrate was treated with 1% hydrofluoric acid and rinsed with deionized water (DI water) just before each experiment. The thin native oxide layer on the silicon surface was thus removed and a hydrogen-terminated hydrophobic surface was obtained [21, 22].

The droplets were an aqueous solution of 0.1 mol/l sulfuric acid. The volume of the droplets was 2 µl and a micro-dispenser was used to place the droplets on the substrate.

2.2. Methods

Figure 3 shows the schematic diagram of the experimental set-up. Current–potential measurements were conducted with three electrodes [18]. These consisted of a working electrode (the silicon substrate), a counter electrode, and a reference electrode. For the experiments, the counter electrode was a Pt needle and the reference electrode was a Pt wire coated with a fluororesin. These electrodes were immersed in the droplet located on the silicon substrate. The Pt wire was exposed at the tip, and the potential of the solution was measured at the tip. The tip of the reference electrode was set in the center of the droplet, and the gap between the tip of the reference electrode and the substrate surface was 200 µm. A commercial potentiostat system (ALS/CHI model 750A, CH Instruments, Austin, TX, USA) was used to control the potential of the substrate *versus* the reference electrode as well as to measure the current. The target values of the potential were set on the computer, and the computer controlled the potentiostat system. The

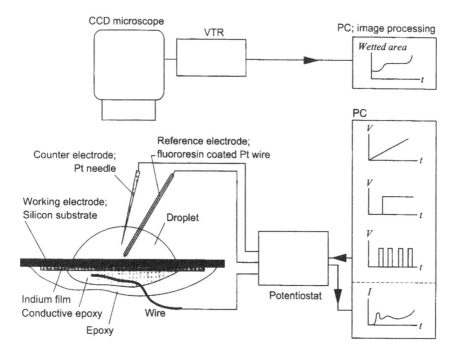

Figure 3. Schematic of the experimental setup using three electrodes for electrically-induced wetting of a droplet on a silicon substrate.

values of the current measured by the potentiostat were stored on the computer hard drive, together with the target values of the potential.

Three kinds of experiments using different applied potential waveforms were conducted. In the first experiment, the potential of the substrate *versus* the reference electrode was swept at a fixed rate, starting from the open circuit potential. The rate of potential sweep was from 0.01 to 0.2 V/s and the maximum potentials were 4.5 V for the positive-type substrates and 6.0 V for the negative type. Second, single-potential-step measurements were carried out. The potential of the substrate was changed instantaneously from the open circuit potential to the target potential in a single step. The maximum target value of the potential was 5.0 V. Finally, an experiment on the pulse response to a short-time rectangular wave was carried out. The pulse amplitude of the potential was 1.7 V with respect to the open-circuit potential. The pulse width and the pulse cycle were 0.01 s and 1 s, respectively.

The magnitude of the wetted area accompanying spreading of the droplets was observed directly from above *via* a CCD microscope, and was recorded with a digital video recorder at a rate of 30 frames/s. After the experiment, the wetted area was calculated by image processing using a personal computer.

3. RESULTS

3.1. Potential sweep

Figure 4 shows the current–potential curves when the potential of the substrate was swept in the anodic direction at a fixed rate, and the change in the wetted area accompanying electrically-induced spreading of the droplets was recorded. The potential sweep was started from the open circuit potential. The open circuit potentials were about –0.7 V for the positive-type silicon substrate and –0.9 V for the negative type. Generally, since the chemical potential of electrons in a solid is different from that in an electrolytic solution, when the solid and the solution are in contact with each other an electrical potential difference arises between them in the equilibrium state. In this experiment, the open circuit potential implies the value obtained by adding the electrical potential difference between the solution and the platinum of the reference electrode to the electrical potential difference between the silicon and the solution. That is, when the substrate was set at the open circuit potential, the original thermodynamic equilibrium between the substrate and the droplet was maintained and the current did not flow.

After the potential sweep started, the first increase in current occurred before the potential of 0 V. At this time, the hydrogen atoms terminating the surface of the silicon were replaced by silanol groups. Subsequently, an anodic oxide film was formed along the interface between the substrate and the droplet. The spreading of the droplet started at 0 V for the negative-type substrates and at 0.4 V for

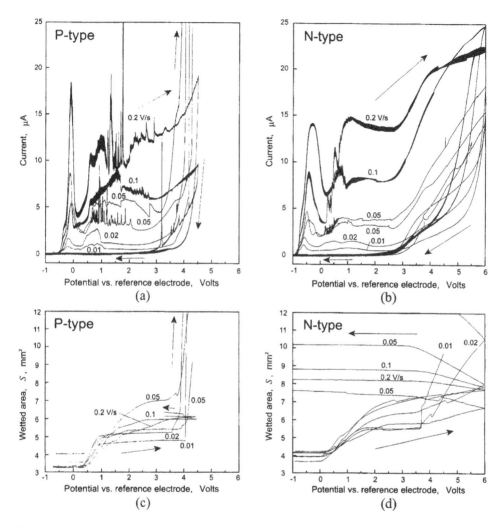

Figure 4. Current-potential relation when a potential sweep was performed for electrically-induced wetting by anodic polarization of the substrates: (a) p-type [111] Si and (b) n-type [111] Si; and the change in the wetted area by spreading of the droplet: (c) p-type [111] Si and (d) n-type [111] Si. The potential sweep rate was constant and is indicated in the figures (unit V/s). The volume of the droplet was 2 µl and the solution was 0.1 mol/l sulfuric acid.

the positive type, and stopped by the time the applied potential was 3 V. The increase in the wetted area accompanied by spreading of the droplet resulted in increase in current. This increase in current was due to the anodic oxidation of the new interface formed by spreading of the droplet. When the sweep rate was especially low on the positive-type substrates (0.01 and 0.02 V/s), the start and cessation of spreading were clearly observed at 0.4 and 1 V, as shown in Fig. 4c. From these results, it could be recognized that the increase in current was accompanied

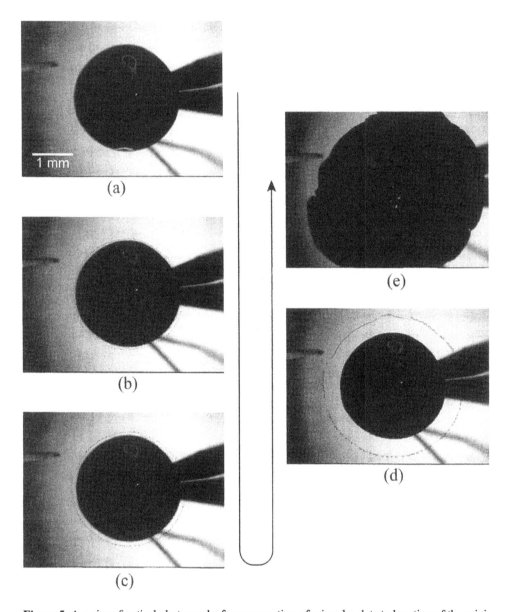

Figure 5. A series of optical photographs from generation of microdroplets to bursting of the original droplet on a p-type [111] silicon substrate; (a) after limited spreading (3.32 V), (b) generation of microdroplets around the original droplet (3.79 V), (c) starting of microdroplets movement (3.81 V), (d) microdroplets located radially away from the original droplet (4.05 V) and (e) bursting of the droplet (4.06 V). The potential sweep rate was 0.01 V/s. The droplet was 0.1 mol/l sulfuric acid with a volume of 2 µl.

by the spreading of the droplet. The droplets spread intermittently at this time and this intermittent spreading caused a sharp change in current in Fig. 4.

At 3 V applied potential and more positive potential, the current increased further, and the droplet began to spread continuously. This spreading continued until the applied potential was turned down to 3 V or less. The contact angle of the droplet became almost zero during the spreading, but the spreading still continued. For a positive-type substrate and a sweep rate of 0.05 V/s or less, at about 3–4 V, the droplets exploded and spread instantaneously, and the current increased rapidly. This current made an anodic oxide film on the new interface formed with the explosion of the droplets. On the other hand, when the potential sweep rate was high, the droplets did not burst but the current increased gradually. The increase in current was partially due to the anodic oxidation of the new interface. However, this increase in current is believed to be due either to electrolysis of water along the interface between the droplet and the substrate at more than 3 V or a change in the reaction mechanism for the anodic oxidation. Further study is required to clarify the relationship between the change in current at 3 V and the change in spreading behavior.

For the negative-type substrate, the current was saturated although a similar increase in current took place above 3 V. Generally, when anodic polarization of a semiconductor is carried out in solution, since holes behave as a carrier inside the semiconductor, the interface between the negative-type semiconductor and the solution shows the rectification effect. At this time, the potential drop becomes dominant in the space charge layer within the semiconductor near the interface, and the potential drop in the electrical double layer on the solution side of the interface becomes small compared with a positive-type semiconductor arrangement. For this reason, in the negative-type substrates, the current was limited. For a negative-type substrate with a low sweep rate, it was also observed that the droplet spread momentarily at about 4 V, however the subsequent droplet explosion was small compared to that in the positive-type substrate experiment.

Although it is mentioned above that the droplet exploded on the positive-type substrate with a low sweep rate of 0.05 V/s or less; however, before the burst of the droplet numerous microdroplets appeared around the original droplet at a potential of 3.5 to 4 V. The appearance of such microdroplets has been reported as one of the electrowetting effects on a dielectric material also [14]. Furthermore, with an increase in potential during the experiment, it was observed that the microdroplets moved radially away from the original droplet. A series of these situations are shown in Fig. 5. Generation of microdroplets was seen at the same potential as when the wetting stage changed from limited wetting to continuous wetting as mentioned above. In the expectation that degree of anodic oxidation and accumulation of electric charge on the interface between the substrate and the droplet, as well as the potential, would affect generation of microdroplets, the electric charge density was calculated, as shown in Fig. 6. This electric charge density is the value calculated by carrying out time integration of the current and dividing the result by the wetted area at each time. This electric charge density is the sum of the current consumed in order to form the anodic oxide film and the amount of electric charges that actually accumulate at the interface. For the low

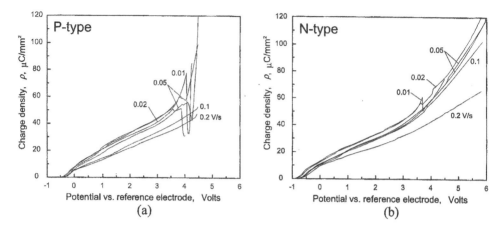

Figure 6. Charge density as calculated by carrying out time integration of the current divided by the wetted area; (a) p-type [111] Si and (b) n-type [111] Si. The potential sweep rate was constant and is indicated in the figures (unit V/s). The droplet was 0.1 mol/l sulfuric acid with a volume of 2 μl.

potential sweep rate (0.01–0.05 V/s), the electric charge density did not depend on the potential sweep rate, and it appeared to be saturated. But, before the burst of the droplet, the electric charge density increased rapidly at about 3.5 V. Overall, however, the electric charge density was low when the sweep rate was fast (0.1, 0.2 V/s). The difference in the charge density suggests that the microdroplets would be generated around the original droplet when anodic oxidation is saturated at about 3 V.

3.2. Step and pulse responses

The results of step response are shown in Fig. 7. These results clearly show the three stages of spreading that were seen for the potential sweep and classified in terms of the electrical potential. There was no spreading of the droplet up to an applied potential of 0 V from the open circuit potential. From 0–3 V, limited spreading of the droplet appeared and the increase in the wetted area was dependent on the applied potential. At this time, the flow of current was in agreement with the spreading of the droplet, and it simultaneously became almost zero as the spreading stopped. With an applied potential above 3 V, a two-step behavior of spreading appeared. First, during the first part of the potential application ($t = 0$–20 s) the spreading behavior was the same as that seen when applying a potential of 0 to 3 V. Second ($t > 20$ s), the droplet spread continuously with a constant current. In the experiments, there seemed to be no limit to spreading during the second step. Since the first and second steps of spreading were entirely different in current-time relation and spreading behavior, it was expected that these two spreading steps would have different spreading mechanisms. With a potential step, the droplet did not burst.

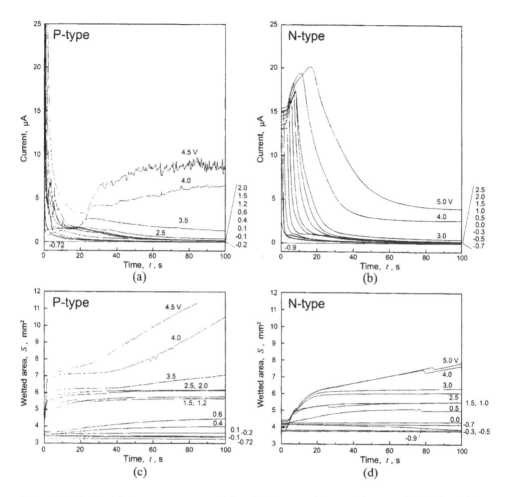

Figure 7. Relationship between current and time for the potential step: (a) p-type [111] Si and (b) n-type [111] Si; and the change in the wetted area by spreading of the droplet: (c) p-type [111] Si and (d) n-type [111] Si. The applied potential is indicated in the figures (unit V). The droplet was 0.1 mol/l sulfuric acid with a volume of 2 µl.

For the negative-type silicon, the rectification effect on the substrate–solution interface was clearly seen at the start of the potential application, and the current was limited to about 15 µA. Due to the current limit there was a several seconds delay before spreading of the droplet. It was believed that the current for several seconds after application of the potential step was unrelated to the spreading of the droplet and was consumed in forming the anodic oxide film on the original interface. However, for the positive-type silicon, because there was no current limit and the current surged with the potential step, the spreading of the droplet started almost concurrently with the potential step.

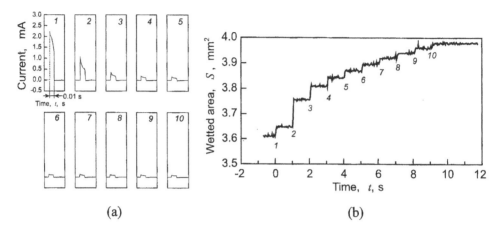

Figure 8. Pulse response for a p-type silicon substrate. 1.7 V pulse amplitude from the open circuit potential, 0.01 s pulse width and 1 Hz cycle: (a) current–time relations and (b) change in the wetted area by spreading of the droplet. The droplet was 0.1 mol/l sulfuric acid with a volume of 2 μl.

The pulse–response experiment was performed using a positive-type substrate. The pulse amplitude of the potential was 1.7 V with respect to the open circuit potential. The pulse width and the pulse cycle were 0.01 s and 1 s, respectively. Figure 8 shows the current–time relations and the increase in the wetted area. The increase in the wetted area became saturated, and simultaneously the current also decreased continuously with every pulse. These current–time relations and the spreading results were almost the same as the result of a step response with a potential of 0–3 V. The spreading occurred within 67 ms (equivalent to two video frames) or less.

4. CONCLUSIONS

To conclude, the different stages of electrically-induced wetting on the silicon substrates are summarized with respect to the potential and the rate of potential sweep based on the experimental results in Fig. 9. The open circuit potential of the silicon substrates with respect to the reference electrode immersed in the droplets was –0.7 to –0.9 V. When the potential of the silicon substrate was swept from the open circuit potential in the anodic direction, the spreading of the droplets occurred beyond the threshold of about 0 V. This spreading was intermittent, periodically stopping until 3 V. Above 3 V, droplet spreading re-started, and it appeared to be continuous. When the sweep rate was 0.05 V/s or less, the droplet burst at a potential of 3–4 V. Furthermore, just before the burst of the droplet, numerous microdroplets were generated around the original droplet. Subsequently, the microdroplets combined and moved radially away from the original droplet, and the droplet burst.

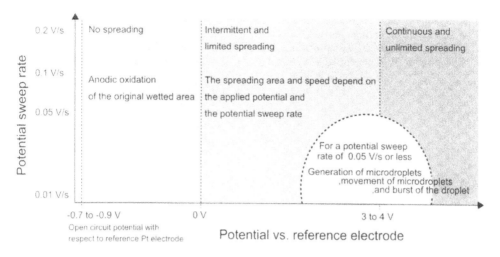

Figure 9. Summary of the stages of the wetting behavior induced by anodic polarization of the silicon substrate with respect to the potential and the potential sweep rate; the droplet was 0.1 mol/l sulfuric acid with a volume of 2 μl.

When the substrate was kept at a constant potential, three stages of wetting behavior were classified in terms of the applied potential: no spreading of the droplet below a potential of about 0 V, limited spreading at a potential of 0–3 V, and a continuous and unlimited spreading above 3 V. The continuous and unlimited spreading occurred following the limited spreading.

REFERENCES

1. E. Colgate and H. Matsumoto, *J. Vac. Sci. Technol.* **A8**, 3625–3633 (1990).
2. M. G. Pollack and R. B. Fair, *Appl. Phys. Lett.* **77**, 1725–1726 (2000).
3. H. Moon, S. K. Cho, R. L. Garrell and C. J. Kim, *J. Appl. Phys.* **92**, 4080–4087 (2002).
4. G. Lippmann, *Ann. Phys.* **149**, 546–561 (1873).
5. G. Lippmann, *Ann. Chim. Phys.* **5**, 494–549 (1875).
6. B. Janocha, H. Bauser, C. Oehr, H. Brunner and W. Goupel, *Langmuir* **16**, 3349–3354 (2000).
7. D. Hegemann, H. Brunner and C. Oehr, *Plasmas Polymers* **6**, 221–235 (2001).
8. J. H. Lee and C. J. Kim, *J. Microelectromech. Syst.* **9**, 171–180 (2000).
9. P. K. Wong, T. H. Wang, J. H. Deval and C. M. Ho, *IEEE/ASME Trans. Mechatron.* **9**, 366–376 (2004).
10. S. K. Cho, H. Moon and C. J. Kim, *J. Microelectromech. Syst.* **12**, 70–80 (2003).
11. R. A. Hayes and B. J. Feenstra, *Nature* **425**, 383–385 (2003).
12. V. Peykov, A. Quinn and J. Ralston, *Colloid Polym. Sci.* **278**, 789–793 (2000).
13. A. Quinn, R. Sedev and J. Ralston, *J. Phys. Chem.* **B107**, 1163–1169 (2003).
14. F. Mugele and S. Herminghaus, *Appl. Phys. Lett.* **81**, 2303–2305 (2002).
15. D. Kamiya and M. Horie, in: *Contact Angle, Wettability and Adhesion*, Vol. 2, K. L. Mittal (Ed.), pp. 507–520. VSP, Utrecht (2002).
16. R. Memming, *Semiconductor Electrochemistry.* Wiley-VCH, Weinheim (2001).
17. H. Gerischer and W. Mindt, *Electrochim. Acta* **13**, 1329–1341 (1968).
18. J.–N. Chazalviel, *Electrochim. Acta* **37**, 865–875 (1992).

19. R. Memming and G. Schwandt, *Surface Sci.* **4**, 109–124 (1966).
20. W. Kern, *J. Electrochem. Soc.* **137**, 1887–1892 (1990).
21. V. Lehmann, *Electrochemistry of Silicon.* Wiley-VCH, Weinheim (2002).
22. H. Gerischer, P. Allongue and V. C. Kieling, *Ber. Bunsenges. Phys. Chem.* **97**, 753–757 (1993).

Contact Angle, Wettability and Adhesion, Vol. 4, pp. 295–305
Ed. K.L. Mittal
© VSP 2006

Criteria for ultralyophobic surfaces

C. W. EXTRAND*

Entegris Inc., 3500 Lyman Blvd., Chaska, MN 55318, USA

Abstract—Surfaces that exhibit super-repellency combine inherent lyophobicity and tortuous topography. The combination of these traits is a necessary but not sufficient requirement for super repellency, two additional criteria must be met to invoke ultralyophobicity. First, interaction of a liquid with asperities must direct the surface forces at the contact line upward. Second, the asperities must be tall enough that liquid protruding between them does not contact the underlying solid. Based on these criteria, a model has been developed and tested using experimental data available in the literature. The criteria were found to correctly predict suspension of small water drops on model rough surfaces with a variety of asperity sizes and spacings.

Keywords: Wetting; wettability; contact angles; rough surfaces; composite surfaces; ultrahydrophobicity; ultralyophobicity; super-repellency.

1. INTRODUCTION

If a small liquid drop is deposited on a rough surface, in most cases it will spread, engulfing surface asperities. Figure 1a shows such a surface where the contact liquid has penetrated the spaces between the asperities. Here, interaction of the liquid with the asperities can lead to large values of contact angle hysteresis [1–11] and, consequently, substantial forces may be required to initiate drop movement [12]. Less frequently, drops are suspended atop inherently lyophobic asperities leaving air (or vapor) between them [2, 13–26]. This suspension, depicted in Fig. 1b, leads to very poor liquid–solid adhesion and very large apparent contact angles (140° to 180°) that are characteristic of super-repellent or ultralyophobic surfaces.

Studies of ultralyophobic surfaces first appeared in the scientific literature in the 1940s [2]. While this subject has received continued attention since that time, most investigators have focused on the relation between contact angles and surface geometry [2, 5, 9, 13, 18–20, 23–26]. More recent investigations have targeted the conditions that lead to ultralyophobicity [10, 27–30]. This paper describes the conditions required for the creation of ultralyophobic surfaces.

*Tel.: (1-952) 556-8619; Fax: (1-952) 556-8023; e-mail: chuck_extrand@entegris.com

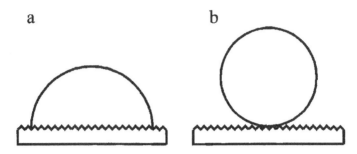

Figure 1. Small sessile drops on rough surfaces. (a) A collapsed drop on rough surface; the contact area is composed solely of a liquid/solid interface. (b) A suspended drop on an ultralyophobic surface.

Quantitatively, two criteria must be met: a contact line density criterion and an asperity height criterion. After derivation and explanation of these two criteria, their ability to correctly predict experimental observations on model rough surfaces with regular, periodic asperities is demonstrated.

2. THEORETICAL BASIS

In order for surfaces to show ultralyophobic behavior two conditions must be met. First, interaction of a liquid with asperities must direct the surface forces at the contact line upward; the surface forces must be of sufficient magnitude to suspend

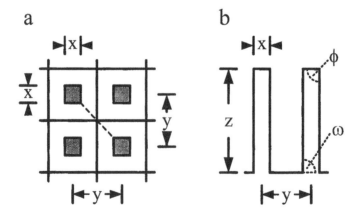

Figure 2. Schematic depiction of a surface covered by a regular array of square pillars of top width, x, and height, z. (a) Plan view. Each unit cell has a linear dimension of y. The shaded areas are the tops of asperities. The length of the dashed diagonal line connecting adjacent asperities is 2b. (b) Side view. The angle subtended by the top edges of the asperities is ϕ and the rise angle of their sides is ω. As drawn, $\phi = \omega = 90°$.

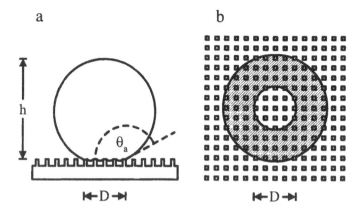

Figure 3. A small liquid drop suspended on an ultralyophobic surface consisting of a regular array of square pillars. (a) Side view. (b) Plan view. The apparent contact area is a composite of gas/liquid and liquid/solid interfaces. θ_a is the apparent advancing contact angle and D is the apparent diameter of the drop.

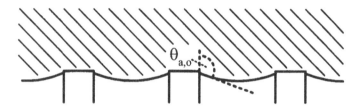

Figure 4. A magnified side view of a suspended liquid drop on a model ultralyophobic surface. The liquid exhibits its true advancing value, $\theta_{a,o}$, on the sides of the pillars.

the liquid against the downward pull of gravity (or other body forces). Second, the asperities must be tall enough that liquid protruding between them does not contact the underlying solid. The first condition can be described quantitatively by a contact line density criterion [27], the second by an asperity height criterion [30].

Consider the model rough surface shown schematically in Fig. 2. Surface asperities take the form of a regular array of square pillars. The pillars have a top width x and height z. The spacing between them is y. The angle subtended by the top edge of the asperities is ϕ, the rise angle of the sides of the asperities is ω and

$$\phi = 180° - \omega. \tag{1}$$

As drawn in Fig. 2, $\phi = \omega = 90°$. The area density, δ, of asperities is

$$\delta = 1/y^2. \tag{2}$$

Contact line density, Λ, which is the length of asperity perimeter per unit area that could potentially suspend a liquid drop, is defined as the product of δ and the feature perimeter, p,

$$\Lambda = p\delta. \tag{3}$$

For the regular array of square pillars described in Fig. 2, Λ is

$$\Lambda = 4x/y^2. \tag{4}$$

Figure 3 shows side and plan views of a liquid drop on a model ultralyophobic surface covered with a regular array of square pillars. The drop is suspended atop the pillars and exhibits a very large apparent advancing contact angle, θ_a. Its apparent contact area (defined by the contact diameter D) is composite in nature, consisting of both gas–liquid and liquid–solid interfaces. The fraction of liquid–solid contact area within this composite interface, α_p, can be estimated as

$$\alpha_p = (x/y)^2. \tag{5}$$

A magnified side view of the composite interface is shown in Fig. 4. Suspended drops protrude between the pillars and exhibit their true advancing value, $\theta_{a,o}$, on the sides of the asperities. The true advancing contact angle, $\theta_{a,o}$, is the value measured on a chemically-equivalent, smooth, horizontal surface. The longest linear distance, 2b, between adjacent pillars is a diagonal line (depicted in Fig. 2),

$$2b = 2^{\frac{1}{2}}(y-x). \tag{6}$$

2.1. Contact line density criterion for small drops

A contact line density criterion can be derived for a small liquid drop on a rough surface by examining the interplay of body and surface forces [27]. Body forces, F, associated with a gently deposited drop are determined from the density, ρ, and the unsupported volume, V_u, of the liquid drop,

$$F = \rho g V_u, \tag{7}$$

where g is the acceleration due to gravity. The unsupported or suspended volume can be approximated from the total volume of the liquid drop, V, and the fractional area of contact between the liquid and the pillar tops, α_p, as

$$V_u = V - \alpha_p h A, \tag{8}$$

where h is drop height and A is its apparent contact area. On the other hand, the surface forces, f, depend on surface tension of the liquid, γ, the contact angle on the sides of the asperities with respect to vertical plane, θ_s, the perimeter of each asperity, p, the area density of the asperities, δ, and the apparent contact area of the drop, A,

$$f = -\rho\delta A\gamma \cos\theta_s. \tag{9}$$

The contact angle on the sides of the asperities, θ_s, is related to $\theta_{a,o}$ by ω,

$$\theta_s = \theta_{a,o} + \omega - 90°. \tag{10}$$

The contact line density has a critical value, Λ_c,

$$\Lambda = \Lambda_c, \tag{11}$$

where the body and surface forces are equal. It is assumed that the ultralyophobic surface is sufficiently porous so that gases are not trapped between asperities and, in turn, do not contribute to the suspension forces.

$$F = f. \tag{12}$$

Combining equations (3) and (7)–(12) leads to the critical value of Λ_c for small drops:

$$\Lambda_c = -\rho g V(1 - \alpha_p hA/V)/A\gamma \cos(\theta_{a,o} + \omega - 90°). \tag{13}$$

To simplify the analysis, we use the maximum height of the drop at the apex, equation (8). Away from the apex, its height varies with position. If the drop is small and the apparent contact angle is large, then the contact area should be small and the height of the liquid supported by the contact patch will not vary much. If sufficiently small, drops retain spherical proportions, which allows both drop height and apparent contact area, A, to be written in terms of the apparent advancing contact angle, θ_a, and the drop volume, V [27]:

$$h = ((3V/\pi)(1 - \cos\theta_a)/(2 + \cos\theta_a))^{1/3} \tag{14}$$

and

$$A = \pi^{1/3}(6V)^{2/3}(\tan(\theta_a/2)(3 + \tan^2(\theta_a/2)))^{-2/3}. \tag{15}$$

Substituting equations (14) and (15) into (13) allows estimation of Λ_c values without knowledge of drop dimensions,

$$\Lambda_c = -\rho g V^{1/3}(1 - k)(\tan(\theta_a/2)(3 + \tan^2(\theta_a/2)))^{2/3}/((36\pi)^{1/3}\gamma \cos(\theta_{a,o} + \omega - 90°)), \tag{16}$$

where k is a correction factor that accounts for liquid supported by pillars or asperities,

$$k = \alpha_p(96(1 - \cos\theta_a)/(2 + \cos\theta_a))^{1/3}(\tan(\theta_a/2)(3 + \tan^2(\theta_a/2)))^{-2/3}. \tag{17}$$

If α_p is small and θ_a is large, then k is effectively zero and equation (16) reduces to

$$\Lambda_c = -\rho g V^{1/3}(\tan(\theta_a/2)(3 + \tan^2(\theta_a/2)))^{2/3}/((36\pi)^{1/3}\gamma \cos(\theta_{a,o} + \omega - 90°)). \tag{18}$$

Error in Λ_c values from neglecting k is small and is discussed elsewhere [30].

2.2. Critical asperity height criterion

Because suspended drops establish true advancing contact angles on the sides of asperities, surface curvature may cause liquid to protrude downward between asperities, as depicted in Figs 4 and 5. Figure 5 shows a detailed view of a liquid protrusion. The protrusion depth, d, depends on the distance between the asperities, 2b, and the depth angle, θ_d, between the horizontal plane and the liquid protrusion. Assuming that the cross-section of the liquid protrusion can be described as a segment of a circle, then d can be calculated as [30]

$$d = b \tan(\theta_d/2), \tag{19}$$

where

$$\theta_d = \theta_{a,o} + \omega - 180°. \tag{20}$$

If the protrusion depth equals the critical value of asperity height, z_c,

$$z_c = d, \tag{21}$$

then liquid will just touch the floor between asperities, causing collapse. Combining equations (19)–(21) gives an expression for estimating the critical asperity height,

$$z_c = b \tan((\theta_{a,o} + \omega - 180°)/2). \tag{22}$$

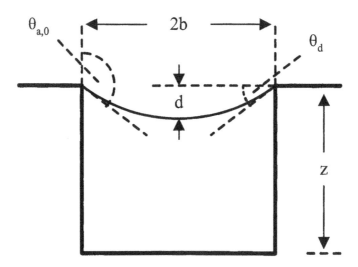

Figure 5. A magnified side view of a suspended liquid protruding to a depth, d, between two model asperties. The asperities have a height z and a maximum distance between them 2b. The liquid exhibits its true advancing value, $\theta_{a,o}$, on the sides of the asperities. θ_d is the angle between the horizontal plane and the protruding liquid.

3. RESULTS AND DISCUSSION

3.1. Attributes of the model

The reported size of asperities required to create ultralyophobic surfaces has been controversial [23]. The problem lies in defining surfaces solely in terms of asperity height or spacing. While these are important considerations, other parameters such as liquid density, surface tension, contact angles and asperity shape also play a role. From equation (18), if ρ, V, or θ_a increase, higher Λ_c will be required to suspend drops. On the other hand, if γ, ω, or $\theta_{a,o}$ increase, a lower Λ_c will suffice. The edge angle is extremely important in determining ultralyophobicity. In order for a surface to support a liquid drop, ω (and $\theta_{a,o}$) must be sufficiently large to direct the surface forces upward against the body force (a positive Λ_c value). Marshalling the surface forces against gravity requires that $\theta_{a,o} + \omega - 90° > 90°$. In terms of the contact line density criterion, Λ must be $> \Lambda_c$. If the surface and body forces are both directed downward, Λ_c will be negative and drops will collapse regardless of the magnitude of Λ.

High contact-line density alone is not a sufficient condition for ultralyophobicity. Not only must $\Lambda > \Lambda_c$, but asperity height must be greater than the protrusion, $z > d$ (or $z > z_c$). Otherwise, the protruding liquid will contact the base of the solid surface, instigating collapse.

Consider a hypothetical ultrahydrophobic surface covered with a regular array of square pillars where x = 8 µm, y = 16 µm, z = 40 µm, $\omega = 90°$, $\theta_{a,o} = 120°$ and $\theta_a = 170°$. The contact line density of this model surface is $\Lambda = 1.3 \times 10^5$ m^{-1} (equation (4)). On the other hand, equation (18), which captures the interplay of surface and body forces, suggests that the minimum Λ value for suspension of small water drops is $\Lambda_c \approx 1 \times 10^4$ m^{-1}. From equations (19) and (20), water suspended atop these square pillars would protrude downward 1.5 µm. Because $\Lambda > \Lambda_c$ and $z > z_c$, this surface would be expected to suspend small, gently-deposited water drops.

3.2. Comparison with experimental data

In this section, results from experimental studies on model ultralyophobic surfaces [10, 26] are used to test the proposed criteria. Öner and McCarthy [10] prepared hexagonal arrays of square pillars on silicon wafers with $\omega = 90°$ (Fig. 6) using photolithography and then treated them with hydrophobic silane agents. Table 1 lists the dimensional parameters and apparent advancing contact angles on dimethyldichlorosilane (DS)-treated surfaces. (Based on geometry, Λ for hexagonal arrays of square pillars is $4x/y^2$.) The hydrophobic surfaces created by Öner and McCarthy had pillar heights that were greater than the penetration depth of the liquid, i.e., $z > z_c$; therefore, water drop suspension was determined by contact line density. Based upon examination of the experimental data, the critical value of Λ for these water/surface combinations was found to lie between 1×10^4 m^{-1} and

a   b

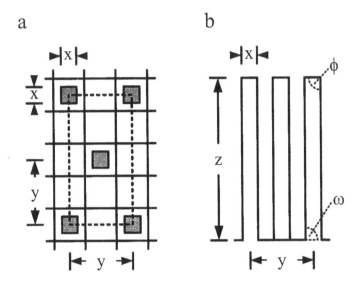

Figure 6. Schematic depiction of a surface covered by a hexagonal array of square pillars of top width x and height z. (a) Plan view. The square grid has a linear dimension of ½y. The dashed rectangle defines a single unit cell. The shaded areas are the tops of asperities. (b) Side view. The angle subtended by the top edges of the asperities is ϕ and the rise angle of their sides is ω. As drawn here, $\phi = \omega = 90°$.

3×10^4 m^{-1}. This observation is in general agreement with the Λ_c value of 3×10^4 m^{-1} calculated from equation (18).

Yoshimitsu and colleagues [26] also examined the wettability of regular arrays of square pillars (Fig. 2). They used a diamond saw to sculpt silicon wafers and then treated them with a fluorosilane to produce super-repellency. Table 2 lists data for their surfaces. Pillar width and spacing were held constant at x = 50 μm and y = 150 μm, corresponding to $\Lambda = 8.9 \times 10^3$ m^{-1}. Pillar height was varied between z = 0 μm and 282 μm. The critical Λ value from equation (18) for a 1-μl water drop was $\Lambda_c = 1.6 \times 10^3$ m^{-1}. Even though $\Lambda > \Lambda_c$, not all of these surfaces showed ultrahydrophobic behavior. Water was suspended atop the tallest pillars and exhibited very large apparent advancing contact angles, $\theta_a = 155°$. However, drops collapsed over the shortest pillars, z = 10 μm, producing a lower apparent value, $\theta_a = 138°$. Water could have protruded as deep as 15 μm between the pillars. For the shortest pillars (z = 10 μm), this much protrusion would have caused water to touch the floor between the pillars and thereby triggered the collapse. The exact shape of the interfaces between pillars is not clear. Curvature, if it exists, could produce contact that would trigger collapse. If interfaces are relatively flat due to capillary pressure constraints, then other factors (inertial flows associated with drops impacting a surface, injection of liquid into an existing sessile

Table 1.

Geometric parameters and apparent advancing contact angles for suspended or collapsed water drops on hydrophobic silicon surfaces covered with hexagonal arrays of square pillars, z = 40 μm and ω = 90°

x (μm)	y (μm)	x/y (μm)	δ (μm^{-2})	Λ (m^{-1})	α_p	2b (μm)	d (μm)	θ_a (°)	Suspended or collapsed?
2	4	0.5	6.3×10^{-2}	5.0×10^5	0.25	3.6	0.3	176	Suspended
8	16	0.5	3.9×10^{-3}	1.3×10^5	0.25	14.4	1.1	173	Suspended
16	32	0.5	9.8×10^{-4}	6.3×10^4	0.25	28.8	2.2	171	Suspended
32	64	0.5	2.4×10^{-4}	3.1×10^4	0.25	57.7	4.3	168	Suspended
64	128	0.5	6.1×10^{-5}	1.6×10^4	0.25	115	8.6	139	Collapsed
128	256	0.5	1.5×10^{-5}	7.8×10^3	0.25	231	17.2	116	Collapsed
8	23	0.35	1.9×10^{-3}	6.1×10^4	0.13	14.4	1.1	175	Suspended
8	32	0.25	9.8×10^{-4}	3.1×10^4	0.069	24.2	1.8	173	Suspended
8	56	0.14	3.2×10^{-4}	1.0×10^4	0.023	36.9	2.8	121	Collapsed

The surfaces were treated with a hydrocarbon silane (DS), $\theta_{a,o}$ = 107° [10]. x is pillar top width and y is the width of each unit cell. δ is the area density of asperities (equation (2)). Λ is contact line density (equation (4)). α_p is the fractional contact area between the liquid and pillar tops. 2b is the longest distance between asperities and d is the depth a suspended liquid protrudes between asperities (equation (19)). θ_a is the apparent advancing contact angle.

Table 2.

Geometric parameters and apparent advancing contact angles for suspended or collapsed water drops on silicon surfaces covered with regular arrays of square pillars, x = 50 μm, y = 150 μm and ω = 90°

z (μm)	δ (μm^{-2})	Λ (m^{-1})	α_p	2b (μm)	d (μm)	θ_a (°)	Suspended or collapsed?
10	4.4 × 10^{-5}	8.9 × 10^3	0.11	141	15	138	Collapsed
36	4.4 × 10^{-5}	8.9 × 10^3	0.11	141	15	155	Suspended
148	4.4 × 10^{-5}	8.9 × 10^3	0.11	141	15	151	Suspended
282	4.4 × 10^{-5}	8.9 × 10^3	0.11	141	15	153	Suspended

Surfaces were rendered hydrophobic with a fluorosilane, $\theta_{a,o}$ = 114° [26]. z is asperity height, δ is the area density of asperities (equation (2)); Λ is contact line density (equation (4)); and α_p is the fractional contact area between the liquid and pillar tops (equation (5)). 2b is the longest distance between asperities and d is the depth a suspended liquid protrudes between asperities (equation (19)). θ_a is the apparent advancing contact angle.

drop, or liquid flowing over a surface, etc.) could drive liquids between pillars, thereby destroying super-repellency.

This work has focused on the interaction of surface forces and gravity. Other body forces, such as the inertia, could also be important [27, 31]. Allowing drops to fall onto surfaces [29] or pressing down on suspended drops [25] can cause collapse.

4. CONCLUSIONS

The criteria developed and explored in this paper constitute a set of design rules for creating ultralyophobic surfaces. In order for a liquid–solid combination to show super-repellency, a contact line density and an asperity height criterion must be met. Surface forces acting around the perimeter of asperities must be greater than body forces and must be directed upward. Also, asperities must be tall enough that liquid protruding between them does not contact the base of the solid causing the liquid to be drawn downward, leading to collapse. Quantitatively, this requires that values of contact line density and asperity height determined from surface geometry must exceed critical values calculated from properties of the liquid, solid and their interfacial interactions. If $\Lambda > \Lambda_c$ and $z > z_c$, then the liquid will be suspended by the asperities, producing an ultralyophobic surface. If $\Lambda < \Lambda_c$ or $z < z_c$, then the liquid will collapse and the contact interface will be solely liquid/solid. These criteria were tested using data from several experimental studies of model rough surfaces. In all cases, they correctly predicted either suspension or collapse.

Acknowledgements

Entegris management is acknowledged for supporting this work and allowing publication. Also, thanks to Prof. T. M. McCarthy for providing details of their experimental work on ultralyophobic surfaces and to Dr. J. M. Miller for insightful discussions and helpful comments.

REFERENCES

1. R. N. Wenzel, *Ind. Eng. Chem.* **28**, 988 (1936).
2. A. B. D. Cassie and S. Baxter, *Trans. Faraday Soc.* **40**, 546 (1944).
3. R. Shuttleworth and G. L. J. Bailey, *Disc. Faraday Soc.* **3**, 16 (1948).
4. J. J. Bikerman, *J. Colloid Sci.* **5**, 349 (1950).
5. F. E. Bartell and J. W. Shepard, *J. Phys. Chem.* **57**, 211 (1953).
6. J. F. Oliver, C. Huh and S. G. Mason, *Colloids Surfaces* **1**, 79 (1980).
7. H. J. Busscher, A. W. J. van Pelt, P. de Boer, H. P. de Jong and J. Arends, *Colloids Surfaces* **9**, 319 (1984).
8. C. W. Extrand and A. N. Gent, *J. Colloid Interface Sci.* **138**, 431 (1990).
9. J. D. Miller, S. Veeramasuneni, J. Drelich, M. R. Yalamanchili and G. Yamauchi, *Polym. Eng. Sci.* **36**, 1849 (1996).
10. D. Öner and T. M. McCarthy, *Langmuir* **16**, 7777 (2000).
11. M. Tanaguchi and G. Belfort, *Langmuir* **17**, 6465 (2001).
12. G. MacDougall and C. Okrent, *Proc. Roy. Soc. (London)* **180A**, 151 (1942).
13. R. E. Johnson Jr. and R. H. Dettre, in: *Contact Angle, Wettability, and Adhesion*, Adv. Chem. Ser. No. 43, p. 112. Am. Chem. Soc., Washington, DC (1964).
14. R. H. Dettre and R. E. Johnson Jr., in: *Contact Angle, Wettability, and Adhesion*, Adv. Chem. Ser. No. 43, p. 136. Am. Chem. Soc., Washington, DC (1964).
15. R. E. Johnson Jr. and R. H. Dettre, in: *Surface and Colloid Science*, E. Matijević (Ed.), Vol. 2, p. 85. Wiley, New York, NY (1969).

16. M. Morra, E. Occhiello and F. Garbassi, *Langmuir* **5**, 872 (1989).
17. Y. Kunugi, T. Nonaka, Y.-B. Chong and N. Watanabe, *J. Electroanal. Chem.* **353**, 209 (1993).
18. T. Onda, S. Shibuichi, N. Satoh and K. Tsujii, *Langmuir* **12**, 2125 (1996).
19. S. Shibuichi, T. Onda, N. Satoh and K. Tsujii, *J. Phys. Chem.* **100**, 19512 (1996).
20. K. Tadanaga, N. Katata and T. Minami, *J. Am. Ceram. Soc.* **80**, 1040 (1997).
21. K. Tadanaga, N. Katata and T. Minami, *J. Am. Ceram. Soc.* **80**, 3213 (1997).
22. A. Hozumi and O. Takai, *Thin Solid Films* **303**, 222 (1997).
23. W. Chen, A. Y. Fadeev, M. C. Hsieh, D. Öner, J. P. Youngblood and T. M. McCarthy, *Langmuir* **15**, 3395 (1999).
24. J. P. Youngblood and T. M. McCarthy, *Macromolecules* **32**, 6800 (1999).
25. J. Bico, C. Marzolin and D. Quéré, *Europhys. Lett.* **47**, 220 (1999).
26. Z. Yoshimitsu, A. Nakajima, T. Watanabe and K. Hashimoto, *Langmuir* **18**, 5818 (2002).
27. C. W. Extrand, *Langmuir* **18**, 7991 (2002).
28. N. A. Patankar, *Langmuir* **19**, 1249 (2003).
29. B. He, N. A. Patankar and J. Lee, *Langmuir* **19**, 4999 (2003).
30. C. W. Extrand, *Langmuir* **20**, 5013 (2004).
31. A. Lafuma and D. Quéré, *Nature Mater.* **2**, 457 (2003).

Contact Angle, Wettability and Adhesion, Vol. 4, pp. 307–320
Ed. K.L. Mittal
© VSP 2006

Highly hydrophobic textile surfaces obtained by thin-layer deposition

THOMAS BAHNERS,* KLAUS OPWIS, TORSTEN TEXTOR and
ECKHARD SCHOLLMEYER

Deutsches Textilforschungszentrum Nord-West e.V., Adlerstr. 1, D-47798 Krefeld, Germany

Abstract—The potential of thin-layer deposition by physical processes for the surface modification of polymer films and synthetic fibers was investigated. A new approach to photochemical surface modification using monochromatic UV excimer lamps is described. The experiments have shown that it is possible to initiate grafting or even cross-linking of functional thin layers on the substrate surface by treatment in reactive atmospheres. The general condition to achieve such reactions is a marked difference in the absorbance of a low- or non-absorbing atmosphere and a strongly absorbing substrate, which leads to radical processes at the boundary between the atmospheric medium and the activated substrate. Using perfluoro-4-methylpent-2-ene as a medium the contact angle on poly(ethylene terephthalate) surfaces increased up to 116°. In case of bi-functional substances such as 1,5-hexadiene and diallylphthalate, the increased resistance against chemical degradation as well as infrared spectroscopy provided evidence for thin layer generation. Well-known and widely applied in several industries, surface modifications by gas discharge (plasma) processes are based on activation of the substrate mainly by electrons. The surface radicals generated then react with radicals in the plasma leading to substitution, grafting or cross-linking reactions. With regard to treatment of textiles, the atmospheric pressure plasma concept allows the use of rather simple machinery and a continuous treatment. In order to deposit functional thin layers for highly hydrophobic surfaces, fluorocarbons can be used as process gases. In the work described here fluorinated thin layers with a total fluorine concentration on the surface more than 50% could be deposited. An oil repellence rating, according to AATCC, of up to 7 to 8 was achieved.

Keywords: Thin-layer deposition; polymer; hydrophobicity; photochemistry; atmospheric plasma.

1. INTRODUCTION

In the field of technical textiles, both highly hydrophobic and oleophobic finishes are increasingly required to create barrier coatings or to attain self-cleaning properties. A number of physical surface modifications have been described in the literature, examples of which can be found, e.g., in Refs [1, 2]. All these processes are based on activation of the polymer surface with energetic particles. With

---

*To whom correspondence should be addressed. Tel.: (49-2151) 843-156;
Fax: (49-2151) 843-143; e-mail: bahners@dtnw.de

Table 1.
Possible reactions following UV lamp irradiation

Type		Reaction scheme	Effect
I	Recombination	$A_1^* + {}^*A_1 \rightarrow A_1\text{-}A_1$	None
II	Reaction with radical(s) of a neighboring chain	$A_1^* + {}^*A_2 \rightarrow A_1\text{-}A_2$	Cross-linking
III	Reaction with reactive atmosphere	$A_1^* + Z \rightarrow A_1\text{-}Z^*$	Addition
IV	Reaction with bi-functional material	$A_1^* + Z \rightarrow A_1\text{-}Z^*$ $A_1\text{-}Z^* + {}^*A_2 \rightarrow A_1\text{-}Z\text{-}A_2$	Cross-linking, thin-layer deposition

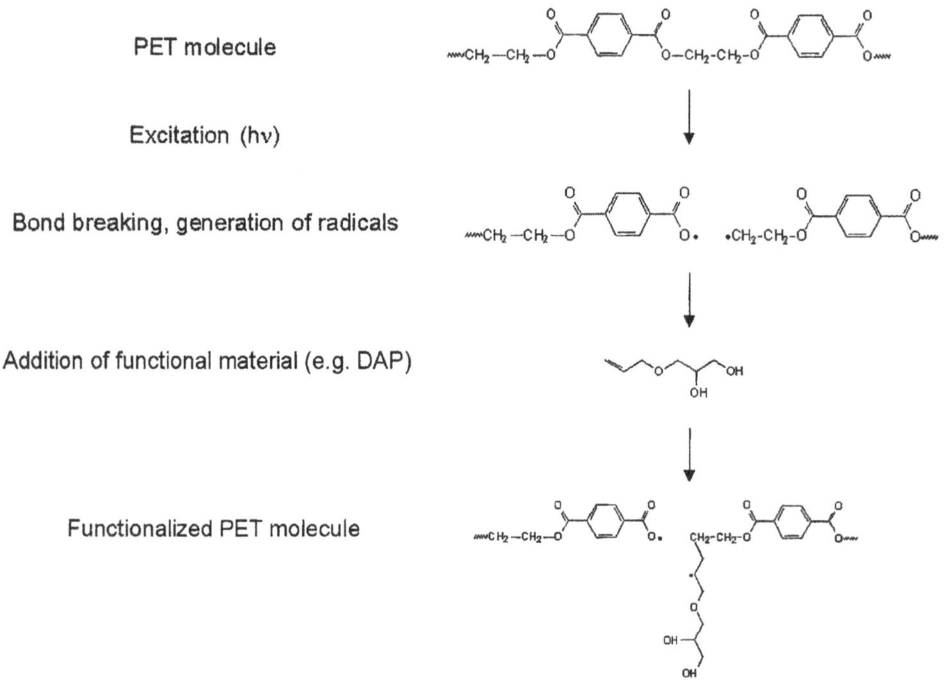

PET molecule

Excitation (hv)

Bond breaking, generation of radicals

Addition of functional material (e.g. DAP)

Functionalized PET molecule

Figure 1. Scheme of a photo-chemical addition reaction.

respect to achievable surface modification, considering low technical effort, and absence of disposable chemicals, gas discharge based processes and photo-induced surface modifications seem to offer a great potential.

The irradiation of a fibrous polymer using UV lamps can effect photochemical surface modifications, if the photons are sufficiently absorbed. Conventional UV sources, however, are not specific to the polymer due to their broad spectra, including a marked emission in the infrared. Accordingly, the radiation penetrates the substrate and can affect the polymer structure thermally by altering the orien-

cross-linked layer

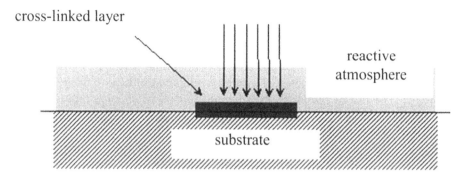

reactive
atmosphere

substrate

Figure 2. Photochemically-induced generation of a cross-linked layer on the surface of a strongly absorbing substrate (reaction type III). The reactive 'atmosphere' does not absorb the ultraviolet radiation, which reaches and activates the substrate initiating radical cross-linking at the interface.

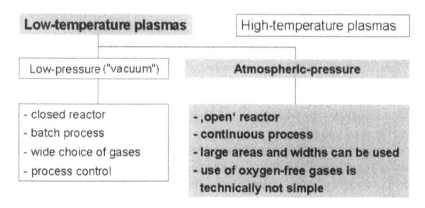

Figure 3. Definition and characteristics of technical plasma processes used for material processing.

tation or degree of crystallinity. The need to restrict the intensity of the radiation in order to avoid thermal damage accordingly reduces the yield for photochemically initiated reactions. A new approach is offered by monochromatic excimer lamps which emit in the UV region. At present, excimer lamps are available with wavelengths of 172, 222 and 308 nm. The bandwidth is of the order of 5 nm.

Effective surface modifications can be expected from treatments in reactive atmospheres [3, 4]. This is quite similar to plasma-based processes. The general condition to achieve such reactions is a marked difference in the absorbance of a low or non-absorbing atmosphere and a strongly absorbing substrate. The actual reaction takes place at the boundary between the atmosphere and the activated substrate, where radical processes are initiated. Here, it should be noted that backbone radicals have a higher probability to be generated due to the absorptive properties of the individual bonds, but have, in comparison to side-chain radicals, an extremely short lifetime in polymers with high crystallinity such as poly(ethylene terephthalate) (PET) fibers. Basically, four different types of reactions are possible (Table 1): (I) Recombination of radicals, (II) cross-linking of polymer chains, (III) addition of radicals from the reactive atmosphere (an example is given in Fig. 1) and (IV) addition of bi-functional molecules with ensuing cross-linking between the functional groups, which can induce thin-layer deposition (Fig. 2). The use of monochromatic lamps has the advantage that it does not induce cross-linking in the atmosphere itself. Also, unlike in plasma processes, the 'atmosphere' can be gaseous or liquid.

Gas-discharge processes are well known in materials science. As in photo-based processes, the activation of the substrate surface by energetic particles initiates radical reactions which may lead to oxidation, etching, grafting, as well as plasma polymerization. In the last case, the plasma gas needs to contain radicals which allows cross-linking and deposition of a thin layer. In contrast to a photo-chemical process, the co-polymerization of the layer on the surface, i.e. with actual bonding between the substrate and the polymerized layer, is in competition with homo- or gas-phase polymerization. Due to the bonding to the substrate only co-polymerization can be expected to establish durable thin layers.

Gas-discharge processes are broadly classified as shown in Fig. 3. Low temperature, low pressure processes are widely employed in various industrial applications, but pose problems when applied to large-width products, such as technical textiles, which may be as wide as 10 m.

Based on the above, technical (and scientific) approaches to the treatment of textiles more and more favor an atmospheric pressure plasma. A promising process design is the dielectric barrier discharge (DBD), which requires rather simple machinery without the need for low-pressure reactors and offers a continuous treatment. Latest developments have been the use of oxygen-free gases for long-lasting hydrophilic, as well as hydrophobic finishes.

The objective of the work presented here was to study the potential of photo-chemical modification using UV lamps and atmospheric pressure plasma (dielectric barrier discharge) for thin-layer deposition on polymeric surfaces with the aim to increase their liquid repellence.

2. EXPERIMENTAL

2.1. UV lamp irradiation

Poly(ethylene terephthalate) (PET), polyetherimide (PEI), m-aramide and poly(ether ether ketone) (PEEK) films were irradiated in a closed reaction chamber which was equipped with a quartz window and allowed irradiation in both gaseous or liquid atmospheres. The reactive media used are shown in Table 2. Where the reactive medium was applied in a liquid form, the samples were immersed in the liquid and dried afterwards. The irradiation was done at 222 nm, after which the samples were extracted (Soxhlet, 4 h in water/methanol followed by

Table 2.
Reactive media used for photochemical modification

Medium	Formula	Structure
1,5-Hexadiene	C_6H_{10}	
Perfluoro-4-methylpent-2-ene	C_6F_{12}	
Diallylphthalate (DAP)	$C_{14}H_{14}O_4$	

4 h in petroleum ether) in order to remove residual, i.e. not bonded, reactive medium.

2.2. Plasma treatment

The plasma treatments were performed at the University of Stuttgart, Institut für Plasmaforschung (IPF) in a closed chamber, which allowed defined process conditions. While the gas was at atmospheric pressure in all experiments, several plasma gases (fluorocarbons such as C_3HF_7 or perfluorocyclobutane (c-C_4F_8)) and carrier gases (Ar, N_2) were used. The oxygen content was kept at 0% in all cases. In this part of the work standardized (with regards to yarn, fabric construction and surface finish) fabrics made of PET served as samples.

2.3. Sample characterization

Surface chemical properties of the treated samples were determined either by X-ray photoelectron spectroscopy (XPS) or Fourier-transform infrared (FT-IR) spectroscopy. FT-IR measurements were made using a Biorad FTS-45 spectrometer with an MTEC-300 photoacoustic (PAS) detector unit.

The resulting wettability of the samples was determined by measurement of advancing contact angles of water. Because of the problems encountered in the measurement of contact angles on porous and textured fabrics, this was done only on film samples. In the case of textiles, a drop penetration test was used.

For several samples the oil repellence was measured in addition to the (water) wettability. This was done using the drop test according to AATCC 118-1972. This test uses 8 different oils, namely n-alkanes of varying chain length (ranging from paraffin to n-heptane) and appropriately designated numbers '1' to '8'. Beginning with oil no. '1', i.e. paraffin which has the smallest chain length, droplets are placed on the fabric. If the droplets do not penetrate the sample, the oil with the higher chain length will be used, i.e. oil number '2' and so on. The number of the highest n-alkane, i.e., longest chain, that does not penetrate the sample gives a ranking. If all oils penetrate, the ranking will be '0'. Accordingly, '0' signifies a non-repellent and '8' a totally repellent sample.

3. RESULTS AND DISCUSSION

3.1. Surface modification by irradiation with monochromatic UV excimer lamps

As examples of photochemical reaction type III (cf., Table 1), irradiations of various films were performed in the presence of (bi-functional) 1,5-hexadiene and perfluoro-4-methylpent-2-ene in gaseous, as well as liquid forms. The resulting changes in the contact angles (advancing) are given in Table 3. In the case of PET, the water contact angle increased to more than 95° after irradiation in 1,5-hexadiene, while the irradiation in perfluoro-4-methylpent-2-ene lead to a contact angle of 116°. The dependence of the contact angle on the irradiation time is shown for 1,5-hexadiene as the atmosphere in Fig. 4. A steep increase in hydrophobicity is observed already after a short irradiation time with a saturation behavior at longer treatments, indicating an increasing thin layer deposition on the polymer surface.

From XPS analysis it could be shown that the reactive substances were bound to the substrate surface after irradiation (Fig. 5). The spectra of PET films irradiated in the presence of 1,5-hexadiene showed an increase in the signal at the binding energy of 285 eV (C–H and C–C bonds), which was up to nearly 70% in the

Table 3.
Advancing contact angles of water (deg) on films made of several polymers which were irradiated in the presence of 1,5-hexadiene and perfluoro-4-methylpent-2-ene (gaseous)

Polymer	Untreated	After irradiation in	
		1,5-Hexadiene	Perfluoro-4-methylpent-2-ene (gaseous)
PET	75	> 95	116
Meta-aramide	65	> 90	–
PEI	70	95	–
PEEK	75	95	–

case of the liquid medium. In case of perfluoro-4-methylpent-2-ene as the medium, the XPS analysis revealed signals related to fluorine (690 eV) and to –CF$_3$ (294 eV) carbons. The F/C-ratio increased to 0.30 when using the liquid medium. This should correspond to a decrease of oxygen at the sample surface. In some cases, however, it was found that the resolved C$_{1s}$ spectrum, as well as the oxygen signals (532 and 534 eV) were in contrast to this assumption. One reason might be a re-organization of the macromolecules, but one has also to consider insufficient purification of the reactive media.

The comparison of the spectra recorded from PET films which were irradiated in liquid and gaseous atmospheres of perfluoro-4-methylpent-2-ene and 1,5-hexadiene showed that the surface modification was stronger in case of a liquid atmosphere.

The FT-IR spectrum (PAS) of the 1,5-hexadiene treated PET sample showed new hydrocarbon peaks at 2927 and 2854 cm^{-1} caused by aliphatic C–H stretching vibrations (Fig. 6). There was no evidence of unsaturated carbon bonds at wavenumbers around 1600 cm^{-1}. AFM studies of PET film surfaces after 10 min of irradiation in gaseous atmospheres (cf., Ref. [4]) showed a smoother surface after the treatment, which might indicate further cross-linking. An additional indication of cross-linking is provided by the observation that the modification following irradiation in gaseous 1,5-hexadiene also showed some resistance against chemical attacks.

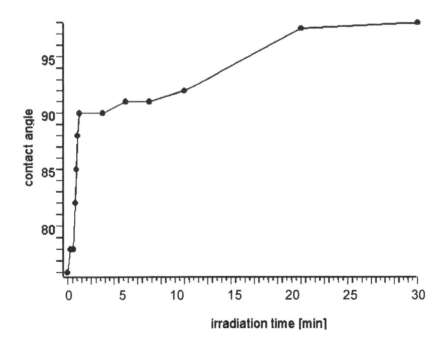

Figure 4. Advancing contact angle (distilled water, in deg) on PET films after irradiation under 1,5-hexadiene as a function of irradiation time.

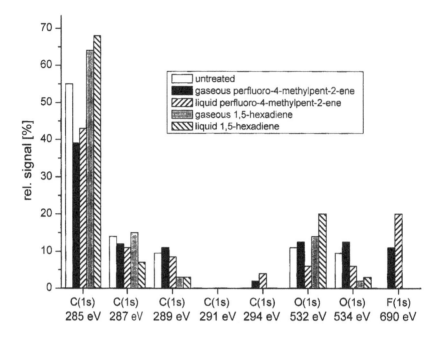

Figure 5. Comparative XPS studies of PET surfaces after irradiation (10 min) in 1,5-hexadiene and perfluoro-4-methylpent-2-ene atmospheres (in both liquid and gaseous forms).

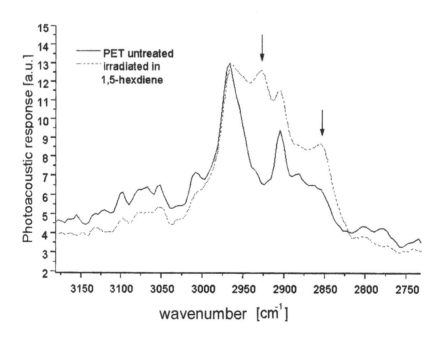

Figure 6. FT-IR spectra (PAS) of PET film untreated and after irradiation in 1,5-hexadiene.

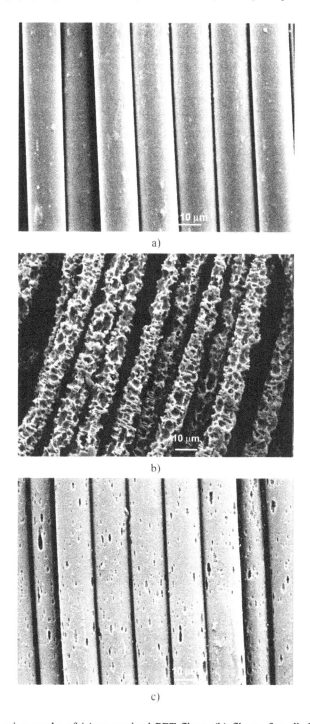

Figure 7. SEM micrographs of (a) as-received PET fibers, (b) fibers after alkaline hydrolysis in 25% NaOH-solution (90 min at 60°C) and (c) photochemically-modified fibers after alkaline hydrolysis (irradiation in DAP).

The bi-functional material diallylphthalate (DAP) was used for further studies on thin layer generation. Again, the deposition of the material was confirmed in FT-IR spectra (not shown here). In order to prove the formation of a thin layer, an experiment was devised to evaluate the barrier function of this layer (cf., Ref. [5]). The samples, as-received and after photochemical treatment in DAP, were stored for 90 min. in 25% NaOH-solution at 60°C. The SEM micrographs presented in Fig. 7a–c show that the untreated (as-received) sample was totally destroyed through alkaline hydrolysis (Fig. 7b). The loss in specific mass was determined to be 86%. In comparison, the sample treated in DAP (Fig. 7c) showed local surface defects, but still maintained considerable mechanical strength. In this case the mean loss in specific mass was only 25%. These experimental observations were interpreted as a proof for a cross-linked layer acting as a barrier against chemical attacks.

3.2. Surface modification using dielectric barrier discharge (DBD)

The objective of this part of this work was to study the potential of an atmospheric plasma treatment to deposit a low-surface-energy thin layer by plasma polymerization in order to establish high water, as well as oil repellence. Given this objective, the choice of the process gases (fluorocarbons) and the design of the experiment had to meet several criteria.

With regard to the process gas, the F/C-ratio is a prime parameter, as the occurrence of CF_2 radicals in the plasma is strongly dependent on this parameter (Fig. 8, taken from Ref. [6]). Mainly CF_2 radicals promote polymerization. In

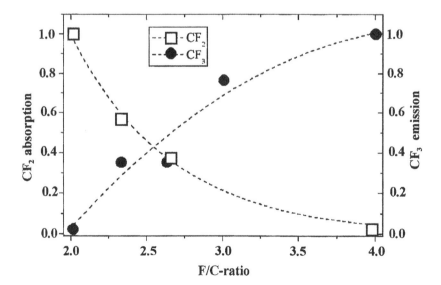

Figure 8. Influence of F/C-ratio of a fluorocarbon process gas on the occurrence of CF_2 and CF_3 radicals in the plasma [6]. According to their spectroscopic properties, CF_2 radicals can best be detected in the plasma by UV absorption, while CF_3 radicals are detected by UV emission.

addition, since competitive reactions with oxygen may work against the growth of a hydrophobic thin layer, appropriate measures have to be taken to avoid diffusion of air and water and/or their transport into the reactor by the porous textile. With regard to the latter, it is also well known that certain materials, e.g., polyamides or cellulose, have rather high water content. Thus, the laboratory experiments reported here were performed in a closed reactor, although at atmospheric pressure. It should be noted here that atmospheric plasma treatments of moving textile fabrics were performed in 'open' reactors with an effective oxygen content of less than 0.5% (cf., Ref. [7]).

The analysis of a large number of samples which were treated under varying plasma parameters, i.e., process gas (e.g., C_3HF_7 or perfluorocyclobutane (c-C_4F_8)), mixture of plasma and carrier gas and duration of the treatment, revealed that there was a linear correlation between the resultant oil repellence of the samples and surface fluorine content which was determined by XPS (Fig. 9). In the case of samples with high oil repellence, the XPS spectra also showed CF_2 and CF_3 signals. The best effects were achieved when c-C_4F_8 was employed. With regard to changes in the hydrophobic properties of the PET fabrics, a correlation between the wettability and the surface content of carbon-bonded oxygen (sum of signals due to C–O and C–OOR) was found.

The experiments showed that a general problem was the low resistance of the thin layers against mechanical stress (abrasion) and extraction. Two reasons can

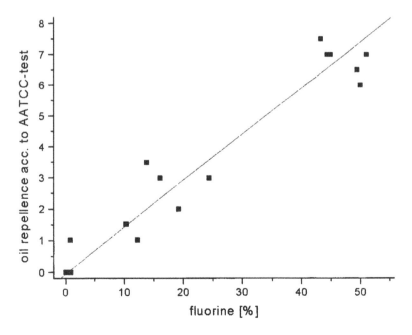

Figure 9. Correlation between oil repellence according to AATCC test and (total) fluorine content determined by XPS as taken from a number of samples treated under varying plasma parameters.

Table 4.
Chemical composition (in %) of the PET fabric (a) as-received and (b) after extensive Soxhlet extraction (4 h in water/methanol followed by 4 h in petroleum ether)

	PET fabric		According to stoichiometry of PET $(C_{10}H_8O_4)_n$
	As received	After extraction	
C–C/C–H	61.4	46.0	42.8
C–O	11.6	15.7	14.3
COOR	3.5	12.1	14.3
C total	76.5	73.8	71.4
O total	14.4	25.6	28.6
F total	0.8	–	–
N total	1.9	–	–
Si total	4.4	0.7	–

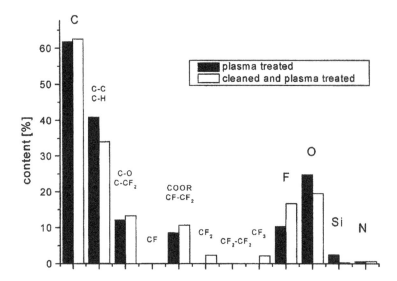

Figure 10. Comparative XPS studies of PET fabrics which were treated by atmospheric plasma using perfluorocyclobutane gas (c-C_4F_8) and underwent extraction. One sample was cleaned before the plasma process.

be considered here: (1) All processes based on plasma polymerization include (wanted) co-polymerization – i.e. covalent bonding, as well as competitive homo-polymerization in the gas phase, where the latter will not promote high thin layer adhesion. (2) In case of technical samples, residual finishes with low adhesion might cover the surface and prohibit covalent bonding. Especially this latter case (2) has to be considered in the case of textile samples. In order to investigate the actual chemistry of the sample surface, as-received fabrics and samples which were extensively extracted (Soxhlet, 4 h in water/methanol followed by 4 h in petroleum ether) were analyzed by XPS. The data obtained (Table 4) show that the surface chemistry of the as-received sample clearly deviates from the stoichiometry of PET, indicating the presence of finishing agents from the production process. Following thorough extraction, on the other hand, the composition of the same sample was in good agreement with the chemical formula of PET within analysis error.

Based on the above, plasma treatments of samples which were cleaned before thin layer deposition were performed. The exemplary XPS data on the samples given in Fig. 10 show that the sample which was cleaned before plasma treatment had a higher content of (total) fluorine as well as of CF_2 and CF_3 groups.

4. CONCLUSIONS

The objective of this work was to study the potential of photochemical modification using UV lamps, as well as an atmospheric pressure plasma (dielectric barrier discharge) for thin-layer deposition on polymeric surfaces with the aim to increase liquid repellence.

Monochromatic UV excimer lamps were employed to initiate photochemical modifications of the sample surface. It was shown that radicals generated during irradiation reacted with molecules from the atmospheric medium. Depending on the reactive atmosphere, grafting and thin layer formation was shown by FT-IR, XPS and studies of resistance against chemical degradation. Using perfluoro-4-methylpent-2-ene as the atmosphere the water contact angle on PET-surfaces increased to 116°. In order to study the possible effect of thin layer formation by bifunctional materials experiments using diallylphthalate (DAP) were performed. The observed barrier effect against chemical attack was interpreted as a proof for the generation of a cross-linked layer.

With regard to the plasma process studied, we can conclude that an atmospheric plasma process employing fluorocarbons as process gases allows to deposit hydrophobic as well as oleophobic thin layers. The results showed a linear correlation between the achieved oil repellence of the samples and the surface fluorine content. In the case of samples with high oil repellence, the XPS spectra also showed CF_2 and CF_3 signals. Surface fluorine contents of up to 50% were achieved with perfluorocyclobutane (c-C_4F_8) process gas. One important aspect for future work will be to establish appropriate cleaning procedures in order to promote good adhesion of the thin layer formed.

Acknowledgements

The authors wish to thank the Forschungskuratorium Textil e.V. for financial support of part of this work (AiF-nos. 11290 and 12864). This support is granted from resources of the Bundesministerium für Wirtschaft und Arbeit (BMWA) as a

supplementary contribution by the Arbeitsgemeinschaft Industrieller Forschungs-
vereinigungen "Otto-von-Guericke" e.V. (AIF). In addition, we acknowledge the
financial support from the Bundesministerium für Wirtschaft und Arbeit
(BMWA) (project no. 16IN0036) and the experimental help by the University of
Stuttgart, Germany, Institut für Plasmaforschung, in conducting plasma treatments.

REFERENCES

1. K. L. Mittal (Ed.), *Polymer Surface Modification: Relevance to Adhesion*, Vol. 2. VSP, Utrecht
   (2000).
2. K. L. Mittal (Ed.), *Polymer Surface Modification: Relevance to Adhesion*, Vol. 3. VSP, Utrecht
   (2004).
3. D. Praschak, T. Bahners and E. Schollmeyer, *Appl. Phys.* **A66**, 69 (1998).
4. T. Bahners, T. Textor and E. Schollmeyer, in: *Polymer Surface Modification: Relevance to Adhesion,* Vol. 3, K. L. Mittal (Ed.), pp. 97–124. VSP, Utrecht (2004).
5. K. Opwis, T. Bahners and E. Schollmeyer, *Chem. Fibers Int.* **54**, 116 (2004).
6. R. d'Agostino, F. Cramarosa and F. Fracassi, in: *Plasma Deposition, Treatment and Etching of
   Polymers*, R. d'Agostino (Ed.). Academic Press, San Diego, CA (1990).
7. T. Bahners, W. Best, J. Erdmann, Y. Kiray, A. Lunk, T. Stegmaier and N. Weber, *Techn. Textil.*
   44, 147 (2001).

Contact Angle, Wettability and Adhesion, Vol. 4, pp. 321–333
Ed. K.L. Mittal
© VSP 2006

Wettability of surface-modified keratin fibers

R. MOLINA, [1,*] E. BERTRÁN,[2] M. R. JULIÀ[1] and P. ERRA[1]

[1]*Surfactant Technology Department, Institute of Chemical and Environmental Research, (IIQAB-CSIC), 08034 Barcelona, Spain*
[2]*Applied Physics and Optics Department, Faculty of Physics, University of Barcelona, 08034 Barcelona, Spain*

Abstract—Keratin fibers are hydrophobic in nature due to the presence of an outermost monolayer of fatty acids, which are covalently bonded to proteins of the fiber cuticle surface membrane *via* ester or thioester linkages, with the methylene or ethylene groups oriented towards the outermost fiber surface. This characteristic exerts considerable influence on the shrinkage of wool fabrics submitted to aqueous washing process, on the adhesion to other materials, and on the dye diffusion into the fiber bulk. It is known that the hydrogen peroxide at alkaline pH, or low-temperature plasma (LTP) treatments provide hydrophilic properties to the fiber surface and, consequently, the shrinkage is reduced and the adhesion and dye diffusion are enhanced. This study focuses on the relationship between surface composition and wetting behavior of keratin fibers. For this purpose, wetting force measurements on single fibers according to the Wilhelmy procedure and XPS analysis were carried out on chemically-modified or plasma-treated keratin fibers. Whereas the scale direction of untreated keratin fibers does not exert any influence on the advancing adhesion tension values, it exerts a considerable influence on the receding ones. This behavior was found to be related to the different chemical compositions existing between the dorsal and the frontal parts of the fiber scales. The oxygen relative atomic concentration determined by XPS analysis is well correlated to advancing water contact angle values and it can be used to predict wetting behavior of keratin fibers.

Keywords: Wool; hair; plasma; bleaching; contact angle; XPS.

1. INTRODUCTION

Keratin fibers, like wool or the human hair, consist of cortical and cuticular cells. The cuticular cells are located on the outermost part of the fiber surrounding the cortical cells. The cuticle consists of endocuticle, A and B exocuticles and an exterior thin membrane called epicuticle. According to the model proposed [1], the epicuticle has a thickness of 5–7 nm and it is chemically composed of an external fatty acid monolayer (F-layer) and a protein layer with hydrophilic groups. The fatty acids are covalently bonded to proteins via ester or thioester linkages [2]. This fatty acid monolayer confers hydrophobicity to keratin fiber surface owing to

---

*To whom correspondence should be addressed. Tel.: (34-93) 400-6165;
Fax: (34-93) 204-5904; e-mail: rmmqst@iiqab.csic.es

the presence of methylene or ethylene groups on the outermost part of the fiber surface. This characteristic exerts considerable influence on the shrinkage of wool fabrics submitted to aqueous washing process, on the adhesion to other materials, and on the dye diffusion into the fiber bulk [3–5]. However, hydrophobicity can be decreased by the formation of hydrophilic groups on the fiber surface and/or by removing the fatty acid monolayer by means of plasma or chemical treatments [6, 7]. If the fatty acid monolayer is removed the hydrophilic groups of the protein matrix can be brought on the outermost part of the fiber surface.

A low temperature plasma or glow discharge is a partially ionized gas, generated by means of the application of an electrical discharge. The plasma is formed by neutral particles that can be molecules, excited atoms, free radicals and metastable particles and by charged particles, electrons and ions. Plasma active species affect exclusively the substrate surface at a depth of 30 nm, generating new functional groups and/or etching surface material [4].

In this study wool or human hair fibers were treated with chemical reagents (at alkaline pH with or without hydrogen peroxide) or subjected to plasma treatments in order to study the wetting properties of differently modified keratin fibers. Wettability was estimated by dynamic contact angle determination on single fibers according to the Wilhelmy procedure. The solid–liquid contact angle, which is sensitive to the chemical composition of the outermost 0.5 nm of the fiber [8], is a useful technique for evaluating the surface modification as a result of plasma or chemical treatments. Also, the XPS technique was used in order to confirm and to evaluate the chemical composition changes promoted by different treatments on the fiber surface. Finally, contact angle measurements and XPS analysis results were compared in order to find the relationship between contact angle and chemical composition.

2. EXPERIMENTAL

2.1. Materials

2.1.1. Untreated keratin fibers
Merino wool and dark brown European human hair fibers were used in this work. Before and after treatments, the samples were stored in a conditioned room (20°C, 65% relative humidity).

2.1.2. Chemically-treated keratin fibers
Hair fibers were treated under conventional experimental conditions for wool bleaching (70°C, pH 9, 1 h) with either 0.6% (w/v) of hydrogen peroxide (H_2O_2) or without hydrogen peroxide. The ratio of treatment solution to hair fiber was 50:1 (v/w).

2.1.3. Plasma-treated keratin fibers

Wool and human hair fibers were submitted to glow discharge treatments in a radiofrequency reactor built in the Applied Physics and Optics Department of the University of Barcelona [9]. The pressure before introducing plasma gas (10 Pa), power (100 W), electrode distance (8.5 cm) and plasma gas pressure (100 Pa) were kept constant in all experiments; plasma gases were air, nitrogen or water vapor and the treatment times were 10, 40 and 120 s.

2.1.4. Palmitoyl-chloride-treated fibers

Hair fibers were treated with palmitoyl chloride (synthesis reagent grade, Merck) in order to introduce new hydrophobic groups on the plasma-treated fiber surface. The palmitoyl chloride treatment was performed for a few seconds after taking out the samples from the plasma reactor. A freshly prepared solution of palmitoyl chloride (0.1 M) in hexane was used at a solution to keratin fiber ratio of 20:1 (v/w), at room temperature for 2 min.

2.1.5. Wetting liquids

Milli-Q water (pH 6.5), with surface tension of 72.8 mN/m was used. Decane (purum, Fluka) with a surface tension of 23.8 mN/m was employed as the zero contact angle liquid to determine the fiber perimeter.

2.2. Methods

2.2.1. Dynamic wetting force measurements

Advancing contact angles were calculated from dynamic wetting force (F_w) measurements carried out in an electrobalance, KSV Sigma70 Contact Angle Meter, using water as the wetting liquid. A single keratin fiber was mounted, overhanging 2–3 mm from an aluminum support in order to keep the fiber straight and rigid. This helps to avoid the buoyancy effect and to obtain constant fiber perimeter during wetting measurements. The fiber was scanned for 1 mm at a velocity of 0.5 mm/min in both advancing and receding modes. One hysteresis cycle was carried out for each fiber. The forces acting on a solid immersed in a liquid are the wetting force (F_w), the buoyancy force (F_b) and the solid weight force (F_g). The resulting force (F) is given by equation (1).

$$F = F_w - F_b + F_g \tag{1}$$

When the solid is, for instance, a single human hair fiber, the buoyancy force (F_b) is negligible, compared to the wetting force. The solid weight (F_g) is zeroed before the solid touches the liquid surface. Thus, the resulting force can be considered equivalent to the wetting force, i.e.,

$$F = F_w = \gamma_L \cdot L \cdot \cos\theta, \tag{2}$$

where L is the fibre perimeter, γ_L the liquid surface tension and θ the fiber–liquid contact angle.

The fiber perimeter can be calculated by applying equation (2) for liquids showing a total wettability (zero contact angle). From the fiber perimeter determined and by applying equation (2), the fiber-liquid contact angle can be determined.

2.2.2. Determination of fiber perimeter

Perimeters of the scanned fibers were estimated from the wetting force measured in a total wetting liquid, decane (γ_L = 23.8 mN/m), where cos θ is assumed to be unity. The measurement was carried out in a closed chamber in order to equilibrate the fiber with the vapor of the wetting liquid. The average perimeter value of the fibers can be calculated from equation (2).

2.2.3. Determination of fiber swelling

The samples were observed with a Reichert Polyvar 2 optical microscope (supplied by Leica) equipped with video. Swelling measurements were carried out after exposure of the fiber to the swelling liquid (water at pH 6.5) for 30 min. Wetting force measurements in water were corrected for the swelling effect.

2.2.4. Determination of scale direction of the fiber

The root–tip or tip–root cuticular direction of immersion of each fiber was visualized by a Wild Heerbrug (Switzerland) optical microscope after the fiber was mounted on an aluminum support.

2.2.5. X-ray photoelectron spectroscopy (XPS)

XPS was used to monitor the modifications produced in the outermost (5–10 nm) wool fiber surface. The wool fabric samples were analyzed using a PHI Model 5500 Multitechnique System with an Al K_α monochromatic X-ray source operating at 350 W. The pressure inside the analysis chamber was 6×10^{-7} Pa. The measurements were taken using a take-off angle of 0°. Survey scans were taken in the range 0–1100 eV, with pass energy of 187.85 eV and acquisition time of 10 min. High-resolution scans were obtained on the C_{1s}, N_{1s}, O_{1s} and S_{2p} peaks, with a pass energy of 23.5 eV and an acquisition time of 3 min. Binding energies of all photopeaks were referenced to the C_{1s} photopeak position for C–C and C–H species at 285.0 eV.

3. RESULTS AND DISCUSSION

3.1. Wetting hysteresis cycles

Force over length or adhesion tension (F/L) values as a function of the immersion depth in water wetting liquid for untreated wool and hair fibers are shown in Fig. 1. Adhesion tension and thus contact angle hysteresis is attributed to several factors such as surface chemical heterogeneity, roughness, surface deformability,

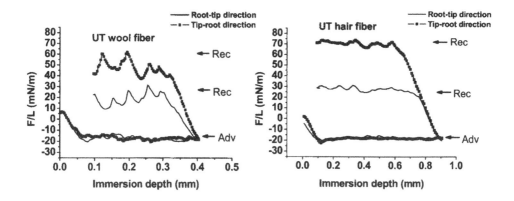

Figure 1. Adhesion tension (*F/L*) for untreated (UT) wool and human hair fibers in the root–tip and tip–root fiber scale directions. Abbreviations: advancing (Adv), receding (Rec) (data from Ref. [18]).

surface configuration change and adsorption or desorption [10–13]. In accordance with the model of Johnson and Dettre [14], advancing contact angles are associated with regions of low wettability (low-energy regions) and receding contact angles with regions of high wettability (high-energy regions).

Both wool and human hair keratin fibers exhibit similar wetting behaviors, attributed to their similar surface chemical compositions. The negative advancing adhesion tension value can be attributed to the presence of the fatty acid monolayer in the outermost fiber surface. Whereas the advancing adhesion tension values are independent of the fiber scale direction, the receding ones are strongly dependent. As had been reported by Kamath *et al.* [15], wettabilities for the untreated (UT) fibers in the receding mode are higher for the tip–root cuticular direction than for the root–tip cuticular direction.

Chemical or plasma treatments can alter the wetting properties of keratin fibers. Thus, the advancing adhesion tension values for keratin fibers treated at alkaline pH (pH 9) without hydrogen peroxide, close to 0 mN/m (Fig. 2a), or treated with hydrogen peroxide at alkaline pH, close to 10 mN/m (Fig. 2b), are higher than the corresponding value for untreated fibers (Fig. 1). It can be observed for both treated fibers, as in the case of untreated fibers, that the advancing adhesion tension values are independent of the fiber scale direction; however, the receding values show some differences. The receding adhesion tension values obtained in the tip–root direction are always slightly higher than those for the root–tip direction. The increase in the advancing adhesion tension values with respect to the untreated keratin fibers clearly indicates a surface modification [6].

Advancing adhesion tension values for air-plasma-treated keratin fiber (Fig. 2c) increase up to 50 mN/m, indicating a noticeable increase in the hydrophilicity of keratin fibers, which can be attributed to the fatty acid monolayer oxidation and/or removal. Also, both advancing and receding adhesion tension values are

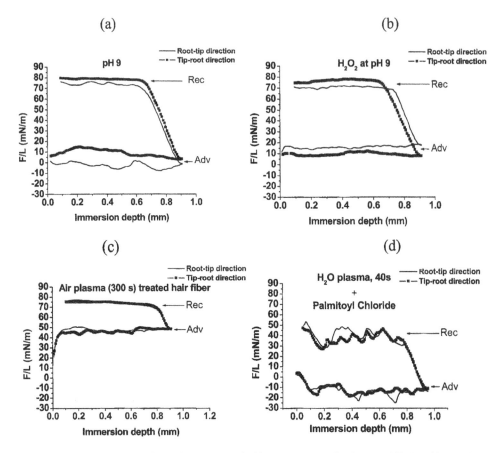

Figure 2. Adhesion tension (*F/L*) for human hair fibers treated at alkaline pH (pH 9) without (a) or with hydrogen peroxide (b), air plasma (c) and water vapor plasma and post-treated with palmitoyl chloride (d) in the root–tip and tip–root fiber scale directions. Abbreviations: advancing (Adv), receding (Rec) (c and d, data from Ref. [18]).

independent of the fiber scale direction. However, the advancing adhesion tension values decrease after the palmitoyl chloride treatment (Fig. 2d), reaching values similar to untreated fibers. This shows the presence of hydrophobic groups on the fiber surface. It should be noticed that both advancing and receding adhesion tension values are independent of the fiber scale direction. The receding adhesion tension values are very similar to those obtained for the untreated hair in the root–tip direction.

The wetting behavior observed for the differently treated keratin fibers is in accord with the model proposed by Kamath *et al.* [15] for untreated keratin fibers to explain the different wetting behaviors in relation to the fiber scale direction. This model assumes that the frontal part of the scales is more hydrophilic than the dorsal part. Thus, in the receding mode, in the root–tip direction, wetting liquid interacts with the dorsal part, which is more hydrophobic. In the tip–root direction, wetting liquid contacts the frontal part of the scales, which is more hydrophilic and thus the receding adhesion tension values are greater in this scale direction.

Receding adhesion tension values in the root–tip direction for alkaline treated fibers increase noticeably with respect to untreated fibers, suggesting that the dorsal part of scales becomes more hydrophilic. Since the dorsal part has now become more hydrophilic and the frontal part is already hydrophilic, little differences in receding adhesion tension values with respect to fiber scale direction were found. The receding adhesion tension values for plasma treated fibers are very high and independent of fiber scale direction. This suggests that plasma treatment modifies both dorsal and frontal parts of scales, generating new hydrophilic groups. On the contrary, the receding adhesion tension values for palmitoyl-chloride-after-treated fibers are lower than those obtained for plasma-treated fibers and they are independent of fiber scale direction. This clearly indicates that both dorsal and frontal parts of the fiber scales are hydrophobic. Evidently, the differences in receding adhesion tension values with respect to the fiber scale direction are mainly due to different chemical nature (hydrophobic or hydrophilic) of the dorsal and frontal parts of the fiber scales.

3.2. XPS analysis

In order to assess the surface chemical changes obtained with chemical or plasma treatments, the fiber surfaces were analyzed by X-ray photoelectron spectroscopy. The relative elemental concentrations (at%) and the C/N and O/C atomic ratios for untreated keratin fibers are shown in Table 1. Both wool and human hair keratin fibers have similar surface chemical composition. The high value of C/N atomic ratio in Table 1 for the untreated keratin fiber and the value reported from amino-acid analysis of the epicuticle (C/N=3.4) [16] suggests the presence of an excess of carbon on the fiber surface, which can be attributed to the presence of the fatty acid monolayer. Since the C/N atomic ratio can be related to the

Table 1.
XPS elemental concentrations (at%) for untreated keratin fibers (wool and human hair) and human hair treated at alkaline pH with or without hydrogen peroxide

Sample	C	N	O	S	Si	Na	C/N	O/C
Untreated wool	76.7	6.4	13.5	2.7	0.7	0	12.0	0.18
Untreated hair	73.4	6.2	15.9	1.3	3.0	0	11.8	0.22
Hair treated at pH 9	74.0	7.9	14.8	2.3	0.8	0.2	9.4	0.20
Hair treated with H_2O_2 at pH 9	71.6	7.2	17.1	0.8	3.4	0	9.9	0.24

The error associated with each measurement is < 5% of the reported value.

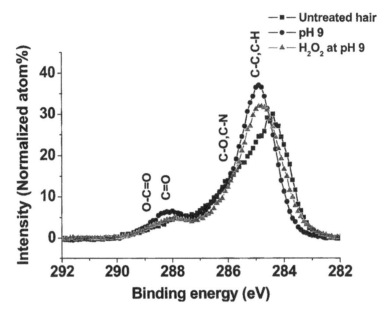

Figure 3. High-resolution XPS C_{1s} spectra for untreated and treated at alkaline pH (pH 9) without or with H_2O_2 at pH 9 human hair fibers.

Table 2.
XPS elemental concentrations (at%) for untreated and plasma-treated wool with different plasma gases

Sample	C	N	O	S	Si	C/N	O/C
Untreated wool	76.7	6.4	13.5	2.7	0.7	12.0	0.18
N_2 plasma, 40 s	62.9	7.7	27.0	1.3	1.1	8.2	0.43
Air plasma, 40s	64.9	7.4	26.1	1.2	0.4	8.8	0.40
H_2O plasma, 40 s	68.0	7.9	22.9	1.2	0	8.6	0.34

The error associated with each measurement is < 5% of the reported value.

hydrocarbon chain length of the fatty acid monolayer, the lower the C/N ratio the lower the length of hydrocarbons chain of the fatty acid monolayer.

Human hair treated at alkaline pH without hydrogen peroxide shows an increase in the N and S relative atomic concentrations and a decrease in the C/N atomic ratio with respect to untreated hair (Table 1). This suggests a partial removal of the fatty acid monolayer which can be attributed to the hydrolysis of some of the acylated bonds between fatty acids and the protein of the epicuticle. The XPS results for human hair treated with hydrogen peroxide at alkaline pH show an increase in the relative N and O concentrations, as well as a decrease in the C/N atomic ratio and an increase in the O/C atomic ratio. These ratios point

Table 3.
XPS elemental concentrations (at%) for untreated (UT) and water-vapor-plasma-treated wool

Sample	C	N	O	S	Si	C/N	O/C
Untreated wool	76.7	6.4	13.5	2.7	0.7	12.0	0.18
H_2O plasma, 10 s	71.4	4.9	21.5	0.8	1.3	14.6	0.30
H_2O plasma, 40 s	68.0	7.9	22.9	1.2	0	8.6	0.34
H_2O plasma, 120 s	58.8	12.3	25.5	2.1	0.3	4.8	0.43

The error associated with each measurement is < 5% of the reported value.

out a partial removal of the fatty acid monolayer due to the alkaline pH and the oxidation process by hydrogen peroxide. It is known that the oxidation process with hydrogen peroxide promotes cysteic acid residue formation as a consequence of disulfide bond oxidation [17].

The high-resolution spectra corresponding to the carbon photoelectron peaks (Fig. 3) reveal that the peak corresponding to the carbonyl groups is more resolved after alkaline treatments, pointing out that the protein matrix can be detected more clearly as a consequence of a partial removal of the fatty acid monolayer.

The elemental concentrations and the C/N and the O/C atomic ratios for untreated and plasma-treated keratin fibers with different gases are shown in Table 2. Whereas the relative atomic oxygen concentration and O/C atomic ratio increase, the relative atomic carbon concentration and the C/N atomic ratio decrease, irrespective of the plasma gas used. These changes could be explained by the oxidation and partial removal of hydrocarbon chains belonging to the fatty acid monolayer. Unexpectedly, the nitrogen plasma also promotes oxidation of the fatty acid monolayer, indicating a possible contamination of the plasma gas with water present either in the keratin fibers or in the reactor chamber walls [18]. For this reason, treatments with water vapor plasma were carried out and the influence of the treatment time on the surface composition of the keratin fibers was studied (Table 3). The elemental chemical composition of the fiber surface changed noticeably after 10 s of water plasma treatment. The relative O atomic concentration and the O/C atomic ratio increased considerably with respect to untreated sample. The C/N atomic ratio decreased and the O/C atomic ratio increased with increasing treatment time. This suggests a progressive oxidation of the fatty acid monolayer by increasing the treatment time. However, the oxidation of the hydrocarbon chains prevailed over their removal in the early stages of the plasma treatment. The fatty acid monolayer removal is more evident at long treatment times.

The high-resolution spectra of carbon photopeaks for plasma treated wool with different gases (Fig. 4a) suggest that both air and nitrogen plasmas modify the keratin fiber surface in a similar way. However, the water vapor plasma seems to

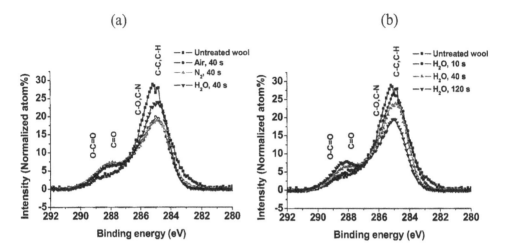

Figure 4. (a) High-resolution XPS C_{1s} spectra for untreated and plasma treated with different gases and (b) water vapor-plasma-treated wool at different treatment times.

be less efficient. The high-resolution spectra of the carbon photopeaks as a function of the water vapor plasma treatment time (Fig. 4b) reveal a progressive decrease of the peak located at 285 eV and a progressive increase of carbonyl and carboxyl groups. This confirms that the fatty acid monolayer is progressively oxidized and removed with treatment time.

3.3. Contact angle vs. XPS

Regardless of plasma gas used, the advancing contact angle increases with time elapsed after the plasma treatment, as a consequence of the reorientation of hydrophilic groups towards the fiber bulk [6, 19–21]. Thus, in order to correlate contact angle measurements with the XPS results (obtained 1 month after the plasma treatment), advancing contact angles corresponding to aged samples were used (Table 4) [6, 7, 21]. The advancing contact angles for fibers treated with or without hydrogen peroxide at alkaline pH are also included.

Figure 5 shows the advancing contact angle as functions of the C/N and O/C ratios. As can be observed, that whereas the fatty acid monolayer removal, expressed as the C/N ratio, is not directly related to the advancing contact angle (Fig. 5a), the degree of functionalization expressed as the O/C ratio seems to be directly related to the advancing contact angle (Fig. 5b).

The advancing contact angles as function of the relative atomic concentrations of carbon and oxygen are plotted in Fig. 6. The advancing contact angle tends to decrease with decreasing relative atomic carbon concentration (Fig. 6a) and with increasing relative atomic oxygen concentration (Fig. 6b). Two different regions can be distinguished in Fig. 6a. Between 103° and 70°, the advancing contact angle decay is linear with the decrease in carbon concentration. This behavior could

Table 4.
Average advancing water contact angles (θ_{Adv}) on chemical or aged-plasma-treated human keratin fibers

Sample	θ_{Adv} (deg)
Untreated wool	105±4
Untreated hair	103±1
Hair treated at pH 9	90±4
Hair treated with H_2O_2 at pH 9	84±3
H_2O plasma, 10 s	79±8
H_2O plasma, 40 s	68±6
N_2 plasma, 40 s	58±9
Air plasma, 40 s	55±4
H_2O plasma, 120 s	54±6

Errors are indicated as 95% confidence level. For comparison, θ_{Adv} on untreated wool is also given.

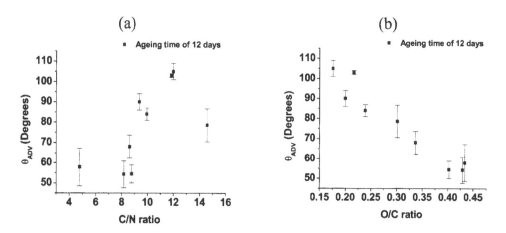

Figure 5. (a) Water advancing contact angles on fibers treated with or without hydrogen peroxide at alkaline pH and plasma-treated samples with an ageing time of 12 days after the plasma treatment as a function of the C/N and (b) O/C atomic ratios.

be attributed to the progressive oxidation and removal of the fatty acid monolayer. After that, the advancing contact angle remains approximately constant, whereas the carbon concentration decreases from 65 to 57%. Similar behavior can be observed in Fig. 6b; however, when the contact angle remains approximately constant, the oxygen concentration increases from 25 to 27%. Therefore, the relative oxygen concentration in the keratin fiber surface is more related to the advancing contact angle values than the relative carbon concentration.

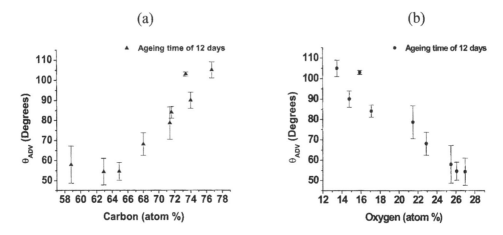

Figure 6. (a) Water advancing contact angles on fibers treated with or without hydrogen peroxide at alkaline pH and plasma-treated samples with an ageing time of 12 days after the plasma treatment as a function of the C and (b) O relative atomic concentrations.

4. CONCLUSIONS

It is shown here that contact angle determination and XPS analysis are complementary techniques for chemically characterizing the surfaces of solid materials. These techniques provide information about the changes produced on the surface of keratin fibers by chemical or plasma treatments. From wetting hysteresis cycles it has been demonstrated that the differences in receding adhesion tension values with respect to fiber scale direction are mainly due to the differences in the chemical nature (hydrophobic or hydrophilic) of the dorsal and frontal parts of the fiber scales. The C/N atomic ratio seems to be related to the hydrocarbon chain length in the fatty acid monolayer and it is not directly related to the advancing contact angle. Instead, the O/C atomic ratio, which can be attributed to the oxidation degree of the hydrocarbon chains of the fatty acid monolayer, apparently is related to the advancing contact angle. Nevertheless, the advancing contact angle values are dependent on both the carbon and oxygen atomic concentrations. However, the oxygen atomic concentration seems to be more related to the advancing contact angle. Therefore, the oxygen atomic concentration can be used to predict wetting behavior of keratin fibers.

Acknowledgements

The financial support by the MCYT (MAT 2002-02613 project) and the Generalitat de Catalunya (2001SGR00357 and Xarxa Tematica n° 2003/Xt/00025 "Plasma Amb Materials Polimerics") is acknowledged. This research has been carried out within the FEMAN and INQUISUP partner units.

REFERENCES

1. A. P. Negri, H. J. Cornell and D. E. Rivett, *Textile Res. J.* **63**, 109–115 (1993).
2. A. P. Negri, H. J. Cornell and D. E. Rivett, *Textile Res. J.* **62**, 381–387 (1992).
3. P. Erra, R. Molina, D. Jocic, M. R. Julià, A. Cuesta and J. M. D. Tascón, *Textile Res. J.* **69**, 811–815 (1999).
4. H. Hesse, H. Thomas and H. Höcker, *Textile Res. J.* **65**, 335–361 (1995).
5. R. H. Kienle, G. L. Royer and H. R. McCleary, *Rayon Textile Monthly* **26**, 408–412, 482–485, 541–542 (1945).
6. R. Molina, P. Jovancic, F. Comelles, E. Bertran and P. Erra, *J. Adhesion Sci. Technol.* **16**, 1469–1487 (2002).
7. R. Molina, F. Comelles, M. R. Julià and P. Erra, *J. Colloid Interface Sci.* **237**, 40–46 (2001).
8. E. A. Vogler, in: *Wettability*, J. C. Berg (Ed.), pp. 237–239. Marcel Dekker, New York, NY (1993).
9. E. Bertran, J. Costa, G. Sardin, J. Campmany, J. L. Andújar and A. Canillas, *Plasma Sources Sci. Technol.* **3**, 348–354 (1994).
10. C. W. Extrand and Y. Kumagai, *J. Colloid Interface Sci.* **191**, 378–383 (1997).
11. J.-H. Wang, P. M. Claesson, J. L. Parker and H. Yasuda, *Langmuir* **10**, 3887–3897 (1994).
12. A. Marmur, *J. Colloid Interface Sci.* **168**, 40–46 (1994).
13. H. S. Van Damme, A. H. Hoght and J. Feijen, *J. Colloid Interface Sci.* **114**, 167–172 (1996).
14. R. E. Johnson Jr. and R. H. Dettre, *J. Phys. Chem* **68**, 1744–1750 (1964).
15. Y. K. Kamath, C. J. Dansizer and H. D. Weigmann, *J. Appl. Polym. Sci.* **22**, 2295–2306 (1978).
16. N. L. R. King and J. H. Bradbury, *Aust. J. Biol. Sci.* **21**, 375–384 (1968).
17. J. A. Maclaren and B. Milligan, *Wool Science. The Chemical Reactivity of the Wool Fibre*, pp. 64–65. Science Press, Marrickville (1981).
18. R. Molina, P. Jovančić, L. Julià, E. Bertran, D. Jocić and P. Erra, in *Polymer Surface Modification: Relevance to Adhesion*, Vol. 3. K. L. Mittal (Ed.), pp. 51–68. VSP, Utrecht (2004).
19. M. R. Wertheimer, L. Martinu, J. E. Klemberg-Sapieha and G. Czeremuszkin, in: *Adhesion Promotion Techniques: Technological Applications*. K. L. Mittal and A. Pizzi (Eds.), pp. 139–173. Marcel Dekker, New York, NY (1999).
20. J. Nakamatsu, L. F. Delgado-Aparicio, R. Da Silva and F. Soberon, *J. Adhesion Sci. Technol.* **13**, 753–761 (1999).
21. R. Molina, P. Jovančić, D. Jocić, E. Bertran and P. Erra, *Surface Interface Anal.* **35**, 128–135 (2003).

Contact Angle, Wettability and Adhesion, Vol. 4, pp. 335–350
Ed. K.L. Mittal
© VSP 2006

Ammonia plasma-simulating treatments and their impact on wettability of PET fabrics

KENTH S. JOHANSSON*

Institute for Surface Chemistry, P.O. Box 5607, SE-114 86 Stockholm, Sweden

Abstract—Ammonia plasma treatments were performed on both thermoplastic plates and fabrics made of poly(ethylene terephthalate) (PET). The plates became more hydrophilic with improved adhesion properties as expected, whereas the fabrics became more hydrophobic, yet positively charged. Plasma treatments of PET fabrics using gas mixtures such as NH_3/N_2 and H_2/N_2 were performed in order to simulate pure ammonia plasma treatments since such treatments are not always applicable in industrial applications, due to environmental and safety reasons. It was shown that the recommended and allowed gas compositions, 15% NH_3 in N_2 and 5% H_2 in N_2, did not show any similarity with pure ammonia plasma treatments with respect to surface charge, wettability and chemical surface composition. At least 80% NH_3 in N_2 or 80% H_2 in N_2 is needed to simulate an ammonia-like plasma treatment of PET fabrics.

Keywords: Ammonia plasma simulation; H_2/N_2 plasma; water wettability; surface charge; chemical composition; XPS; PET fabric.

1. INTRODUCTION

Nitrogen-containing plasmas are widely used to improve wettability, printability, bondability and biocompatibility of polymer surfaces [1, 2]. Several applications have been investigated and they all show that wettability and adhesion properties of various polymer surfaces improve after ammonia or allylamine plasma treatment. Holmes and Schwartz, for instance, treated ultra-high-strength polyethylene fibers using ammonia plasma [3] and found that fabrics treated with ammonia plasma became completely epoxy resin wettable and the interfacial adhesion strength increased by 40%.

Li *et al.* [4] investigated a similar system and also found that ammonia plasma treatment improved the wettability of the fibers and also the adhesion to epoxy resins. Tusek *et al.* [5] performed a careful surface characterisation study of ammonia plasma-treated polyamide 6 films, confirming the introduction of various N-functionalities, such as amino, imino, cyano and others on the polyamide sur-

---

*Tel.: (46-8) 5010-6052; Fax: (46-8) 208-998; e-mail: kenth.johansson@surfchem.kth.se

face. Yang *et al.* [6] investigated the effects of ammonia plasma treatment of poly (D,L-lactide) (PDLLA), followed by collagen anchorage. It was found that the surface energy of the PDLLA increased by the ammonia plasma treatment and the collagen attachment and cell affinity were greatly improved [6]. The introduction of amino groups on the surface of polystyrene films with ammonia plasma treatment was also reported to improve cell affinity [7]. Loh *et al.* [8] introduced nitrogen-containing groups onto the surface of carbon fibers in a cold plasma process using mixtures of ammonia and N_2/H_2. They found that the surface concentration of functional groups containing nitrogen decreased with time and proposed several mechanisms for this degradation. The results suggested that plasma-modified carbon fibers had excellent potential for improving adhesion between fibers and the matrix through physical compatibility and/or chemical bonding. Waldman *et al.* [9] have concluded that although ammonia gas is not polymerizable, it can provide reactive sites including polar groups when mixed with, e.g., ethylene. Ammonia plasma treatment was reported to increase the peel strength between poly(tetrafluoroethylene) and nitrile rubber when a phenol-type adhesive was used [10].

The present study concerns ammonia plasma and ammonia plasma-simulating treatments of PET fabrics. For comparison, ammonia and oxygen plasma treatments of smooth PET and other thermoplastic plates have also been performed.

Ammonia plasma treatment of fabrics gives the fabric surface interesting properties. However, one important drawback of ammonia plasma treatment is that it cannot be used in a large-scale industrial plasma process without special precautions due to environmental and safety reasons. It is, therefore, of great interest to find a plasma process that can be run at large scale, giving the fabric surface the same properties as does ammonia plasma treatment. In search for such an alternative, two different approaches have been evaluated in this study:

NH_3/N_2 plasmas: Ammonia diluted with nitrogen. How much can the ammonia be diluted without losing its interesting effect on the fabric surface? According to studies at HTP Unitex (Caronno Pertusella, Italy), an industrial partner in the project, it is possible to use up to 15% ammonia in nitrogen without violating the safety regulations.

H_2/N_2 plasmas: H_2 and N_2 mixed at various ratios. According to HTP Unitex, it is possible to use up to 5% hydrogen in nitrogen without violating the safety regulations. Is it possible to obtain ammonia-like plasma treatments at such conditions?

2. EXPERIMENTAL

2.1. Materials

2.1.1. Thermoplastics
ABS (acrylonitrile-butadiene-styrene), PMMA (poly(methyl methacrylate)) and PET (poly(ethylene terephthalate)) thermoplastic plates were supplied by Arla Plast (Borensberg, Sweden). The plates were ultrasonically cleaned in isopropanol and rinsed with ethanol prior to plasma treatments.

2.1.2. Fabrics
Polyester (PET) microfiber fabric samples weighing 90 g/m^2 were used in all experiments. The fabric was produced for high-quality workwear fabrics. Comprehensive information on microfiber products, their production and properties is given in Ref. [11].

2.2. Plasma surface modifications

All plasma treatments were performed in YKI's large plasma reactor, which is a commercially available batch reactor (Technics Plasma, model 3027-E). The process chamber is made of aluminum and its inner dimensions (width × height × depth) are 650 × 600 × 700 mm. The reactor has been described in more detail elsewhere [11].

2.2.1. NH_3 and O_2 plasma treatments of PET thermoplastic plates
ABS, PMMA and PET thermoplastic plates samples were oxygen and ammonia plasma-treated at two different power-to-pressure ratios (W/P):
- $W/P = 1.6$ (1000 W and 600 mTorr) (O_2 plasma 1 and NH_3 plasma 1)
- $W/P = 5$ (1000 W and 200 mTorr) (O_2 plasma 2 and NH_3 plasma 2)

The treatment time was always 30 s.

2.2.2. NH_3/N_2 plasma treatment of PET fabrics
PET fabric samples were plasma treated at 8 different NH_3/N_2 ratios ranging from pure NH_3 to pure N_2. Such series of samples were prepared at four different W/P ratios:
- $W/P = 1.6$ (1000 W and 600 mTorr)
- $W/P = 3.3$ (2000 W and 600 mTorr)
- $W/P = 5$ (1000 W and 200 mTorr)
- $W/P = 10$ (2000 W and 200 mTorr)

2.2.3. H_2/N_2 plasma treatment of PET fabrics
Samples of PET fabrics were plasma treated at 8 different H_2/N_2 ratios ranging from pure H_2 to pure N_2. Such series of samples were prepared at four different W/P ratios:
- $W/P = 1.6$ (1000 W and 600 mTorr)

- $W/P = 3.3$ (2000 W and 600 mTorr)
- $W/P = 5$ (1000 W and 200 mTorr)
- $W/P = 10$ (2000 W and 200 mTorr)

2.3. Adhesion measurements

The plasma-treated and untreated thermoplastic plates were laminated with glass fibre mats immersed in an unsaturated polyester resin (Crystic U904, Scott Bader, UK), using the vacuum injection method. The adhesion between the thermoplastic plates and the polyester resin was determined using the ASTM method 3807-92.

2.4. Characterization of plasma-treated fabric surfaces

2.4.1. XPS analyses

The chemical composition of plasma-treated and untreated PET fabric surfaces was determined by X-ray Photoelectron Spectroscopy (XPS) using a Kratos AXIS HS X-ray photoelectron spectrometer (Kratos Analytical, Manchester, UK). The samples were analysed in fixed analyser transmission (FAT) mode using a mono-chromatic Al K_α X-ray source operated at 300 W (15 kV/20 mA). The analysis area was below 1 mm^2. The take-off angle of the photoelectrons was perpendicu-lar to the sample. Detailed spectra for C_{1s}, O_{1s}, N_{1s} and S_{2p} were acquired with a pass energy of 80 eV. The sensitivity factors used were 0.25 for C_{1s}, 0.66 for O_{1s}, 0.42 for N_{1s} and 0.54 for S_{2p} (supplied by Kratos). The high-resolution C_{1s} spectra were acquired with a pass energy of 20 eV. Using the curve-fitting programme supplied with the spectrometer, Gaussian curves were fitted for deconvolution of the carbon (C_{1s}) peak.

2.4.2. Surface charge measurements

The surface charge was measured using a static sensor, 3M Static Sensor, model 709. The sensor is designed to detect and measure the electrostatic field strength (in V/m) of the charges generated at the surfaces. The Static Sensor Method thus gives an indirect measure of the surface charge as it is anticipated that the electro-static field strength is proportional to the surface charge. The method is contact-less and can be used to distinguish between positively and negatively charged sur-faces. It should be noted that the method is only indicative and not precise. However, it is possible to obtain relative levels of the surface charge (positive as well as negative). All calibration measurements performed at the 3M Austin Static Calibration Laboratory comply with MIL-STD-45662A.

The surface charge measurements were performed directly after the plasma treatments while the fabric samples were still hanging on the sample holder. The sensor was kept approximately 20 mm from the fabric and the charge was meas-ured at three different locations on each sample film and the measurements were repeated at least three times. The sensor was re-zeroed between each set of meas-urement.

2.4.3. Dynamic contact angle measurements

The wettability measurements were performed in a Dynamic Absorption Tester (DAT, Fibro Systems, Hägersten, Sweden). The spreading and absorption of small water droplets into treated and untreated PET fabric samples were recorded with a high speed CCD camera, with a measurement every 30 ms. This results in plots where the advancing contact angle is plotted *versus* time.

3. RESULTS AND DISCUSSION

3.1. NH₃ and O₂ plasma treatments of thermoplastic plates

The ammonia and oxygen plasma treatments of the various thermoplastic plates were evaluated with respect to wettability (water contact angles) and adhesion to an unsaturated polyester resin, and the results are shown in Tables 1 and 2, respectively.

Each result in Table 1 is an average of four measurements. As shown in Table 1, all untreated thermoplastics have a water contact angle around 70±3°. After the various plasma treatments, the contact angle decreases to 24–56 degrees, depending on the type of plasma treatment and thermoplastic. In all cases, the surface energy of the thermoplastics increased, which normally leads to an improvement in adhesion strength [12]. This is also the case, as shown in Table 2. The adhesion

Table 1.
Wetting properties of oxygen and ammonia plasma-treated thermoplastic plates

Thermoplastic	Water contact angle (θ_{adv}, deg)				
	Untreated	O_2-plasma 1	O_2-plasma 2	NH_3-plasma 1	NH_3-plasma 2
ABS	72	27	24	49	55
PMMA	67	51	50	52	56
PET	70	24	28	45	32

Table 2.
Adhesion properties of oxygen and ammonia plasma-treated thermoplastic plates to an unsaturated polyester resin

Thermoplastic	Adhesion force (N)				
	Untreated	O_2-plasma 1	O_2-plasma 2	NH_3-plasma 1	NH_3-plasma 2
ABS	6	20	22	30	24
PMMA	7	15	16	15	16
PET	Cohesive failure	Cohesive failure	Cohesive failure	Cohesive failure	Cohesive failure

force between the plasma-treated thermoplastics and an unsaturated polyester resin is significantly increased, and the ammonia plasma treatment gives just as good or even better adhesion than the oxygen plasma treatment.

The PET thermoplastic plates had unfortunately a poor cohesive strength and, thus, cohesive failures were obtained both for the untreated and plasma-treated samples. The evaluation of the plasma-treated samples shows that both the oxygen and ammonia plasma treatments of thermoplastic plates give very similar results, i.e., improved wettability, as well as improved adhesion.

3.2. NH₃ and O₂ plasma treatments of PET fabrics

Exactly the same plasma conditions and same reactor were used in the oxygen and ammonia plasma treatments of PET fabrics as in the treatment of the thermoplastic plates. As shown in Fig. 1, oxygen plasma treatment of the fabric gives similar results, i.e., the wettability is significantly improved.

Ammonia plasma treatment of the PET fabric, on the other hand, gives a totally different result. As shown in Fig. 2, the fabric becomes hydrophobic instead of hydrophilic. Thus, identical ammonia plasma conditions give hydrophilic thermoplastic surfaces and hydrophobic PET fabric surfaces. As of today we have no simple explanation for these results.

The oxygen and ammonia plasma treatments were also applied to fluorocarbon (FC)-coated PET fabric in order to improve its adhesion properties. As shown in Fig. 3, the oxygen plasma treatment gives a drastic improvement of the water wetting properties. It is obvious that the oxygen plasma treatment destroys the

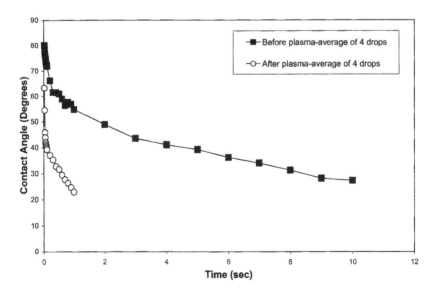

Figure 1. Wettability of oxygen plasma-treated PET fabric, illustrated as advancing water contact angle as a function of time.

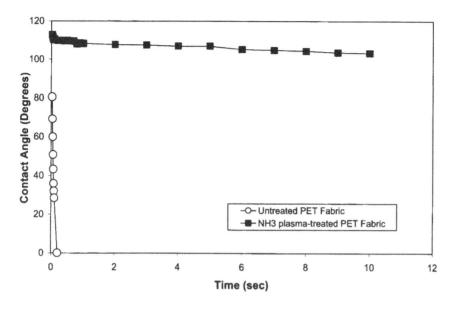

Figure 2. Wettability of ammonia plasma-treated PET fabric, illustrated as advancing water contact angle as a function of time.

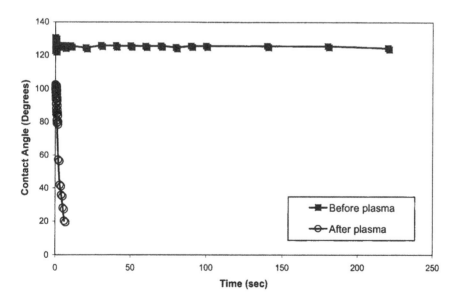

Figure 3. Wettability of FC-coated PET fabric, before and after oxygen plasma treatment, illustrated as advancing water contact angle as a function of time. Average of three measurements.

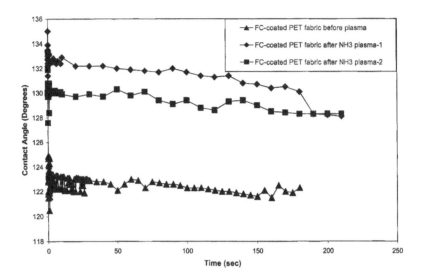

Figure 4. Wettability of FC-coated PET fabric, before and after two different ammonia plasma treatments, illustrated as advancing water contact angle as a function of time. Plasma conditions NH₃ plasma-1 and NH₃-plasma-2 here refer to 1000 W, 800 mTorr, 30 s, and 1000 W, 200 mTorr, 30 s, respectively.

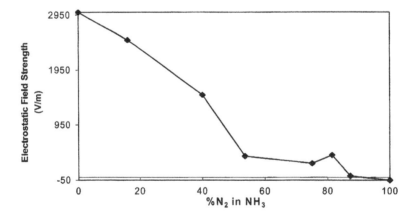

Figure 5. Surface charge characteristics of NH₃/N₂ plasma-treated PET fabrics expressed as electrostatic field strength as a function of the gas composition at plasma power 2000 W and pressure 200 mTorr.

hydrophobic fluorocarbon layer. This will certainly improve the adhesion properties of the fluorocarbon-coated PET fabric. However, if an additional water-based coating is applied on top of the modified fluorocarbon coating, that coating will be soaked into the bulk of the fabric, making it very stiff upon drying. This is an unwanted effect as the additional coating is expected to stay on top of the fluoro-carbon coating.

As shown in Fig. 4, the ammonia plasma treatment here also gives a hydropho-bic character to the PET fabric. The already hydrophobic FC-coated PET fabric becomes even more hydrophobic. At the same time, electrostatic field strength measurements (an indirect measure of the surface charge) show that the PET fab-ric becomes positively charged (Fig. 5). The combination of hydrophobicity and high positive surface charge gives the PET fabric interesting properties. The hy-drophobic character of the surface ensures that an additional coating applied on top of the modified FC-coated PET fabric will stay there and not be soaked into the bulk of the fabric. The high positive surface charge gives a sufficiently good adhesion between the additional coating and the FC coating (data not shown).

However, due to the problems mentioned above, it is of great interest to find a process that can substitute the ammonia plasma process.

3.3. NH₃/N₂ plasma treatment of PET fabrics

3.3.1. Surface charge characteristics of NH₃/N₂ plasma-treated PET fabrics
The surface charge (or the electrostatic field strength) of the NH_3/N_2 plasma-treated PET fabrics is shown in Fig. 5 as a function of the gas composition at W/P = 10. It is shown that fabrics treated with a pure ammonia plasma attain relatively high positive field strengths around 3000 V/m (untreated fabrics have no field strength, i.e., 0 V/m). The positive charge then gradually decreases as the ammonia

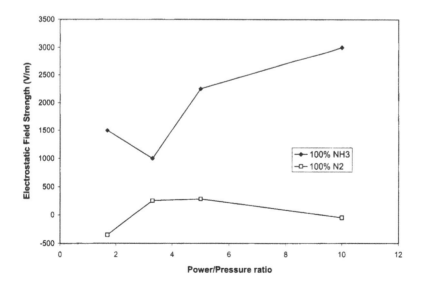

Figure 6. Influence of two operational parameters, plasma discharge power and pressure, on the electrostatic field strength of PET fabrics at 100% NH_3 and 100% N_2 plasmas.

is diluted with nitrogen in the plasma. The surface charge of the fabrics after a pure N_2 plasma treatment becomes slightly negative. The same trend was observed also at the other W/P ratios.

Figure 6 shows the influence of two operational parameters, plasma power and pressure, on the surface charge at the two extremes: 100% NH_3 and 100% N_2. The surface charge is only slightly affected by the operational parameters in pure N_2 plasma, whereas the charge increases significantly with increasing W/P ratio in the pure ammonia plasma.

3.3.2. Wettability of NH_3/N_2 plasma-treated PET fabrics

The water wettability of the NH_3/N_2 plasma-treated PET fabrics, ranging from pure nitrogen to pure ammonia plasma, is displayed in Fig. 7. It is shown that fabrics treated with pure N_2 plasma exhibit very good wetting properties. As the concentration of ammonia in the gas feed increases, the wettability of the fabrics gradually decreases. At an NH_3 concentration of 60% or more, an advancing water contact angle of 105–115 degrees is obtained, giving the fabric surface a very hydrophobic character.

In conclusion, PET fabrics exposed to pure ammonia or ammonia-rich plasmas attain a high positive surface charge and a high surface hydrophobicity. The positive surface charge disappears and the water wettability increases significantly when ammonia is diluted with nitrogen. It is thus obvious that the interesting sur-

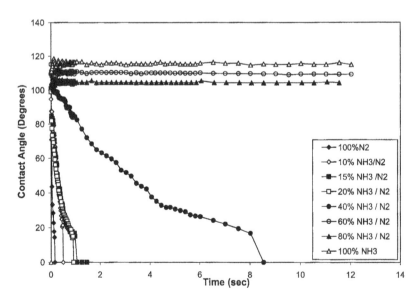

Figure 7. Wettability of NH_3/N_2 plasma-treated PET fabrics, illustrated as the advancing water contact angle as a function of time. The plasma power and pressure were 1000 W and 200 mTorr, respectively.

face properties of ammonia plasma-treated PET fabrics cannot be preserved if it is diluted to less than 60% ammonia in nitrogen.

3.4. H₂/N₂ plasma treatment of PET fabrics

3.4.1. Surface charge characteristics of H_2/N_2-plasma-treated PET fabrics

The surface charge (read electrostatic field strength) of the H_2/N_2-plasma-treated PET fabrics is shown in Fig. 8 as a function of the gas composition at various plasma conditions. It is shown that fabrics treated with a 80%H_2/20% N_2 plasma attain the highest positive field strengths, around 3000 V/m, at 1000 W and 600 mTorr ($W/P = 1.6$), which is the same field strength level as after pure ammonia plasma treatment. The surface charge seems to decrease with increasing W/P ratio for a given gas composition. It is also indicated that a mixture of the gases is needed in order to obtain a positively charged surface, since both pure H_2 and pure N_2 plasmas give slightly negatively charged surfaces.

3.4.2. Wettability of H_2/N_2-plasma-treated PET fabrics

The water wettability of the H_2/N_2-plasma-treated PET fabrics, ranging from pure nitrogen to pure hydrogen plasma at three different plasma conditions, is displayed in Figs 9–11. It is shown that fabrics treated with pure N_2 plasma exhibit very good wetting properties. As the concentration of hydrogen in the gas feed

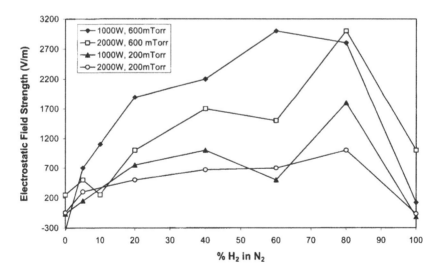

Figure 8. Surface charge characteristics of H_2/N_2 plasma-treated PET fabrics expressed as electrostatic field strength as a function of the gas composition at various plasma conditions.

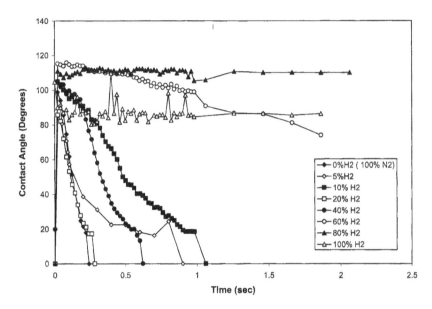

Figure 9. Wettability of H_2/N_2 plasma-treated PET fabrics, illustrated as the advancing water contact angle as a function of time. The plasma power and pressure were 1000 W and 600 mTorr, respectively ($W/P = 1.6$).

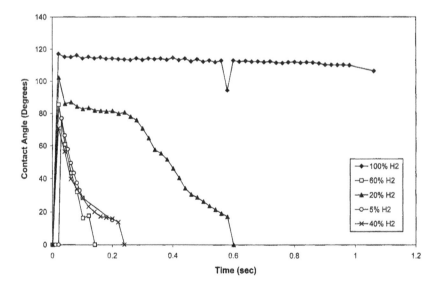

Figure 10. Wettability of H_2/N_2 plasma-treated PET fabrics, illustrated as the advancing water contact angle as a function of time. The plasma power and pressure were 1000 W and 200 mTorr, respectively ($W/P = 5$).

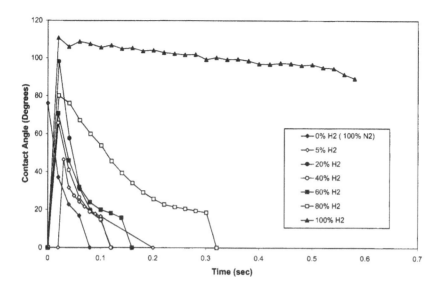

Figure 11. Wettability of H_2/N_2 plasma-treated PET fabrics, illustrated as the advancing water contact angle as a function of time. The plasma power and pressure were 2000 W and 200 mTorr, respectively ($W/P = 10$).

increases, the wettability of the fabrics gradually decreases. At $W/P = 1.6$ and an NH_3 concentration of 60% or more, an advancing water contact angle of 105–115 degrees is obtained, at least within the investigated time interval, giving the fabric surface a hydrophobic character. Figures 9–11 also show that the general wettability of the H_2/N_2-plasma-treated fabrics increases with increasing W/P ratio.

3.5. Comparison between pure NH3 and H2/N2-plasma-treated PET fabrics

The similarity between pure NH_3 and H_2/N_2-plasma-treated PET fabrics has been investigated with respect to surface charge, water wettability and chemical surface composition.

3.5.1. Surface charge and water wettability
It was found that H_2/N_2 plasmas with 80% H_2 gave the best correlation with pure ammonia plasma treatment. A comparison of the water wettability is shown in Fig. 12. It is shown that the water repellence properties of the H_2/N_2 plasma-treated fabric are not quite as good as those of the NH_3 plasma-treated fabric. However, they are significantly better than those of the untreated PET fabric. It is also shown in Figs 5 and 8 that the surface charge (read electrostatic field strength) after the two treatments is very similar: 2950 V/m after NH_3 plasma and 2800 V/m after the H_2/N_2 plasma treatment.

It is also shown that treatment at the recommended hydrogen concentration, i.e., 5% in nitrogen, does not give the desired properties. The fabric surface

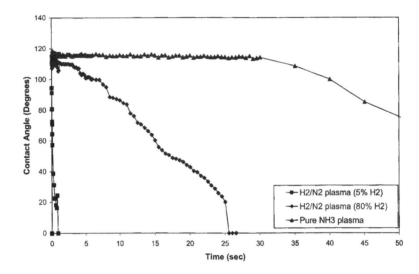

Figure 12. Wettability of pure NH₃ plasma-treated, H₂/N₂ plasma-treated (with 80% H₂) and H₂/N₂ plasma-treated (with 5% H₂) PET fabrics, illustrated as the advancing water contact angle as a function of time. The plasma power and pressure were 1000 W and 600 mTorr, respectively ($W/P = 1.6$) during the H₂/N₂ plasma treatments.

becomes very hydrophilic and the field strength is only 700 V/m, i.e., significantly lower than after the ammonia plasma treatment.

3.5.2. Chemical surface composition

The chemical composition of the plasma-treated fabric surfaces according to XPS analyses is shown in Table 3.

NH₃ plasma treatment is known to give rise to various N-containing functionalities, such as amino (–NH₂), imino (–CH=NH), cyano (–C≡N) and others, on polymers [6]. In addition, also oxygen-containing groups, such as amido (–CONH₂) are formed due to post-discharge atmospheric oxidation [13]. It is shown in Table 3 that the NH₃ plasma-treated PET fabric sample contains less O than the untreated sample. It appears as if the ammonia plasma introduces N atoms at the expense of O atoms in the PET molecule. In the H₂/N₂ plasma-treated

Table 3.
XPS elemental concentrations (in %) of H₂/N₂ and ammonia plasma-treated PET fabrics

PET sample	C	O	N
Untreated, Reference	78.2	21.8	–
Pure NH₃ plasma-treated	79.5	18.5	2.0
H₂ /N₂ plasma-treated (80:20)	76.5	21.6	1.9
H₂ /N₂ plasma-treated (5:95)	71.1	26.5	2.4
Pure N₂ plasma-treated	69.5	28.1	2.4

Table 4.
XPS C_{1s} peak deconvolution results (in %) of H_2/N_2- and ammonia-plasma-treated PET fabrics

PET sample	C1 (C–C, C–H)	C2 (C–O, C=N)	C3 (N–C–O, C=O)	C4 (–C–O–C*=O)
Untreated (Reference)	58.1	33.9	–	8.0
Pure NH_3 plasma-treated	66.8	22.0	3.0	8.2
H_2/N_2 plasma-treated (80:20)	63.9	21.7	5.3	9.0
H_2/N_2 plasma-treated (5:95)	54.9	26.9	5.7	12.6
Pure N_2 plasma-treated	52.6	25.6	5.2	16.6

and pure N_2 plasma-treated samples, the N atoms are introduced on the surface without any loss of O atoms. In fact, the O concentration increases with increasing N_2 concentration in the gas feed. It is a common phenomenon that oxygen is incorporated onto polymer surfaces after and during non-oxygen plasma treatments. The reason for this is that free radicals created on a polymer surface can react with small amounts of water vapour that are almost always present in most plasma reactors, even during a non-oxygen plasma treatment. In addition, free radicals that remain on a polymer surface after a plasma treatment will react with oxygen when the surface is exposed to the atmosphere.

As shown in Tables 3 and 4, the chemical surface compositions of the H_2/N_2 (80% H_2) and ammonia plasma-treated PET fabrics are not identical but the resemblance is quite good. One small but still significant difference is that the H_2/N_2 plasma-treated fabric contains more oxygen than the NH_3 plasma-treated sample. Table 4 shows that the relative concentration of the ester carbon (C4) is more or less unchanged after NH_3 plasma treatment. After the H_2/N_2 plasma treatments, on the other hand, the C4 concentration increases and it increases with increasing N_2 content in the gas feed. The PET fabric treated with a pure N_2 plasma attained a C4 concentration twice as high as in untreated and ammonia-plasma-treated fabrics. It is obvious that the increase of total O concentration in Table 3 is mainly due to the increase in C4 and C2 components of the C_{1s} peak. The difference in C4 concentration between the NH_3 plasma-treated and the H_2/N_2 (80:20) plasma-treated fabric samples (8.2 and 9.0 atom%, respectively) could be the main reason for the difference in wetting properties observed in Fig. 12.

It is not obvious why the O concentration increases with increasing N_2 content in the gas feed. One reason could be that the excited nitrogen atoms create more radicals on the PET surface, i.e., have a higher capability to abstract hydrogen atoms from the PET chain, and that the radicals react with oxygen upon exposure to the atmosphere. The H_2/N_2 plasma with only 5% H_2 does not give a surface chemical composition similar to an ammonia plasma-treated PET fabric surface.

4. CONCLUSIONS

Both oxygen and ammonia plasma treatments improve the wetting and adhesion properties of thermoplastic plates.

Ammonia plasma treatments of PET thermoplastic plates and PET fabrics, performed at identical plasma conditions in the same plasma reactor, give totally different results with respect to wettability. The PET (and other thermoplastic) plates become more hydrophilic, whereas the fabrics become hydrophobic. As of today we have no simple explanation for this difference.

H_2/N_2 plasma treatments can give PET fabrics surface properties (surface charge, water wettability and chemical composition) which are similar to those obtained by ammonia plasma treatments if the hydrogen concentration in the gas mixture is about 80% or near the stoichiometric ratio of ammonia (67%). Plasma treatments performed with the recommended gas mixtures, i.e., 15% NH_3 in N_2 and/or 5% H_2 in N_2, did not yield such surface properties.

Acknowledgements

Dr. Paolo Canonico at M & H Srl is thanked for valuable discussions. Dr. Marie Ernstsson and Mr. Mikael Sundin are acknowledged for help with the XPS analyses. Mrs. Katarina Wiklander is thanked for performing the plasma treatments, dynamic absorption and surface charge measurements at YKI. The figure editing by Ms. Karin Hallstenson is gratefully acknowledged. This work was partially funded by the European Community under the 'Competitive and Sustainable Growth' Programme (G1RD-CT-2000-00234).

REFERENCES

1. M. S. Strobel, C. S. Lyons and K. L. Mittal (Eds.), *Plasma Surface Modification of Polymers: Relevance to Adhesion.* VSP, Utrecht (1994).
2. C.-M. Chan, T.-M. Ko and H. Hiraoka, *Surface Sci. Rep.* **24**, 1–54 (1996).
3. S. Holmes and P. Schwartz, *Composites Sci. Technol.* **38**, 1 (1990).
4. Z.-F. Li, A. N. Netravali and W. Sachse, *J. Mater. Sci.* **27**, 4625–4632 (1992).
5. L. Tusek, M. Nietschke, C. Werner, K. Stana-Kleinschek and V. Ribitsch, *Colloids Surfaces A,* **195**, 81–95 (2001).
6. J. Yang, J. Bei and S. Wang, *Biomaterials* **23**, 2607 (2002).
7. Y. Nakayama, T. Takahagi, F. Soeda, K. Tahada, S. Nagaoka, J. Suzuki and A. Ishitani, *J. Polym. Sci. A* **26**, 559 (1988).
8. I. H. Loh, R. E. Cohen and R. F. Baddour, *J. Mater. Sci.* **22**, 2937 (1987).
9. D. A. Waldman, Y. L. Zou and A. N. Netravali, *J. Adhesion Sci. Technol.* **9**, 1475 (1995).
10. N. Inagaki, S. Tasaka and H. Kawai, *J. Adhesion Sci. Technol.* **3**, 637 (1989).
11. A. Nihlstrand, T. Hjertberg and K. S. Johansson, in: *Mittal Festschrift on Adhesion Science and Technology*, W. J. van Ooij and H. R. Anderson Jr. (Eds.), pp. 285–305. VSP, Utrecht (1998).
12. K. L. Mittal, *Polym. Eng. Sci.* **17**, 467 (1977).
13. P. Favia, M.-V. Stendardo and R. d'Agostino, *Plasmas Polym.* **1**, 91 (1996).

Contact Angle, Wettability and Adhesion, Vol. 4, pp. 351–367
Ed. K.L. Mittal
© VSP 2006

Study of the effect of acidic species on wettability alteration of calcite surfaces by measuring partitioning coefficients, IFT and contact angles

K. A. REZAEI GOMARI, A. A. HAMOUDA,* T. DAVIDIAN and
D. A. FAGERLAND

Petroleum Department, University of Stavanger, P. O. Box 8002 Ullandhaug, NO-4036 Stavanger, Norway

Abstract—The effect of the pH and salt content on partitioning coefficient (K) and interfacial tension (IFT) for fatty acids dissolved in the n-decane/water system has been investigated. The fatty acids investigated here are divided into three categories: short chain (heptanoic acid), long chain (stearic and oleic acids), and naphthenic acids (cyclohexane-pentanoic and decahydronaphthalene-pentanoic acids). A similar trend is obtained for both K and IFT when systems of these acids are tested at different pH values and salt contents. Minimum changes in K and IFT are shown for systems containing short-chain fatty acids. Divalent cations (Mg^{2+}) showed a larger effect in reducing both IFT and K as a function of pH than the monovalent Na^+. An estimation is made for the acid dissociation constant (pK_a) of stearic acid using the partitioning coefficient (K-model) and IFT data (ionized surface group (ISG) model). The estimated pK_a values for stearic acid are found to be 7.5 and 7 from K and ISG models, respectively. The humidified (presence of water film) calcite surface is shown to be altered to a more oil-wet at lower pH, and as the pH increases the contact angle decreases. The degree of alteration of wettability is dependent on the structure of the fatty acid and, hence, on K, IFT and pH.

Keywords: Wettability alteration; contact angle; fatty acid; calcite; salt; pH.

1. INTRODUCTION

The wetting state of most reservoir minerals is considered oil-wet or mix-wet, while originally they are strongly water-wet. This wettability alteration is found to be due to adsorption of polar components in crude oil onto the mineral surfaces [1, 2]. The adsorption of polar components can take place directly from the oil phase, or through the water film [3]. The study of the adsorption of organic components onto carbonate minerals shows that fatty acids are the main components that adsorb strongly on the surface and alter the wettability to strongly oil-wet [4].

---

*To whom correspondence should be addressed. Tel.: (47-51) 832-271;
e-mail: aly.hamouda@uis.no

In the oil reservoirs where three phases of oil/water/rock coexist, the partitioning of oil components through, water as well as the interfacial tension between these three phases are important determining factors in the wetting state of the reservoir rock.

Several authors have studied the partition coefficients of carboxylic acids between oil and water phases as a function of pH and salinity [5–7]. They found a strong dependency for partitioning between oil and water on the pH and salinity of the water phase. The type and structure of the acids are also important parameters for the distribution of acids between the two phases [8].

Lowering of the IFT of oil/water system in presence of carboxylic acids by increasing the pH has been investigated by many researchers [9–14].

In this work a simple model oil (n-decane) is used. The model system used in this work includes different fatty acids that are present in the oil phase. The aim of this study was to address the possible relations between partitioning of carboxylic acids and their interfacial activity and the wetting state of calcite surface, measured by contact angle, at different pH values and salinities of water phase.

2. MATERIALS AND METHODS

2.1. Materials

A simple model oil, n-decane, was selected here to investigate the partitioning behavior, and interfacial activity of dissolved fatty acids in the model oil in contact with an aqueous phase. This simple model oil was also used to check the wettability of calcite caused by the selected fatty acids. The n-decane was supplied by Chiron (Trondheim, Norway) with +99% purity. The properties of all components are listed in Table 1. In this study, n-decane was considered to be the oil phase and hereafter will be termed "oil phase" unless otherwise stated.

Table 1.
Fatty acids tested

Organic component	Supplier and purity		Structural formula
Heptanoic acid	Fluka	> 99%	$CH_3(CH_2)_5COOH$
Oleic acid	Aldrich	> 99%	$C_8H_{17}CH{=}CH(CH_2)_7COOH$
Stearic acid	Aldrich	> 98.5%	$CH_3(CH_2)_{16}COOH$
Cyclohexane-pentanoic acid	Chiron AS	99%	[COOH]
Decahydronaphthalene-pentanoic acid	Chiron AS	–	[COOH]

The type of calcite crystals used in contact angle measurements was "Island-spar" calcite from India supplied by J. Brommeland (Norway). Pieces of the chunks were formed to the correct shape by filing (29×26.2×2 mm and 39.6×14×2 mm) to fit to the surface holder of the contact angle cell. The loose calcite particles from filing (using silicon carbide grinding paper) were blown off the surface using air, followed by a distilled water wash. The water-wetness was then checked for each piece by placing a drop of water on the surface. This was done before and after filing and cleaning the calcite surface to assure that there was no obvious change in the wettability during the process of preparation.

2.2. Methods

2.2.1. pH measurements
The pH measurements were performed by a pH meter model PHM 92 (supplied by Radiometer Analytical, Denmark). The aqueous solutions were adjusted to the desired pH values using drops of 0.01 M NaOH or 0.01 M HCl. The pH values in this study were between 5 and 10. At higher pH values, the solutions became unstable.

2.2.2. Partitioning coefficient determination
The partitioning coefficient was determined by subtracting the measured concentration of the fatty acids equilibrated with water at the desired pH and salinity from the initial concentration of 0.01 M in the n-decane. The value obtained represents the concentration of the fatty acid in the aqueous phase. The measurements were done using a Perkin-Elmer spectrophotometer in the IR region, at a wavenumber of 1713 cm^{-1} [15]. The n-decane and water weight ratio was constant at 1:1 (5 g of n-decane to 5 g of aqueous phase in a glass bottle). The bottle was then placed in an agitator and shaken at ambient temperature of 25±2°C for 1 day. Standard deviation of these measurements (log K) is estimated to be about ±0.15 (based on 5 measurements).

2.2.3. Interfacial tension measurements
The interfacial tension (IFT) measurements were performed by the ring method, using a Krüss tensiometer type 8451, having an accuracy of ±2 mN/m.

2.2.4. Contact angle measurements
The organic components were initially dissolved in the oil phase. The filed cleaned pieces of calcite were pre-wetted by dipping in the aqueous phase for 5–7 min. The water pre-wetted calcite was immediately inserted into the oil phase and aged for 1 day. After this period of time, calcite was taken out and washed with distilled water, n-heptane and dried with air. The air-dried surface was then placed in the contact angle cell where pure n-decane was introduced by a needle onto the calcite surface, which was surrounded by distilled water. The advancing contact angle was then viewed and measured accurately by a microscope. In some cases, as indicated in the Discussion, the oil droplets were not stable on the calcite surface surrounded by water, so the contact angle was then measured at the air interface.

2.2.5. Zeta potential measurements

The zeta potential for CaCO$_3$ (powder) was measured by AcoustoSizer Is, Colloidal Dynamics (Australia), between pH 7 and pH 12. In this experiment solution of 2% by wt. of CaCO$_3$ in distilled water was used. The test temperature was set to 25°C and the mixer was set to 360 rpm for all the experiments carried out in this work.

3. RESULTS AND DISCUSSION

3.1. Partitioning coefficients of fatty acids

The effect of the pH and salt content/type on the partitioning of the fatty acids between the oil (n-decane) and aqueous phases is addressed in this section. The partitioning, in general, for acids can be expressed as follows:

$$HA_o \longleftrightarrow HA_w, \tag{1}$$

where, HA$_o$ and HA$_w$ are the undissociated forms of the fatty acid in the oil and water phases, respectively. The concentration of the acid in the oil phase after equilibration period of 1 day was measured by IR spectra at a wavenumber of 1713 cm^{-1} which corresponds to C=O stretching vibrations of the undissociated acid [15]. Thereafter, the measured acid concentration was used in the following equation to obtain the concentration of acid transferred to the water phase at equilibrium:

$$C_{w,eq} = \left[C_{o,i} - C_{o,eq} \right] \frac{w_o}{w_w} \tag{2}$$

where, $C_{o,i}$ and $C_{o,eq}$ are the concentrations of fatty acid (mol/kg) in the oil phase initially and at equilibrium, respectively. w_o and w_w are, respectively, the weights of the oil and water phases in the system. In our experiments for simplicity equal weights of the two phases were used; thus, equation (2) is reduced to

$$C_{w,eq} = C_{o,i} - C_{o,eq} \tag{3}$$

Thus, the partitioning coefficient can then be expressed as:

$$K = \frac{\left[HA \right]_{o,eq}}{\left[HA \right]_{w,eq}} = \frac{C_{o,eq}}{C_{w,eq}} \tag{4}$$

Partition coefficients (K) for stearic acid (saturated fatty acid), oleic acid (unsaturated fatty acid), heptanoic acid (short chain) as a function of pH are shown in Fig. 1. This figure demonstrates the dependence of K on the pH. Jafvert *et al.* [5] divided the dissociation curve into three distinct parts, where at the low pH, the

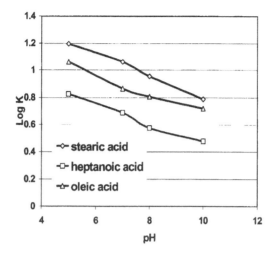

Figure 1. Partitioning coefficient of different fatty acids in n-decane/water system.

neutral species are dominant in both phases. At intermediate pH values, the neutral species are dominant in the oil phase, the anionic species become dominant in the aqueous phase, and the distribution ratio becomes more pH dependent. At high pH (above pH 9), the anion predominates in both phases, and the distribution ratio is again independent of pH. Reinsel *et al.* [6] investigated the effect of pH, temperature and organic acid concentration on partition coefficients for short-chain organic acids in crude oil/water system. They concluded that temperature and acid concentration had only small effects on *K* compared to the effect of pH.

Rudin and Wasan [9], Standal *et al.* [7] and Havre *et al.* [13] have investigated the partitioning coefficient of different acid components in an oil/water system. Their results demonstrate that the reduction of *K* with pH is related to the acid dissociation K_a. Havre *et al.* [13] combined the two parameters, equilibrium constant and partition coefficient, and proposed an expression for the total acid concentration of naphthenic acid in the water phase as:

$$C_{w,tot} = C_{ini,o} \frac{V_{oil}}{V_{water}} \times \left(1 - \frac{1}{1 + \frac{1}{K}\left(\frac{V_{water}}{V_{oil}}\right) \times \left[1 + \left(\frac{K_a}{[H^+]}\right)\right]} \right), \tag{5}$$

where $C_{ini,o}$ is the initial concentration of the stearic acid in the oil phase in our work, V_{oil} and V_{water} are the volumes of the oil and water phases, respectively, $[H^+]$ is the concentration of H^+, and *K* and K_a are the partition coefficient and the acid dissociation constant, respectively. The experimentally determined results from

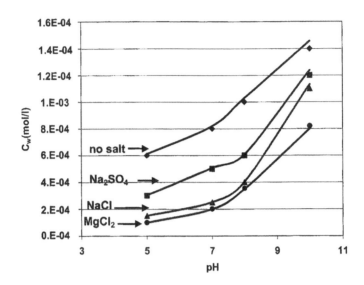

Figure 2. Calculated concentration of stearic acid in the water phase as a function of pH in presence and absence of salts. Equation 5 (line) is fitted to the experiment by setting K_a. The estimated pK_a for stearic acid based on the K-model is found to be 7.5.

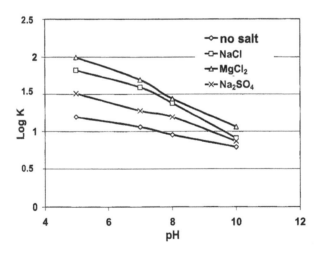

Figure 3. Effect of salts on the partitioning coefficient of n-decane-stearic acid/water system at different pH values.

the stearic acid concentration in water at different pH values were fitted to equation (5) by adjusting the K_a as shown in Fig. 2. The estimated pK_a is 7.5 for stearic acid. As can be seen that the experiment and model are in good agreement except at high pH, where a slight deviation is observed. A similar deviation was also reported by Havre *et al.* [13].

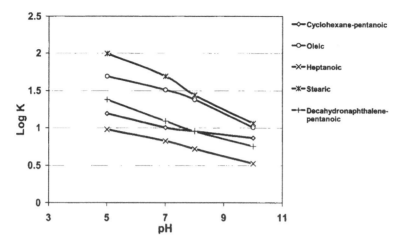

Figure 4. Effect of pH on partitioning of stearic acid, oleic acid, saturated 1-ring naphthenic acid (cyclohexane-pentanoic acid), saturated 2-ring naphthenic acid (decahydronaphthalene-pentanoic acid) and short chain (heptanoic acid) acid   between n-decane and water in presence of 0.1 M $MgCl_2$ at different pH values.

The effect of the salt content in the aqueous phase on the partition coefficient (K) of stearic acid between n-decane and water is shown in Fig. 3. As can be seen the addition of salt results in increasing K. This may be explained by the salting-out effect. This was also observed by Price [16], who investigated the effect of NaCl on the solubility of benzene, pentane, toluene and methyl cyclopentane in an aqueous phase.

A significant increase in K is observed when magnesium chloride is added to the aqueous phase compared to the system with sodium chloride. This is in agreement with the work of Jafvert *et al.* [5], who showed different contributions of the divalent ($MgCl_2$ and $CaCl_2$) and monovalent cations (NaCl and KCl) salts. The difference in partitioning coefficients caused by the added salts diminishes with pH.

Investigation of the effect of the pH on the partitioning of fatty acids in presence of Mg^{2+} is shown in Fig. 4. As can be seen, the partition coefficients for all acids decrease with increasing pH of the aqueous phase in the following order: stearic acid > oleic acid > saturated 2-ring naphthenic acid (decahydronaphthalene-pentanoic acid) > saturated 1-ring naphthenic acid (cyclohexane-pentanoic acid) > short chain (heptanoic acid).

3.2. Interfacial activity of fatty acids

The effects of the pH and salt content on the interfacial tension of n-decane-fatty acid/water systems are investigated in this section. The K_a for fatty acids can be represented by the point where the IFT decreases sharply with increasing H^+ concentration. The dissociation of acids at the interface can be expressed by equations (6) and (7):

$$HA \rightleftarrows H^+ + A^-$$ (6)

$$K_a = \frac{\left[H^+\right]_{int.}\left[A^-\right]}{\left[HA\right]}$$ (7)

The determination of K_a (both theoretically and experimentally) for fatty acids using the IFT data as a function of pH has been investigated for many years. Peters [17] studied the change in IFT as a function of the pH for different fatty acids. He obtained from the graph a pK_a value of about 7.5, where a sharp change in IFT was observed. Cratin [11] developed an approach to calculate the interfacial pK_a of fatty acids. In his model he assumed that the value of pK_a was equal to the H^+ concentration when the highest measured IFT dropped to half of its original value. Cratin [11] reported a pK_a value of 10.5 for stearic acid from his approach. In another study, Joos [18] reported an interfacial pK_a of 8.2 for stearic acid.

Healy and White [19] proposed a model (ionized surface group, ISG) to explain the double-layer properties of clays, inorganic oxides, polymer latex colloids and bio-surfaces in aqueous electrolyte solutions. Takamura and Chaw [20] modified the ISG model and applied it to liquid/liquid interfaces. The equations of this model used in our work to estimate the interfacial pK_a are given below.

The basic equation for this model is expressed as equation (7). This equation can simply be applied for dissociation of an acid at the interface using the reduced surface potential equation given by:

$$\left[H^+\right]_{int.} = \left[H^+\right]_b \exp\left(y_0\right),$$ (8)

where $[H^+]_{int.}$ and $[H^+]_b$ are the interface and bulk concentrations of hydrogen ions respectively, $y_0 = \dfrac{e\Psi}{kT}$ is the reduced surface potential. In the reduced surface potential, e, ψ, k and T are electron charge, potential at the interface, Boltzmann distribution constant, and temperature in Kelvin, respectively. Ionization of an acid at the interface charges the surface. The surface charge density is expressed as:

$$\sigma_0 = -e\left[A^-\right]$$ (9)

The sum of ionized and un-ionized forms of the acid gives the total number of acidic sites at the interface ($N_{int.}$):

$$N_{int.} = \left[A^-\right] + \left[HA\right]$$ (10)

Combination of equations (7)–(10) results in:

$$\sigma_0 = \frac{eN_{int.}}{1+\left(\dfrac{\left[H_b^+\right]}{K_a}\right)\exp(-y_0)} \qquad (11)$$

In the presence of a monovalent salt in the bulk phase, the Grahame equation can be applied to express the charge density in the diffuse part of the electrical double layer as:

$$\sigma_d = (8kTn_{mv}\varepsilon\varepsilon_0)^{0.5}\sinh(\frac{-y_0}{2}) \qquad (12)$$

where n_{mv} is the monovalent ion density in the solution (here Na^+), and ε and ε_0 are the dielectric constant of the medium (water) and vacuum permittivity, respectively. The Grahame equation for the divalent cation (Mg^{2+}) is:

$$\sigma_d = (8kT\varepsilon\varepsilon_0)^{0.5}\left\{n_{dv}(2+\exp(-y_0))\right\}^{0.5}\sinh(\frac{-y_0}{2}) \qquad (13)$$

In equation (13), n_{dv} is the density of divalent ion in the bulk solution. Assuming equivalency, $\sigma_d = -\sigma_0$, $N_{int.}$ is obtained from the experimental IFT data. K_a is, then, obtained by matching σ_0 as a function of pH with the experimental IFT data at various pH values.

To calculate the number of sites at the interface ($N_{int.}$), the general equation of state for neutral films at an oil/water interface is given as:

$$\pi(A - A_0) = kT, \qquad (14)$$

where k is the Boltzmann constant and T is temperature in Kelvin, π is the surface pressure of the adsorbed film, which may be calculated from the difference between the interfacial tension (γ_0-γ) of the n-decane/water systems in the absence(γ_0) and presence (γ) of the fatty acids. The surface pressure in this work was calculated from the difference between the interfacial tension of water (0.1 M NaCl)/n-decane (0.01 M stearic acid) system (γ) and (0.1 M NaCl)/n-decane system (γ_0). Interfacial tensions γ_0 and γ at pH 5 are found to be 44.5±1 and 30.1±1 mN/m, respectively. The area occupied by the adsorbed molecules, A, of 48.5 Å^2 is calculated from the limiting area A_0 for the adsorbed molecule in a closely packed film at the interface of 20 Å^2 [12, 21]. This corresponds to $N_{int.} = 2.056\times10^{18}$ molecules/m^2. Spildo and Høiland [12] following a similar approach obtained $N_{int.} = 2\times10^{18}$ for 4-heptylbenzoic acid at pH 4 for the acidic species (unionized). In this work, IFT at pH 5 was used to calculate π and estimate $N_{int.}$.

The value of $N_{int.}$ obtained is then used in equation (11) and the K_a is obtained by iteration to match the σ_0 vs. pH and the interfacial tension as a function of the corresponding pH. In order to obtain the best fit, the surface charge was slightly adjusted. The pK_a of stearic acid, following this process, was found to have a value of 7. The curve fits are shown in Figs 5–7 for the three cases of 0.1 M NaCl, Na_2SO_4 and $MgCl_2$. Another method was also used to check the obtained pK_a. This consisted in applying a nonlinear fit to the interfacial tension/pH data and solving the above equations for K_a in terms of σ_0. In this method, Mathematica software program was used. The pK_a obtained by this method was 7.2. This is almost the same as the value of 7 obtained above. This (pK_a) is also in good agreement with the value of 7.5 obtained by Peters [17]. The values of pK_a obtained by the ISG model and the K-model for the partitioning coefficient, see previous section, are found to be in a good agreement with a difference of only about 0.3/0.5.

The IFT for the n-decane/water system without fatty acids is shown in this work to be 45.5 ± 1 mN/m. This is in agreement with the values reported in the literature, where values of 47 and 46 ± 1 mN/m were reported by Morrow et al. [22] and Spildo and Høiland [12], respectively. It is shown in Fig. 8 that heptanoic acid and naphthenic acid (1 or 2 rings) decreased the IFT from initial value of 45.5 to 37 and 35 mN/m at pH 5, respectively. The IFT of these acids (heptanoic and naphthenic acids) are not affected by the pH.

For long chain fatty acids, a significant change was observed as a function of pH. Oleic acid and stearic acid decreased the IFT from 45.5 (obtained from n-decane/water without fatty acid) to 30 and 32 mN/m at pH 5, respectively. Further increase in the pH beyond 7, shows a significant decrease of the IFT. This reduction in the IFT may be explained by acid dissociation at pH ≥ 7.

The less effect on the IFT obtained from the short chain fatty acids is expected since increasing the CH_2 group increases the adsorption energy of molecules at an interface (Traube's rule). The reported magnitude of adsorption energy for each CH_2 group at an oil/water interface is 800 cal/mol [23].

The effect of adding 0.1 M NaCl to n-decane/water system in presence of 0.01 M heptanoic, cyclohexane-pentanoic, decahydronaphthalene-pentanoic and stearic acids is shown in Fig. 9. A large decrease in IFT for all the acids, with different degrees, occurs at a pH > 7, as shown in Fig. 9. The decreasing order in IFT caused by the acids is as follows: stearic acid > decahydronaphthalene-pentanoic acid > cyclohexane-pentanoic acid > heptanoic acid. This is also in accord with the previous study of the partitioning coefficients of these acids between n-decane and water (see previous section), where the short-chain acids had lower K values compared to the long-chain acids, as discussed above. This demonstrates the importance of the structure of the organic acids on the IFT, and hence on wettability.

The effect of different salts, 0.1 M NaCl, Na_2SO_4 and $MgCl_2$, on the IFT in n-decane/water system at pH 8 with 0.01 M heptanoic, cyclohexane-pentanoic,

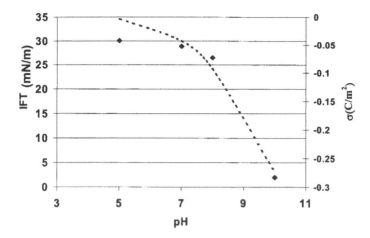

Figure 5. Estimation of interfacial dissociation constant K_a using the ISG approach. Experimental IFT (♦) and calculated interfacial charge, σ (----) for n-decane containing 0.01 M stearic acid and aqueous phase with 0.1 M NaCl as a function of pH.

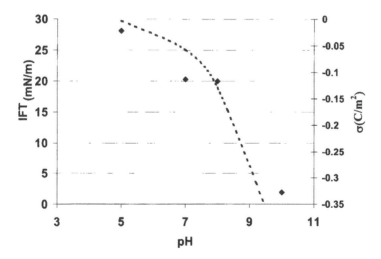

Figure 6. Estimation of interfacial dissociation constant K_a using the ISG approach. Experimental IFT (♦) and calculated interfacial charge, σ (----) for n-decane containing 0.01 M stearic acid and aqueous phase with 0.1 M Na_2SO_4 as a function of pH.

decahydronaphthalene-pentanoic, stearic and oleic acids is shown in Table 2. The IFT does not seem to be affected by the salts used in the systems that contain heptanoic and cyclohexane-pentanoic acids. However, for decahydronaphthalene-pentanoic, stearic and oleic acids, a large reduction of IFT was observed going from NaCl, Na_2SO_4 to $MgCl_2$, where $MgCl_2$, showed the highest effect on the IFT, while NaCl had the lowest effect. As shown also in the table the stearic acid

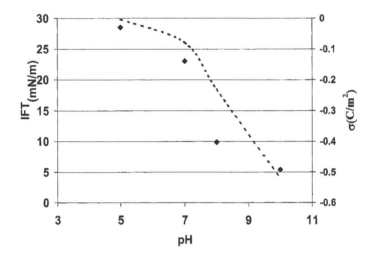

Figure 7. Estimation of interfacial dissociation constant K_a using the ISG approach. Experimental IFT (♦) and calculated interfacial charge, σ (---) for n-decane containing 0.01 M stearic acid and aqueous phase with 0.1 M MgCl₂ as a function of pH.

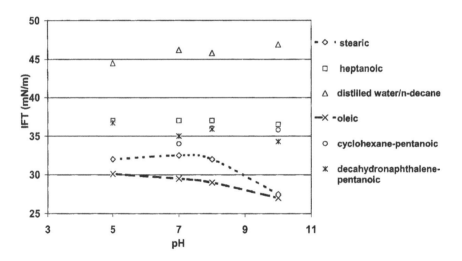

Figure 8. Effect of pH on IFT for systems containing stearic acid, oleic acid, saturated 1-ring naphthenic acid (cyclohexane-pentanoic acid), saturated 2-ring naphthenic acid (decahydronaphthalene-pentanoic acid) and short chain (heptanoic acid) acid without salts.

has the highest effect followed by oleic acid, then decahydronaphthalene-pentanoic acid. The first two acids have the longest chains among the acids tested. This is consistent with the results obtained on the partitioning coefficients for these acids in presence of Mg^{2+} compared to that, for example, for Na^+. The large

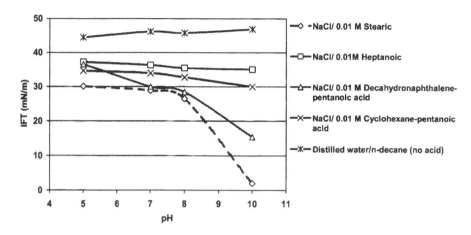

Figure 9. Effect of pH on IFT for systems containing stearic acid, saturated 1-ring naphthenic acid (cyclohexane-pentanoic acid), saturated 2-ring naphthenic acid (decahydronaphthalene-pentanoic acid) and short chain (heptanoic acid) acid in presence of 0.1 M NaCl.

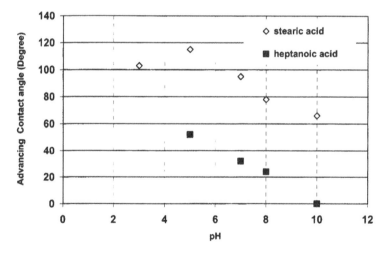

Figure 10. Comparison between measured advancing contact angles of stearic/heptanoic acid in n-decane/calcite/air system as a function of pH.

effect of the Mg^{2+} ions compared to Na^+ ions on the IFT may be explained based on the larger change in the charge density at the interface caused by Mg^{2+} ions compared to that resulted in presence of Na^+ ions, promoting more acid dissociation (driving the reaction in equation (6) towards the right side); hence, reduction of the IFT.

Table 2.
IFT (mN/m) values for n-decane (fatty acid)/water system at pH 8

Salt content	Type of acid				
	Heptanoic acid	Cyclohexane-pentanoic acid	Decahydronaphthalene-pentanoic acid	Stearic acid	Oleic acid
NaCl (0.1 M)	36.3	31.9	28.4	26.5	22.5
Na$_2$SO$_4$ (0.1 M)	34.8	30	26	19	20.9
MgCl$_2$ (0.1 M)	35.2	32.7	23	9.9	15.7

3.3. Contact angles

The present investigation was carried out to understand the observation made earlier on the reversed wettability alteration (RWA), where in presence of water the calcite surface was altered to be oil-wet by fatty acids [24]. In the work carried out here, contact angle measurements were performed on the same systems as reported earlier in the partitioning coefficient and IFT sections of this paper in order to draw a correlation between the three techniques used here and the effect of the fatty acid structure on the wettability.

A summary of the results of the contact angle measurements in air is shown in Table 3. The results show that stearic acid is by far the most influential among the acids tested in altering the surface to a more oil-wet as indicated by the high advancing contact angle. Other acids, heptanoic, cyclohexane-pentanoic and decahydronaphthalene-pentanoic acids as shown in Table 3 at pH 7 did not alter the calcite surface. This is in agreement with the results reported by Zullig and Morse [25] where no adsorption of short chain fatty acids on the calcite surfaces was observed. The stated stable/unstable condition in Table 3 is to reflect whether the released oil droplet stays on the calcite surface (stable) or it is rolled away (unstable) from the surface to the bulk liquid (water) in the contact angle cell.

So far, in this work it has been shown that an acid with a high partitioning coefficient shows a high interfacial activity at the oil/water interface and its adsorption on the calcite surface is stronger than that of the acid with low partitioning coefficient and low interfacial activity.

In order to draw a correlation between the partitioning of the fatty acid in oil (n-decane)/water system and the contact angle, as well as addressing the importance of the acid structure, Fig. 10 shows a comparison between stearic (long chain) and heptanoic (short chain) acids as a function of pH in the range 5–10. The measured contact angle on the calcite treated with stearic acid at neutral pH is found to be 96° in this work, which is comparable to the reported advancing contact angle of 100° by McCaffery and Mungan [26] (at the same concentration as in our work here). A contact angle of 98° has also been reported by Hamouda and co-workers [24]. The effect of the pH on the calcite surface in presence of water is related to the concentration of the potential determining ions Ca^{2+} and CO$_3{}^{2-}$ at the calcite surface, the degree of acid dissociation and the preferential affinity of carboxylate ions to Ca^{+2} ions. Long-chain saturated stearic acid, as well as

Table 3.
Advancing contact angles (degrees) in n-decane-fatty acid/calcite/air system. Stable/unstable indicates whether the oil droplet adheres to the calcite surface or rolls off

Component	Advancing contact angle in air	n-decane droplet stability on calcite in water
Stearic acid	95	Stable
Heptanoic acid	32	Unstable
Cyclohexane-pentanoic acid	47	Unstable
Decahydronaphthalene-pentanoic acid	70	Unstable

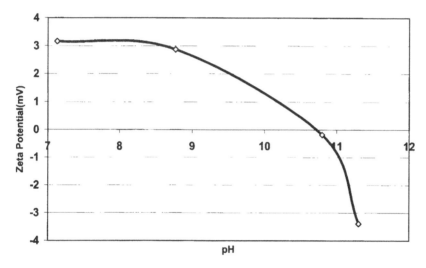

Figure 11. Smoluchowski zeta potential of surface measured in solution of 2 wt% CaCO₃ in distilled water; the zero point charge of calcite is found to be at pH 10.6.

long-chain unsaturated oleic acid are shown to alter the calcite surface to a more oil-wet [4, 25–27]. Zullig and Morse [25] have shown that at pH values of 7.8 to 8.4, the adsorption isotherms of stearic and oleic acids have the same isothermal adsorption affinity to the calcite surface.

Lebell and Lindström [28], Thompson and Pownall [29], Schramm *et al.* [30] and Legens *et al.* [2] have studied the change in zeta potential on calcite surfaces. They found calcite to be positively charged almost in all pH range. Thomson and Pownall [29] concluded that the major surface ions on calcite surface were Ca^{2+} and CO_3^{2-} in the range of pH 7–12. Lebell and Lindström [28] found that the zero point charge of calcite when immersed in water was 10.1. In our work zeta potential for $CaCO_3$ as a function of pH was measured. The zero point charge is found to be 10.6, as shown in Fig. 11. Figures 11 and 12 show a decrease in zeta potential of calcite surface that corresponds to a decrease in contact angle.

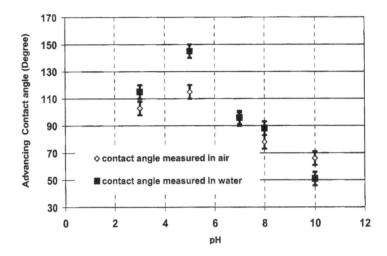

Figure 12. Measured advancing contact angles at the oil/water/calcite and air/water/calcite interfaces as a function of pH. The calcite surface was aged for 1 day in n-decane including stearic acid (0.01 M).

4. CONCLUSIONS

The following conclusions are drawn from this study:

1. The partition coefficient (K) of fatty acids between two liquid phases strongly depends on the pH, salinity and acid structure. Partition coefficients for all acids decrease with increasing pH in the following order: stearic acid > oleic acid > saturated 2-ring naphthenic acid (decahydronaphthalene-pentanoic acid) > saturated 1-ring naphthenic acid (cyclohexane-pentanoic acid) > short chain (heptanoic acid). Dissolved salts, as expected, reduce K as a function of pH. A significant reduction in K is observed in presence of Mg^{2+} ions with pH. This may be explained by the increased charge density at the interface.

2. IFT measurements have shown the same trends as the partitioning coefficient with the pH, as well as with the salt content and the type of the salt, i.e., mono/divalent ions.

3. The pK_a values for stearic acid using the partitioning coefficient (K-model) and the ISG-model are found to be 7.5 and 7, respectively.

Acknowledgements

The authors would like to thank the University of Stavanger for the financial support of this project. The authors also wish to thank Professor Tanja Barth, Postdoc. Sylvi Høiland Standal (Staff of Bergen University) and senior researcher Kristine Spildo for helpful discussions during the project. We are indebted to

Mehmed Nazecic for his assistance with the experimental work and Unni Hakli for her positive attitude and help in planning to get the chemicals needed in time.

REFERENCES

1. W. G. Anderson, *J. Petroleum Technol.* 1125–1144 (October 1986).
2. C. Legens, H. Toulhoat, L. Cuiec, F. Villieras and T. Palermo, *paper No. 49319 presented at the SPE Annual Technical Conference and Exhibition held in New Orleans, LA* (1998).
3. L. E. Cuiec, paper presented at the *21st Intersociety Energy Conversion Engineering Conference*, San Diego, CA (1986).
4. M. M. Thomas, J. A. Clouse and J. M. Longo, *Chem. Geol.* **109**, 201–213 (1993).
5. C. T. Jafvert, J. C. Westall, E. Grieder and R. P. Schwarzenbach, *Environ. Sci. Technol.* **24**, 1795–1803 (1990).
6. M. A. Reinsel, J. Borkowski and J. T. Sears, *J. Chem. Eng. Data* **39**, 513–516 (1994).
7. S. H. Standal, A. M. Blokhus, J. Haavik, A. Skauge and T. Barth, *J. Colloid Interface Sci.* **212**, 33–41(1999).
8. A. Leo, C. Hansch and D. Elkins, *Chem. Rev.* **71**, 525 (1971).
9. J. Rudin and D. T. Wasan, *Colloids Surfaces* **68**, 67–79 (1992).
10. J. Rudin and D. T. Wasan, *Colloids Surfaces* **68**, 81–94 (1992).
11. P. D. Cratin, *Colloids Surfaces* **89**, 103–108 (1994).
12. K. Spildo and H. Høiland, *J. Colloid Interface Sci.* **209**, 99–108 (1998).
13. T. E. Havre, J. Sjöblom and J. E. Vindstad, *J. Dispersion Sci. Technol.* **24**, 789–801 (2003).
14. Y. Touhami, V. Hornof and G. H. Neale, *J. Colloid Interface Sci.* **177**, 446–455 (1996).
15. K. William, *Organic Spectroscopy.* Macmillan, London (1975).
16. L. C. Price, *Am. Ass. Petrol. Geol. Bull.* **60**, 213 (1976).
17. R. A. Peters, *Proc. Roy. Soc. London Ser. A* **133**, 140 (1931).
18. P. Joos, *Bull. Soc. Chim. Belg.* **80**, 277 (1971).
19. T. W. Healy and L. R. White, *Adv. Colloid Interface Sci.* **9**, 303–345 (1978).
20. K. Takamura and R. S. Chow, *Colloids Surfaces* **15**, 35–48 (1984).
21. A. W. Adamson, *Physical Chemistry of Surfaces*, 3rd ed. Wiley, New York (1976).
22. N. R. Morrow, P. J. Cram and F. G. McCaffery, *Society of Petroleum Engineers of AIME J.* **13**, 221–232 (1973).
23. A. K. Chatterjee and D. K. Chattoraj, *J. Colloid Interface Sci.* **26**, 1 (1968).
24. G. Hansen, A. A. Hamouda and R. Denoyel, *Colloids Surfaces* **172**, 7–16 (2000).
25. J. J. Zullig and J. W. Morse, *Geochim. Cosmochim. Acta* **52**, 1667–1678 (1988).
26. F. G. McCaffery and N. Mungan, *J. Can. Petrol. Technol.* 185–196 (July–Sept. 1970).
27. L. Madsen, L. G. Madsen, C. Grøn, I. Lind and J. Engell, *Org. Geochem.* **24**, 1151–1155 (1996).
28. J. C. Lebell and L. Lindström, *Finn. Chem. Lett.* 134–138 (1982).
29. T. W. Thompson and P. G. Pownall, *J. Colloid Interface Sci.* **131**, 74–82 (1988).
30. L. L. Schramm, K. Mannhardt and J. J. Novosad, *Colloids Surfaces* **55**, 309–331 (1991).

Contact Angle, Wettability and Adhesion, Vol. 4, pp. 369–384
Ed. K.L. Mittal
© VSP 2006

Surfactant-induced modifications in interfacial and surface forces and multiphase flow dynamics in porous media

SUBHASH C. AYIRALA and DANDINA N. RAO[*]

*The Craft and Hawkins Department of Petroleum Engineering, Louisiana State University,
3516 CEBA Bldg., Baton Rouge, LA 70803-6417, USA*

Abstract—The use of surfactants to enhance recovery of oil from petroleum reservoirs has been well documented in the literature. The two main mechanisms for oil recovery enhancement by surfactants are modifications in interfacial forces (interfacial tension) and surface forces (rock wettability). However, the ability of these surfactants to alter wettability, as well as the relative contributions from the two mechanisms on multiphase flow characteristics in porous media and enhanced oil recovery remain largely unexplored. Hence, flow experiments through porous media were carried out in two different rock–fluids systems using a nonionic ethoxy alcohol surfactant. The first system consisted of Berea rock, Yates synthetic brine and n-decane, in which wettability effects were assumed to be negligible; and the second system consisted of Berea rock, Yates synthetic brine and crude oil from the Yates reservoir in West Texas, with its accompanying wettability effects. The multiphase flow characteristics are represented by oil–water relative permeabilities. A coreflood simulator has been used to calculate oil–water relative permeabilities by matching the recovery and the pressure drop data obtained during these flow tests. The relative permeability variations have then been interpreted to characterize the wettability alterations induced by the surfactant. The significantly higher incremental oil recoveries, observed due to surfactant in the rock–fluids system containing Yates crude oil when compared to that consisting of n-decane, clearly demonstrated the wettability-altering capability of the surfactant in addition to confirming the hypothesis that the wettability alteration was the principal mechanism for enhancing oil recovery by the surfactants. In addition, this study has provided evidence that the surfactant was able to render a special kind of heterogeneous wettability, known as mixed wettability, that enables preferential draining of oil phase through the formation of continuous wetting films of oil on the rock surface. This study has also demonstrated the possibility of making erroneous wettability interpretations from relative permeability characteristics due to the strong impact of oil–water emulsions on flow dynamics.

Keywords: Capillary forces; interfacial tension; mixed wettability; oil–water emulsion; surfactants; relative permeabilities; wettability; contact angle.

---

[*]To whom correspondence should be addressed. Tel.: (1-225) 578-6037; Fax: (1-225) 578-6039;
e-mail: dnrao@lsu.edu

1. INTRODUCTION

Hirasaki [1] has described wettability and its dependence on surface forces. The thermodynamics of wettability can be understood using the region of three-phase contact line, where the three phases of rock, oil and water come together. At the three-phase contact line, interactions take place between the pairs of interfaces separated by a distance called thickness. This thickness can be considered as a thermodynamic variable, as it affects the system energy. Disjoining pressure is the change in energy per unit area with respect to the change in distance between the pair of interfaces. This is the force that tends to separate the two interfaces and is dependent on intermolecular forces such as van der Waals, electrical and structural forces. All the various types of surface forces have been lumped together in the term "wettability" in rock–fluids systems.

In the bulk liquid phase, each molecule is surrounded by similar molecules and hence the net force on the molecule will be zero. On the other hand, a molecule at the interface between two immiscible bulk fluid phases is acted upon by different forces from the molecules on either side of the interface. This asymmetrical force field exerts a tension at the interface, which is called interfacial tension. It is represented by a vector tangential to the interface at the line of contact and can be defined as the free energy per unit area or as the force per unit length. Molecular origins of interfacial tension can be understood in terms of differences in interaction between molecules in the bulk phase and at the interface.

Nearly two-thirds of original oil in place found in petroleum reservoirs is left behind at the end of primary and secondary recovery processes. This is due to capillary forces that prevent oil from flowing within the pores of reservoir rock, trapping large amounts of residual oil in reservoirs. The capillary force (capillary pressure) can be defined as the force per unit area resulting from the interaction of the surface and interfacial forces and the geometry of the porous medium in which they exist and is given by:

$$P_c = \frac{2\sigma \cos\theta}{r} \tag{1}$$

where P_c is the capillary pressure, σ is the oil–water interfacial tension, θ is the contact angle and r is the capillary pore radius.

From equation (1), it is evident that the capillary pressure can be reduced either by reducing the interfacial forces (σ) or by altering the surface forces (θ), which means an alteration of wettability of the rock–fluids system. However, most of the research work done in this area has ignored the effect of surface forces on capillary force by assuming contact angle to be equal to zero. This is equivalent to assuming strongly water–wet conditions in all the reservoirs ignoring the fact that there are more non-water–wet reservoirs than water–wet ones [2].

Surfactants have long been considered in oil industry for oil recovery enhancement. There are mainly two mechanisms behind the use of surfactants, namely reduction in oil–water interfacial tension (interfacial forces) and alteration of wettability (surface forces). The ability of surfactants to reduce oil–water interfacial tension is well known and widely practiced. However, the other aspect, namely the ability of surfactants to alter wettability, has largely been ignored. For significant enhancements in oil recovery by interfacial tension reduction mechanism, 4–5 orders of magnitude reduction in interfacial tension is required [3]. Surfactants capable of providing such interfacial tension reduction are expensive and are required in large quantities, rendering them uneconomical for field application. This is the major reason behind the continuous decline and subsequent extinction of chemical EOR projects in the world over past few decades [4].

Surfactants can alter wettability from water–wet to oil–wet or from oil–wet to water–wet, depending on the type of surfactant orientation on rock surface [5]. Austad *et al.* [6] demonstrated using laboratory experiments on spontaneous imbibition and oil recovery that certain dilute surfactants at low concentrations could alter the wettability of nearly oil–wet chalk to water–wet conditions. Standnes and Austad [7] conducted imbibition and contact-angle measurements using chalk and reservoir dolomite cores and showed the surfactant-induced wettability alterations from oil–wet to water–wet. Chen *et al.* [8] performed dilute surfactant imbibition tests on initially strongly oil–wet dolomite cores of Yates field and measured United States Bureau of Mines (USBM) wettability indices to demonstrate the ability of surfactants to shift the wetting characteristics of Yates rock towards less oil–wet. Thus, it appears that wettability alterations could be induced by low-cost surfactants at moderate concentrations. This opens another avenue for using low-cost chemicals in enhancing oil recovery through wettability alteration.

Therefore, the objectives of this study were (1) to experimentally investigate the wettability altering capability of low-cost surfactants and at low concentrations by measuring oil–water relative permeability characteristics and (2) to discern the relative contributions of the two surfactant-induced mechanisms of interfacial tension reduction and wettability alteration on enhanced oil recovery. For this purpose, two series of coreflood experiments have been conducted using Berea rock and nonionic surfactant (ethoxy alcohol) in varying concentrations. The first series used n-decane as the oil phase to quantify the effect of interfacial tension reduction on oil recovery, while making a reasonable assumption that wettability effects in the decane–brine–Berea system were negligible. The second series used Yates crude oil (from Yates field in West Texas) in place of n-decane to quantify the effects of interfacial tension reduction as well as wettability alteration on enhanced oil recovery.

2. EXPERIMENTAL APPARATUS AND PROCEDURE

Figure 1 shows schematic diagram of coreflood apparatus used in this study. The Berea core used in the experiments was 3.81 cm in diameter and 7.62 cm in length. The absolute permeability was 400 milliDarcies (1 Darcy = 0.98692 μm^2)

and porosity was 22%. All the coreflooding experiments were conducted at ambient temperature (25°C) with the outlet open to atmosphere. The outlet pressure was always atmospheric (101.325 kPa) and, hence, the pressure at the inlet provided the absolute pressure drop across the core. Synthetic brine matching the Yates reservoir brine composition was used in all the experiments. The compositions and pH of Yates reservoir brine, as well as the synthetic brine used were given in Ref. [9].

The Rapaport and Leas [10] criterion was used to calculate the stable volumetric flow rates for use in all the experiments, since it was essential to conduct all the floods in a flow regime where the recovery was independent of injection flow rate. This scaling causes the capillary pressure gradient in the flow direction to be small compared to the imposed pressure gradient, enabling a simplified analytical calculation of relative permeabilities.

At first the core was saturated with brine to determine its porosity and the absolute permeability. Then oil (n-decane or Yates crude oil) was injected at a flow rate of 2.0 ml/min for 2 pore volumes and 6.0 ml/min for 5 pore volumes to bring the core to initial water saturation (S_{wi}). Then, a waterflood was conducted at a flow rate of 2.0 ml/min for 2 pore volumes. The core was then brought back to S_{wi} by flooding with oil. Then the effect of surfactant concentration on oil recovery was studied by carrying out several floods with synthetic brine containing different concentrations (500, 1500, 3500 and 5000 ppm) of nonionic (ethoxy alcohol) surfactant. During each of these floods, pressure drop and oil and brine productions were continuously monitored and recorded.

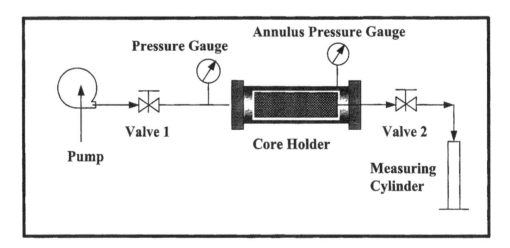

Figure 1. Schematic diagram of coreflood apparatus used in this study.

3. CALCULATION OF RELATIVE PERMEABILITIES

A coreflood simulator was used to simulate the experimentally measured pressure-drop and recovery histories using the relative permeabilities (k_r) described by:

$$k_{ro} = (1 - S)^{e_o} k_{rom} \qquad (2)$$

$$k_{rw} = S^{e_w} k_{rwm} \qquad (3)$$

where the subscripts o and w denote the variables with reference to oil and water phases, respectively. k_{ro} is the oil relative permeability, k_{rw} is the brine relative permeability, k_{rom} is the oil relative permeability at S_{wi}, k_{rwm} is the brine relative permeability at S_{or} and e_o and e_w are the Corey exponents. S, being a ratio of instantaneous change to maximum change in water saturation, represents dimensionless water saturation and is given by:

$$S = \frac{(S_w - S_{wi})}{(S_{wm} - S_{wi})} \qquad (4)$$

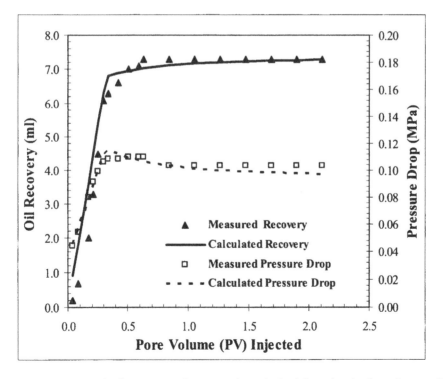

Figure 2. History match of recovery and pressure drop obtained from the simulator for waterflood of Yates crude oil in Berea core at 0 ppm surfactant concentration.

Table 1.
Rules-of-thumb used in this study for wettability interpretation

S. No	Criterion	Water–wet	Oil–wet	Reference
1	Initial water saturation, S_{wi}	>0.25	<0.15	[13]
2	Water saturation at cross-over point	>0.5	<0.5	[13]
3	End-point water relative permeability at S_{or}	<0.3	>0.5	[13]
4	End-point oil relative permeability at S_{wi}	>0.95	<0.7–0.8	[14, 15]

Here S_w is the brine saturation, S_{wi} is the initial brine saturation and S_{wm} is the maximum brine saturation or $(1\text{-}S_{or})$.

This semi-analytical model was developed by Okazawa [11] for cases where the capillary pressure data were unavailable. However, this limitation has been taken care of by using Rapaport and Leas scaling criterion [10] to calculate the stable flow rates for the coreflood experiments conducted. The model uses the fractional flow theory of Buckley and Leverett [12] to calculate the recovery and pressure drop data at any given time after the start of the displacement and then calculates the relative permeabilities by minimizing the sum-of-squares of weighted deviations of the experimental pressure and production histories from the calculated values. The history match of recovery and pressure drop obtained from the simulator during the relative permeability calculations for waterflood of Yates crude oil in Berea rock at 0 ppm surfactant concentration is shown in Fig. 2 as an example. The relative permeability data obtained from the simulator were interpreted using Craig's rules-of-thumb [13] to discern the wettability alterations induced by the surfactant solutions of varying concentrations. The rules-of-thumb used in this study to characterize wettability are summarized in Table 1.

4. RESULTS AND DISCUSSION

4.1. Rock–fluids system without wettability effects

Oil–water emulsions were observed during the coreflood experiments in this rock–fluids system to strongly affect the multiphase flow dynamics and relative permeability characteristics. These emulsions, formed at all surfactant concentrations, blocked the flow or caused very high pressure drops. The high pressure drops during primary drainage resulted in low end-point oil permeabilities, which raises concerns about the applicability of rules-of-thumb (Table 1). The summary of experimental and simulator results for waterflood of n-decane in Berea rock at various surfactant concentrations are shown in Table 2. These results indicate only minor adjustments in end-point water relative permeabilities in the simulator to obtain acceptable history match. The ratio of pressure drops ($\Delta p_{surfactant}$ / $\Delta p_{no\ surfactant}$) at each surfactant concentration is plotted against pore volume in-

Table 2.
Comparison between the experimental and simulator results for waterflood of n-decane in Berea core at various surfactant concentrations

Case	Experimental					Simulator			
	Recovery (%OOIP)	S_{wi}	S_{or}	k_{ro}	k_{rw}	S_{wi}	S_{or}	k_{ro}	k_{rw}
Brine	48	0.400	0.320	0.700	0.0840	0.39	0.32	0.700	0.0750
500 ppm	45	0.450	0.332	0.554	0.0905	0.45	0.30	0.554	0.0262
1500 ppm	47	0.365	0.362	0.425	0.0951	0.35	0.33	0.425	0.0255
3500 ppm	54	0.300	0.365	0.225	0.0900	0.28	0.29	0.225	0.0225
5000 ppm	54	0.285	0.375	0.208	0.0650	0.28	0.30	0.208	0.0275

% OOIP = percent original oil in place; S_{wi} = initial water saturation; S_{or} = residual oil saturation; k_{ro} = end-point oil relative permeability at S_{wi}; k_{rw} = end-point water relative permeability at S_{or}.

jected in Fig. 3. Figure 4 presents the effect of surfactant concentration on cross-over-point water saturation.

From these experimental results, the following important observations can be made.

(1) There is a significant change in the initial water saturation and it gradually decreases from 39% to 28% in these floods as the surfactant concentration is increased from 0 ppm to 5000 ppm, except for the case of 500 ppm surfactant concentration, where the initial water saturation is 45%. Thus, the initial water saturation is always greater than 25% for all these floods, indicating water-wet characteristics according to Craig's rules.

(2) The end-point water relative permeability at residual oil saturation gradually decreases from 7.5% to 2.75% in these floods as the surfactant concentration is increased from 0 ppm to 5000 ppm and is always less than 30% of the end-point oil relative permeability for all these floods. These results also suggest water–wet characteristics.

(3) There is a significant drop in the end-point oil relative permeability at initial water saturation from 70% to 21% in these floods as the surfactant concentration is increased from 0 ppm to 5000 ppm. This significant drop in the end-point oil permeabilities appears to indicate a shift from the initial water–wet/intermediate-wet to oil–wet nature as the surfactant concentration is increased. This shift is attributed mainly to high pressure drops caused by the oil–water emulsion formation at higher surfactant concentrations and not to wettability alteration.

(4) There is a significant shift to the left in the water saturation at cross-over point from 55% to 45% in these floods as the surfactant concentration is increased from 0 ppm to 5000 ppm. This shift appears to indicate a shift from water–wet to less water–wet or intermediate-wet condition as the surfactant concentration is increased. However, emulsion formation is the reason for this shift and not wettability change.

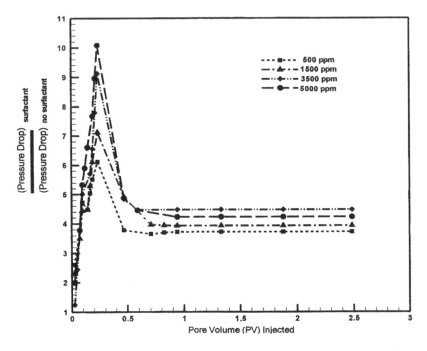

Figure 3. Effect of surfactant concentration on pressure drop ratio in n-decane containing rock–fluids system.

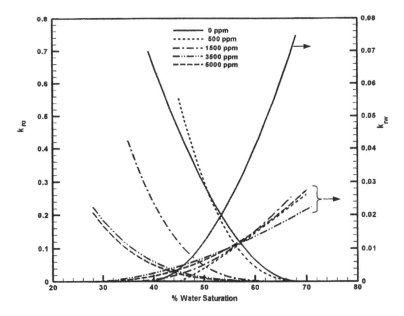

Figure 4. Effect of surfactant concentration on water saturation at cross-over point in n-decane containing rock–fluids system.

From all the above observations, it can be concluded that the rock–fluids system of n-decane, Yates synthetic brine and Berea rock is insensitive to wettability alterations, as originally assumed, in the presence of surfactant. The ratio of pressure drop at each surfactant concentration to that at no surfactant is always greater than 1 (Fig. 3). The ratio gradually increases as the surfactant concentration is increased. This can be attributed to the formation of oil–water emulsion observed during the experiments. The emulsions formed at all surfactant concentrations appear to be unstable due to high pressure drops and visual observations of emulsion droplets in the collected production streams. Furthermore, the surfactant concentrations used are below the critical micelle concentration (cmc) and, hence, it is reasonable to assume that the surfactant does not impose a stabilizing barrier between the emulsion drops at these concentrations.

From Fig. 4, it can be seen that water saturation at cross-over point is gradually shifting to the left from 55% to 45% as the surfactant concentration is increased. This is due to oil–water emulsion formation. Therefore, it can be concluded that the oil–water emulsion has the characteristic of shifting the cross-over point water saturation to the left. From Table 2, it is observed that the oil recovery gradually increases from 45% to 54% as the surfactant concentration is increased from 0 ppm to 5000 ppm. These modest enhancements in oil recovery (from 45% to 54%) obtained in this system are attributed solely to interfacial tension (IFT) reduction caused by the surfactant. 3500 ppm surfactant concentration appears to be the optimum as there is no further increase in oil recovery above 3500 ppm (Table 2).

4.2. Rock–fluids system with wettability effects

In this system consisting of Yates reservoir crude oil, brine and Berea rock, oil–water emulsion was not observed at any surfactant concentration in contrast with the n-decane containing rock–fluids system discussed previously. Hence, the changes observed in relative permeability characteristics can be attributed to wettability shifts. The summary of experimental and simulator results for waterflood of Yates crude oil in Berea rock at various concentrations of the surfactant is shown in Table 3. The pressure drop ratio ($\Delta p_{surfactant} / \Delta p_{no\ surfactant}$) at each surfactant concentration is plotted against pore volume injected in Fig. 5. Figure 6 depicts the effect of surfactant concentration on cross-over point water saturation. The important observations are:

(1) There is a significant change in the initial water saturation and it gradually increases from 39% to 65% in all these floods as the surfactant concentration is increased from 0 ppm to 5000 ppm. It is always greater than 30% in all the floods.

(2) No significant change is observed in the end-point water relative permeability at residual oil saturation for all these floods. It remains almost the same and is always less than 30% of end-point oil relative permeability.

Table 3.
Comparison between the experimental and simulator results for waterflood of Yates crude oil in Berea core at various surfactant concentrations

Case	Experimental					Simulator			
	Recovery (%OOIP)	S_{wi}	S_{or}	k_{ro}	k_{rw}	S_{wi}	S_{or}	k_{ro}	k_{rw}
Brine	56	0.400	0.290	0.970	0.0710	0.39	0.250	0.970	0.0280
500 ppm	62	0.500	0.210	0.830	0.1435	0.49	0.191	0.830	0.0430
1500 ppm	86	0.520	0.085	0.945	0.0950	0.50	0.077	0.945	0.0410
3500 ppm	94	0.650	0.038	1.000	0.1322	0.65	0.012	1.000	0.0350
5000 ppm	94	0.630	0.039	0.886	0.1483	0.63	0.013	0.886	0.0550

% OOIP = percent original oil in place; S_{wi} = initial water saturation; S_{or} = residual oil saturation; k_{ro} = end-point oil relative permeability at S_{wi}; k_{rw} = end-point water relative permeability at S_{or}.

(3) There is a significant change (especially because of absence of emulsion) in the end-point oil relative permeability at initial water saturation in these floods and it varies from 97% to 88% as the surfactant concentration increases from 0 ppm to 5000 ppm.

(4) There is a significant shift to the right in the water saturation at crossover-point from 60% to 86% in these floods as the surfactant concentration is increased from 0 ppm to 5000 ppm.

From Fig. 5, it can be seen that the ratio of pressure drop at each surfactant concentration to that at no surfactant is less than one (except at 1500 ppm). This is due to absence of oil–water emulsion observed in this particular rock–fluids system. It was noted earlier that the formation of oil–water emulsion resulted in the pressure drop ratio to be greater than unity in n-decane containing rock–fluids system (Fig. 3). From Table 3, it is observed that the oil recovery significantly increases from 56% to 94% as the surfactant concentration is increased from 0 ppm to 5000 ppm. The surfactant concentration of 3500 ppm appears to be the optimum also for this system, as the oil recovery remains unchanged above 3500 ppm.

The very high oil recoveries observed in this rock–fluids system at higher surfactant concentrations indicate that the system is neither water–wet nor oil–wet. Therefore, Craig's rules-of-thumb [13] are not applicable to infer wettability shifts for this system, as these rules are used only to distinguish between strongly water–wet and oil–wet systems based on relative permeability characteristics. Hence, the shifts in cross-over point water saturations are used to interpret the wettability states for this system. From the plots of relative permeabilities against water saturation for different surfactant concentrations (Fig. 6), it can be seen that

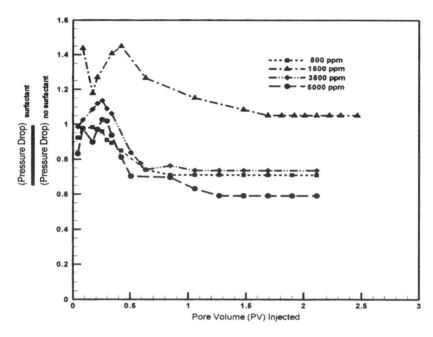

Figure 5. Effect of surfactant concentration on pressure drop ratio in Yates crude oil containing rock–fluids system.

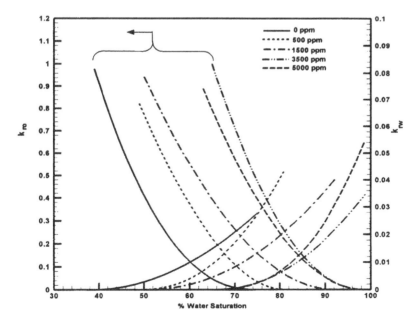

Figure 6. Effect of surfactant concentration on water saturation at cross-over point in Yates crude oil containing rock–fluids system.

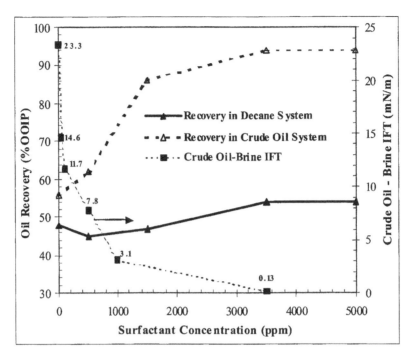

Figure 7. Effect of surfactant concentration on oil recovery and crude oil–brine interfacial tension.

the cross-over point water saturation is gradually shifting to the right as the surfactant concentration is increased. This type of relative shift in cross-over point water saturation appears to indicate development of a mixed-wettability condition [16, 17]. The very high oil recoveries, very low residual oil saturations and steady increase in initial water saturation at higher surfactant concentrations (Table 3) also indicate the development of Salathiel [18] type mixed-wettability due to surfactant. The possible mechanisms responsible for development of mixed-wettability condition are discussed below.

The system is initially strongly water–wet when surfactant is not present. Hence, the rock surface is covered with water and the oil exists in the form of globules in the middle of the bigger pores. Thus there would be a thin film of water between the rock surface and oil globules. In the presence of a water-soluble surfactant, this thin water film could become unstable [19], which will cause surfactant molecules to migrate and adsorb on the rock surface. Somasundaran and Zhang [20] recently reported that there were four regions to describe the surfactant adsorption on a rock surface. Region 1 corresponds to the absence of surfactant in the rock–fluids system and, hence, the rock surface exhibits its original wetting characteristics in this region (hydrophilic or water–wet in our current study). In region 2, the surfactant molecules adsorb on the rock surface with their hydrophobic tail groups toward the bulk solution making the surface hydrophobic (oil–wet). This will occur at surfactant concentrations below the critical micelle concentration (cmc) of the surfactant. The hydrophobicity reaches a maximum at the end of region 2 and then the surface is electrically neutralized. In region 3, the neutralization of the surface causes adsorption of some of surfactant molecules on the surface with their hydrophilic heads orienting towards the aqueous phase. This makes the rock surface less oil–wet. Region 4 corresponds to bilayer adsorption rendering the surface hydrophilic (water–wet) and this will happen at the surfactant concentrations above the cmc. Since all the surfactant concentrations used in our study are below the cmc (as can be seen in Fig. 7 where the measured IFT values still continue to decline even at high surfactant concentrations of 3500 ppm), the surfactant adsorption of region 2 is most likely to be applicable to our study resulting in strong oil–wet characteristics for the surface. This will enable the formation of a continuous film of oil on the rock surface. This oil film not only provides a path of least resistance to the flow of oil but also mitigates the competing effect of simultaneous flow of water. These effects of mixed-wettability development resulted in increased oil recoveries of up to 94% observed in the corefloods of this particular rock–fluids system.

5. CONTROLLING MECHANISM FOR OIL RECOVERY ENHANCEMENT BY SURFACTANTS

The oil recoveries for the two rock–fluids systems studied are summarized in Fig. 7. The oil recovery at each surfactant concentration in the rock–fluids system consisting of Yates crude oil is significantly higher than the recovery at the same surfactant concentration in the n-decane containing rock–fluids system. Furthermore, smaller slope of decane recovery curve in Fig. 7 indicates a relatively smaller effect of IFT reduction mechanism on oil recovery.

The significantly higher oil recoveries in the Yates crude oil containing rock–fluids system as observed in this study would be associated with 4–5 orders of magnitude reduction in the interfacial tension if there were no wettability alteration. The oil–water interfacial tension measured between Yates crude oil and Yates synthetic brine at various surfactant concentrations is also shown in Fig. 7. From Fig. 7, it can be seen that the interfacial tension reduction observed with the surfactant is only of two orders of magnitude (from 23.3 mN/m at 0 ppm to 0.13 mN/m at 3500 ppm).

In order to further substantiate the absence of surfactant-induced wettability alterations in n-decane containing rock–fluids system, capillary number N_c ($N_c = v\mu/\sigma$, where v is the displacing fluid velocity, μ is the displacing fluid viscosity and σ is the interfacial tension between the two fluids) is plotted against $S_{or}/S_{or,wf}$ (S_{or} is the oil saturation after EOR and $S_{or,wf}$ is the oil saturation after waterflood) in the recovery of residual oil *versus* capillary number plot of Klins [3] and is shown as Fig. 8. This plot was originally developed for strongly water–wet systems ignoring the effect of wettability on capillary number by setting contact angle equal to zero. Two surfactant concentrations of 0 ppm (no surfactant) and

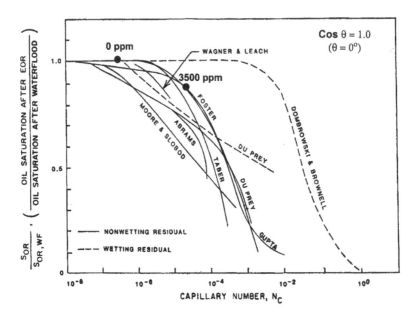

Figure 8. Recovery of residual oil *versus* capillary number [3] for strongly water–wet systems. Wettability effects on multiphase flow dynamics are ignored in this plot by assuming zero contact angle.

Table 4.
Summary of capillary number (N_c) calculations at 0 and 3500 ppm surfactant concentrations for n-decane containing rock–fluids system

Surfactant concentration (ppm)	S_{or}	Velocity (m/s)	Viscosity (kg/ms)	IFT (mN/m)	N_c	$S_{or}/S_{or,wf}$
0	0.32	3.06E-05	9.24E-04	53.54	5.28E-07	1.00
3500	0.29	3.06E-05	9.24E-04	0.42	6.73E-05	0.90

Velocity and viscosity are for displacing fluid (brine) and $S_{or,wf}$ is the oil saturation after brine flood.

3500 ppm (optimum) were used in the calculations. The results of calculations are summarized in Table 4. The measured oil–water IFT values at these two surfactant concentrations can be seen in Table 4. From Fig. 8, it can be seen that both the 0 ppm and 3500 ppm points of this study exactly fall on the two correlations by Foster [21] and by Gupta [22] in the Klins plot. This clearly substantiates the absence of wettability alteration in n-decane containing rock–fluids system used in our current study.

From all the above cited reasons, it can be concluded that wettability alteration is the controlling mechanism and the significant enhancement in oil recovery observed in Yates crude oil containing rock–fluids system is mainly due to alteration of wettability from water–wet to mixed-wet state.

6. SUMMARY AND CONCLUSIONS

This study has pointed out the relative benefits of the two main mechanisms for oil recovery enhancement by surfactants. The use of two different rock–fluids systems to explore the relative contributions from IFT reduction and wettability alteration mechanisms due to surfactants on enhanced oil recovery was found to be successful.

Only marginal increment in oil recovery (from 45% to 54%) was observed in the rocks–fluids system having no wettability effect (n-decane containing system). This is due to absence of wettability alteration and only two orders of magnitude reduction in IFT observed with the surfactant in this system (from 53.54 mN/m at 0 ppm to 0.42 mN/m at 3500 ppm).

Contrarily, significant enhancement in oil recovery (from 56% to 94%) was observed in the rocks–fluids system having wettability effect (Yates crude oil containing system). For such enhancement in oil recovery by IFT reduction mechanism, 4–5 orders of magnitude reduction in IFT would be required. However, the interfacial tension reduction observed with the nonionic surfactant in this rock–fluids system was only of two orders of magnitude (from 23.3 mN/m at 0 ppm to 0.13 mN/m at 3500 ppm). Therefore, the major portion of significant enhancement in oil recovery observed in this system was due to alteration of wettability from water–wet to mixed-wet condition.

The conventional surfactant floods operating mainly by IFT reduction mechanism have become almost extinct, since the surfactants capable of providing large reductions in IFT are uneconomical for filed applications. However, this study has experimentally demonstrated the beneficial use of low cost surfactants at low concentrations to alter wettability resulting in significant enhancements in oil recovery. Therefore, this study offers a new avenue for the use of surface-active chemicals mainly to alter wettability instead of the conventional approach of utilizing them for only IFT reduction. Thus surface force modification by surfactants is demonstrated to be far more effective than altering interfacial forces in influencing multiphase flow dynamics in porous media.

Acknowledgements

Financial support of this project by the Louisiana Board of Regents through the Board of Regents Support Fund Contract LEQSF (2000-03)-RD-B-06 and Marathon Oil Company is greatly appreciated. Sincere thanks are due to Dan Lawrence and Chandra S. Vijapurapu, Department of Petroleum Engineering (LSU) for the technical help and IFT measurements. Sincere thanks are also due to Eugene Wadleigh, Steve Bourgeois, Brian Policky and Edward Yang of Marathon Oil Company for their support and encouragement.

REFERENCES

1. G. J. Hirasaki, *SPE Form. Eval.* **6**, 217 (1991).
2. W. G. Anderson, *J. Petrol. Technol.* **38**, 1125 (1986).
3. M. A. Klins, *Carbon Dioxide Flooding*, International Human Resources Development Corporation, Boston, MA (1984).
4. G. Moritis, *Oil Gas J.*, 47 (Apr. 15, 2002).
5. E. A. Spinler and B. A. Baldwin, *Surfactants: Fundamentals and Applications in the Petroleum Industry*, pp. 159–202. Cambridge University Press, Cambridge (2000).
6. T. Austad, B. Matre, J. Milter, A. Saevareid and L. Qyno, *Colloids Surfaces A* **137**, 117 (1998).
7. D. C. Standnes and T. Austad, *Colloids Surfaces A* **218**, 161 (2003).
8. H. L. Chen, L. R. Lucas, L. A. D. Nogaret, H. D. Yang and D. E. Kenyan, *SPE Reservoir Eval. Eng.* **4**, 16 (2001).
9. C. S. Vijapurapu and D. N. Rao, *Colloids Surfaces A* **241**, 335 (2004).
10. L. A. Rapaport and W. J. Leas, *Trans. AIME* **198**, 139 (1953).
11. T. Okazawa, *User's Manual for ANRPM and EXRPM Relative Permeability Simulators*, Petroleum Recovery Institute, Calgary, Alberta (1983).
12. S. E. Buckley and M. C. Leverett, *Trans. AIME* **146**, 107 (1942).
13. F. F. Craig, *The Reservoir Engineering Aspects of Waterflooding*, pp. 12–44, Monograph Series. SPE, Richardson, TX (1971).
14. E. C. Donaldson and R. D. Thomas, Paper No. SPE 3555, *Proc. 46th SPE Annual Fall Meeting*, New Orleans, LA (1971).
15. W. W. Owens and D. L. Archer, *J. Petrol. Technol.* **23**, 873 (1971).
16. D. N. Rao, M. Girard and S. G. Sayegh, *SPE Reservoir Eng.* **7**, 204 (1992).
17. W. G. Anderson, *J. Petrol. Technol.* **39**, 1453 (1987).
18. R. A. Salathiel, *J. Petrol. Technol.* **25**, 1216 (1973).
19. C. K. Lin, C. C. Hwang and W. H. Uen, *J. Colloid. Interface Sci.* **231**, 379 (2000).
20. P. Somasundaran and L. Zhang, *Proc. 8th International Symposium on Evaluation of Reservoir Wettability and its Effect on Oil Recovery*, Houston, TX (2004).
21. W. R. Foster, *J. Petrol. Technol.* **25**, 205 (1973).
22. S. P. Gupta and S. P. Trushenski, *Soc. Petrol. Eng. J.* **19**, 116 (1979).

Part 4

Surface Characteristics and Adhesion

Contact Angle, Wettability and Adhesion, Vol. 4, pp. 387–405
Ed. K.L. Mittal
© VSP 2006

Surface modification of banana-based lignocellulosic fibres

NEREIDA CORDEIRO,[1,*] LÚCIA OLIVEIRA,[1] HELENA FARIA,[1]
MOHAMED NACEUR BELGACEM[2] and JOÃO C. V. P. MOURA[3]

[1]*CEM and Departamento de Química, Universidade da Madeira, 9000-390 Funchal, Portugal*
[2]*LGP2, Ecole Française de Papeterie et des Industries Graphiques, BP65,
38402 St Martin D'Hères, France*
[3]*Departamento de Química, Universidade do Minho, 4700-320 Braga, Portugal*

Abstract—Lignocellulosic raw materials were isolated from rachis of *Musa acuminata* Colla var. *cavendish* and characterised before and after chemical modification. The rachis was submitted to different mechanical treatments, milling and defibration, resulting in rachis powder and rachis fibres, respectively. The chemical composition of these two samples was established and it was shown that rachis fibres exhibited higher polysaccharide and lignin contents and lower amounts of ash and extractives components, as compared with the rachis powder. The effects of solvent extraction, alkali treatment and chemical modification using phenyl isocyanate, maleic anhydride, alkenyl succinic anhydride and alkyl ketone dimer as grafting agents were studied. The materials were characterized in terms of chemical structure by ATR–FT-IR and ¹³C-CP-MAS-NMR spectroscopy, morphology by scanning electron microscopy and surface energies by inverse gas chromatography and contact angle measurements. The surface energy of these materials was found to be very close to other similar lignocellulosic materials. Finally, the water absorption of these materials before and after treatment was ascertained. The modified fibres showed considerable changes in ATR–FT-IR and ¹³C-CP-MAS-NMR spectra and surface properties, providing very convincing evidence that chemical grafting had occurred.

Keywords: "Dwarf Cavendish"; banana plant; rachis fibres; surface modification; morphology characterization; surface properties.

1. INTRODUCTION

The use of lignocellulosic fibres as a reinforcing agent in composite materials has spurred renewed interest recently, as evidenced by numerous publications from various laboratories and reviews [1–4]. The incorporation of these natural fibres into polymeric matrices has been shown to be interesting, because of the certain excellent characteristics of these fibrous materials, such as their good mechanical properties, wide availability directly from numerous vegetal species, renewable

---

*To whom correspondence should be addressed. Tel.: (351) 291-705107;
Fax: (351) 291-705149; e-mail: ncordeiro@uma.pt

character, low density compared with, for example, glass fibre, and their low cost in most instances [1–4].

The quality and the suitability of a composite material obviously depend on the appropriate choice of its two basic components (matrix and fibre), depending on specific applications envisaged for it. Thus, in order to prepare high-performance composite materials, one should ascertain the properties of each constituent.

Cellulose-based composites are gaining increasing interest because of the many advantages of this renewable natural polymer. Moreover, recycling of such composites is much easier to accomplish when compared to glass-fibre-based counterparts. Cellulose is a very highly polar macromolecule, which makes it poorly compatible with non-polar matrices commonly used for composite materials. Moreover, this hydrophilic character induces high affinity for water molecules in the surrounding atmosphere, which negatively affects the mechanical properties and the dimensional stability of the resulting materials. These are the main reasons for submitting cellulose to chemical modification before its incorporation into a given matrix. In the literature, many different approaches are documented for treating cellulosic material to make it compatible with common matrices. These approaches were recently reviewed [5].

The island of Madeira currently has a reduced forest area of pine and eucalyptus; nevertheless, it has an extensive area of banana plants. After harvesting the bunch of bananas from each plant, the pseudo-stems are cut giving rise to large amounts of agricultural waste. Before commercialisation, the rachis is cut from the bunch of banana, which is presently unused and is an undesirable waste. In this context, we recently started a research program to investigate banana plant as a source of lignocellulosic fibres, in view if its value as a renewable resource for papermaking [6, 7] and composites fabrication [8].

Here we present the main bulk and surface properties of the fibrous materials studied, in order to predict their suitability for composite preparation and to help in the choice of the appropriate matrix or in the selection of the most suitable chemical modification for a given combination. In this study we have carried out chemical modification of the surface of the banana plant derived fibrous material, using mono-functional molecules, with a view to use the modified banana rachis with polymeric matrices such as polypropylene, styrene–butadiene latex, or polyethylene. Thus, chemical modification of rachis and rachis fibres from banana plants was carried out and the resulting modified fibres fully characterised.

2. MATERIALS AND METHODS

2.1. Lignocellulosic material

Rachis was harvested from mature banana plants *Musa acuminata* Colla var. *cavendish* in Funchal, Portugal and kindly supplied by Cooperativa Agrícola dos Produtores de Fruta da Madeira. Two kinds of material were used: (i) the raw material (labeled rachis) and (ii) mechanically isolated fibres of rachis obtained from a prototype defibration machine (labeled rachis fibres).

After air-drying for two weeks, the samples were milled and sieved to 40–60 mesh fractions, in order to determine their chemical composition.

2.2. Chemical composition

The ash content was determined by calcination at 600°C for 6 h. The different materials were extracted in a Soxhlet extractor for 8 h sequentially with n-hexane, dichloromethane, ethyl acetate, ethanol and water. The lignin content was determined by Klason's method according to Tappi standard T222 om-88. The holocellulose content was quantified by the peracetic acid method [9] whereas the cellulose content was determined by the Kurschner–Hoffner procedure [9]. In order to obtain the respective amounts of hemicelluloses A and B and cellulose, holocelluloses were submitted to fractionation by extraction with KOH under a nitrogen atmosphere as described elsewhere [10]. All chemical analysis and fractionation experiments were carried out, at least, in duplicate.

In alkali treatment the material was dipped in 2% and 10% aqueous solutions of NaOH for 30 min and washed in a very dilute hydrochloric acid to remove the unreacted alkali. Washing continued until the fibres were alkali free. The washed fibres were then dried in an oven at 70°C for 3 h.

2.3. Chemical modification

For grafting reactions the samples were thoroughly dried in vacuum at 80°C for 3 days before carrying out surface modification. They were then placed in dry toluene under a nitrogen atmosphere at 80°C and the grafting agents (Fig. 1), phenyl isocyanate (PI) and maleic anhydride (MA) (commercial high purity products); alkenyl succinic anhydride (ASA) and alkyl ketone dimer (AKD) (industrial products), were added, in large excess, and stirred under reflux in an inert atmosphere. In the reaction with PI a small amount of triethylamine was used as a catalyst (1% of the added NCO). The suspensions were left under magnetic stirring to complete the reaction, i.e., at least for 10 days under reflux, taking particular care to prevent any spurious inlet of moisture [11].

The treated lignocellulosic materials were then recovered by filtration and washed with dichloromethane. In order to eliminate all the ungrafted species, the modified materials were submitted to exhaustive Soxhlet extraction for 48 h using methylene chloride. Then the materials were dried at 40°C. The modified products were characterized by spectroscopic, morphological and surface energy methods, in order to evaluate the efficiency of these modifications.

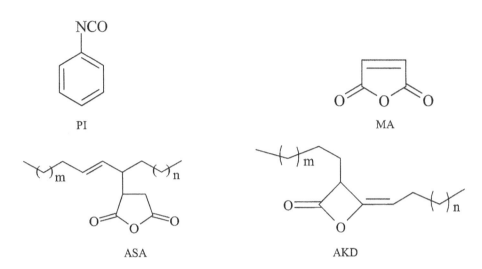

Figure 1. Structures of grafting agents: phenyl isocyanate (PI), maleic anhydride (MA), alkenyl succinic anhydride (ASA); alkyl ketone dimer (AKD).

2.4. Characterization of the modified samples

ATR spectra were obtained on a Bruker IFS 55 spectrophotometer equipped with a horizontal micro-ATR Golden Gate unit (SPECAC). Solid-state ^{13}C-NMR spectra were recorded at 100.6 MHz (9.4 T) on a Bruker MSL-400 spectrometer.

The surface morphology of the materials was analysed by scanning electron microscopy (SEM) using a Jeol JSM-6100 microscope.

Inverse gas chromatography was performed using an HP 5890 Series II chromatograph equipped with a flame ionisation detector and a 100 cm × 4 mm stainless steel column. The practical details as well the calculation methods and the probes used were reviewed recently [12]. Contact angles were measured with a home-made apparatus designed to acquire up to 200 images per second as described in a previous publication [13].

The water absorption was determined for lignocellulosic materials. A minimum of three samples were tested for each material. After drying overnight at 70°C, the samples were cooled and weighed, and then stored at 25°C. The water absorption in the samples was determined by conditioning the sample in a high moisture atmosphere. The samples were stored at 25°C in desiccators containing sodium sulphate to ensure an atmosphere with RH of 98%. The samples were removed at specified intervals and weighed using a four-digit balance. The water absorption (WA) was determined using equation (1):

$$WA\,(\%) = \frac{(M_t - M_0)}{M_0} \times 100 \tag{1}$$

where: M_t and M_0 are the weights of the sample after exposure to 98% RH atmosphere for time t and before exposure, respectively.

3. RESULTS AND DISCUSSION

3.1. Chemical characterization

The quantitative results on the chemical composition of "Dwarf Cavendish" rachis samples are shown in Table 1, from which several remarks can be made when rachis and rachis fibres are compared. A high amount of ashes (*ca.* 48%) is extracted during rachis defibration, as confirmed by the decrease of the ash content in rachis fibres (Table 1). The ash content of rachis is quite high when compared to other fast growing plants [14–17]. However, the ash content of rachis fibres is similar to that obtained, for example, for the stalk fibre of rice straw (15–20%) [15]. The extractives content for rachis (18.1%) is similar to those obtained for other fast growing plants [14–17]. The defibration process reduced the extractives content. Lignin was found in a slightly higher amount in rachis fibres (11.8%) than in rachis (10.5%). The high lignin content indicates that the mechanical treatment did not extract any appreciable amount of lignin, as expected due to the high molecular weight and cross-linked character of this macromolecule. The low lignin content is a good feature, since it does not contribute as well as do cellulose macromolecules in the mechanical performance of these fibres.

Holocellulose and cellulose contents were found to be higher in rachis fibres than in rachis (Table 1). The cellulose content of the rachis fibres (56.1%) is in agreement with that reported for other stalk fibres (43–66%) [15]. These results can be explained in terms of significant removal of ashes and extractives in the juice plant during the defibration process. Hemicelluloses and cellulose contents differ significantly for these two materials. Cellulose was the major component of holocelluloses, representing *ca.* 77 and 75% for rachis fibres and rachis, respectively. The cellulose content of rachis fibres (51.5%) is quite high when compared with the values obtained for other stalk fibres (26–45%) [15]. The hemicelluloses

Table 1.
Chemical composition (% in dry material) of rachis and rachis fibres of "Dwarf Cavendish"

Component	Rachis	Rachis fibres
Ash	26.8	14.0
Extractives[a]	18.1	10.2
Lignin[a]	10.5	11.8
insoluble	9.6	11.1
soluble	0.9	0.7
Holocellulose[a]	41.8	67.0
Cellulose	31.5	51.5
Hemicellulose A	4.3	8.5
Hemicellulose B	4.0	6.2
Cellulose[a]	34.9	56.1

[a] Corrected for ash content.

content in rachis fibres was higher than that of the starting rachis, probably because of the extraction procedure. In order to confirm the above arguments, the analysis of the extracted juice during defibration is in progress.

3.2. Spectroscopic characterization

Figures 2 and 3 show the ATR–FT-IR spectra of the rachis and rachis fibres, before and after solvent extraction. All these spectra are characterised by common features, namely (i) a strong band around 3400 cm^{-1}, attributed to hydrogen-bonded –OH stretching vibration; (ii) a band at *ca.* 2900 cm^{-1} due to CH stretching vibration of aliphatic moieties; (iii) a peak near 1250 cm^{-1} associated with C–C, C–O and C=O stretchings in the lignin; and (iv) a prominent band in the range 1200–900 cm^{-1}, attributed to the presence of polysaccharides. The high intensity of this band in all samples shows that polysaccharides are present in large amounts. The absorptions at 1157 and 1033 cm^{-1} are attributed to the C–O–C and C–O stretchings in polysaccharides, respectively, whereas the C–H deformation in cellulose can be seen at 898 cm^{-1}.

Nevertheless, some peaks are present only in the rachis (Fig. 2a and 2b), namely those near 1731 and 1600 cm^{-1}. The signal of the non-cellulose components at 1731 cm^{-1} is characteristic of the carboxyl-containing substances present in extractives, such as fatty acids widely present in the rachis [10]. The peak at

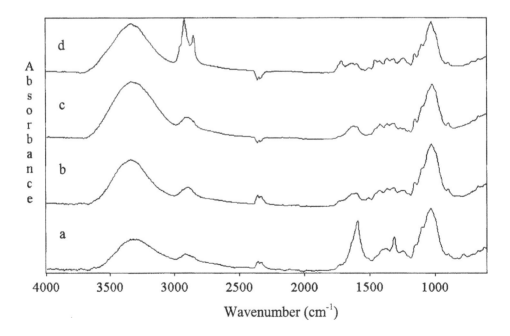

Figure 2. ATR–FT-IR spectra of (a) rachis, (b) extracted rachis, (c) rachis treated with 10% NaOH, and (d) rachis chemically modified with AKD.

1600 cm^{-1} can be assigned mainly to the C=C bond of extractives and appears with much more pronounced intensity in the rachis spectrum, before extraction (Fig. 2a). In the ATR–FT-IR spectra of rachis fibres the signals attributed to extractives decreased significantly after the defibration process (Fig. 3Ia). This is in agreement with the results obtained for the chemical composition (Table 1). The ATR–FT-IR spectrum of rachis before extraction presents also some differences in the region of $1500–1300 \text{ cm}^{-1}$ due to extractives removal (Fig. 2b).

The ATR–FT-IR spectrum of the rachis submitted to alkali treatment with NaOH is shown in Fig. 2c. The spectrum of the treated rachis with 2% NaOH (not presented here) was very similar to that of the extracted material, which indicates that even a small alkali concentration is capable to extract high quantities of extractives and ashes. The increase of NaOH to a level of 10% (Fig. 2c) caused disappearance of the C=O band at *ca.* 1739 cm^{-1}. However, high amounts of polysaccharides were removed with 10% NaOH solution, as seen from the weakening of the band at $1200–900 \text{ cm}^{-1}$.

The occurrence of chemical modification of the samples, when submitted to reactions with grafting agents was verified by ATR–FT-IR spectroscopy, as illustrated in Figs 2 and 3. Despite the chemical modification taking place mainly on the material surface, some changes were detected in the ATR–FT-IR spectrum of the extracted fibres, namely, two peaks at 1720 cm^{-1} and 1540 cm^{-1}, characteristic of the carbonyl groups, with higher intensity in the case of MA-modified fibres, and NH moieties of the urethane group (R–O–(CO)–(NH)–) in the case of PI-modified materials. This observation is also verified by the ATR–FT-IR spectroscopy for PI-, MA- and ASA-modified unextracted as well as extracted material. In all these cases, the chemical modifications were more pronounced in the extracted material as compared with the non-extracted one. Figure 2d shows a significant alteration in the ATR–FT-IR spectrum of the AKD-modified original

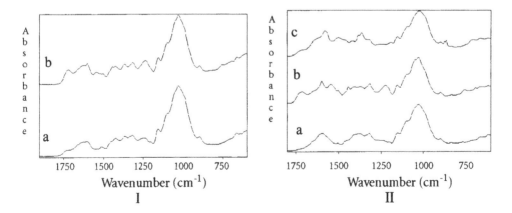

Figure 3. ATR–FT-IR spectra of **I**: (a) extracted rachis and (b) extracted rachis chemically modified with PI; **II**: (a) rachis fibres and rachis fibres modified with: (b) MA and (c) ASA.

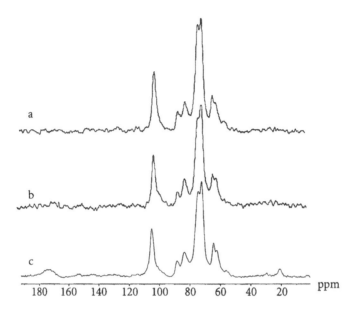

Figure 4. ^{13}C-CP-MAS-NMR spectra of (a) rachis fibres, (b) rachis and (c) rachis modified with ASA.

rachis, namely the intensity increase of the peaks at 2922 and 2851 cm^{-1} attributed to the aliphatic moiety of AKD and the appearance of three strong peaks at 1720, 1460 and 1370 cm^{-1} characteristics of the carbonyl and CH groups.

^{13}C-CP-MAS-NMR spectra of rachis and rachis fibres are shown in Fig. 4. The main difference is with respect to the peak of methoxy at 56 ppm, attributed to the presence of both hemicellulose and lignin, which was stronger in the rachis fibres (Fig. 4a). This fact can be explained by the removal of high amount of components, principally extractives, during the defibration process (Table 1). The peaks between 61 and 105 ppm, which arose from the overlapping signals of carbohydrates and lignin aliphatic carbons, are very similar in both spectra. However, different relative intensities of the signals at 62, 74 and 88 ppm in rachis fibres spectrum suggested that during the defibration process some sugars were extracted. The sugar analysis [10] showed that the predominantly removed sugars were glucose-containing.

The occurrence of chemical modifications in rachis can also be observed by ^{13}C-CP-MAS-NMR. As an example, in Fig. 4c is presented the spectrum of the modified rachis with ASA. As expected, two signals appear in the rachis spectrum after reaction with ASA: one signal at 21 ppm characteristic of the aliphatic moiety of the grafting agent and a large signal at 175 ppm, indicative of the presence of the carbonyl groups.

3.3. Morphological characterization

The cross-sectional and longitudinal views of rachis, before and after extractives removal, are shown in Figs 5 and 6. The original rachis showed some irregularities and defects (Fig. 5a–c) at the surface of the fibres, which does not allow observation of fibrillar structure of the material in longitudinal view. However, in cross section we see that original rachis has a circular and fibrillar structure

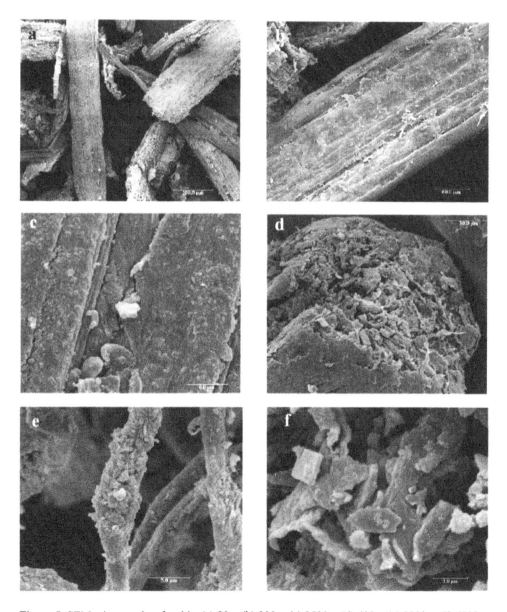

Figure 5. SEM micrographs of rachis: (a) 80×; (b) 300×; (c) 3500×; (d) 600×; (e) 3000× ; (f) 6000×.

(Fig. 5d). Particular attention should be paid to Fig. 5e and 5f, which show very well organised mineral crystals in rachis. Mineral elemental analysis revealed the presence of different constituents, mainly potassium, calcium and silicium [10].

The location of these minerals was studied and it was found that they were grown not only at the surface of the fibres, but also inside the channels, which

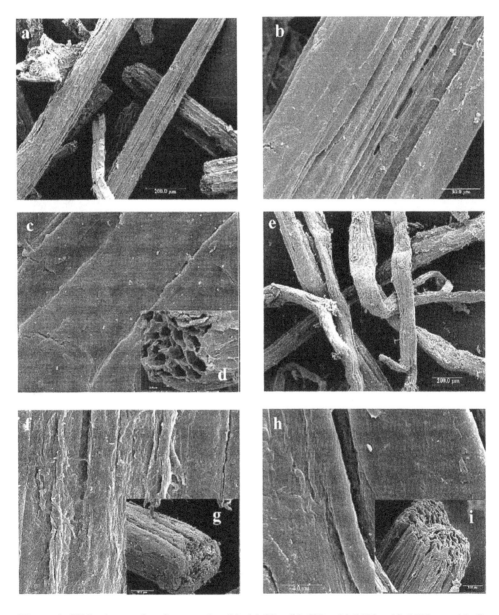

Figure 6. SEM micrographs of extracted rachis (a) 80×, (b) 600×, (c) 1500×, (d) 1500×; rachis fibres (e) 80×, (f) 1500×, (g) 600×; extracted rachis fibres (h) 4000×, (i) 600×.

explains that even after exhaustive water extraction, a significant amount of ashes remained in the material [10]. After extraction (Fig. 6), the rachis surface became smoother, displaying a multi-layer organisation (Fig. 6b, c). A large number of surface cracks, or pit formations, were observed when compared to virgin samples. These cracks might occur because of the partial removal of wax or fatty substances. The cross-section of the extracted rachis shows that the extraction removed a high amount of wax or fatty substances leaving a tubular structure (Fig. 6d).

The rachis fibres were also investigated by SEM (Figs 6e to 6i) and showed similar features as those found for rachis, except for two main differences: (i) fibres seem to be thinner than in the rachis (Fig. 6a and 6e), due to compression during the defibration process and (ii) the surface of fibres shows a fibrillar structure of this material, which indicates that defibration removed a significant amount of components, principally ash, as observed previously from the chemical composition data (Table 1).

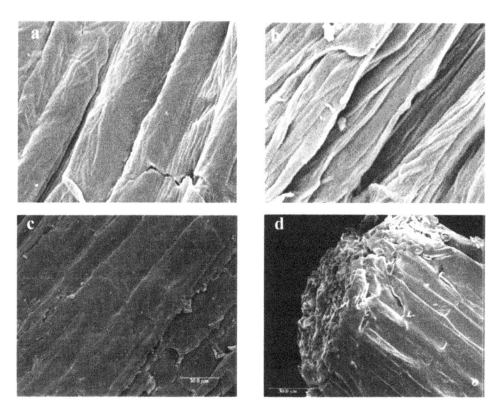

Figure 7. SEM micrographs of rachis treated with: (a) 2% NaOH, 4000×; (b) 10% NaOH, 4000×; (c) ASA, 600×; (d) AKD, 600×.

The rachis extracted with different alkaline solutions was also analysed by SEM (Fig. 7a and 7b). Micrographs of the two samples confirmed the ATR-FTIR observations, i.e., the use of a higher concentration of NaOH was more efficient in the components removal, where the border between individual fibres can be easily observed in the extracted rachis with 10% NaOH (Fig. 7b). Similar observation was also made in rachis extracted with 2% NaOH, but to a smaller extent.

Alkali treatment induced extraction of the carbohydrate–lignin complex and led to breaking down of the composite fibre bundle into smaller fibres by breaking linkages between the constituents in the cell wall. This increase of the surface area available for contact with the polymeric matrix will play a beneficial role. In other words, alkali treatment reduces the fibre diameter, enhances the development of surface roughness, which, in principle, will offer better fibre–matrix interface adhesion and, consequently, higher mechanical properties.

Finally, the SEM micrographs of the materials showed the presence of longitudinally-oriented unit cells with more or less parallel orientations. The intercellular spaces are probably filled up by cellulose macromolecules, hemicellulose as a compatibiliser, and lignin as a binder.

The SEM micrographs of the grafted materials did not show significant differences, principally in cases of PI and MA agents (small molecules). In the ASA and AKD cases the surface of the material showed a slight coverage (Fig. 7c and 7d, respectively).

3.4. Surface characterization

The surface characterization of the samples was carried out by using the IGC technique. The excellent linearity relative to n-alkane probes provides a reliable method for the calculation of the dispersion component of the surface energy of the samples. With this method, the dispersion component of the surface energy, γ_S^D as a function of temperature was obtained, as summarised in Table 2. The relative retention of the characteristic acidic (chloroform) and basic (tetrahydrofuran) probes provided an estimate of the Lewis acid–base properties of the surfaces and is reflected by the ratio K_a/K_b (acidic and basic parameters, respectively [18]).

Table 2 shows that both the dispersion component and acidic character (K_a/K_b ratio) of surface energy in extracted materials were higher than for original material. These could be attributed to the high amount of extractives removal (with high percentage of ashes), which increases the dispersion component of surface energy and makes the lignocellulosic –OH groups more accessible, increasing the acidic character. Similar results were previously reported for other lignocellulosic materials [19–21]. The mechanical defibration increased the dispersion component of surface energy and the K_a/K_b ratio.

The chemically-modified materials showed that grafting increased the dispersion component of surface energy in particular in the case of AKD. The observed increase of K_a/K_b ratio can be attributed to the introduction of nucleophilic sites

Table 2.

Values of the dispersion component of surface energy, γ_S^D, at different temperatures, K_a/K_b ratio, obtained by IGC, and water contact angles (°) at 25°C

Sample	Grafting agent	γ_S^D (mJ/m²)			K_a/K_b	Contact angle (°)
		60°C	70°C	80°C		
Rachis						
original	None	34	31	29	0.7	71
	Phenyl isocyanate	38	36	34	1.7	67
	Maleic anhydride	38	35	33	2.4	56
	ASA	42	38	32	1.9	72
	AKD	40	39	37	0.5	85
extracted	None	40	34	29	3.7	69
	Phenyl isocyanate	44	40	35	1.8	63
Rachis fibres						
original	None	35	31	27	0.9	71
	Phenyl isocyanate	40	36	32	2.1	73
	Maleic anhydride	38	35	32	3.1	63
extracted	None	41	38	31	2.2	46
	Phenyl isocyanate	54	49	40	2.2	72
Rachis treated with 10% NaOH			n.d.			50
Cellulose from rachis fibres			n.d.			37

n.d., not determined.

by the urethane, ester and acid moieties resulting from the grafting reaction. Thus, a higher K_a/K_b ratio for MA and a smaller value for AKD (due to the long aliphatic chain) are observed.

The γ_S^D values for the materials under study were found to be close to those reported in the literature for similar materials. For example, the Avicell cellulose before and after purification had values of 38 and 42 mJ/m², respectively, and after modification with ASA the values did not change significantly (41–43 mJ/m²). In the case of Whatman paper before and after treatment with AKD, the K_a/K_b ratios were 1.6 and 0.7, respectively [20, 21].

The contact angles on the rachis fibres were measured with five different liquids, in order to determine both the dispersion and polar components of surface energy of these materials according to the Owens–Wendt method [22]. The evolution of contact angles for different liquids used with time for rachis fibres is shown in Fig. 8. The calculated total surface energy of rachis fibres gives a value of 41 mJ/m², with 33 and 8 mJ/m² for γ_S^D and γ_S^P, respectively. The dispersion

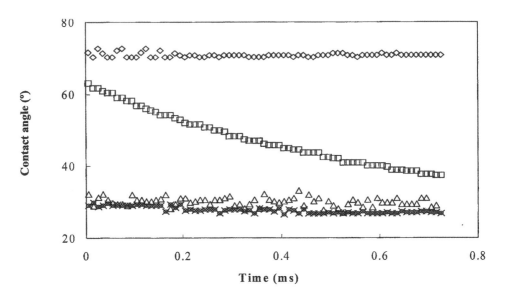

Figure 8. Variation of contact angle as a function of time for various liquids on rachis fibres: (◊) water; (□) formamide; (Δ) diiodomethane; (○) bromonaphthalene; (+) hexadecane.

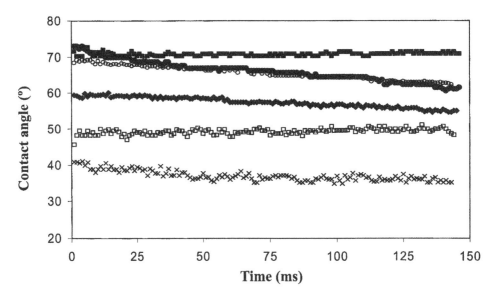

Figure 9. Short-time evolution of water contact angles on the surfaces of: (●) rachis; (○) extracted rachis; (♦) rachis treated with 10% NaOH; (■) rachis fibres; (□) extracted rachis fibres; (×) cellulose from rachis fibres.

component of the rachis fibres was lower than that obtained by IGC (44 mJ/m^2 at 25°C), as already observed by different authors [18, 20, 21, 23].

The contact angles of water drops deposited on the surface of the investigated materials are presented in Table 2 and Fig. 9. We can remark that the extracted materials seem to be most capable of establishing hydrogen bonds (high hydrophilicity) since they have the lowest initial contact angle. This could be attributed, as referred above, to the removal of the compounds from the surface of the material, which makes the fibre more exposed (as observed by SEM micrographs) and, consequently, the –OH groups of cellulose fibres are more accessible.

The surface of rachis treated with NaOH has a lower value of θ. This is due to the higher extraction of the components from the surface of fibres (as shown in Fig. 7a and 7b), which is also found for other alkali-treated cellulose fibres [24].

As expected, the rachis fibres (high content of cellulose) have a smaller water contact angle. The water contact angle decreases by 35% for rachis fibres after a simple extraction. This could indicate that defibration process made the hydrophilic component present in the surface of the material more accessible for extraction.

The values of water contact angle for chemically modified materials are also given in Table 2 and Fig. 10. We can remark that, in general, the contact angles of water decrease in the case of reaction with MA, due to the presence of new acid groups on surface of the material. In the case of the surface modified with ASA and AKD the contact angles have higher values due to the presence of non-polar long aliphatic hydrophobic chains (lower density of polar groups). This effect is more pronounced in the case of AKD, because its aliphatic chains are saturated and longer.

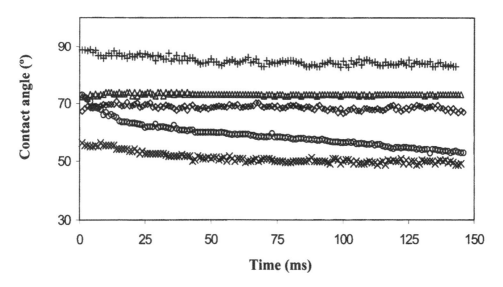

Figure 10. Short-time evolution of water contact angles on the surfaces of (○) rachis and rachis modified with: (◊) PI; (×) MA; (△) ASA; (+) AKD.

3.5. Water absorption

The samples were conditioned in a high moisture atmosphere (98% RH). The water absorption of rachis and rachis fibres, before and after extraction, as a function of time, was evaluated as presented in Fig. 11 and Table 3. These absorption data are means of several trials, and reliability of measurements was very good.

The first remark is that all samples absorbed water during the experiment. The second remark is the existence of two well-separated regions: region I at shorter times, where the kinetics of absorption was fast, whereas at extended times the kinetics of absorption slowed down and reached a plateau, i.e., region II. Region II corresponds to equilibrium water absorption. The initial absorption of water results in hydrogen bonds with the remaining –OH groups on the material surface and increases until it reaches a point of saturation. In general, in the materials studied the rate of water absorption was high during the first 16 h, and after that it decreased. An exception was observed for the unextracted materials, where the equilibrium was reached at longer times (*ca.* 75 h), probably because of the presence of extractives (wax or fatty substances), which made water penetration difficult into the material. The water absorption at equilibrium is presented in Table 3.

The extracted materials show two times less absorption of water than the respective original materials. These observations are in apparent discord with the water contact angles, which decrease with the extraction. The same contradiction is observed for alkali-treated material. However, if on the one hand the water absorption is dependent on the polar component in the material; on the other hand, the compressed or more crystalline material makes water diffusion difficult, consequently the water absorption decreases. The effect of compression or

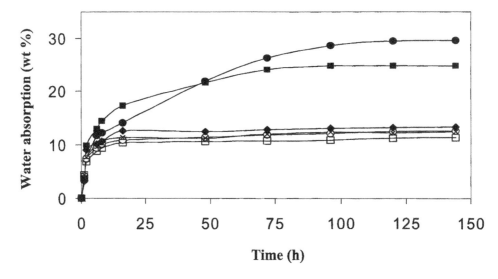

Figure 11. Water absorption *versus* time for rachis: (●) original; (○) extracted; (♦) treated with 10% NaOH; and from rachis fibres: (■) original; (□) extracted; (×) cellulose from rachis fibres.

crystallinity is more pronounced for water absorption measurements than for water contact angles measured in very short times.

It is observed that the original rachis absorbs 30 wt% of water (Table 3). The water content at equilibrium decreased with the solvent extraction to only 13 wt% (57% with respect to extracted material). For the rachis fibres the water absorption decreased by 54% for the extracted fibres. The smaller value in the rachis fibres is probably overcome by the high crystallinity which reduces water absorption capacity. Extracted materials have very similar values of water absorption when compared to cellulose. This could be attributed to fibre compression, making water absorption more difficult for the fibre.

The absorption of water by different lignocellulosic materials is largely dependent on the availability of free –OH groups on the surface. The difference in absorption of water between modified and original materials is due principally to blocking of –OH groups by urethane formation and/or esterification. The water absorption on exposure to 98% RH of different lignocellulosic materials before chemical modification as a function of time was evaluated and is presented in Fig. 12.

Water absorption by original material is reduced by chemical modification with PI, MA and ASA but the maximum reduction was 22% when compared to

Table 3.
Water absorption (wt%) for the samples conditioned at 98% RH

Material	Grafting agent	Water absorption
Rachis		
original	None	30
	Phenyl isocyanate	28
	Maleic anhydride	28
	ASA	23
	AKD	10
extracted	None	13
	Phenyl isocyanate	13
Rachis fibres		
original	None	25
	Phenyl isocyanate	23
	Maleic anhydride	25
	ASA	17
extracted	None	12
	Phenyl isocyanate	10
	Maleic anhydride	11
Rachis treated with 10% NaOH		13
Cellulose from rachis fibres		12

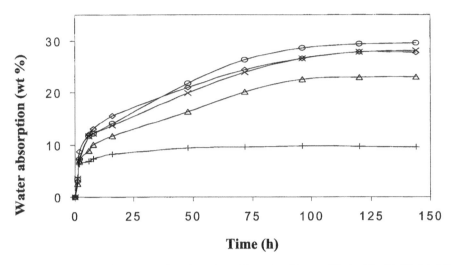

Figure 12. Water absorption *versus* time for (o) rachis and rachis modified with: (◊) PI; (×) MA; (Δ) ASA; (+) AKD.

original material (Table 3). Therefore, in the case of modification with AKD the reduction is around 30%. This phenomenon was ascribed to the long aliphatic chain in AKD, which prevented water absorption by the material. In conclusion, the minimum absorption is observed for the AKD agent, followed by extracted rachis. The order of absorption in rachis is: original rachis > modification with MA > PI > ASA > AKD. The order of absorption in rachis fibres is the same as for rachis. These results are in agreement with the chemical structures of the grafting agents used and provide evidence for the occurrence of modification.

4. CONCLUSIONS

In this paper data concerning the chemical composition, morphology and surface energy of "Dwarf Cavendish" are presented. The influence of both mechanical treatments and solvent extraction of these materials was also established. These structural changes induced a decrease of water absorption. These findings will be useful in the application of "Dwarf Cavendish" in different areas, such as paper-making and cellulose-based composite materials.

The chemical modification of rachis surface with grafts resulted in loss of hydrophilic character, leaving essentially the modified surface with a dispersion character. Thus, these results promise a better surface compatibility between the fibrous materials and polymer matrices possessing a predominantly dispersion character, such as polypropylene, natural rubber or latices. This work is in progress.

Acknowledgements

The Portuguese Foundation for Science and Technology (FCT) has sponsored this work. H. F. and L. O. also thank the FCT for awarding of Ph.D grants (BD/4745/2001 and BD/4749/2001, respectively). The authors also wish to thank the Cooperation programme GRICES/EGIDE.

REFERENCES

1. S. J. Eichhorn, C. A. Baillie, N. Zafeiropoulos, L. Y. Mwaikambo, M. P. Ansell, A. Dufresne, K. M. Entwistle, P. J. Herraro-Franco, G. C. Escamilla, L. Groom, M. Hughes, C. Hill, T. G. Rials and P. M. Wild, *J. Mater. Sci.* **36**, 2107 (2001).
2. T. Peys and G. Baillie (Eds.), *Composites Science and Technology*, Volume 63, issue 9. Elsevier, Amsterdam (2003).
3. A. K. Mohanty, M. Misra and G. Hinrichsen, *Macromol. Mater. Eng.* **1**, 276 (2000).
4. J. Gassan and A. K. Bledzki, *Prog. Polym. Sci.* **24**, 221 (1999).
5. M. N. Belgacem and A. Gandini, *Composite Interfaces*, in press.
6. N. Cordeiro, M. N. Belgacem, I. C. Torres and J. V. C. P. Moura, *Industrial Crop Prod.*, **19**, 147 (2004).
7. N. Cordeiro, M. N. Belgacem, I. C. Torres and J. V. C. P. Moura, *Cellulose Chem. Technol.* **39**, 517–529 (2006).
8. H. Faria, L. Oliveira, N. Cordeiro, M. N. Belgacem and A. Dufresne, *Macromol. Mater. Eng.* **291**, 16–26 (2006).
9. B. L. Browning (Ed), in: *Methods of Wood Chemistry*, Vol. II, p. 406. Wiley, New York, NY (1967).
10. L. Oliveira, C. S. F. Freire, A. J. D. Silvestre, N. Cordeiro, I. C. Torres and D. Evtuguin, *Industrial Crop Prod.* **22**, 187–192 (2005).
11. H. Faria, L. Oliveira, N. Cordeiro and M. N. Belgacem, *Proc. International Congress on Production, Processing and Use of Natural Fibres*, p. 94, Potsdam, Berlin (2002).
12. M. N. Belgacem and A. Gandini, in: *Interfacial Phenomena in Chromatography*, E. Pefferkorn (Ed.), p. 41. Marcel Dekker, New York, NY (1999).
13. P. Aurenty, V. Lanet, A. Tessadro and A. Gandini, *Rev. Sci. Instrum.* **68**, 1801 (1997).
14. C. Pascoal Neto, A. Seca, A. M. Nunes, M. A. Coimbra, F. Domingues, D. Evtuguin, A. Silvestre and J. A. S. Cavaleiro, *Industrial Crop Prod.* **6**, 51 (1997).
15. J. E. Atchison, in: *Pulp and Paper Manufacture*, F. Hamilton and B. Leopold (Eds.), Vol. III, p. 157. TAPPI Press, Atlanta, GA (1993).
16. C. Pascoal Neto, A. Seca, D. Fradinho, M. A. Coimbra, F. Domingues, D. Evtuguin, A. Silvestre and J. A. S. Cavaleiro, *Industrial Crop Prod.* **5**, 189 (1996).
17. A. Antunes, E. Amaral and N. M. Belgacem, *Industrial Crop Prod.* **12**, 85 (2000).
18. N. Cordeiro, C. Pascoal Neto, A. Gandini and M. N. Belgacem, *J. Colloid Interface Sci.* **174**, 246 (1995).
19. S. Katz and D. G. Gray, *Svensk Papperstidning* **8**, 226 (1980).
20. M. N. Belgacem, G. Czeremuszkin, S. Sapieha and A. Gandini, *Cellulose* **2**, 145 (1995).
21. M. N. Belgacem, A. Blayo and A. Gandini, *J. Colloid Interface Sci.* **182**, 431 (1996).
22. J. Schultz and L. Lavielle, in: *Inverse Gas Chromatography*, D. R. Lloyd, T. C. Ward and H. P. Schreiber (Eds.), ACS Symposium Series, No. 391. Am. Chem. Soc., Washington, DC (1989).
23. J. Felix and P. Gatenholm, *Nordic Pulp Paper Res. J.* **2**, 200 (1993).
24. A. Bismarck, J. Springer, A. K. Mohanty, G. Hinrichsen and M. A. Khan, *Colloid Polym. Sci.* **229**, 278 (2000).

Contact Angle, Wettability and Adhesion, Vol. 4, pp. 407–436
Ed. K.L. Mittal
© VSP 2006

XeCl excimer laser treatment of ultra-high-molecular-weight polyethylene fibers

J. ZENG and A. N. NETRAVALI[*]

Fiber Science Program, Cornell University, Ithaca, NY 14853-4401, USA

Abstract—Ultra-high-molecular-weight polyethylene (UHMWPE) fibers (Spectra® 1000) were treated using pulsed XeCl excimer laser (308 nm) to improve their adhesion to epoxy resin. The laser treatments were carried out in air and in diethylenetriamine (DETA) environments with varying numbers of pulses and fluences. The effects of the laser treatments on the fiber surface topography, chemistry and wettability were investigated. The interfacial shear strength (IFSS) with epoxy resin was measured using a single fiber pull-out test. The surface roughness was characterized using scanning electron microscopy (SEM) and atomic force microscopy (AFM). The laser treatment introduced grooves along the fiber length and increased the fiber surface roughness up to 6-times the control value, as measured by AFM. The X-ray photoelectron spectroscopy (XPS) data indicated that oxygen and/or nitrogen were incorporated on the fiber surface after the laser treatments depending on the environment in which the treatments were carried out. For some treatments the oxygen/carbon ratio increased by up to 2.5-times the control value. The dynamic wettability data showed that the laser treatments significantly increased the acid-base component of the surface energy and the work of adhesion. The UHMWPE fiber/epoxy resin IFSS increased by up to 400% for certain treatment conditions. The introduction of polar groups and higher surface roughness were found to be the two main factors that contributed to the significant increase in the adhesion of the UHMWPE fiber/epoxy resin system.

Keywords: Ultra-high-molecular-weight polyethylene (UHMWPE) fiber; excimer laser; adhesion; surface roughness; fiber/resin interface; wettability; surface energy; interfacial shear strength.

1. INTRODUCTION

Ultra-high-molecular-weight polyethylene (UHMWPE) fibers possess high specific strength and fracture toughness [1]. These fibers also have excellent chemical resistance because of their chemical composition consisting solely of methylene groups. UHMWPE fibers are currently used in various commercial and military applications, such as ropes and ballistic materials. However, applications of these fibers in advanced composites are limited because of their poor adhesion

---

[*]To whom correspondence should be addressed. Tel.: (1-607) 255-1875; Fax: (1-607) 255-1093; e-mail: ann2@cornell.edu

to most polymeric resins due to their nonpolar nature, low surface energy and smooth surface.

The performance of fiber reinforced composites depends not only on the properties of fiber and resin, but also on the fiber/resin interfacial adhesion [2]. Good interfacial adhesion helps resin to transfer the load from broken fibers to intact fibers, which maintains the integrity of the composites and results in better mechanical properties. Interfacial adhesion in composites is primarily controlled by three factors: (i) chemical interaction, (ii) mechanical interaction (interlocking) and (iii) physicochemical interaction [3]. Enhancement in any of these three factors can improve the fiber/resin interface bonding.

Various chemical treatments have been employed, in the past, to improve the adhesion between UHMWPE fibers and different polymeric resins. These treatments include oxidation in chromic acid [4] and sulfonation in chlorosulfonic acid [5]. However, these treatments require long immersion times in acids and are often accompanied by undesirable fiber strength loss. Corona discharge and flame treatments have also been used to increase the surface polarity of UHMWPE fibers, and thus to improve the interlaminar shear strength in composites [6].

A variety of radiofrequency (rf) plasma treatments have been employed to modify UHMWPE surfaces to improve their adhesion to various resins [7–12]. These include ammonia plasma treatment [10], and polymer forming allylamine plasma treatment [7, 8, 11]. Researchers have also used atmospheric pressure helium plasma treatment [13]. These treatments have shown an increase between 100% and 300% in the IFSS between UHMWPE fibers and epoxy resins. Although plasma treatments can increase the interfacial shear strength (IFSS) values, the process conditions may be difficult to control. Also, most rf plasma treatments require vacuum.

Cohen *et al.* [14] used 1,2,3,4-tetrahydronaphthalene (tetralin) solvent to treat both UHMWPE fiber and polyethylene resin to improve adhesion between them. They observed that the solvent increased the molecular mobility in both the fiber and the resin resulting in molecular entanglement. This entanglement significantly increased the interfacial adhesion. Nam and Netravali [15] used tetralin treatment followed by ammonia plasma treatment to modify the UHMWPE fiber surface, which increased the IFSS of the fibers with epoxy resin by more than 350%. Tetralin solvent increased the fiber surface roughness and the ammonia plasma treatment incorporated amine groups on the fiber surface, increasing acid–base interaction between the fibers and epoxy resin. The increase in IFSS was attributed to the combination of these two factors.

Intense pulsed high-power ion beam (HPIB) was used to modify UHMWPE fiber surface [16]. In this case, increases in both surface roughness and surface polarity were achieved simultaneously by only a single pulse of argon ions. Both factors contributed to improve its IFSS with epoxy resin by 2.5-times that of control fibers.

Netravali and co-workers [17–19] used a XeCl (308 nm wavelength) pulsed excimer laser to modify UHMWPE fiber surface in air, ammonia, argon and helium gas environments. After the laser treatments, UHMWPE (Spectra® 1000 and 2000) fiber/epoxy resin IFSS was found to increase by up to 4-times the control value. These enhancements were again attributed to increased fiber surface roughness and surface polarity. Zeng and Netravali [20] used KrF excimer laser (248 nm wavelength) to modify the UHMWPE fiber surface in air and liquid diethylenetriamine (DETA) environment. DETA was shown to have no significant absorption at 248 nm wavelength. After the laser treatment in DETA, the IFSS was found to increase by up to six times of that obtained for control fiber. X-Ray photoelectron spectroscopy (XPS) analysis of fibers treated in air showed the presence of oxygen containing groups on the fiber surface. Treatment in DETA environment resulted in nitrogen-, as well as oxygen-containing groups on the fiber surface. The amine groups on the fiber surface, although only a few, covalently bonded with epoxy resin, resulting in higher IFSS. Fiber surface roughness measured using atomic force microscopy (AFM) was also observed to increase up to seven times, which would contribute to the increase in mechanical bonding. The combination of oxygen- and nitrogen-containing groups was shown to increase the hydrogen bonding significantly. This considerably increased the work of adhesion between UHMWPE fibers and epoxy resin, and resulted in better spreading of the resin onto the fiber as well.

Several researchers have studied the effects of laser radiation on polymeric materials and several theories have been presented to explain the UV excimer laser interactions with organic polymers [21–27]. At present, it is widely accepted that UV excimer laser commonly causes photodecomposition and/or photoablation of the polymer [21–23]. For fibers which are highly absorbing in the UV range, such as polyesters (e.g., PET) and polyamides (PAs), highly regular roll (wavy) structures in the transverse direction have been observed after UV laser treatments [26]. In the case of non-UV-absorbing polymers, such as UHMWPE fibers, groove structures in the longitudinal direction have been shown, most likely due to the fibrillar structure of the fibers [17, 18]. Additionally, since UHMWPE fibers have low absorption in the UV range, both higher fluence and significantly larger number of pulses were required to achieve similar topographical changes than those required for PET and PA using the same laser source.

In the present research, pulsed XeCl excimer laser (308 nm) has been used to treat UHMWPE (Spectra® 1000) fibers in DETA environment to improve their adhesion to a diglycidyl ether of a bisphenol A (DGEBA)-based epoxy resin. For comparison, laser treatment was also carried out in air. The effects of fluence, number of pulses and environments of the excimer laser treatment on the fiber surface and interfacial properties with the epoxy resin have been investigated. These results are also compared with the corresponding results obtained using KrF (248 nm) laser.

2. EXPERIMENTAL

2.1. Materials

The UHMWPE fibers (Spectra® 1000) used in this study were obtained from Honeywell. The DGEBA-based epoxy resin (DER 331) with an average epoxy equivalent weight of 18.7 and tetraethylenepentamine (DEH 26) curing agent with amine hydrogen equivalent weight of 27.1 were obtained from Dow Chemical. HPLC grade water, DETA, formamide, and methylene iodide were obtained from Aldrich. Quartz coverslips, 2.54 cm × 2.54 cm, were obtained from Electron Microscopy Sciences (www.emsdiasum.com).

2.2. Specimen preparation and laser treatment

A schematic diagram of the experimental setup for laser treatment is shown in Fig. 1. A Lambda Physik LPX 200 apparatus was used for the laser treatment. Xenon (Xe) and chlorine (Cl) gases were used to obtain XeCl laser beam. Fibers were separately irradiated in DETA and air environments. Different fluences, 200, 400 and 600 mJ/(pulse·cm^2) were used in this study. The laser pulse repetition rate was fixed at 10 Hz and each pulse was of 20 ns duration. The fluence, repetition rate and the number of pulses were controlled by a computer linked to the laser apparatus. For every laser treatment, the fiber specimens were first exposed to half the number of pulses from one side and the remaining half from the opposite side. The laser fluence was measured with a photodiode calibrated using the Si crystal melt technique [28]. To carry out treatment in air, fibers were mounted in a parallel array between two cardboard pieces with 7 mm diameter circular windows in the center. For laser treatment in DETA, fibers were placed in a parallel array between two quartz coverslips and the gap between the coverslips was filled with DETA, immersing the fibers (Fig. 1).

The intensity of laser beam was measured prior to passing through the quartz coverslips and as it emerged from the coverslips. The beam intensity was found to reduce by less than 10% due to coverslip reflection. In the second set of experiments DETA was added between the coverslips and the laser intensity was

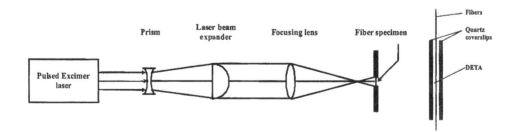

Figure 1. A schematic diagram of the experimental setup for laser treatment.

measured again. The intensity of the laser beam was again reduced by approximately 10%. This indicated that the small amount of DETA present between the cover slips did not absorb significantly at 308 nm wavelength. In addition, the UV spectrum of DETA was also measured and it showed negligible absorption. The absorption by quartz coverslips was taken into account while exposing the fibers to the laser beam. After the laser treatment in DETA, the fibers were observed again to make sure that they were still immersed in DETA. The fibers treated in DETA were carefully washed in acetone and water successively to remove any residual DETA before performing various surface analyses.

2.3. Fiber surface characterization

The surface topographies of control and treated fibers were characterized using a scanning electron microscope (SEM, Leica, Model 440X) and an AFM (Digital Instruments, Model Dimension^{TM} 3100). Wettability properties of treated and control fibers were measured by the Wilhelmy technique using the TRI/Princeton wettability apparatus. In order to obtain the acid–base and dispersion components of the fiber surface energy, three different probe liquids: water, formamide and methylene iodide were used. The three-probe liquid wettability method has been described in detail by Good *et al.* [29] and Kamath *et al.* [30]. The chemical changes on the UHMWPE fiber surfaces were characterized using XPS (Surface Science Instruments, Model SSX100).

2.4. Fiber/resin interface characterization

The single fiber pull-out test was used to estimate the IFSS between the UHMWPE fiber and epoxy resin. The specimen geometry and other experimental details have been explained elsewhere [7, 20]. The fiber was embedded into epoxy resin, keeping the length of the embedment constant at 1.5 mm. Epoxy resin DER 331 and the curing agent DEH 26 were used in stoichiometric ratio. The resin was cured at 80°C for 3 h and the oven was turned off to allow the specimens to cool down over 12 h. The single fiber pull-out test was carried out using an Instron Tensile Tester (model 1122) at a cross-head speed of 2 mm/min.

The average IFSS, τ, in MPa, was calculated using the following equation:

$$\tau = \frac{F_p}{\pi d l} \tag{1}$$

where F_p is the fiber pull-out force in N, l is the length of embedment = 1.5 mm and d is the fiber diameter in mm, which was measured using an optical microscope. It is assumed that F_p is uniformly distributed along the entire length of the fiber embedment.

3. RESULTS AND DISCUSSION

3.1. Fiber surface topography

Figures 2 and 3 show SEM micrographs of control and laser-treated fibers at a fluence of 600 mJ/(pulse·cm^2) in air and in DETA, respectively. The micrographs of control fibers are included in both figures for comparison. The control fibers shown in Figs 2a and 3a, exhibit a smooth surface with a few grooves along their axes, indicative of their fibrillar nature.

Figure 2 shows laser-treated fibers at a fluence of 600 mJ/(pulse·cm^2) in air. After 100 laser pulses, the fiber surface was slightly rougher with narrow and shallow grooves along the longitudinal direction, as shown in Fig. 2b. When the number of pulses increased up to 200, the surface roughness increased and the grooves became deeper and better defined, as seen in Fig. 2c. When the number

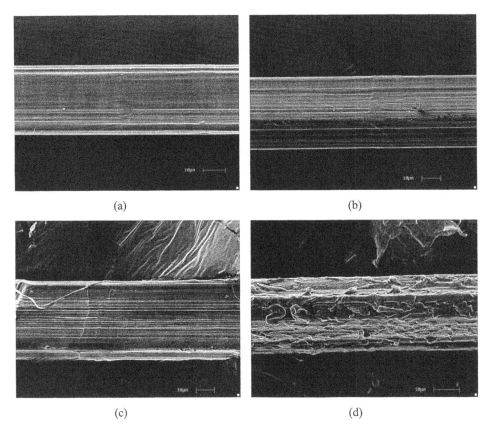

(a)   (b)

(c)   (d)

Figure 2. SEM micrographs of UHMWPE fibers: (a) control; (b) laser-treated with 100 pulses; (c) laser-treated with 200 pulses; (d) laser-treated with 400 pulses. Fibers in panels b–d were treated at a fluence of 600 mJ/(pulse·cm^2) in air.

of pulses increased up to 400, the fiber surface showed a much rougher scale structure and at the same time the small grooves disappeared, as seen in Fig. 2d.

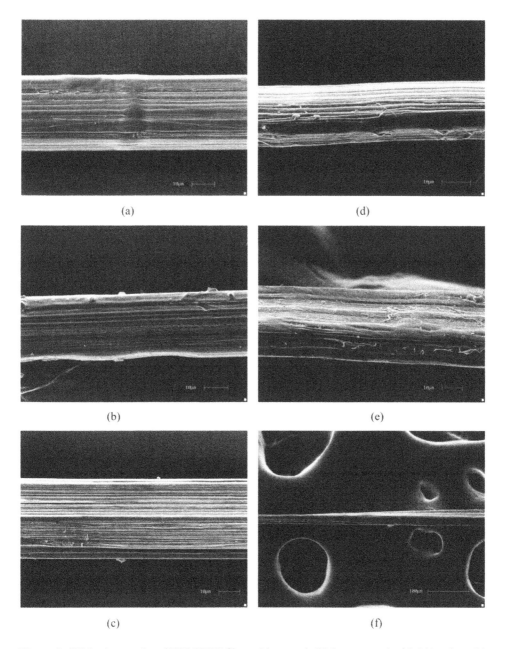

(a)

(d)

(b)

(e)

(c)

(f)

Figure 3. SEM micrographs of UHMWPE fibers: (a) control; (b) laser-treated with 200 pulses; (c) laser-treated with 400 pulses; (d) laser-treated with 600 pulses; (e, f) laser-treated with 800 pulses. Fibers in panels b–f were treated at a fluence of 600 mJ/(pulse·cm^2) in DETA.

Figure 3 shows fiber treated in DETA at a fluence of 600 mJ/(pulse·cm²). With 200 laser pulses, some small cracks and shallow grooves appeared, as seen in Fig. 3b. As the number of pulses increased up to 400, the grooves became deeper and their edges became sharper (Fig. 3c). After 600 pulses, the depth of the grooves on the surface became higher, but their edges became blunt, suggesting melting or softening at the fiber surface. Since the melting temperature of UHMWPE is below 152°C, it is not difficult to achieve it in spite of its low UV absorption [31]. When the number of pulses increased up to 800, narrow grooves disappeared and were replaced by an irregular furrowed structure containing some texture on the surface (Fig. 3e). This again suggests UHMWPE fibers melting at the surface, resulting in smoother grooves.

Figure 3f shows the same fiber at the same location as Fig. 3e, but at a lower magnification. The left part of the fiber underwent the laser treatment, while the right part remained untreated. It is clear that the diameter of the left half of the fiber is significantly smaller than the right half, indicating that a significant part of the fiber, when irradiated, was ablated. The diameters of the fibers were measured before and after the treatment. Figure 4a and 4b shows the fiber diameter remaining, in percent, after the laser treatment as a function of number of pulses at different fluences for fibers treated in air and in DETA, respectively. From these plots, it is clear that both laser fluence and number of pulses have significant effect on ablation. Furthermore, the ablation thresholds at different fluences are different. Higher fluence has a lower ablation threshold, as expected. For example, for fibers treated in air at the fluence of 200 mJ/(pulse·cm²), the threshold is about 200 pulses. In other words, there is no change in measured diameter up to 200 pulses and for fibers treated at the fluence of 400 mJ/(pulse·cm²), there is no change in fiber diameter up to 100 pulses. At the fluence of 600 mJ/(pulse·cm²), the threshold is close to zero, i.e., the ablation starts almost from the very first pulse. Similar trend is also seen for fibers treated in DETA. These results clearly show that the diameters of the UHMWPE fibers decrease when treated with XeCl laser as a result of ablation and that the ablation is a function of both laser fluence and number of pulses (total energy).

The energy of XeCl laser (388.4 kJ/mol) is higher than the bond energy of the C–C bond (343 kJ/mol), which forms the backbone of the polyethylene molecules [32]. As a result, the XeCl laser can break these covalent bonds near the fiber surface and create a number of much smaller molecular species as well as free radicals. Simultaneously, a part of the released energy, which is the energy difference between the photon energy of XeCl laser and the C–C bond energy, could convert to kinetic energy of these polymer fragments, resulting in these fragments escaping from the fiber surface [32]. The UHMWPE fibers have a fibrillar structure, as seen in the SEM micrograph in Fig. 2c. Also, UHMWPE fibers have been shown to easily fibrillate when fractured in tension indicating weak interfibrillar interactions [33]. After the laser irradiation, the longitudinal grooves on the fiber surface deepen and widen, as mentioned earlier, resulting in increased fiber surface roughness. One reason for this, perhaps, is the weak interfibrillar interaction

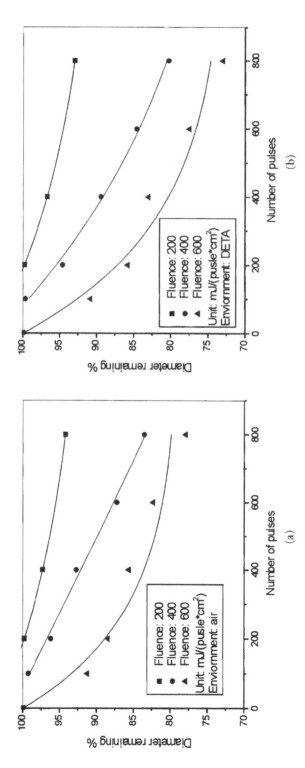

Figure 4. UHMWPE fiber diameter remaining (%) after laser treatment of fiber *vs.* number of pulses: (a) in air; (b) in DETA.

which allows them to separate, especially at higher temperature. The increase in the fiber temperature could be due to the oxygen containing groups introduced at the fiber surface during storage creating species with higher absorption coefficient at 308 nm. The radicals generated at the fiber surface by the laser treatment also allow oxygen from air to combine with the fiber. The XPS results, which will be discussed later in Section 3.3, confirm the presence of oxygen on both control and treated fiber surfaces. The photoablation process is enhanced by the higher temperature at the fiber surface [23, 34].

During the laser treatment, the higher temperature can damage/melt the highly crystalline structure at the fiber surface, increase the chain to chain distance and thus decrease the van der Waals forces among the molecules. These factors facilitate escaping of the molecular fragments from the fiber surface created by bond breakage. Moreover, the higher surface temperature also helps the free radicals to react with oxygen from the air incorporating oxygen containing groups on the fiber surface. For the treatments carried out in DETA, the free radicals can react with the reactive amine groups in DETA and incorporate them onto the fiber surface. Again, the heat generated during the process can enhance the reaction rate. The excess heat also raises the temperature of the fiber above the melting point of the UHMWPE fiber (144–152°C), resulting in melting of a thin layer on the fiber surface [31]. The molten material can flatten the edges of the sharp peaks reducing the surface roughness of the fiber. For fibers such as polyesters (e.g., PET) and polyamides (PAs), which are highly absorbing in the UV range and melt easily, highly wavy structures in the transverse direction have been observed after exposing to a significantly lower number of laser pulses [26]. Since UHMWPE fibers have very low absorption in the UV range, both the fluence and the number of pulses required to achieve equivalent topographical changes are significantly higher than those required for PET and PA.

AFM was used to quantitatively characterize fiber surface roughness at the nanoscale level. The scan area in AFM measurements was 3 µm×3 µm, much smaller than the scan area in the SEM measurements. With significantly higher resolution, AFM images reveal much smaller structural features of the fiber surface than possible with SEM.

Figure 5 shows the AFM images of surfaces of fibers treated at a fluence of 600 mJ/(pulse·cm^2) in air with different number of pulses. Figure 5a shows a control fiber surface which has a smooth surface with some nodular structures on it. With increased number of laser pulses, the surface became rougher. After 400 laser pulses, the surface became irregular with sharp peaks (Fig. 5b). When the number of pulses was increased to 800, the fiber surface became smoother again, as seen in Fig. 5c. This, again, confirms the melting and flattening of sharp peaks on the fiber surface.

Figure 6 shows the AFM images of surfaces of fibers treated at a fluence of 600 mJ/(pulse·cm^2) in DETA with varying number of pulses. A typical AFM image of control fibers presented in Fig. 6a, again, illustrates its smooth surface with a few shallow grooves running along the fiber length. Figure 6b shows the fiber

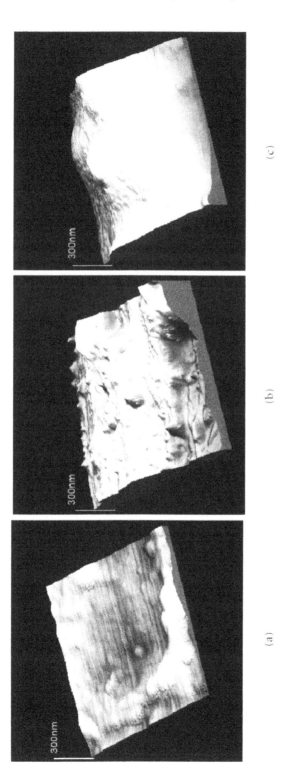

Figure 5. AFM images of surfaces of UHMWPE fibers treated in air: (a) control, RMS = 11.6 nm; (b) laser-treated with 800 pulses, RMS = 42.0 nm. Fibers in panels b and c were treated at a fluence of 600 mJ/(pulse·cm^2). The scan area of AFM measurements was 3 μm×3 μm and the scan rate was 1 Hz. The RMS values above are averages of measurements on at least 5 specimens.

surface after 100 pulses of laser beam with several shallow grooves in the longitudinal direction. With more pulses, the grooves became wider and deeper, and the surface roughness of the fibers increased as well (Fig. 6c–6e). When the number of pulses increased to 600, some furrowed structures appeared on the fiber surface, which indicated some melting at the surface, as can be seen in Fig. 6f. With 800 pulses, the grooves on the fiber surface disappeared and the surface became smoother implying that the fiber surface underwent melting. Similar features have also been seen in corresponding SEM micrographs for fibers treated in air and have been discussed earlier.

To quantify the roughness of the fiber surfaces that underwent different laser treatments, three separate roughness parameters RMS, R_a and R_{max} were calculated from AFM measurements. The RMS (root mean square) value measures the standard deviation of the height values within the scanned area and is given by the following equation [35]:

$$RMS = \sqrt{\frac{\sum (z_i - z_{av})^2}{n}} \tag{2}$$

where z_i is the height value for point i, z_{av} is the average height value in the area and n is the number of points within the given area. R_a measures the arithmetic average of the absolute values of the surface height deviations measured from the mean plane and is given by the following equation [35]:

$$R_a = \frac{\sum |z_i - z_{av}|}{n} \tag{3}$$

R_{max} signifies the difference in height between the highest (peak) and the lowest (valley) points within the image.

Table 1 presents the RMS, R_a and R_{max} values for control fibers, as well as fibers laser treated in air and DETA environments. Figure 7 presents two three-dimensional plots which show the relationships between RMS, fluence and number of pulses for fibers treated in air and in DETA, respectively. From these two plots, it can be seen that both fluence and number of pulses are significant factors in determining laser-treated fiber surface roughness. In general, higher fluence and larger number of pulses provide greater surface roughness value. For control fibers, the RMS roughness value is 11.6 nm. For treated fibers, all RMS values are significantly higher than control values, indicating much higher surface roughness. For fibers treated in air, with 600 pulses at the fluence of 600 mJ/(pulse·cm^2), the highest RMS value of 72.3 nm is about 6.2 times the control value. Beyond that, for 800 pulses, the RMS value decreased. At the same time R_{max}, which indicates the maximum peak-to-valley distance, and R_a values also decreased. As explained earlier, the decrease in the roughness is caused by melting of the fiber surface. Similarly, for fibers treated in DETA, the highest RMS value of 57.1 is about 5-times that obtained for control fibers. This highest value

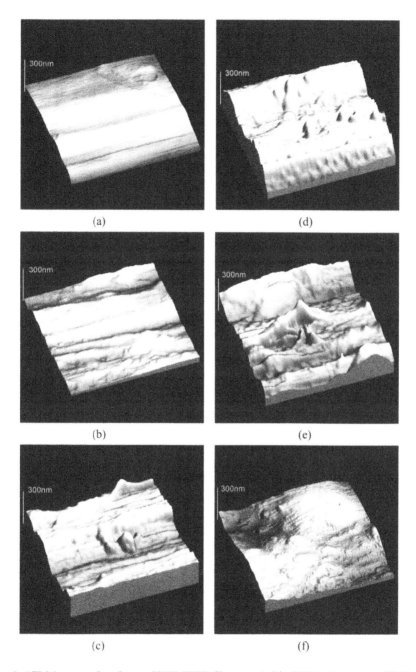

Figure 6. AFM images of surfaces of UHMWPE fibers treated in DETA: (a) control, RMS = 11.6 nm; (b) 100 pulses, RMS = 24.0 nm; (c) 200 pulses, RMS = 41.4 nm; (d) 400 pulses, RMS = 57.1 nm; (e) 600 pulses, RMS = 56.7 nm; (f) 800 pulses, RMS = 38.2 nm. Fibers in panels b–f were treated at a fluence of 600 mJ/(pulse·cm^2). The scan area of AFM measurements was 3 μm×3 μm and the scan rate was 1 Hz. The RMS values above are averages of measurements on at least 5 specimens.

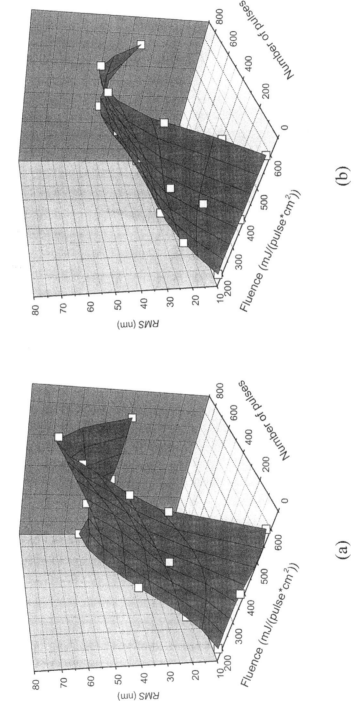

Figure 7. 3-D plots of RMS surface roughness *vs.* fluence and number of pulses: (a) laser treatment in air; (b) laser treatment in DETA.

Table 1.
Effect of XeCl laser treatments on UHMWPE fiber surface roughness

Fluence (mJ/(pulse·cm^2))	Number of pulses	Environment	RMS (nm)	R_a (nm)	R_{max} (nm)
Control			11.6	9.1	74.4
200	200	Air	19.6	16.5	120.5
200	400	Air	35.3	28.5	238.4
200	800	Air	55.2	46.1	269.2
400	100	Air	20.4	15.8	149.1
400	200	Air	33.2	28.1	180.3
400	400	Air	43.5	34.4	260.9
400	600	Air	58.2	49.7	273.7
400	800	Air	45.5	36.0	224.5
600	100	Air	42.1	31.5	266.0
600	200	Air	53.1	44.9	320.3
600	400	Air	66.0	53.4	394.2
600	600	Air	72.3	56.0	362.5
600	800	Air	42.0	31.5	266.6
200	200	DETA	20.7	16.9	115.9
200	400	DETA	25.9	20.4	151.2
200	800	DETA	28.8	23.9	191.3
400	100	DETA	22.7	16.5	137.5
400	200	DETA	32.7	27.5	189.5
400	400	DETA	39.5	31.2	284.5
400	600	DETA	45.8	35.4	367.0
400	800	DETA	50.1	41.8	389.2
600	100	DETA	24.0	19.8	208.5
600	200	DETA	41.4	32.4	251.0
600	400	DETA	57.1	46.1	293.4
600	600	DETA	56.7	43.8	342.3
600	800	DETA	38.2	32.6	261.4

was observed for fibers treated with 400 pulses at the fluence of 600 mJ/(pulse·cm^2). Beyond 400 pulses, the fiber surface roughness decreased due to melting. Comparing the effect of different environments, it can be concluded that significantly higher surface roughness is achieved for fibers treated in air than for those treated in DETA. One reason may be that a part of the laser energy is absorbed by the DETA. With large number of laser pulses, the temperature of DETA liquid increases to its boiling point (approx. 209°C), which is higher than the melting point of UHMWPE fiber. Since the fiber was immersed in DETA, the fiber softens, thereby reducing the surface roughness. In the present study, the fiber surface roughness after 800 pulses was significantly lower than that after 400 pulses at the fluence of 600 mJ/(pulse·cm^2) in both air and DETA environments.

From these results, it can be concluded that the UHMWPE fiber surface changed from smooth, prior to UV excimer laser treatment, to rough after the laser treatment up to a certain number of pulses. Any additional pulses decreased the surface roughness as a result of melting The increase in roughness varied with the fluence, the number of pulses and the treatment environment.

The surface topography and surface roughness of UHMWPE fibers treated by KrF (248 nm) laser were significantly different from the results obtained in the current study [20]. When irradiated with KrF laser, a wavy structure at the UHMWPE fiber surface appeared after merely 400 pulses at a fluence of 200 mJ/(pulse·cm^2) in DETA [20]. However, no wavy structure appeared at the fiber surface under any treatment condition in DETA when irradiated with XeCl laser. Under KrF laser irradiation with 400 pulses at a fluence of 200 mJ/(pulse·cm^2) in DETA, the RMS value was 79.0 nm, whereas the corresponding RMS value was only 25.9 nm with XeCl laser. In addition, none of the SEM micrographs of KrF laser treated fibers exhibited surface melting, as clearly seen in Fig. 3e. These phenomena reveal that the mechanisms of laser-polyethylene interactions are different when irradiated with XeCl than with KrF laser. The XeCl laser energy of 388.4 kJ/mol can break C–C bonds (343 kJ/mol), but not C–H bonds (394 kJ/mol) in polyethylene [32]. The KrF laser energy of 482.4 kJ/mol, however, is much larger than that of the XeCl laser and can break C–H bonds as well. Compared to XeCl laser, KrF laser provides larger kinetic energy to the molecular fragments and makes it easier for them to escape from the fiber surface. In other words, the polymer surface may ablate before it can melt. As a result, no melting of the fiber surface could be seen when irradiated with KrF laser. Additionally, the absorption of polyethylene at 248 nm is slightly higher than that at 308 nm, which means polyethylene absorbs more energy at 248 nm than at 308 nm at the same fluence [36]. As a result, the extent of photoablation of polyethylene at 248 nm is higher than at 308 nm, at the same fluence.

3.2. Wettability properties and acid–base interaction measurements

The work of adhesion, contact angles and surface energies for both control and treated fibers were estimated by the Wilhelmy technique using three probe liquids [29, 30]. The work of adhesion was obtained for both advancing and receding modes. Hysteresis, which is caused by metastable states at the solid/liquid/vapor interface, was obtained as the advancing work of adhesion minus the receding work of adhesion [37]. While chemical heterogeneity at the surface is the major cause of hysteresis, surface roughness also leads to some hysteresis. Some probe liquid molecules may remain attached and be difficult to be removed from the fiber surface, particularly from deep valleys, as the fiber is pulled out of the probe liquid [37]. The typical plots of work of adhesion, in water, for control and for laser-treated fibers are presented in Fig. 8. The lower curve in every plot is for the

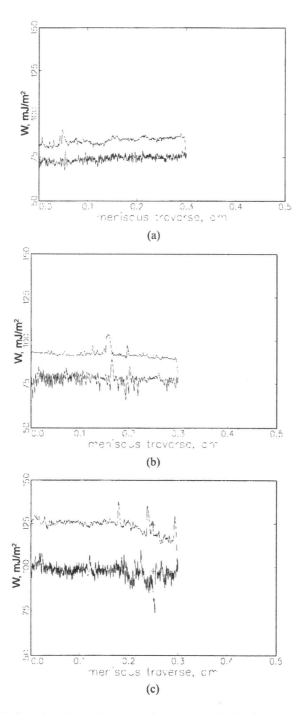

Figure 8. Typical advancing (bottom) and receding (top) work of adhesion plots for control and treated fibers in water: (a) control; (b) laser-treated fiber (fluence: 400 mJ/(pulse·cm^2) pulses: 200 in air); (c) laser-treated fiber (fluence: 400 mJ/(pulse·cm^2) pulses: 600 in DETA).

J. Zeng and A. N. Netravali

Table 2.
Effect of XeCl laser treatments on work of adhesion, water contact angle and surface energies (UHMWPE fibers treated in air)

Experimental conditions		Work of adhesion		Contact angle (Adv mode, θ_a, deg)	Dispersion (γ^d, mJ/m²)	Acid–base (γ^{AB}, mJ/m²)	Acid parameter (γ^+, mJ/m²)	Base parameter (γ^-, mJ/m²)	Surface energy (γ, mJ/m²)	Hysteresis (W_{adv}–W_{rec}, mJ/m²)
Fluence (mJ/(pulse·cm²))	Pulses	W_{adv} (mJ/m²)	W_{rec} (mJ/m²)							
Control		68.7	78.5	92.3	33.8	0.03	0.0	2.3	33.8	-9.7
200	200	68.8	81.3	93.2	31.4	0.9	0.1	1.7	32.3	-12.5
200	400	69.7	83.6	92.4	30.2	1.3	0.2	1.9	31.5	-13.9
200	800	77.2	92.5	86.6	29.4	2.4	0.4	4.1	31.8	-15.4
400	100	71.7	84.2	90.9	29.5	1.5	0.2	2.5	31.0	-12.5
400	200	74.2	88.9	88.9	30.2	2.2	0.5	2.5	32.4	-14.7
400	400	86.3	103.9	79.3	29.8	4.1	0.6	7.6	33.9	-17.6
600	100	79.2	94.2	85.0	29.2	2.3	0.2	5.6	31.5	-15.0
600	200	88.0	105.2	78.0	29.5	3.9	0.4	9.2	33.4	-17.3
600	400	91.9	111.9	74.8	29.6	5.6	0.8	10.1	35.2	-20.0

W_{adv} is work of adhesion in advancing mode, W_{rec} is work of adhesion in receding mode.

Table 3.
Effect of XeCl laser treatments on work of adhesion, water contact angle and surface energies (UHMWPE fibers treated in DETA)

Experimental conditions		Work of adhesion		Contact angle (Adv mode, θ_a, deg)	Dispersion (γ^d, mJ/m²)	Acid–base (γ^{AB}, mJ/m²)	Acid parameter (γ^+, mJ/m²)	Base parameter (γ^-, mJ/m²)	Surface energy (γ, mJ/m²)	Hysteresis (W_{adv}-W_{rec}, mJ/m²)
Fluence (mJ/(pulse·cm²))	Pulses	W_{adv} (mJ/m²)	W_{rec} (mJ/m²)							
Control		68.7	78.5	92.3	33.8	0.03	0.0	2.3	33.8	-9.7
200	200	80.3	93.2	84.1	31.6	0.9	0.4	0.6	32.5	-12.9
200	400	84.3	98.4	81.0	30.4	1.3	0.5	0.9	31.7	-14.2
200	800	89.8	106.6	76.5	29.8	2.5	0.9	1.8	32.3	-16.8
400	100	81.2	94.2	83.4	30.9	0.9	0.3	0.8	31.8	-12.9
400	200	86.8	102.0	79.0	28.1	2.2	0.5	2.4	30.4	-15.2
400	400	94.5	113.2	72.7	26.7	4.6	0.7	8.0	31.3	-18.7
400	600	100.1	120.5	68.0	26.5	5.5	0.8	9.1	32.0	-20.4
400	800	103.5	124.5	65.1	26.2	5.8	0.9	9.3	32.0	-21.0
600	100	83.3	98.4	81.8	28.3	2.1	0.5	2.2	30.4	-15.2
600	200	91.7	110.3	74.9	28.2	4.7	0.7	7.5	32.9	-18.6
600	400	98.1	119.4	69.7	27.8	6.4	0.8	12.5	34.2	-21.3
600	600	102.3	124.7	66.1	27.9	6.8	0.9	13.5	34.7	-22.4
600	800	103.3	126.4	65.2	27.8	7.0	0.9	13.7	34.8	-23.1

W_{adv} is work of adhesion in advancing mode, W_{rec} is work of adhesion in receding mode.

advancing mode and the upper curve is for the receding mode. Comparing the work of adhesion curves for the control fiber shown in Fig. 8a with those for the two laser-treated fibers shown in Fig. 8b and 8c, it is clear that the laser-treated fibers show significantly higher work of adhesion than the control fibers. The control UHMWPE fibers have highly non-polar surface consisting of methylene groups which give low work of adhesion in water [7, 8]. After the laser treatments, as the surface polarity of the fiber improves as a result of incorporation of oxygen and nitrogen containing groups on the surface, the work of adhesion in water increases. It is also clear from Fig. 8 that the absolute value of hysteresis for control fibers in water is smaller than that obtained for the laser-treated fibers. Control fibers with smoother surface as seen from SEM micrographs and AFM images, and mostly non-polar surface result in small hysteresis in water. Laser-treated fibers show larger hysteresis due to increasingly polar and rougher surface. Introduction of the polar groups increases the chemical heterogeneity of the fiber surfaces.

It can also be seen from the work of adhesion plots, in the advancing mode, that there are two types of fluctuations. The first type of fluctuation with high amplitude but low frequency is caused by the surface roughness [17–20]. The second type of fluctuation consisting of low amplitude but high frequency is a result of the surface heterogeneity. Both types of fluctuations are much larger in the case of treated fibers compared to the control fibers, confirming that the laser treatment significantly increases both the fiber surface roughness and surface heterogeneity as expected [17–20].

Tables 2 and 3 list the work of adhesion in advancing and receding modes, water contact angles, surface energy and its components for laser-treated fibers (in advancing mode) in air and in DETA, respectively. From Tables 2 and 3, it is evident that the total surface energies for control and laser-treated fibers are close and not statistically different. However, the work of adhesion, in advancing mode, in water (W_{adv}) and the acid–base component of surface energy γ^{AB} increased significantly after the laser treatment. The increase in the acid–base component of surface energy after the laser treatment confirms that the fiber surface has become more polar. The enhancement of fiber surface polarity can be mainly attributed to oxygen containing groups at the surface. However, not all oxygen containing groups contribute to γ^{AB} to the same extent. For example, R–COO–R (ester) is a non-polar group and C–O–C is less polar than C–OH. These groups cannot contribute much to increasing the γ^{AB}. It is widely accepted that the acid–base interaction can form hydrogen bonding between the fibers and the epoxy resin [33, 38]. Therefore, it can be expected that the IFSS between UHMWPE fibers and epoxy resin would increase after laser treatment in DETA. Since γ^{AB} is a critical parameter that determines the extent of acid–base interaction or hydrogen bonding between the fiber and the epoxy resin, the γ^{AB} data are shown in three-

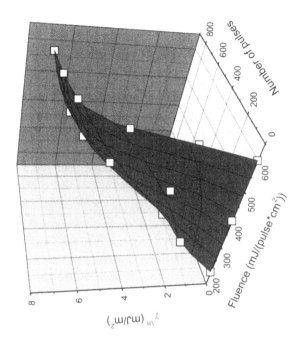

(B)

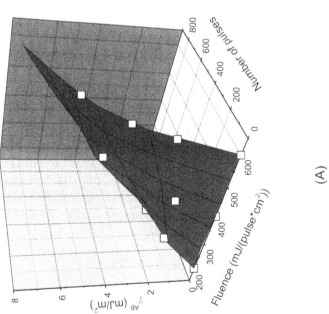

(A)

Figure 9. 3-D plots of acid–base component of surface energy *vs.* fluence and number of pulses: (A) laser treatment in air; (B) laser treatment in DETA.

dimensional plots as a function of fluence and number of pulses (Fig. 9) for fibers treated in air and in DETA. From these plots, it can be seen that both fluence and number of pulses are significant factors in increasing the γ^{AB}. At the same fluence, especially at fluences of 400 and 600 mJ/(pulse·cm^2) in DETA, with increasing the number of pulses the γ^{AB} increased at a slower rate. When the number of pulses increased to 800 at these two fluences, in DETA, γ^{AB} increased only slightly over that obtained for 600 pulses. This indicates that the acid-base component of the surface energy almost achieves the maximum value at the corresponding fluences with 800 pulses and that the rate of photoablation is approximately the same as introduction of the polar groups onto the fiber surface. As a result, further increase in the number of pulses may not be expected to increase γ^{AB} any further. For control fibers, W_{adv} was 68.7 mJ/m^2, the water contact angle was 92.3° and the hysteresis was -9.7 mJ/m^2. With 200 pulses at the fluence of 200 mJ/(pulse·cm^2) in DETA, W_{adv} increased to 80.3 mJ/m^2, the water contact angle decreased to 84.1° and the hysteresis increased to -12.9 mJ/m^2. For fibers treated with 400 and 800 pulses at the fluence of 200 mJ/(pulse·cm^2) in DETA, W_{adv} increased further to 84.3 and 89.8 mJ/m^2, the water contact angles decreased to 81.0° and 76.5°; and the hysteresis increased to -14.2 and -16.8 mJ/m^2, respectively. Similar trends were also observed for fibers treated at 400 and 600 mJ/(pulse·cm^2) fluence in air. Overall, these observations further confirm that the fiber surface polarity, heterogeneity and roughness all increased significantly after the laser treatments. However, there is a limit to this increase which has been reached in these experiments.

DETA is a liquid amine which contains a high density of amine groups compared to ammonia gas environment used in earlier studies [17, 18]. Laser treatment generates certain amount of free radicals on the polyethylene fiber surface. DETA environment can provide additional opportunity for these free radicals to combine with amine groups. Thus, it can be expected that the acid–base component of surface energy of the fibers treated in DETA would be larger than that of fibers treated in air. However, when compared with γ^{AB} values in Table 3 with corresponding data for laser-treated fibers in air (Table 2), it can be seen that they are not significantly different. The reason is believed to be that DETA absorbs little energy at 308 nm wavelength. As a result, only small numbers of amine groups are generated during XeCl laser irradiation to combine with the fiber surface. This limited amount of amine groups cannot increase the acid–base component of the fiber surface energy significantly.

3.3. Surface chemical composition analysis by XPS

The surface elemental composition of control and a few laser-treated fibers was investigated using XPS. The fiber specimens were scanned using Al Kα X-rays between 0 and 1000 eV and 55° take-off angle. Figure 10 shows XPS spectra of a control and laser-treated fibers in air and in DETA. The XPS spectrum for the control fiber shows four distinct peaks, corresponding to C_{1s} and O_{1s}. Small peaks for Si_{2s} and Si_{2p} can also be seen. The O_{1s} peak indicates that the control fiber has oxygen at the surface. The presence of oxygen can be attributed to surface oxidation due to prolonged exposure to air [39]. The Si peaks indicate that some impurities such as dust particles (SiO_2) present in the air may settle on the fiber during handling. These impurities may also increase the oxygen content on the fiber surface.

The elemental composition data for various binding energies are presented in Table 4. The data in Table 4 indicate that the oxygen content of the laser-treated fiber surface increases significantly. The O/C ratio also increases about 2-fold after the laser treatment. The free radicals formed during the laser treatment react with the oxygen in the air, creating oxygen containing groups on the topmost layer of the fiber surface. However, the O/C ratio increased only slightly with increased number of pulses at the same fluence for both environments. As discussed earlier, this is because the oxidation and photoablation occur simultaneously and, thus, the initially oxidized layer ablates and a fresh surface which is exposed to the laser starts to oxidize [20]. With more laser pulses, the surface temperature increases and accelerates the oxidation process. As a result, the oxygen content on the fiber surface increases only slightly with larger number of pulses. When the UHMWPE fibers were treated in DETA environment, a small peak of N_{1s}, at a binding energy of 399.5 eV, was observed. As discussed earlier, the DETA environment provides an opportunity for incorporating $-NH_2$ groups on the fiber surface. The radicals generated at the fiber surface can also combine with the oxygen adsorbed on the fiber surface or when the fiber is exposed to the air after the treatment. As a result, the UHMWPE fibers irradiated in DETA environment contained both nitrogen and oxygen.

The binding energies of deconvoluted peaks for C_{1s} and O_{1s}, and the corresponding relative intensities (%) are also listed in Table 4. In general, for C_{1s} the 285 ± 0.5 eV peak is assigned to C–H and C–C groups. The 286 ± 0.5 eV peak is assigned to groups in which carbon forms single bonds to one oxygen or nitrogen, such as C–O and C–N. The 290 ± 0.5 eV peak is assigned to groups in which carbon makes double bond to one oxygen, such as C=O. The satellite peaks at binding energies of 295 ± 0.5 eV commonly result from non-monochromatic Al Kα X-ray source [12, 17–20, 40]. The O_{1s} peaks were deconvoluted into two peaks at 531 ± 0.5 eV and 533 ± 0.5 eV. The two positions correspond to the C=O and C–O bonds, respectively. From Table 4, it can be seen that most of the oxygen is singly bonded to carbon. This is consistent with the C_{1s} analysis, which suggests that C–O is the main form of oxygen bonding.

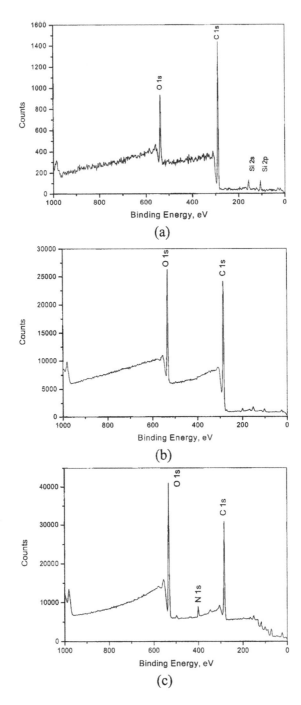

Figure 10. XPS spectra of UHMWPE control and laser treated fibers: (a) control; (b) 200 pulses at a fluence of 600 mJ/(pulse·cm²) in air; (c) 600 pulses at a fluence of 600 mJ/(pulse·cm²) in DETA.

Table 4.
Effect of XeCl laser treatment on surface elemental composition of UHMWPE fibers

Experimental condition			C_{1s} peak (%)					O_{1s} peak (%)			N_{1s} (%)	O/C ratio
Environment	Fluence	Pulses	285±0.5 (C–C/C–H)	286±0.5 (C–O/C–N)	290±0.5 (C=O)	295±0.5 (satellite)	Carbon (%)	531±0.5 (C=O)	533±0.5 (C–O)	Oxygen (%)	399.5±0.5	
Control			58	29	0	0	87.4	1	12	12.6	0.0	0.14
Air	600	200	54	18	3	1	76.2	6	18	23.8	0.0	0.31
Air	600	600	52	17	3	1	73.1	5	22	26.9	0.0	0.37
DETA	600	200	49	24	2	0	74.1	3	22	24.5	1.4	0.33
DETA	600	600	47	21	2	0	70.4	4	22	26.0	3.6	0.37

Fluence in mJ/(pulse·cm^2), binding energy in eV.

3.4. Interfacial shear strength with epoxy resin

Figure 11 presents the 3-D plots showing the effects of fluence and number of pulses on the IFSS values for control fibers and laser-treated fibers with the epoxy resin treated in air and in DETA. The data were obtained using the single fiber pull-out technique [8]. The average IFSS value for control fibers with epoxy resin was calculated to be 1.09 MPa using equation (1). It is clear that after the laser treatment, the adhesion between the fiber and epoxy resin increased significantly and that higher fluence and more pulses result in higher IFSS values. The maximum IFSS value of 3.73 MPa, for fiber treated in air, was achieved at a fluence of 600 mJ/(pulse·cm^2) with 400 pulses in air, which is about 350% of the control value 1.09 MPa. The maximum IFSS value of 4.35 MPa for fibers treated in DETA was obtained at a fluence of 600 mJ/(pulse·cm^2) with 600 pulses, which is about 400% of the control value. However, with additional laser pulses, the fiber surface melts and, thus, the surface roughness of the fibers decreases significantly. At the same time, γ^{AB} increases only by a small amount. The combined effects of these changes result in a reduction in IFSS.

As discussed previously, the IFSS value depends on both surface roughness and surface energy, especially the acid–base component of the surface energy [12, 17–20, 40]. Figure 12a and 12b shows the relationships between IFSS and RMS, and between IFSS and γ^{AB}, respectively, for fibers laser-treated in both air and DETA. From these plots, it is seen that both relationships are almost linear. Moreover, these relationships are independent of treatment conditions. Figure 12c shows the relation between RMS and γ^{AB}. It is clear from this linear relation that both RMS and γ^{AB} increase with the laser treatment, in both air and DETA environments. This suggests that the effect seen in both Fig. 12a and 12b is the combined effect of surface roughness and γ^{AB}, since both these quantities change simultaneously. As a result, in the present case, it is difficult to separate the effect of one factor from the other. Earlier, however, Netravali and Song [41] were able to separate the effects of these two factors by treating UHMWPE fibers with XeCl laser in inert environments of argon and helium to alter their surface topography without altering the chemistry. Their results showed an increase of 100% in IFSS due to surface roughness alone. For fibers treated in ammonia, an increase of about 300% in IFSS was observed, as a result of increase in both the surface roughness and the γ^{AB}.

In general, higher UHMWPE fiber/epoxy resin IFSS values can be attributed to several factors: (1) increased surface roughness, which provides better mechanical interlocking as well as larger interface area, (2) increased acid–base component of the surface energy, which provides hydrogen bonding between the fibers and the epoxy resin and (3) low contact angle, which allows better resin spreading and, thus, better fiber/epoxy resin contact.

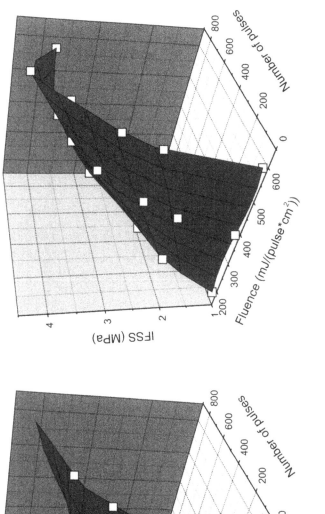

(B)

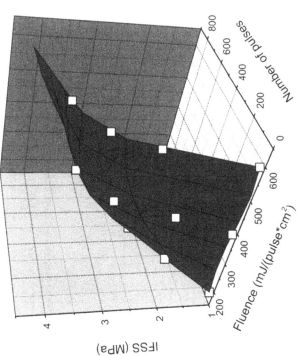

(A)

Figure 11. 3-D plots of interfacial shear strength (IFSS) *vs.* fluence and number of pulses after XeCl laser treatment: (A) in air; (B) in DETA.

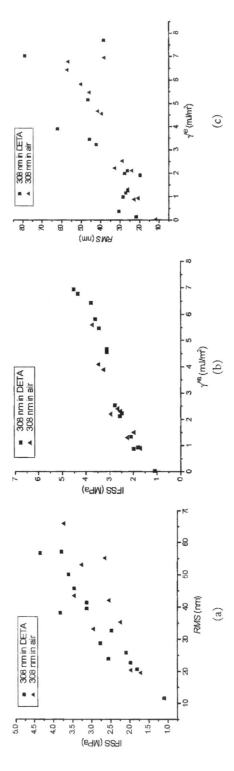

Figure 12. Plots showing relationship between IFSS, RMS and acid–base component of surface energy (γ^{AB}). (a) IFSS *vs.* RMS; (b) IFSS *vs.* acid–base component of surface energy; (c) RMS *vs.* acid–base component of surface energy.

4. CONCLUSIONS

UHMWPE fibers were treated using XeCl excimer laser in air and in DETA with different fluences and numbers of pulses. After laser treatment, the following were observed.

- Fiber surface roughness increased. The extent of roughness was a function of

  (a) Fluence and number of pulses. Higher fluence and larger number of pulses resulted in higher surface roughness due to different photoablation rates at different locations on the fiber (until surface melting) and

  (b) Treatment environment. Higher surface roughness was achieved for fibers treated in air than in DETA. For large number of pulses, the surface melting in DETA was more severe than in air.

- The acid–base component of the surface energy increased significantly after the laser treatment. Higher fluence and larger number of pulses resulted in higher acid–base component values because of more oxygen being incorporated onto the fiber surface.

- Interfacial shear strength (IFSS) increased significantly after the laser treatment. This enhancement is attributed to

  (c) Increase in surface roughness

  (d) Increase in acid–base component of surface energy and

  (e) Lower contact angle and higher work of adhesion.

Acknowledgements

The authors would like to thank Professor Michael Thompson of Cornell University for allowing the use of XeCl excimer laser facility in his laboratory, College of Human Ecology, Cornell University for funding this research, and Cornell Center for Material Research for the use of the facilities.

REFERENCES

1. R. B. Seymour and R. S. Porter, *Manmade Fibers: Their Origin and Development.* Elsevier Applied Science, London (1993).
2. F. L. Matthews and R. D. Rawlings, *Composite Materials: Engineering and Science.* Woodhead, Cambridge (1999).
3. M. Nardin and I. M. Ward, *Mater. Sci. Technol.* **3**, 814–826 (1987).
4. N. H. Ladizesky and I. M. Ward, *J. Mater. Sci.* **18**, 533–544 (1983).
5. A. R. Postema, A. T. Doornkamp, J. G. Meijer, H. Vandervlekkert and A. J. Pennings, *Polym. Bull.* **16**, 1–6 (1986).
6. S. L. Kaplan, P. W. Rose, H. X. Nguyen and H. W. Chang, *SAMPE Q.* **19**, 55–59 (1988).
7. Z. F. Li and A. N. Netravali, *J. Appl. Polym. Sci.* **44**, 319–331 (1992).
8. Z. F. Li and A. N. Netravali, *J. Appl. Polym. Sci.* **44**, 333–346 (1992).
9. S. I. Moon and J. Jang, *Composites Part A* **30**, 1039–1044 (1999).
10. S. Holmes and P. Schwartz, *Composites Sci. Technol.* **38**, 1–21 (1990).

11. A. Netravali and Z. F. Li, in: *Polymer and Fiber Science: Recent Advances*, R. E. Fornes and R. D. Gilbert (Eds.), p. 403. VCH, New York, NY (1992).

12. F. P. M. Mercx, *Polymer* **35**, 2098–2107 (1994).

13. Y. Qiu, C. Zhang, Y. J. Hwang, B. L. Bures and M. McCord, *J. Adhesion Sci. Technol.* **16**, 99–107 (2002).

14. Y. Cohen, D. M. Rein, L. E. Vaykhansky and R. S. Porter, *Composites Part A* **30**, 19–25 (1999).

15. S. Nam and A. Netravali, in: *Contact Angle, Wettability and Adhesion*, Vol. 2, K. L. Mittal (Ed.), p. 147. VSP, Utrecht (2002).

16. A. N. Netravali, J. M. Caceres, M. O. Thompson and T. J. Renk, *J. Adhesion Sci. Technol.* **13**, 1331–1342 (1999).

17. Q. Song and A. N. Netravali, *J. Adhesion Sci. Technol.* **12**, 957–982 (1998).

18. Q. Song and A. N. Netravali, *J. Adhesion Sci. Technol.* **12**, 983–998 (1998).

19. A. Netravali, Q. Song, J. M. Caceres, M. O. Thompson and T. J. Renk, in: *Polymer Surface Modification: Relevance to Adhesion*, Vol. 2, K. L. Mittal (Ed.), p. 355. VSP, Utrecht (2000).

20. J. Zeng and A. Netravali, in: *Polymer Surface Modification: Relevance to Adhesion*, Vol. 3, K. L. Mittal (Ed.), pp. 159–182. VSP, Utrecht (2004).

21. S. Lazare and R. Srinivasan, *J. Phys. Chem.* **90**, 2124–2131 (1986).

22. J. Y. Zhang, H. Esrom, U. Kogelschatz and G. Emig, *J. Adhesion Sci. Technol.* **8**, 1179–1210 (1994).

23. T. Bahners, T. Textor and E. Schollmeyer, in: *Polymer Surface Modification: Relevance to Adhesion*, Vol. 3, K. L. Mittal (Ed.), pp. 97–123. VSP, Utrecht (2004).

24. J. E. Andrew, P. E. Dyer, D. Forster and P. H. Key, *Appl. Phys. Lett.* **43**, 717–719 (1983).

25. Y. Novis, J. J. Pireaux, A. Brezini, E. Petit, R. Caudano, P. Lutgen, G. Feyder and S. Lazare, *J. Appl. Phys.* **64**, 365–370 (1988).

26. T. Bahners and E. Schollmeyer, *J. Appl. Phys.* **66**, 1884–1886 (1989).

27. T. Bahners, *Opt. Quant. Electron* **27**, 1337–1348 (1995).

28. K. K. Dezfulian, J. P. Krusius and M. O. Thompson, *Appl. Phys. Lett.* **81**, 2238 (2002).

29. R. J. Good, M. K. Chaudhury and C. J. van Oss, in: *Fundamentals of Adhesion*, L.-H. Lee (Ed.), p. 454. Plenum Press, New York, NY (1991).

30. Y. K. Kamath, C. J. Dansizer, S. Hornby and H. D. Weigmann, *Textil. Res. J.* **57**, 205–213 (1987).

31. T. Peijs, M. J. N. Jacobs and P. J. Lemstra, in: *Comprehensive Composite Materials*, Vol. 1, T.-W. Chou (Ed.). Elsevier, New York, NY (2000).

32. T. W. G. Solomons, *Organic Chemistry*, Wiley, New York, NY (1984).

33. P. Schwartz, A. Netravali and S. Sembach, *Textil. Res. J.* **56**, 502–508 (1986).

34. R. Srinivasan and V. Maynebanton, *Appl. Phys. Lett.* **41**, 576–578 (1982).

35. *Nanoscope III, Command Reference Manual for Version 2.51*, Digital Instruments (1993).

36. J. Ashok, P. L. H. Varaprasad and J. R. Birch, in: *Handbook of Optical Constants of Solids II*, E. D. Palik (Ed.). Academic Press, Boston, MA (1991).

37. R. E. Johnson Jr. and R. H. Dettre, in: *Wettability*, J. C. Berg (Ed.). p. 531. Marcel Dekker, New York, NY (1993).

38. A. Miller and P. Schwartz, *Plasmas Polym.* **2**, 115–132 (1997).

39. A. Miller, *Effect of aging on plasma treated ultrahigh strength polyethylene and the plasma treated ultrahigh strength polyethylene/epoxy interface*. Ph. D. Thesis, Cornell University, Ithaca, NY (1996).

40. Z. F. Li, A. N. Netravali and W. Sachse, *J. Mater. Sci.* **27**, 4625–4632 (1992).

41. A. N. Netravali and Q. Song, in: *Acid–Base Interactions: Relevance to Adhesion Science and Technology*, Vol. 2, K. L. Mittal (Ed.), pp. 525–537. VSP, Utrecht (2000).

Contact Angle, Wettability and Adhesion, Vol. 4, pp. 437–446
Ed. K.L. Mittal
© VSP 2006

Thin-film coating of textile materials. Part I: Permanent functionalization of textile materials with biopolymers

K. OPWIS,* M. FOUDA, D. KNITTEL and E. SCHOLLMEYER

Deutsches Textilforschungszentrum Nord-West e.V., Adlerstr. 1, D-47798 Krefeld, Germany

Abstract—An easy technique for the immobilization of biopolymers from the carbohydrate family, such as alginates, chitosans, pectins and carrageenans, onto textile materials is described. These biopolymers can be fixed onto the carrier material covalently *via* different anchor molecules. Different analysis procedures are carried out to determine the biopolymer load on the textile materials. The well-known properties of these biomolecules, such as antibacterial behavior, can be shown, even on the fabrics. Moreover, such materials when fixed onto textiles can improve regulation of micro-climate between the skin and the fabric. Several cotton T-shirts were finished and used for derma-tological studies. These shirts were evaluated by patients with atopic eczemas and were found to be very comfortable. The fabrics can be utilized in wound healing or to prevent allergic skin reactions.

Keywords: Biopolymer; carrageenan; chitosan; alginate; pectin; film; gel; cotton.

1. INTRODUCTION

Permanent functionalization of surfaces is of growing interest in textile manufacturing to develop textiles with new, innovative properties. A high functionality is needed, especially for textiles used in contact with the skin (like underwear, bed linen, or bandages). This is of concern both for disposable products, as well as for reusable products that can be found in medical areas. Textiles should have a high biocompatibility and should provide high comfort to the skin. In the case of synthetic fibers an increased hydrophilicity (water retention, sweat transport) is desirable and on natural fibers this could mean anchoring of bacteriostatic or odour binding agents. An advantageous strategy is to rely on existing fiber types (cotton, wool, polyamide, polyester) and to modify their surfaces, thus retaining the desirable mechanical properties of the bulk fiber. In addition such a strategy imparts more flexibility to the textile finishing industry [1].

Biopolymers or their derivatives fixed onto textile materials can offer special physiological properties, such as water retention, hydrogel formation or complex-

---

*To whom correspondence should be addressed. Tel.: (49-2151) 843-205;
Fax: (49-2151) 843-143; e-mail: opwis@dtnw.de

Table 1.
Interactions of biopolymers for permanent immobilization onto different fiber materials

Mode of immobilization	Fiber type				
	CO/CV	WO	PA	PES	PAN
Cross-linking	Y	N	N	N	N
Ionic interaction	N	Y	Y	N	Y
Covalent bonding	Y	Y	Y	N	N
Van der Waals interaction	N	N	Y	Y	Y

Y, possible interaction; N, no or weak interaction; CO, cotton; CV, viscose; WO, wool; PA, polyamide; PES, polyester; PAN, polyacrylonitrile.

Table 2.
Selected properties of biopolymers

Biopolymer	Property (bulk)	Effect of application on textile
Dextrin	Hydrophilicity	Water retention, regulation of micro
Alginate	Gel, film	Water retention, regulation of micro
Pectin	Gel, film	Water retention, regulation of micro
Chitosan	Film formation, hydrophilicity	Antibacterial, antifungal, wound healing
Carrageenan	Protein binding	Antiallergic

ing power. The compatibility of pure biopolymers from the carbohydrate family, such as pectins (from fruits), chitosan(s) from crab shells, or carrageenan(s) from algae, towards the skin, as well as their wound healing effect are well known. So an important R&D task is to develop techniques for fixing the biopolymers permanently onto fiber surfaces in a way that the carbohydrates retain their beneficial properties [2]. The physicochemical interactions used for fiber surface modification are summarized in Table 1 [3]. Some properties of the biopolymers (in bulk) and properties to be achieved by a permanent textile finishing are summarized in Table 2. In this study, the focus was on the use of chitosan because of its wound healing, bacteriostatic and hemostatic properties [4–6] and on carrageenan, which is able to bind proteins and to form hydrogels.

2. EXPERIMENTAL

2.1. Biopolymers and chemicals

The biopolymers κ-carrageenan (Fluka, Buchs, Switzerland), Na-alginate (Fluka), alginate (Satialgine® S 1100, SKW Biosystems, Langhorne, PA, USA), pectin from apple (Sigma-Aldrich, Taufkirchen, Germany) and chitosan (molecular mass 150 kDa, Fluka) were used for the functionalization of textile fabrics. Cyanuric chloride (Degussa, Düsseldorf, Germany), 2,4-dichlor-6-methoxy-s-triazine (synthesized from cyanuric chloride), hydroxydichloro-s-triazine sodium salt (Na-HDCT, Degussa) and butane-1,2,3,4-tetracarboxylic acid (BTCA, Sigma-Aldrich) were used as anchor molecules. For dissolving the anchor molecules the non-ionic detergent Marlipal®013/80 (Degussa) was used. Washing was done using ECE detergent according to ISO 105-C06 (1998). Polyethylenesulfonate sodium salt (PES-Na, BTG-Group, Eclepens, Switzerland) and polydiallyldimethyl ammonium chloride (Poly-DADMAC, BTG-Group) were used for polyelectrolyte titrations. 2,3,5-triphenyltetrazolium chloride (TTC, Merck, Darmstadt, Germany) was used for the TTC-test. Other chemicals such as buffers, ionic materials, or cultivating media used were of highest grade (Merck).

2.2. Fabrics

For the functionalization with biopolymers two cotton fabrics (a, desized, 102 g/m²; b, desized, mercerized, 200 g/m², both from Testex, Bad Münstereifel, Germany) and one Tencel® fabric (183 g/m², Courtaulds, Peterlee, UK) were used.

2.3. Recipes for textile finishing and analyses

A suitable amount of the biopolymer (e.g., chitosan) was dissolved in slightly molar excess of acetic acid, cooled to 0–5°C and after addition of the cyanuric chloride or dichloromethoxytriazine solution (in dioxane or acetone, about 50–100 ml organic solvent/l aqueous solution) the pH value was adjusted to 5.5 by adding NaOH solution. After warming to room temperature, drops of Marlipal® were added and the fabrics padded, squeezed and air dried. Thermal fixation was done for 5–20 min at 140°C. After fixation, the samples were washed twice at 40°C in ECE detergent solution using the Linitest apparatus. Using Na-HDCT as anchor molecule no organic co-solvent is required and in the case of polycarboxylic acids like BTCA as anchor molecules 0.6 mol/l sodium acetate trihydrate to 1 mol anchor molecules was used as a catalyst.

2.4. Characterisation of the biopolymer-treated materials with regard to accessible polymer charges (polyelectrolyte titration)

1–2 g of treated fabric was immersed in 50 ml of polyelectrolyte solution (poly(DADMAC) or PES-Na, with opposite charge to the test sample) and shaken for 2 h. Afterwards the fabric pieces were filtered off. 10 ml of the filtered solution was fed into the measuring cell of the particle charge detector (PCD 03 PH, BTG-Group, Eclepens, Switzerland). The solution was back titrated with an oppositely charged polyelectrolyte solution using the Dosimat 665 (Metrohm, Filderstadt, Germany) titration system. Zero streaming potential served as the endpoint detection. Fabrics with chitosan were pre-conditioned in buffer (pH 4.66).

2.5. Determination of ion binding properties of treated fabrics

For determination of ion binding capacity, 1 g of treated fabric was immersed in 2×10^{-4} mol/l ion solution (CuSO$_4$, ZnSO$_4$, calcium gluconate) of pH 3.5 (to prevent any metal hydroxide formation) and shaken for 5 days. Afterwards the ion content of the filtered solution was determined using conventional titrimetry (Ca^{2+}) and Atomic Absorption Spectroscopy (Cu^{2+}, Zn^{2+}) (SpectrAA-800, Varian, Darmstadt, Germany).

2.6. Microbes and measurement of antimicrobial activity of chitosan

The microbes *Escherichia coli* (*E. coli*) DSM 498, *Candida albicans* DSM 11225 and *C. krusei* ATCC 6258 were used. These actively growing broth cultures were adjusted with sterile saline to a final working concentration of 6×10^7 colony forming units (cfu) per ml.

2.7. Laser nephelometry

To determine the antimycotic influence of chitosan a NEPHELOstar Galaxy® (BMG LABTECH, Durham, NC, USA) Laser Nephelometer was used. This turbidimetric method uses Sabouraud-glucose-bouillon (SGB) as the medium and 3×10^5 cells/ml fungi as an inoculum.

2.8. Measurement of antibacterial activity with chitosan using the TTC test method

The tetrazolium/formazan redox couple (Fig. 1) is well known in biochemistry and the colour reaction can be used for the determination of the viability of microorganisms. For testing the antibacterial activity of chitosan-treated fabrics, circular swatches (diameter 3.8 cm) of treated and untreated fabric material were sterilized at 110°C. The swatches were stacked in 40 ml nutrient broth medium

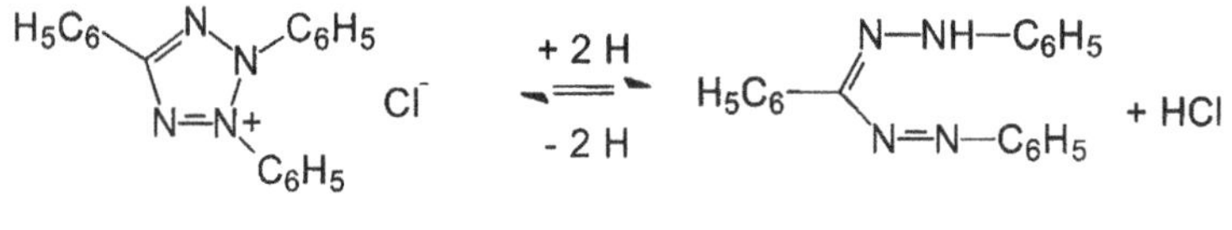

2,3,5-triphenyl-2H-tetrazol-3-ium

no colour

1-phenyl-2-[(Z)-
phenyl(phenyl)hydrazono)methyl]diazene

red

Figure 1. The tetrazolium/formazan redox couple.

containing 10 μl *E. coli* (10^8 cfu/ml) and were shaken at 37°C for 3–4 h. Then 1 ml samples from all flasks containing the test fabrics and the control were added to sterilized test tubes containing 100 μl TTC (0.5 wt%). All tubes were incubated at 37°C for 20 min. The resulting formazan was centrifuged at 4000 rpm for 3 min, followed by decantation of the supernatant. The pellets obtained were re-suspended and centrifuged again in ethanol. The red formazan solution obtained at the end which indicated the activity and viablility of the cells was measured photometrically at 480 nm.

2.9. Protein binding/adsorption

Adsorption measurements of albumin on carragheenan-treated cotton were carried out with a 0.1 wt% protein solution. After immersing the fabric in the solution and after washing, adsorbed protein can be visualized by ninhydrin reagent. Quantitative assay of protein binding was done using the Lowry test and gave about 30–40 mg protein per g biopolymer add-on (about 2 mg protein/g fabric).

2.10. Dermatological evaluation of biopolymer finished cotton

Several cotton T-shirts were finished and used for dermatological studies (2.3 wt% alginate, 4.0–6.2 wt% pectin, 3.7 wt% chitosan, 0.6–1.9 wt% carrageenan). These shirts were evaluated by patients with atopic eczema and were found to be very comfortable. None of the carbohydrate finished fabrics produced skin irritation.

3. RESULTS AND DISCUSSION

3.1. Thermal fixation of biopolymers onto cotton

As described in Table 1 different methods for fixation of biopolymers onto different types of textiles are possible. The most permanent functionalization can be achieved by covalent binding of the biopolymers using an anchor molecule. In the interest of a simple immobilization technique, any multi-step procedures should be avoided. Permanent fixation of biopolymers onto cellulosic material can be done by *in situ* generation of the active, anchorable biopolymer by using cyanuric chloride or other bifunctional spacer molecules such as the sodium salt of hydroxyl-dichloro-triazine into solutions of the biopolymer at temperature of 0–10°C. The resulting solutions can be used without isolation of the modified biopolymer for the impregnation of the fabrics [3–5]. Permanent fixation onto fiber surfaces follows the usual recipes for reactive dyeing of cellulosic material. Other strategies use biopolymer derivatives having hydrophobic chains for the treatment of synthetic fibers (following disperse dyeing conditions).

Table 3.
Thermal fixation of chitosan and carrageenan on cotton fabrics (200 g/m^2) in the presence of cyanuric chloride at 140°C

Biopolymer	c (biopolymer) (g/l)	c (cyanuric chloride) (g/l)	Add-on reached after washing (wt%)
Chitosan 5 (Fluka)	4.4	0.47	0.8
Chitosan 8 (Fluka)	4.4	1.2	2.5
Chitosan 11 (MBP 21)	6.2	2.0	3.2
Chitosan 12 (MBP 21)	12.4	4.0	3.7
Carrageenan 3 (BDC20)	15.0	2.0	1.2
Carrageenan 5-2	12.0	2.0	1.8
Carrageenan 6-1	12.0	2.0	2.2

In Table 3 results are presented for thermal immobilization of the model compounds chitosan (Chi) and carrageenan (Car) on cotton. From the preliminary experiments we found a fixation temperature of 120°C or more as being effective; therefore, for systematic investigations a fixation temperature of 140°C was used.

3.2. Antifungal and antibacterial effects of chitosan-finished fabrics

The antifungal effect of dissolved chitosan was studied by time-dependent nephelometric measurements on *C. krusei* cultures in the presence of different amounts of chitosan. Figure 2 shows the growth of *C. krusei* up to 24 h. Without

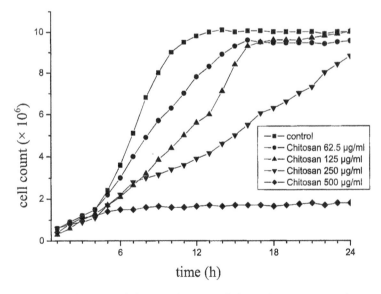

Figure 2. Influence of chitosan on *C. krusei* cultures (nephelometric measurements).

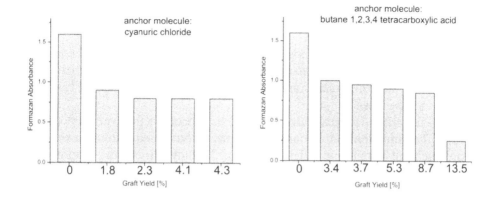

Figure 3. Effect of chitosan grafting onto cotton on the reduction of *E. coli* using the TTC test method.

addition of chitosan the maximal cell population is reached after 12–14 h and does not change significantly after that. With increasing amounts of chitosan the growth of the fungi culture is inhibited: With 62.5 µg/ml or 125 µg/ml the maximum is only achieved after 16–17 h; with 250 µg/ml the growth continues even after 24 h. Using a concentration of 500 µg/ml chitosan no fungi growth could be observed.

The antibacterial effect of cotton fabrics finished with chitosan using different anchor molecules can be evaluated with the TTC test method. Tetrazolium salts and formazans have been well known in biochemistry and histochemistry for many years. The tetrazolium/formazan couple is a special redox system acting as a proton acceptor or as an oxidant. Tetrazolium salts are positively charged yellow dyes and enter the cell where they are reduced to a lipid-insoluble purple formazan by cleavage of the tetrazolium ring due to the activity of dehydrogenase enzymes in the cytosol and the mitochondria of microorganisms. Only living cells are able to reduce the TTC. In the presence of antibacterial agents the bacterial activity is inhibited and thus the formazan generation decreases [7, 8].

Figure 3 illustrates the antibacterial effect of chitosan finished fabrics. The *E. coli* population is reduced because of the smaller amount of formazan generated. Using cyanuric chloride as an anchor molecule, 1.8 wt% chitosan fixed onto cotton yields a reduction of the *E. coli* population of about 50%. Higher chitosan loads do not lead to higher reductions. Using the anchor molecule butane 1,2,3,4-tetracarboxylic acid, the add-on of chitosan (3.4–8.7wt%) strongly decreases the population of *E. coli*. A cotton fabric grafted with 13.5 wt% chitosan reduces the formazan absorbance more than 80%.

3.3. Polyelectrolyte titrations and ion exchange properties of biopolymer finished fabrics

Polyelectrolyte measurements on fabrics finished with biopolymers can be used to assess the availability of ionic charges on the fabric which gives information on the (statistical) sites of anchors and the possible chain segment mobility of the biopolymer on the fiber.

Figure 4 illustrates schematically the analysis procedure for available charges on textile. The immobilized anionic biopolymer is immersed into a known cationic polyelectrolyte solution. After binding of the polycations, the solution with the unused polyelectrolyte is back titrated with an anionic polyelectrolyte.

Figure 5 and Table 4 show some exemplary results of such polyelectrolyte titrations with carragheenan fixed onto cotton and Tencel. The results indicate that through the use of anchoring chemicals (mainly cyanuric chloride) only about one-third of potential charge equivalents remain accessible after fixation onto textile (Table 4). It can be shown that by varying the amount of anchoring molecule used, a higher charge accessibility can be reached using a lower add-on. The charge accessibility related to the amount of permanent add-on can be increased strikingly using less anchoring chemicals but with the risk of only lower coverage and, thus, lower total binding capacity of the treated fabric.

By finishing textiles with ionic carbohydrates an increase in surface humidity can be expected. This can be seen by the change in surface electrical resistivity which is decreased by an order of magnitude vs. the original cotton. Only with a carrageenan treatment a distinctly higher value for water retention is obtained (74% vs. the original value of 39%). This shows good gel forming properties of this biopolymer.

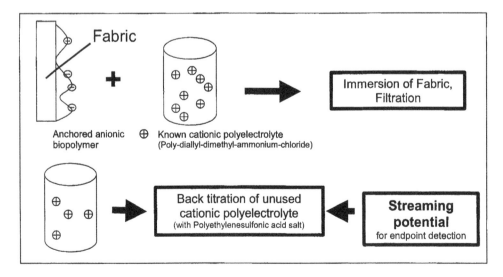

Figure 4. Analysis procedure for available biopolymer charges on textile.

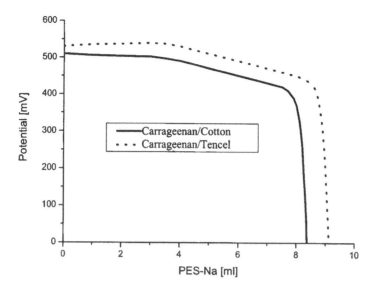

Figure 5. Polyelectrolyte titrations of fabrics finished with carrageenan (without carrageenan = 10 ml polyethylenesulfonic acid sodium salt, PES-Na).

Table 4.
Results of polyelectrolyte titrations on treated fabrics for evaluation of accessible charge of the biopolymer layer

Biopolymer	Fabric	Add-on (wt%)	Milliequivalents per g fabric treated	Milliequivalents per g biopolymer add-on
Alginate	CO	1.4	2.5	1.7
Alginate	CV	1.45	0.8	0.055
Carrageenan	CO	2.1	1.6	0.07

For comparison the maximal available ionic charges in solution of the pure biopolymers used were 4.6 milliequiv/g (alginate) and 2.4 milliequiv/g (carrageenan).

Table 5.
Binding capacity of biopolymer finished cotton fabrics for bivalent cations

Biopolymer	mg biopolymer per g cotton	bound Cu^{2+} per g fabric (μmol/g)	bound Cu^{2+} per g biopolymer add-on (mg/g)
None	–	1.7	–
Alginate	14	24.6	110
Carrageenan	34	14.5	27
Chitosan	47	25.2	34
Pectin	26	35.1	86

Since the biopolymers used for surface modification bear ionic functionalities (depending on the pH value) they may be used for the complexation of heavy metal ions. Some investigations on the exchange capacity of biopolymer finishes have been carried out and the results are presented in Table 5. Using the data for binding of small ions such as Cu^{2+} or Zn^{2+} for 14 mg add-on of alginate per gram cotton fabric 4×10^{-5} equivalents are estimated as possible accessable binding sites per gram.

3.4. Dermatological evaluation of biopolymer finished cotton

Several cotton T-shirts were finished and used for dermatological studies (2.3 wt% alginate, 4.0–6.2 wt% pectin, 3.7 wt% chitosan, 0.6–1.9 wt% carrageenan). These shirts were evaluated by patients with atopic eczemas and were found to be very comfortable. None of the carbohydrate finished fabrics caused skin irritation. Dermatological investigations will be continued and the effect of the molecular weight of the biopolymers used will be elucidated in more detail in a future work.

4. CONCLUSIONS

The method presented here for permanent finishing of fiber surfaces with thin bio-polymer layers offers promising aspects for medical use of textiles near the skin. After a covalent fixation of different biopolymers from the carbohydrate family *via* anchor molecules the treated fabrics have shown antibacterial or ion-exchange properties, as well as an increased water retention. Biopolymer finished fabrics can be used as special clothes for patients with sensitive skin or in wound healing.

Acknowledgements

We would like to thank the Forschungskuratorium Textil e.V. for funding these research projects (Aif-No. 11652 N, 11913 N, 13519). The projects were funded with financial resources of the Bundesministerium für Wirtschaft und Arbeit (BMWA) with a grant of the Arbeitsgemeinschaft industrieller Forschung-svereinigungen "Otto-von-Guericke" e.V. (AiF).

REFERENCES

1. D. Knittel and E. Schollmeyer, *J. Textile Inst.* **91**, 151–165 (2000).
2. D. Knittel, R. Stehr and E. Schollmeyer, German Patent 196 24 170.7 (1996).
3. D. Knittel and E. Schollmeyer, *Textilveredlung* **35**, 29–31 (2000).
4. D. Knittel and E. Schollmeyer, *Adv. Chitin Sci.* **4**, 143–147 (2000).
5. D. Knittel and E. Schollmeyer, *Melliand Textilber.* **83**, 15–16 (2002).
6. E. I. Rabea, M. E. Badawy, C. V. Stevens, G. Smagghe and W. Steurbaut, *Biomacromolecules* **4**, 1457–1465 (2003).
7. F. P. Altman, *Prog. Histochem. Cytochem.* **9**, 6–52 (1976).
8. E. Seidler, *The Tetrazolium-Formazan System: Design and Histochemistry*, pp. 1–86. Gustav Fischer, Stuttgart (1991).

Contact Angle, Wettability and Adhesion, Vol. 4, pp. 447–460
Ed. K.L. Mittal
© VSP 2006

Thin film coating of textile materials. Part II: Enzyme immobilization on textile carrier materials

K. OPWIS,* D. KNITTEL, T. BAHNERS and E. SCHOLLMEYER

Deutsches Textilforschungszentrum Nord-West e.V., Adlerstr. 1, D-47798 Krefeld, Germany

Abstract—Catalase was covalently immobilized on textile carrier fabrics made of cotton, polyamide and poly(ethylene terephthalate) by different techniques using anchor and cross-linking molecules, or photochemical activation. Depending on the procedure used up to 50 mg catalase could be immobilized on 1 g of the carrier material. Even in solutions with high substrate concentrations (e.g., 6.0 g/l H_2O_2) the immobilized enzyme is reusable more than 20 times and the integral activity after all reuses is still much higher than the activity of the free catalase, which can be used only once. Besides their low costs, natural and synthetic fibers have additional advantages compared to other carrier materials such as their easy handling and their flexible construction.

Keywords: Enzyme immobilization; photochemical immobilization; excimer-UV lamp; enzymes; catalase; fabric; cotton; polyamide; PET.

1. INTRODUCTION

Enzymatic processes are of increasing interest in modern catalysis with widespread applications. Enzymes are used on a large scale, e.g., in food and textile industries, as additives in washing agents, also as diagnostic and therapeutic materials for medical applications, as well as in analytical chemistry. The use of enzymes has many advantages compared to conventional, non-enzymatic processes. Enzymes can be used in catalytic concentrations at low temperatures and at pH-values close to neutral. The world-wide sale of industrial enzymes was estimated to be US$ 2.2 billion in 2002 [1] and is expected to reach US$ 3.0 billion in 2008 [2].

An important step for minimizing the costs for enzyme application can be the immobilization of these biocatalysts on suitable carrier materials. In the literature many methods for enzyme immobilization on different carriers are described. The binding can be of adsorptive, ionic or covalent nature [3].

---

*To whom correspondence should be addressed. Tel.: (49-2151) 843-205;
Fax: (49-2151) 843-143; e-mail: opwis@dtnw.de

Here two techniques for immobilization of enzymes on textile carrier materials are studied, taking the iron containing protein catalase as an example. Besides conventional immobilization techniques, using anchor molecules, an easy and inexpensive photochemical method for the fixation of enzymes is reported here.

Oxidoreductase catalase catalyses the disproportionation of hydrogen peroxide into oxygen and water in human and animal metabolism. This enzyme is also of technological interest. Free or immobilized catalase products are used, for example, in textile bleaching process before the dyeing step [4, 5] or after milk sterilisation to remove the excess H_2O_2 [6].

The use of textile carrier materials has several advantages. Textile fabrics are very cheap compared to other commercially available inorganic and organic carrier materials, such as aluminum silicates, porous glass, or various hydrophilic polymers. The flexible and open structure of fabrics guarantees an optimal substrate conversion. It allows reactor construction of any geometry and a quick removal of the catalyst without any residues after the enzymatic reaction.

UV light is able to generate radicals on UV-light-absorbing synthetic polymers, such as poly(ethylene terephthalate) (PET), by homolytic bond cleavage. Because of the small penetration depth the monochromatic light as emitted from excimer-UV lamps causes no significant bulk polymer damage [7]. Suitable functional agents can react with the generated radicals yielding grafted products [7, 8]. Using bifunctional agents, a cross-linked coating can be produced on the polymer surface [9, 10]. Besides their cross-linking properties, the reactive compounds were used as anchor molecules between the textile fabric and the enzyme in this work.

2. EXPERIMENTAL

2.1. Materials

Catalase from bovine liver (Fluka) was used (MW = approx. 240 kg/mol, activity 2500 U/mg). Desized, alkaline-scoured and bleached cotton (CO) fabric (plain weave, 102 g/m^2) and a washed polyamide 6 (PA 6) fabric (plain weave, 65 g/m^2) were used for the conventional immobilization. For the photochemical immobilization, commercial poly(ethylene terephthalate) (PET) and polyamide 6.6 (PA 6.6) fabrics were used. The fabrics were extracted before use (Soxhlet, ethanol/hexane = 20:80 (v/v), 4 h).

2.2. Conventional immobilization on CO and PA 6

2.2.1. Activation of the carrier materials

2.2.1.1. *Cotton.* 8.0 g CO fabric was covered with 30 ml NaOH solution (5 wt%). After 15 min, 270 ml water and 0.05 g sodium dodecyl sulfate (SDS, Fluka) were added. The pH value reached 12.5 to 13. After adding 3.0 g cyanuric chloride (Fluka) the solution was stirred for 3 h at 25°C. Cyanuric chloride was completely dissolved and the pH value decreased to 11.5 to 12. The cotton was washed several times with distilled water and air dried.

2.2.1.2. Polyamide 6. 8.0 g PA 6 fabric was stirred for 2 h at 25°C in 100 ml glutaraldehyde solution (5 wt%, Fluka). The fabric was washed several times with distilled water and air dried.

2.2.2 Enzyme immobilization

2.2.2.1. Adsorptive fixation on non-activated fabrics. Untreated CO or PA 6 fabrics (4.0 g) were stirred for 24 h at 25°C in 100 ml buffer solution (citrate/NaOH, pH 6), which contained 400 mg dissolved catalase. The fabrics were washed several times with distilled water.

2.2.2.2. Covalent enzyme fixation on activated fabrics with or without additional cross-linking. Activated CO or PA 6 (see Section 2.2.1) fabrics (4.0 g) were stirred for 24 h at 25°C in 100 ml buffer solution (citrate/NaOH, pH 6), which contained 400 mg dissolved catalase. For additional cross-linking 1.0 ml glutaraldehyde solution (25 wt%) was added after 2 h. The fabrics were washed several times with distilled water.

2.3. Photochemical immobilization on polyester and polyamide 6.6

2.3.1. Preparation of reactive medium emulsion containing catalase and fabric wetting

Diallylphthalate (DAP, Acros Organics) and cyclohexane-1,4-dimethanoldivinylether (CHMV, Merck) were used as reactive media. Starting with 94.5 ml distilled water, 5.0 ml of the reactive medium was emulsified by adding 0.5 ml Marlipal® surfactant (Hüls) under stirring. 50.0 mg catalase was added to 3.0 ml of the DAP or CHMV emulsion. 1.0 g of the PET or PA 6.6 fabric was wetted with the reactive medium emulsion-catalase system. For comparison, experiments were also carried out without reactive media.

2.3.2. Light source and irradiation

A KrCl*-excimer lamp (BlueLight BLC 222/300, Heraeus Noblelight), emitting nearly monochromatic light at 222 nm, served as the UV-source. The fabrics were irradiated in a steel reactor with a transparent polyethylene window. The irradiation took place in an argon atmosphere to avoid photo-oxidation by air oxygen. The distance between the light source and the sample was 8 cm. The irradiation time was 5.0 min for each fabric side. After the irradiation, the samples were stirred for 0.5 h in 100 ml water containing 0.5 vol% Marlipal. After this treatment the samples were washed five times with 100 ml distilled water.

2.4. Analyses and enzymatic reactions

Activation reactions were analyzed by qualitative and quantitative methods (silver nitrate for chlorine, Schiff reagent for aldehydes [11] and elemental analysis for nitrogen [12]). The immobilization of catalase on textile carrier materials was analyzed qualitatively using the ninhydrin test [13], FT-IR spectroscopy (FTS-45, Bio-Rad) with an ATR-unit (Silver Gate), and UV-Vis spectroscopy (Cary 5E). Scanning electron microscopy (SEM) images of fabrics were acquired using a Topcon microscope ATB-55. The catalase load on the fabrics was determined quantitatively by atomic absorption spectroscopy (SpectrAA 800, Varian) after chemical decomposition in suitable concentrated acids by measuring the iron concentration of these solutions [14]. The enzymatic reactions with immobilized catalase were carried out at 25°C in 50 ml hydrogen peroxide solution ($c = 6.0$ g/l). After 1 min, the H_2O_2 concentration was analyzed quantitatively by high-performance anion-exchange chromatography (HPAEC-PAD, Dionex). The decay of hydrogen peroxide was a measure of the activity of the immobilized catalase [12]. Twenty reuses were carried out (each in fresh hydrogen peroxide solution, 50 ml, $c = 6.0$ g/l).

3. RESULTS

3.1. Conventional immobilization of catalase on textile carrier materials using anchor molecules and cross-linking agents

3.1.1. Conventional immobilization

The covalent immobilization of catalase was studied on fabrics made of cotton and PA 6. In both cases the chemical properties of the materials (reactive groups) were used to activate the carriers by chemical modification with bifunctional anchor molecules.

Cleaned and bleached cotton consists of cellulose, which is constituted of β-1,4-bonded, linear glucose chains. Each glucose unit possesses hydroxyl groups, which can be deprotonated easily in the presence of strong bases. In a second step, this deprotonated cotton reacts with cyanuric chloride (2,4,6-trichloro-1,3,5-triazine). Figure 1a (top) illustrates schematically this nucleophilic substitution of cyanuric chloride (Cy) on cotton. The three chlorine atoms show different reactivities. A quantitative analysis of the residual chlorine amount after the substitution step shows that on each triazine anchor molecule only one chlorine atom remains [12]. An easy qualitative test for the covalent fixation of cyanuric chloride on cotton is the color reaction with silver nitrate solution showing violet or black spots on the fabric in the presence of chloride after light exposure (light sensitivity of silver chloride).

PA 6 is a synthetic fiber with a degree of polymerization between 150 and 200. The amino groups at the chain ends are suitable for activation. For these studies glutaraldehyde (GDA) was used for activation, which reacts by nucleophilic addition with PA 6 (Fig. 1b, top). After this reaction the anchor molecule still carries one free aldehyde function, which is able to react with an amino group of the enzyme [15]. The success of the reaction can be determined by a qualitative test with the Schiff reagent [11].

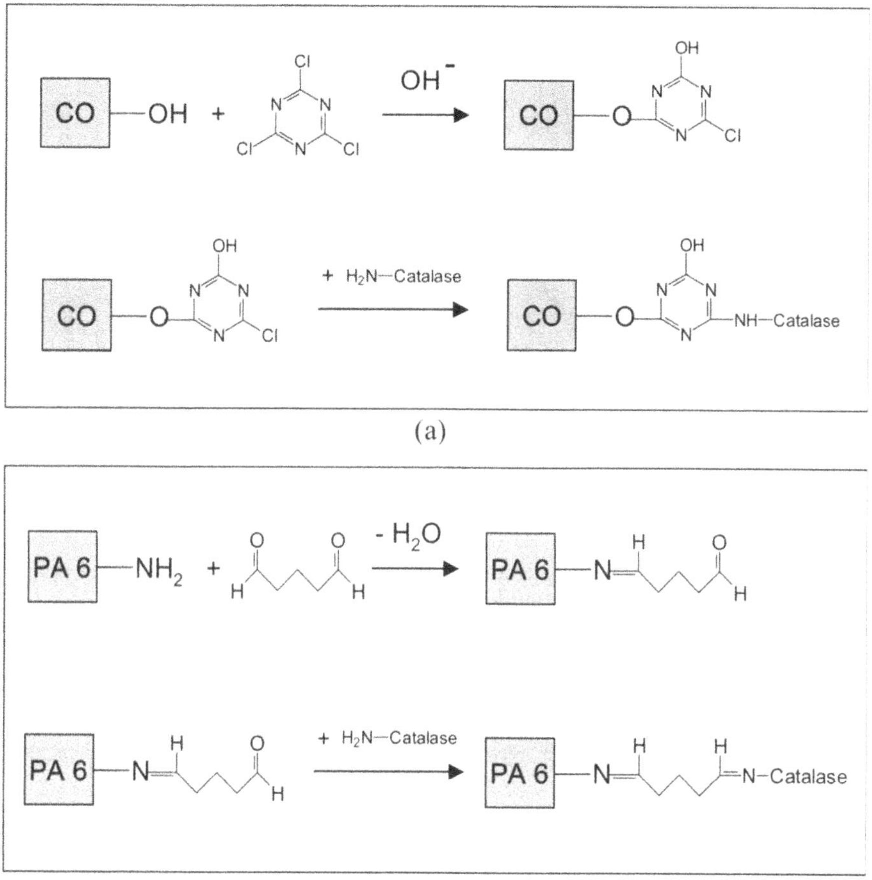

(a)

(b)

Figure 1. (a) Covalent immobilization of catalase on cotton (CO) using cyanuric chloride as anchor molecule (schematic). (b) Covalent immobilization of catalase on polyamide 6 (PA 6) using glutaraldehyde as anchor molecule (schematic).

In a second step, the catalase is fixed covalently to the activated fabrics. Figure 1 shows schematically the reaction of the enzyme with the activated cotton and PA 6. A covalent bond with the anchor molecule (cyanuric chloride or glutaraldehyde) can only be realized in a monomolecular layer on the carrier surface. By additional cross-linking of the immobilized catalase with glutaraldehyde a three-dimensional structure covering the fiber material is created and the total enzyme load is much higher than without cross-linking [16]. Figure 2 shows SEM pictures of cotton fabrics before and after immobilization of catalase using cyanuric chloride as anchor molecule and glutaraldehyde for cross-linking. One can see distinctly the coverage of the fibers with the cross-linked catalase.

Catalase is an iron containing enzyme, so its load can be analyzed quantitatively after chemical decomposition by atomic absorption spectroscopy (AAS)

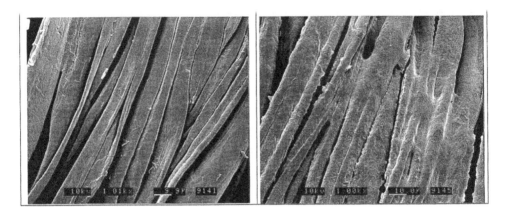

Figure 2. SEM pictures of cotton fibers before (left) and after (right) covalent immobilization of catalase with additional cross-linking.

Table 1.
Load of conventionally immobilized catalase on cotton and PA 6 fabrics using different techniques and loss of catalase by desorption after ultrasonic treatment

Carrier material	Immobilization product	Catalase load (mg/g carrier)	Catalase load after ultrasonic treatment (mg/g carrier)	Loss by desorption (%)
CO	CO-catalase$_{ads}$	9.6	1.9	80.2
	CO-Cy-catalase	17.7	9.6	45.8
	CO-Cy-catalase (cross-linked)	48.7	41.5	14.8
PA 6	PA 6-catalase$_{ads}$	2.1	0.2	90.4
	PA 6-GDA-catalase	22.4	17.0	24.1
	PA 6-GDA-catalase (cross-linked)	50.4	40.7	19.2

[14]. The results are summarized in Table 1. The non-covalent, adsorptive loads are also shown for comparison.

In contrast to PA 6, cotton adsorbs a relatively high amount of catalase, but with covalent fixation the total load can be doubled compared to adsorptive fixation; in the case of PA 6, the covalently bonded amount is even more than 10-times higher. The strength of the different bonds was examined by desorption experiments after ultrasonic treatment [16]. 80% of the catalase, which is adsorptively fixed onto CO, desorbs; in the case of PA 6 even more than 90% is desorbed. If the catalase is fixed covalently on activated fabrics the loss by desorption due to ultrasonic treatment is much lower. With additional cross-linking 50 mg catalase/g fabric can be immobilized on both carrier materials and more than 80% of the catalase load remains on both materials after ultrasonic treatment.

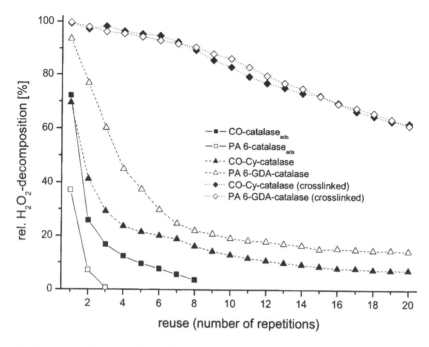

Figure 3. Enzymatic decomposition of hydrogen peroxide with differently immobilized catalase (50 ml H_2O_2 solution, starting concentration 6.0 g/l = 100%, 0.4 g fabric, $T = 25°C$, pH 7, reaction time $t = 1$ min, max. reuse 20 times).

3.1.2. Activity and reuse of conventionally immobilized catalase

The reuse capacity of immobilized enzymes depends on the total load, the relative activity and the strength of fixation (permanence against desorption). The enzyme catalase catalyzes the disproportionation of hydrogen peroxide into oxygen and water. Figure 3 shows the results of reuse experiments on H_2O_2 solutions with catalase immobilized on CO and PA 6 using different immobilization techniques. Adsorptively fixed catalase desorbs very fast and in the case of PA 6 the reactivity after two uses already reaches nearly zero. The covalent fixation of catalase on CO and PA 6 leads to strongly increased activity and both are showing activity even after twenty uses. Using the covalent immobilization technique with additional cross-linking both the total load (approx. 50 mg enzyme per g carrier material; *cf.*, Table 1), and the resistance of the enzyme against desorption increase significantly. This results in very effective immobilizates, which are reusable in both cases (CO and PA 6) more than twenty times with a substrate conversion of more than 60%.

3.2. Photochemical immobilization of catalase on textile carrier materials

3.2.1. Photochemical immobilization

UV light is able to cleave bonds in UV-light-absorbing materials such as PET, enzymes or cross-linking agents yielding radicals. These radicals can react with

neighboring radical species forming new covalent bonds. In the presence of bi
functional agents and catalase a cross-linked structure surrounding the polymer
material is achieved. For these studies, bifunctional cross-linking agents diallyl
phthalate (DAP) and cyclohexane-1,4-dimethanoldivinylether (CHMV) were used
[17]. The procedure for photochemical immobilization is divided into three steps,
which are schematically described in Fig. 4: wetting, irradiation and washing. The
polymers PET and PA 6.6 are wetted with a surfactant based aqueous emulsion of
the reactive cross-linking medium and catalase. After the irradiation, the samples
must be washed to remove non-covalently bonded catalase and cross-linking
agent. The photochemical reactions during irradiation are separately described in
Fig. 5 a–c. In the first step the monochromatic 222 nm UV light generates radicals
on the polymer surface, as well as on the enzyme molecules. Using argon as the
atmosphere the photooxidation of the polymer and also oxidative damage to the
catalase can be avoided (Fig. 5a). In the second step the addition of the radicals to
double bonds of the cross-linking agents takes place (Fig. 5b). In the last step the
bifunctional reactive chemicals form a cross-linked layer covering the polymer
surface (Fig. 5c), which can be seen in the SEM micrographs (Fig. 6). Using different textile carrier materials (PET and PA 6.6) and different cross-linking agents
(DAP and CHMV) in all cases one can see distinctly a three-dimensional layer of
the cross-linked catalase.

   An easy qualitative test for the successful fixation of enzymes, such as catalase,
is the colour reaction with ninhydrin [13]. In the case of treated PET the reaction
is positive. The test does not work for PA 6.6 due to the fact that the amino
groups of this polymer themselves react with ninhydrin. Therefore, UV-Vis spectroscopy was used as a qualitative test for the immobilization of catalase on PA
6.6. Figure 7 shows spectra of different PA 6.6 fabrics. The untreated material and

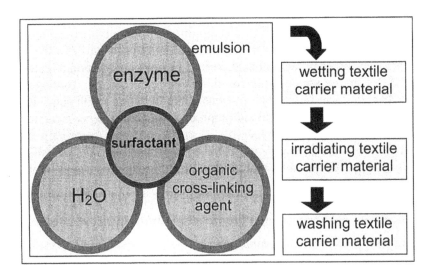

Figure 4. Steps for the preparation of enzyme immobilizates.

the sample treated only with the cross-linking agent CHMV show no absorbance near 410 nm, where the catalase has an absorption maximum. Following irradiation in the presence of catalase a significant signal is visible in this range showing that photochemical immobilization of catalase on PA 6.6 has taken place also without a cross linking agent. A higher yield of immobilized catalase can be achieved using the enzyme in combination with the cross-linking agent; the absorbance of the polymer carrier at 410 nm is strongly enhanced.

Further qualitative evidence for enzyme immobilization on textile carrier materials is offered by FT-IR-spectroscopy using the ATR-technique. Example spectra of PET fabrics before and after UV irradiation are shown in Fig. 8. The irradiation in the presence of the cross-linking agent DAP causes a significant change in the spectrum between 1200 cm^{-1} and 1300 cm^{-1}. The experiment in the presence of catalase without DAP yields no detectable changes in the spectra. Without a cross-linking agent no catalase immobilization on PET takes place. Using a combination of catalase and the cross-linking agent DAP the spectrum shows two characteristic signals of the enzyme between 1500 cm^{-1} and 1700 cm^{-1} (see KBr spectra of catalase powder) and also differs between 1200 cm^{-1} and 1300 cm^{-1} due to bonded DAP.

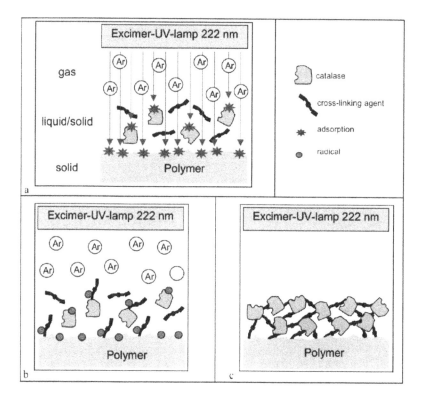

Figure 5. Photochemical immobilization of catalase on textile carrier materials. (a) UV irradiation and radical generation in an inert atmosphere; (b) addition; (c) cross-linking.

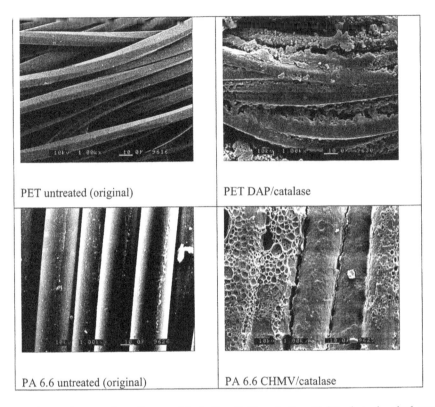

PET untreated (original)

PET DAP/catalase

PA 6.6 untreated (original)

PA 6.6 CHMV/catalase

Figure 6. SEM pictures of textile materials before (left) and after (right) photochemical enzyme immobilization.

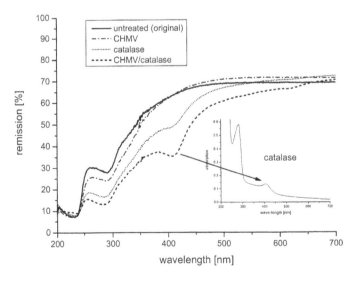

Figure 7. UV-Vis spectra of PA 6.6 fabrics before and after UV irradiation in presence of the cross-linker CHMV, catalase and both together. Inset: UV-Vis spectrum of dissolved free catalase.

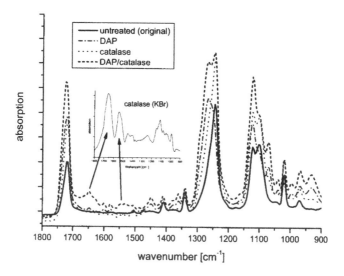

Figure 8. FT-IR spectra (ATR) of PET fabrics before and after UV irradiation in presence of the cross-linker DAP, catalase and both together. Inset: IR spectrum (KBr) of pure catalase.

Table 2.
Load of photochemically-immobilized catalase on PA 6.6 and PET fabrics (as determined from iron analysis by atomic absorption spectroscopy)

Carrier material	Cross-linking agent	Catalase load (mg/g carrier)
PA 6.6	–	6.2
	DAP	20.8
	CHMV	22.0
PET	–	1.3
	DAP	32.2
	CHMV	23.9

In the case of photochemical immobilization on PET or PA 6.6 the fixed catalase can also be analyzed quantitatively by atomic absorption spectroscopy due to its iron content. The results are summarized in Table 2. The data correspond well to the qualitative measurements. 6.2 mg catalase per gram PA 6.6 is immobilized using no cross-linking agent. The yield is more than three times higher using DAP or CHMV. UV-Vis spectroscopy shows similar results (see Fig. 7). Only 1.3 mg catalase is immobilized on 1 g PET. This amount is not high enough for detection by IR-spectroscopy (see Fig. 8). With the cross-linking agent DAP the highest yield was reached (32.2 mg/g PET), this relates to 64% of the catalase input.

3.2.2. Activity and reuse of photochemically-immobilized catalase
Covalent immobilization of enzymes is accompanied by a decrease of activity because of the loss of mobility of the enzyme's conformation due to the fixation of

Table 3.
Relative activity of photochemically immobilized catalase after twenty reuses in comparison to free catalase with an activity of 2300 U/min

Carrier material	Cross-linking agent	Catalase load (mg/g carrier)	Conversion after 20 reuses (mg H_2O_2)	Conversion per mg catalase (mmol/min)	Relative activity (%)
PA 6.6	–	6.2	1.41 (16%)	0.23	10.0
	DAP	20.8	8.73 (99%)	0.42	18.3
	CHMV	22.0	5.74 (65%)	0.26	11.3
PET	–	1.3	0.09 (1%)	0.07	3.0
	DAP	32.2	7.24 (82%)	0.22	9.6
	CHMV	23.9	6.26 (71%)	0.26	11.3
Free catalase				2.30	100.0

1 mg free catalase decomposes 2.3 mmol H_2O_2/min. Start: $n(H_2O_2)$ = 8.82 mmol (300 mg) = 100%.

Table 4.
Integral activity of photochemically immobilized catalase after 20 reuses in comparison to free catalase

Carrier material	Cross-linking agent	Relative activity (%)	Number of reuses	Integral activity (%)
PA 6.6	–	10.0	20	≥ 200
	DAP	18.3	20	≥ 366
	CHMV	11.3	20	≥ 226
PET	–	3.0	20	≥ 60
	DAP	9.6	20	≥ 192
	CHMV	11.3	20	≥ 226
Free catalase		100	1	100

the bonding points to the carrier material. Moreover, the covalent bonds impair the ability of the enzyme to form an enzyme–substrate complex (relative activity = activity of immobilized catalase/activity of free catalase). A successful enzyme immobilization must provide a minimum of activity loss and a maximum of possible reuses. The aim of enzyme fixation in economic terms is an increased activity over all reuses (integral activity = relative activity×number of reuses, assuming rel. activity is approx. constant). By measuring the enzymatic decomposition of hydrogen peroxide as a function of time it is possible to calculate the relative activity of the immobilized enzyme in comparison to free catalase. The enzyme immobilizates obtained with catalase using the cross-linker DAP or CHMV show a significant activity even after 20 reuses (see Table 3). The bonded catalase loses 80–90% of its activity. Nevertheless the integral activity after twenty reuses is much higher than the activity of free catalase, which could only be used once (see

Table 4). The best result was obtained by immobilizing catalase on PA 6.6 using DAP as cross-linking agent. The integral activity increases to 366% after twenty reuses compared to free catalase.

4. CONCLUSIONS

Low-cost textile fabrics, made of cotton, polyester or polyamides, are alternative carrier materials for the immobilization of enzymes. With a low preparative and economic expense fabrics with a high enzyme load, a high relative activity and good permanence against enzyme desorption can be produced using anchor molecules and glutaraldehyde or a photochemical process with excimer-UV lamps in the presence of bifunctional cross-linking agents. The flexible and open construction of fabrics allows a high substrate conversion. Moreover, conventional immobilization products are mostly produced as granulates or pellets, which must be filtered after the enzymatic reactions. Fabrics can be removed very quickly from the reactor without any filtration step and without any enzyme residues after the enzymatic reaction.

The different immobilization techniques were carried out with catalase and up to 50 mg of the enzyme could be fixed per gram textile. Even in solutions with high hydrogen peroxide concentration (6.0 g/l) the immobilized enzyme is reusable more than twenty times, retaining a high activity. Catalase can be used, e.g., in textile manufacturing after bleaching to destroy residual hydrogen peroxide, which otherwise impairs the following dyeing process. The catalase treatment decreases considerably the amount of washing water needed. Unfortunately, the free enzyme can be used only once, because the treated liquor must be rejected before the next manufacturing batch can be started. Immobilized catalase can be taken out of the reaction bath and is, therefore, reusable in the next batch.

The investigations in this field of enzyme immobilization on textile carrier materials are not yet finished and the procedures seem to be transferable to many other enzymes, opening widespread applications in biocatalysis in the future.

Acknowledgements

We wish to thank the Ministry for Science and Research of the country of Northrhine-Westphalia (Germany) for financial support. This support was granted within the project DTNW/Support for attainment of further funds.

REFERENCES

1. www.novozymes.com/library/Downloads/publications/NZ_facts_UK.pdf
2. W. H. Stroh, *Genetic Eng. News* **18**, 11 (1998).
3. A. Wiseman, *Handbook of Enzyme Biotechnology*. Elis Horwood, Chichester (1985).
4. H. Schacht, R. Cronen, E. Cleve, U. Denter, W. Kesting and E. Schollmeyer, *Textilveredlung* **33**, 8–13 (1998).

5. A. Paar, I. Beurer, A. Cavaco-Paulo, M. Gudelj, K. H. Robra and G. M. Gübitz, *AATCC Rev.* **2**, 25–26 (2002).
6. L. Tarhan, *Process Biochemistry* **30**, 623–628 (1995).
7. D. Praschak, T. Bahners and E. Schollmeyer, *Appl. Phys. A* **71**, 577–581 (2000).
8. K. Opwis, T. Bahners, E. Schollmeyer, A. Geschewski, H. Thomas and H. Höcker, *Techn. Textil.* **45**, 215–217 (2002).
9. K. Opwis, T. Bahners and E. Schollmeyer, *Chem. Fibers Int.* **54**, 116–119 (2004).
10. T. Bahners, K. Opwis, T. Textor and E. Schollmeyer, in: *Contact Angle, Wettability and Adhesion*, Vol. 4, K. L. Mittal (Ed.), pp. 307–320. VSP, Leiden (2006).
11. W. Graumann, *Z. Wiss. Mikros.* **61**, 225–226 (1953).
12. K. Opwis, PhD Thesis. University of Duisburg-Essen, Shaker, Aachen (2003).
13. R. West, *J. Chem. Educ.* **42**, 386–387 (1965).
14. K. Opwis, D. Knittel and E. Schollmeyer, *Anal. Bioanal. Chem.* **380**, 937–941 (2004).
15. G. Manecke, E. Ehrenthal and J. Schlüsen, in: *Characterization of Immobilized Biocatalysts*, K. Buchholz (Ed.). Verlag Chemie, Weinheim (1979).
16. K. Opwis, D. Knittel and E. Schollmeyer, *AATCC Rev.* **4**, 25–28 (2004).
17. K. Opwis, D. Knittel, T. Bahners and E. Schollmeyer, *Eng. Life Sci.* **5**, 63–67 (2005).

Contact Angle, Wettability and Adhesion, Vol. 4, pp. 461–469
Ed. K.L. Mittal
© VSP 2006

Biofouling-resistant coatings from low-surface-energy polymers

JOHN TSIBOUKLIS,* THOMAS G. NEVELL and EUGEN BARBU

School of Pharmacy and Biomedical Sciences, University of Portsmouth, Portsmouth PO1 2DT, UK

Abstract—For protecting surfaces against microbial contamination, an alternative to chemical attack (bleaches, detergents) on established colonies of fouling organisms and to the use of toxic surfaces (antibiotics, copper, organotin additives to paints) is to utilise surfaces of environmentally friendly, non-toxic polymeric materials onto which microbial colonisers will not adhere. In this paper, the deployment of low-surface-energy coatings that inhibit settlement by removing the ability of the surface to form a permanent bond with the microorganism is described.

Keywords: Low surface energy; antifouling; fluoropolymer.

1. INTRODUCTION

Bacteria adhere readily to surfaces for survival and propagation. Generally, this brings about the formation of an adherent layer (biofilm) composed of bacteria embedded in an organic matrix (Fig. 1). The biofilm matrix is primarily a glycoprotein, the exopolymer, which is generated by the bacteria, although matter derived from the environment may also be present.

Usually, bacterial adhesion is promoted by the prior adsorption of organic material onto the surface (the conditioning film); for example, polysaccharides and proteins tend to be adsorbed strongly from aqueous solution by most surfaces [1]. Once colonisation has been achieved, formation and subsequent growth of the bacterial biofilm are largely independent of the substrate. Whether or not it is a prerequisite, biofilm formation usually precedes macrofouling. A sequence of four phases is involved:

- phase 1, transport of bacteria to the surface,

- phase 2, reversible attachment of bacteria to that surface; van der Waals interactions overcome repulsive electrostatic forces,

---

*To whom correspondence should be addressed. Tel.: (44-2392) 842-131;
Fax: (44-2392) 843-565; e-mail: john.tsibouklis@port.ac.uk

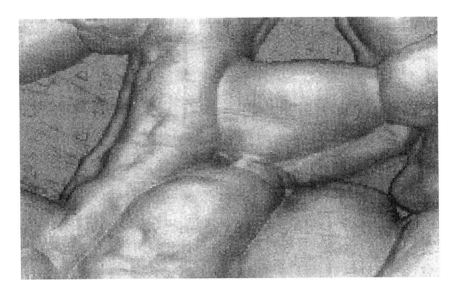

Figure 1. AFM image of typical biofilm in its early stages of development (3 μm×5 μm).

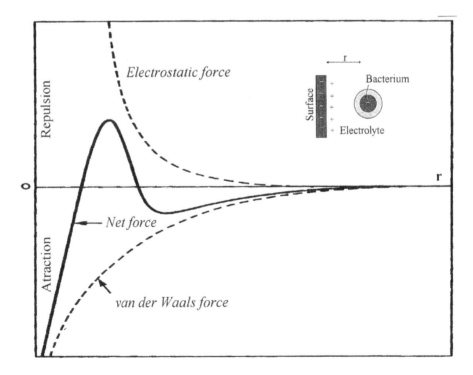

Figure 2. In certain natural aquatic environments, or under physiologically relevant regimes, minimization of the van der Waals interactions creates a formidable barrier between the negatively charged bacterium and the negatively charged surface.

-   phase 3, specific interactions involving chemical (covalent, ionic, hydrogen) bonding between the bacterium and the substrate, and
-   phase 4, colonisation of the surface and formation of a bacterial biofilm.

Biocompatible synthetic materials and fouling-resistant coatings are of great significance in medicine, to industry and in the home [2]. Several types of potentially fouling-resistant materials have been developed, and their performance in preventing the attachment of biopolymers and cells has been investigated [3]. To suppress protein adsorption, molecular design considerations formulated by Harris and co-workers [4–6], Andrade and co-workers [7, 8] and Merrill and Salzman [9] have led to the use of structures that are based on the hydrophilic poly(ethylene glycol) (PEG) backbone, whereas the groups of Marchant [10, 11], Park [12] and Davies [13, 14] have utilised biomimetic hydrophilic surfaces. Furthermore, the quest for readily accessible methods by which the attachment of microorganisms to surfaces may be prevented [15], as well as a full understanding of the underlying mechanisms of bioadhesion [16, 17] so that new anti-fouling strategies can be devise, has stimulated parallel research activities [18–20]. Whilst it has been shown that certain properties of materials, such as surface free energy and roughness, are crucial in determining their *in vivo* biocompatibility [21], the relationships between the chemical functionality of a surface and the extent of bioadhesion are not understood fully [22–24].

Methods derived for passivating synthetic substrates often require aggressive reagents or conditions, which may be unsuitable for treating natural surfaces or those that are introduced into biological environments. Thus, in addition to bio-fouling-resistant surfaces that function through the incorporation of antimicrobial agents into the substrate (e.g., loading of the surface with bacteriocides), two surface-modification strategies have been tried, namely:

-   hydrophilic modification and
-   the low-surface-energy approach.

It is axiomatic to both approaches that surface properties that minimise the initial adsorption processes would also make the substrate unattractive for the direct attachment of colonising organisms. Based on work of adhesion considerations, hydrophilic coatings within the aqueous environment are believed to function by presenting a non-stick surface to bacteria and to other colonising microorganisms [25]. Interest in the low-surface-energy approach dates from the early 1980s, when silicone elastomers were first tried as coating materials; the approach gained added credibility following the observation that gorgonian corals, which have low energy surfaces, were not susceptible to colonisation by marine microbes [26, 27].

The scientific rationale underpinning the low surface energy approach to anti-fouling stems from the DLVO theory: considering that in natural aqueous environments most surfaces, and most bacteria, are negatively charged, the utilisation of ultra-low-surface-energy coatings minimizes the van der Waals interactions operating between the surface and the prospective coloniser, with the result that the repulsive barrier extends further from that surface (Fig. 2).

2. LOW-SURFACE-ENERGY COATINGS

We have tested the low surface energy approach with several classes of readily accessible polymeric coatings [28, 29] (Scheme 1).

Films (30–200 nm in thickness) of these polymers were deposited, from solution or from the melt, onto supporting substrates of glass, Teflon® (PTFE), poly(methyl methacrylate) (PMMA) and glass fibre-reinforced polyester composite (GRP). Contact angles, surface roughnesses and the derived surface energy data for some of these materials are summarised in Table 1.

$$H_3C\!-\!\underset{\displaystyle |}{\overset{\displaystyle \overset{O}{\|}}{Si}}\!-\!(CH_2)_3\!-\!O\!-\!(CH_2)_2\!-\!(CF_2)_n\!-\!CF_3$$

$$H_2C\!=\!\underset{\displaystyle |}{C}R\!-\!\overset{\displaystyle \overset{O}{\|}}{C}\!-\!O\!-\!(CH_2)_2\!-\!(CF_2)_n\!-\!CF_3$$

Scheme 1. Silicones, $n = 3$–9, PFE; acrylates (R = –H), $n = 3$–9; methacrylates (R = –CH$_3$), n = 3–9; itaconates (R = –CH$_2$COOR'; R' = –(CH$_2$)$_2$(CF$_2$)$_n$CF$_3$), $n = 3$–9.

Table 1.
Advancing contact angles

Sample	R_a (nm)	Contact angle, $\theta°$			Surface energy (mJ m^{-2})			
		H$_2$O	DIM	EG	γ_S^{LW}	γ_S^+	γ_S^-	γ_S
Poly(perfluoroalkylacrylate)s								
PFA9	11.0	125 (4)	112 (16)	120 (25)	5.0	0.1	1.6	5.6
PFA7	9.6	117 (8)	112 (12)	108 (17)	5.0	0.1	2.3	6.1
Poly(perfluoroalkylmethacrylate)s								
PFMA9	7.13	125 (6)	109 (13)	107 (34)	5.8	0.3	0.1	6.1
PFMA7	1.17	125 (9)	104 (12)	105 (19)	7.3	0.2	0.1	7.5
Poly(methylpropenoxyfluoroalkylsiloxane)s								
PFE9	3.1	109 (3)	95 (12)	94 (8)	10.6	0.17	2.1	11.8

Advancing contact angles for water (values in mJ m^{-2}: $\gamma_L = 72.8$, $\gamma_L^{LW} = 21.8$, $\gamma_L^+ = 25.5$, $\gamma_L^- = 25.5$), diiodomethane (DIM, $\gamma_L = 50.8$, $\gamma_L^{LW} = 50.8$, $\gamma_L^+ = 0$, $\gamma_L^- = 0$) and ethylene glycol (EG, $\gamma_L = 48$, $\gamma_L^{LW} = 29$, $\gamma_L^+ = 1.92$, $\gamma_L^- = 47$) on acrylate, PFA; methacrylate (PFMA); and silicone, PFE, film structures, and corresponding surface energies, γ_S ($\gamma_S = \gamma_S^{LW} + 2(\gamma_S^+ \gamma_S^-)^{1/2}$ where γ^{LW} is the Liftshitz–van der Waals component, γ^+ is the Lewis acid parameter and γ^- is the Lewis base parameter). The R_a surface roughness parameter, as determined by AFM, is also given. Contact angle hysteresis values, H°, are given in parentheses.

Perfluoro-acrylate and methacrylate polymers exhibited extremely low surface energies, whereas that of the fluorinated silicone was significantly higher; predominantly inward-pointing fluorine groups in the latter account for these differences [28].

3. ANTI-FOULING PERFORMANCE

Polymers were coated onto supporting substrates and tests were conducted with a selection of bacteria and other fouling organisms.

3.1. Bacteria

Resistance to colonisation was evaluated using optical, electron and atomic force microscopies. Sub-cultures of mixed *Pseudomonas* spp. (*P. cepacia, P. paucimobilis* and *P. fluorescens*) and sulphate-reducing bacteria (SRB), strain A1-1 *Desulfovibrio alaskensis* NCIMB 13491, were grown in artificial seawater / Marine Postgate Medium C 4:1 (v/v) [7] or degassed Marine Postgate Medium C. *Pseudomonas* sp. NCIMB 1534 and *Alteromonas* sp. NCIMB 2141 were cultivated in marine broth 2216 (Bacto™, DIFCO Laboratories, Detroit, MI, USA).

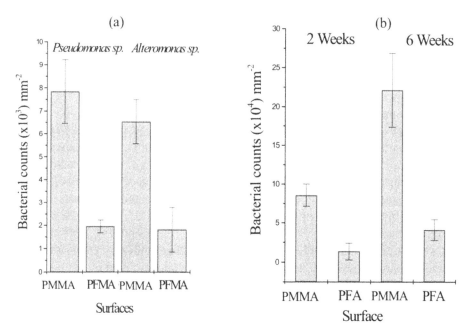

Figure 3. Settlement of bacteria on melt-coated fluoroalkyl polymer films and PMMA control: (a) *Pseudomonas* sp. NCIMB 1534 and *Alteromonas* sp. NCIMB 2141 (after 5 hours, 20°C) on poly(fluoroalkylmethacrylate), PFMA; (b) mixed *Pseudomonas* spp. (after 2 and 6 weeks, 28°C) on poly(fluoroalkylacrylate) (PFA). All experiments were conducted in triplicate.

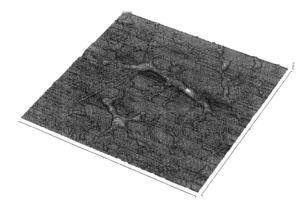

Figure 4. AFM image (10 μm×10 μm) of exopolymer matrix from SRB strain A1-1 *Desulfovibrio alaskensis* NCIMB 13491 on a melt-cast film of poly(fluoroalkylacrylate). The film was exposed to the bacterial culture for two weeks in static experiments; the bacterial colonies were seen to slide off the surface as the coated disc was raised out of the container. The sample was not subjected to any pre-imaging treatment.

Closed batch-culture tests were conducted for long-term settlement on poly(fluoroalkylacrylate)s and poly(acrylate)s, and also for initial settlement on poly(fluoroalkylmethacrylate)s. On all of the controls (glass, PTFE, PMMA, GRP), bacterial settlement was rapid and the cells became strongly attached. For example with *Pseudomonas* NCIMB 1534 and *Alteromonas* NCIMB 2141, cell coverages reached $5.5–8.3×10^3$ mm^{-2} after 5 h; with SRB and *Pseudomonas* spp. cell coverages reached $8.9×10^4$ and $9.5×10^4$ mm^{-2}, respectively, after 2 weeks and mature biofilms were observed after 3–5 months (Fig. 3).

With films of both poly(fluoroalkylacrylate) and poly(fluoroalkylmethacrylate), early settlement of *Pseudomonas* NCIMB 1534 and *Alteromonas* NCIMB 2141 was *ca.* 25% of that on the controls and at this stage there were no significant differences between the fluoropolymers. After exposure for 2 weeks, the corresponding proportions for poly(fluoroalkylacrylate)-coated substrates were 5–20% with *Pseudomonas* spp. and 10–15% with SRB. The poly(fluoroalkylacrylate)s' resistance to settlement was maintained over 5 months. With these surfaces, settled bacterial colonies were weakly attached and very easily displaced, such that only single bacteria or small groups were not removed by gentle rinsing. This is demonstrated in Fig. 4 in which the exopolymeric matrix is seen to spread rather poorly over the poly(fluoroalkylacrylate)-coated surface. Although the non-fluorinated alkylacrylates with long side-chains also showed early resistance to settlement by SRB relative to controls, this was not maintained and mature biofilms were formed.

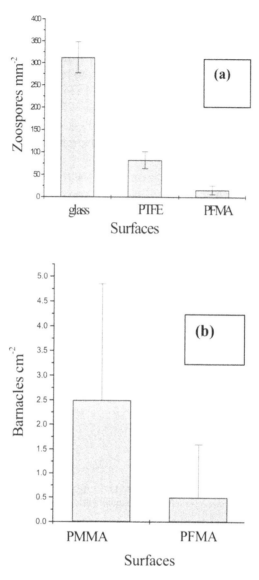

Figure 5. (a) *Enteromorpha* settlement (20°C, dark, 1 h) on glass control, PTFE control and melt-coated film of poly(fluoroalkylmethacrylate); (b) Settlement of barnacle cyprids of *Balanus amphritite* Darwin (14 days, dark, 28°C) on film of poly(fluoroalkylmethacrylate) as compared with PMMA control. All experiments were conducted in triplicate.

3.2. Enteromorpha zoospores and barnacle cyprids

Settlement experiments using zoospores released from *Enteromorpha thalli* [30] were carried out for films of poly(fluoroalkylacrylate) and poly(fluoroalkylmethacrylate) in Petri dishes containing samples of all coated substrates and un-coated controls [31]. Far fewer zoospores settled onto the films of poly(fluoroalkylacrylate) than onto the uncoated controls; with the former, most of the attached zoospores were located over surface imperfections. Crack-free films of poly(fluoroalkylmethacrylate)s showed less settlement. Interestingly, there was also some discrimination between the controls: the relative numbers of zoospores settled were glass /PTFE/ fluoropolymer = 12:4:1 (Fig. 5a).

The settlement over 14 days of cyprid larvae of *Balanus amphritite* Darwin [32] on a poly(fluoroalkylmethacrylate) film [31] is summarised in Fig. 5b. Most of the poly(fluoroalkylmethacrylate)-coated substrates were seen to attract only one or two cyprids.

Strong evidence that the fouling-resistant nature of the low surface energy films is in no way related to toxicity has been provided by animal experiments, performed principally with reference to potential medical applications of the materials. No infection effects could be detected [21].

4. CONCLUSIONS

The principal aim of this work has been to compare the ability of some bacteria, algae and higher fouling organisms to settle and develop on low surface energy fluorinated coating materials. Similar coatings have shown promising resistance to marine fouling [33] but their long-term performance is not yet established. A challenge, with respect to the potential application of these materials as large-scale fouling protection coatings, is that of producing well-adhered and defect-free films.

REFERENCES

1. M. E. Schrader, J. A. Caramone, G. I. Leob and O. P. Arora, in: *Marine Biodeterioration*, M. F. Thomson, R. Sarojini and A. A. Nagabhushanam (Eds), pp. 261–265. Balkema, Rotterdam (1988).
2. L. H. G. Morton and S. B. Surman, *Int. Biodeterior. Biodegrad.* **34**, 203–221 (1994).
3. R. F. Brady, Jr, *Chem. Ind.* 219–222 (1997).
4. J. M. Harris (Ed.), *Polyethylene Glycol Chemistry*. Plenum Press, New York, NY (1992).
5. K. Holmberg, K. Bergström, C. Brink, E. Österberg, F. Tiberg and J. M. Harris, *J. Adhesion Sci. Technol.* 7, 503–517 (1993).
6. E. Österberg, K. Bergström, K. Holmberg, T. P. Schuman, J. A. Riggs, N. L. Burns, J. M. van Alstine and J. M. Harris, *J. Biomed. Mater. Res.* **29**, 741–747 (1995).
7. S. I. Jeon, J. H. Lee, J. D. Andrade and P. G. de Gennes, *J. Colloid Interface Sci.* **142**, 149–158 (1991).
8. J. H. Lee and J. D. Andrade, in: *Polymer Surface Dynamics*, J. D. Andrade (Ed.), p. 119. Plenum Press, New York, NY (1988).
9. E. W. Merrill and E.W. Salzman, *J. Amer. Soc. Artif. Intern. Organs* **6**, 60–72 (1983).
10. N. B. Holland, Y. X. Qiu, M. Ruegsegger and R. E. Marchant, *Nature* **392**, 799–801 (1998).
11. K. Vacheethasanee and R. E. Marchant, *J. Biomed. Mater. Res.* **50**, 302–312 (2000).
12. J. H. Jeong, D. W. Lim, D. K. Han and T. G. Park, *Colloids Surfaces B – Biointerfaces* **18**, 371–379 (2000).

13. R. J. Green, R. A. Frazier, K. M. Shakesheff, M. C. Davies, C. J. Roberts and S. J. B. Tendler, *Biomaterials* **21**, 1823–1835 (2000).
14. R. A. Frazier, G. Matthijs, M. C. Davies, C. J. Roberts, E. Schacht and S. J. B. Tendler, *Biomaterials* **21**, 957–966 (2000).
15. D. L. Elbert and J. A. Hubbell, *Annu. Rev. Mater. Sci.* **26**, 365–394 (1996).
16. R. Bos, H. C. van der Mei and H. J. Busscher, *FEMS Microbiol. Rev.* **23**, 179–230 (1999).
17. Y. H. An and R. J. Friedman, *J. Biomed. Mater. Res.* **43**, 338–348 (1998).
18. Y. L. Ong, A. Razatos, G. Georgiou and M. M. Sharma, *Langmuir* **15**, 2719–2725 (1999).
19. D. Cunliffe, C. A. Smart, J. Tsibouklis, S. Young, C. Alexander and E. N. Vulfson, *Biotechnol. Lett.* **22**, 141–145 (2000).
20. A. Razatos, Y. L. Ong, M. M. Sharma and G. Georgiou, *Proc. Natl. Acad. Sci. USA.* **95**, 11059–11064 (1998).
21. J. Tsibouklis, M. Stone, A. A. Thorpe, P. Graham, V. Peters, R. Heerlien, J. R. Smith, K. L. Green and T. G. Nevell, *Biomaterials* **20**, 1229–1235 (1999).
22. R. F. Brady Jr., *Prog. Org. Coatings* **35**, 31–35 (1999).
23. P. Harder, M. Grunze, R. Dahint, G. M. Whitesides and P. E. Laibinis, *J. Phys. Chem. B* **102**, 426–436 (1998).
24. L. K. Ista, H. Fan, O. Baca and G. P. Lopez, *FEMS Microbiol. Lett.* **142**, 59–63 (1996).
25. L. V. Evans and N. Clarkson, *J. Appl. Bacteriol. Symp. Suppl.* **74**, 119S–124S (1993).
26. A. S. Clare, D. Rittschof, D. J. Gerhart and J. S. Maki, *Invertebr. Reprod. Dev.* **22**, 67–76 (1992).
27. N. H. Vrolijk, N. M. Targett, R. E. Baier and A. E. Meyer, *Biofouling* **2**, 39–54 (1990).
28. J. Tsibouklis, M. Stone, A. A. Thorpe, P. D. Graham, T. G. Nevell and R. J. Ewen, *Langmuir* **15**, 7076–7079 (1999).
29. J. Tsibouklis and T. G. Nevell, *Adv. Mater.* **15**, 647–650 (2003).
30. M. E. Callow, J. A. Callow, J. D. Pickett-Heaps and R. Wetherbee, *J. Phycol.* **33**, 938–947 (1997).
31. P. D. Graham, I. R. Joint, T. G. Nevell, J. R. Smith, M. Stone, A. A. Thorpe and J. Tsibouklis, *Biofouling* **16**, 289–299 (2000).
32. A. S. Clare, *Biofouling* **10**, 141–159 (1996).
33. E. Lindner, *Biofouling* **6**, 193–205 (1992).

Contact Angle, Wettability and Adhesion, Vol. 4, pp. 471–486
Ed. K.L. Mittal
© VSP 2006

Cell adhesion to ion- and plasma-treated polymer surfaces: The role of surface free energy

CRISTINA SATRIANO,[1] GIOVANNI MARLETTA,[1,*] SALVATORE GUGLIELMINO[2] and SANTINA CARNAZZA[2]

[1]*Dipartimento di Scienze Chimiche, Università degli Studi di Catania, Viale Andrea Doria 6, 95125 Catania, Italy*
[2]*Dipartimento di Scienze Microbiologiche, Genetiche e Molecolari, Università di Messina, Salita Sperone, 31, Vill. S. Agata, Messina 98166, Italy*

Abstract—Low-energy ion beams and cold plasma treatments have been used to obtain chemically different altered layers at the surfaces of poly(hydroxymethylsiloxane) and poly(ethylene terephthalate) films, in order to modify their biocompatibility. The cell response, in terms of the number and viability of Normal Human Dermal Fibroblast cells, has been studied for the various surfaces. The different cell–substrate interactions are interpreted in terms of the complex interplay between the chemical structure of the surfaces, the related surface free energy and the cell response. The overall results point towards a critical role of the total surface free energy of the surfaces, which seems to trigger a good cell adhesion above a threshold of about 40 mJ/m^2.

Keywords: Polymer surfaces; ion irradiation; plasma modification; cell adhesion; surface free energy; XPS; poly(hydroxymethylsiloxane); poly(ethylene terephthalate).

1. INTRODUCTION

Polymers are the most commonly used biomaterials due to their compatibile mechanical viscoelastic properties with human tissues [1–6]. Surface modification by high-density energy techniques is a powerful method to tailor the surface composition and the related physico-chemical properties of a polymer [7–9].

In fact, in recent years it has been demonstrated that irradiation of polymers, for instance by ion beams or cold plasma techniques, changes their surface properties, including chemical structure and composition [10], surface free energy [11], electronic structure [12], roughness and morphology at the micro- and nano-scale [13], interaction with biological systems and selective adsorption/adhesion processes [14–25].

In particular, it has also been shown that ion-beam-based processes are able to promote a very significant improvement of adhesion, spreading and/or proliferation processes for such different cells as fibroblasts, endothelial cells, astrocy-

---

*To whom correspondence should be addressed. E-mail: gmarletta@unict.it

tomas, etc., onto very diverse polymeric surfaces as those of polystyrene, segmented polyurethane, polyethylene, polysiloxane, etc., with many different types of ions (for instance, O_2^+, Na^{2+}, Kr^{2+}, N_2^+, N^+, Ar^+ or Ag^-) [26–32].

The exact nature of the ion-beam-induced mechanism for cytocompatibility enhancement is still a matter of debate; however, an increase of surface hydrophilicity or the formation of surface amorphous carbon phases has been demonstrated to be phenomenologically related to improved cytocompatibility effect [31–40]. These two explanations may, in turn, to be related to modification of other more specific factors such as surface polarity [41–45], roughness, as well as topography on the micro- and nano-scale [46–52]. Also, one should take into account the modified electrical properties [53], i.e., trapped charges or conducting domains, and the effect of the long-living radicals or other metastable species [54]. As matter of fact, the irradiated surfaces seem to show complex processes of selective protein attachment, cell membrane reorganization, or they just act as a stimulus for an increased activity of the adhered cells [55]. Thus, it has been shown that the ion-modified polymers alter the membrane protein expression, as well as the mechanism determining the focal areas of the adhering cells [35].

This general picture of the experimental evidence is, however, subject to continuous revision. For instance, recently it has been demonstrated that the capability of ion irradiation to modify cell behaviour may not apply to certain polymers [56]. In this connection, we will show here that irradiation with either inert ions (Ar^+) at relatively high energy (50 keV) or reactive (O_2) cold plasma is unable to enhance the already good PET cytocompatibility, while it dramatically enhances the cytocompatibility of PHMS, so that the irradiated PHMS surfaces exhibit a dramatic enhancement of cell adhesion, proliferation and spreading. In particular, the observed cell response can be satisfactorily related to surface free energy of the various investigated surfaces.

2. MATERIALS AND METHODS

2.1. Preparation and characterization of polymer surfaces

Poly(hydroxymethylsiloxane) (PHMS) and poly(ethylene terephthalate) (PET) (from Aldrich) were deposited as thin films on p-doped silicon (100) wafers by spin coating (3000 rpm, 60 s, room temperature). The thicknesses of the deposited films were 500 ± 30 nm for PHMS and 600 ± 50 nm for PET, as measured with an alpha-step profilometer. The surface irradiation was performed either with 50 keV Ar^+ ions at 1×10^{15} ions/cm^2 fluence (RT, chamber pressure lower than 10^{-5} Pa, current density of 1.5 μA/cm^2) or by O_2-plasma exposure (excitation frequency 13.56 MHz; power 100 W; pressure 53.3 Pa; treatment time 60 s). Ten different samples were irradiated for each modification treatment.

After the irradiation steps, the samples were aged in the laboratory atmosphere for at least 1 week before carrying out surface physico-chemical characterization and adsorption/adhesion experiments.

2.1.1. X-ray photoelectron spectroscopy (XPS)

The chemical structures and compositions of surfaces were investigated by X-ray photoelectron spectroscopy (XPS). XPS analysis was carried out with a PHI 5600 Multi-Technique Spectrometer equipped with a dual Al/Mg anode, a hemispherical analyzer and an electrostatic lens system (Omni Focus III). The electron take-off angle was $45°$ and the analyzer was operated in the FAT mode using the Al $K\alpha 1,2$ radiation with pass energies of 187.85 eV and 11.75 eV for survey and high-resolution scans, respectively.

The spectra were analyzed using an iterative least squares fitting routine based on Gaussian peaks and Shirley background subtraction. Binding energies (BEs) of all the peaks were referenced to the intrinsic hydrocarbon-like C_{1s} peak at 285.0 eV or 284.6 eV for untreated PHMS [31].

2.1.2. Optical microscopy and atomic force microscopy (AFM)

The micro- and nano-scale morphologies were characterized by Optical and atomic force microscopies, respectively. Optical microscopy (OM) was performed with a Leica DM-RME microscope, with 50 to 1000× magnification, equipped with a Polaroid digital camera. The surface micro- and nano-topography, as well as the roughness were measured with a Multimode/ Nanoscope IIIA atomic force microscope (Veeco) in the tapping mode in air with a standard silicon tip. The relative room humidity was 30% and the room temperature was 25°C. Data were acquired on square frames of 10×10 μm^2, 1×1 μm^2 and 350×350 nm^2. Images were recorded using height, phase-shift and amplitude channels with 512×512 measurement points (pixels). Measurements were made at least three times on different zones of each sample. Surface roughness values were determined in five random areas per sample, scanning across areas 10×10 and 1×1 μm^2. The Nanoscope III software was used for image processing and roughness calculation, in terms of average roughness, R_a (the average deviation of the measured z-values from the mean plane), root-mean-square roughness, R_q (standard deviation of an entire distribution of z-values for a large sample size), and maximum roughness, R_{max} (the difference between the largest positive and negative z-values).

2.1.3. Contact angle measurements and surface free energy (SFE) determination

Surface free energies were determined following the Lifshitz–van der Waals/acid–base (LW-AB, or van Oss, Chaudhury and Good) approach [57] from contact angles of water (θ_W), tricresylphosphate (θ_{TCP}) and glycerol (θ_{GL}) onto the untreated and irradiated surfaces. In particular, tricresylphosphate was employed as an apolar liquid, due to its almost negligible polar component (approx. 1.7 mJ/m^2) with respect to the dispersion component (approx. 39.2 mJ/m^2) [58].

Semi-automatic video-based measurements of contact angles were performed at 25°C and 65% relative humidity using an OCA30 instrument (Data Physics). At least five measurements were made for each sample and then averaged. The contact angle data were converted into surface free energies on the basis of the LW-AB approach, which states that the interfacial free energy between the solid (S) and the liquid (L) is given by the following equations:

$$\gamma_{SL} = \gamma_{SL}^{LW} + \gamma_{SL}^{AB} \tag{1}$$

or

$$\gamma_{SL} = 2\left[\left(\sqrt{\gamma_S^{LW}} - \sqrt{\gamma_L^{LW}}\right)^2 + \left(\sqrt{\gamma_S^+} - \sqrt{\gamma_L^-}\right)\cdot\left(\sqrt{\gamma_S^-} - \sqrt{\gamma_L^+}\right)\right] \tag{2}$$

where γ^{LW} and γ^{AB} represent Lifshitz–van der Waals and Lewis acid–base components, respectively. In particular, the term γ^{LW} contains electrodynamic interactions such as due to Keesom, Debye and London forces, while γ^+ and γ^- are used to indicate the parameters of surface free energy which are due to the proton donor (or electron acceptor) and proton acceptor (or electron donor) functionalities, respectively [57].

2.2. Cell adhesion experiments

Normal human dermal fibroblast cell line (NHDF, Clonetics Cambrex Bio Science, Walkersville, MD, USA) was used to test the cell adhesion to the various modified surfaces. The cells were routinely maintained in fibroblasts base medium (FBM, Clonetics) supplemented with 2% fetal bovine serum (FBS), insulin, hFGF-B, gentamycin and anfotericine B (FGM-2 Bullekit, Clonetics). The various PHMS and PET samples, three per treatment, were placed at the bottom of a 6-well polystyrene tissue culture plate (Nunc, Rochester, NY, USA) and 3 ml of cell suspension with a cell concentration of about 1.5×10^5 cells/ml were added to each well. The cell adhesion as well as proliferation and tendency to spreading were evaluated after 48 h of incubation at 37°C in a humidified 5% CO_2 atmosphere. At the end of the culture periods, the samples were washed with PBS (phosphate buffer saline) solution, fixed with 4% p-formaldehyde, then treated with Triton X-100 (0.1% in PBS) and stained with Hoechst 699 and Blue Evans colorants, in order to mark the nucleus and cytoplasm, respectively. The samples were then observed by epifluorescence microscopy (Leica DMRE).

At least 5 different fields per sample were randomly acquired using optical microscopy with 10×, 20× and 40× magnifications with a Leica DC300F Camera and Leica IM50 software. The quantitative evaluation of cell coverage was performed in terms of integrated density (I.D.) values, as described elsewhere [31].

3. RESULTS AND DISCUSSION

3.1. Correlation between surface composition and surface free energy

In this section we will compare the effects of plasma treatment and ion irradiation on the composition and chemical functionalities, morphology at the micrometer scale and SFE of poly(ethylene terephthalate) (PET) and poly)hydroxymethyl-siloxane) (PHMS).

3.1.1. Surface chemical structure

We will now discuss the typical trends in chemical modification for the two polymers as investigated by XPS analysis. Table 1 reports the compositional modifications obtained with both plasma and ion irradiation treatments for PHMS and PET.

In particular, in the case of PHMS both plasma and ion irradiation essentially induce carbon depletion from the surface layer. In fact, the carbon content is about 12% for the Ar^+-irradiated surfaces and about 10% for the plasma-treated samples, *vis-a-vis* the 22% in the untreated polymer [59].

In the case of PET, the stoichiometry changes from the original $C_{2.4}O_1$ of untreated PET to $C_{1.98}O_1$ for the plasma-treated PET, due to the increase of the oxygen content at surface. In contrast, for the ion-irradiated PET surface, the oxygen content decreases to about 15% atomic concentration, i.e., about half of the original value for the untreated PET, corresponding to the change of the stoichiometry from the original $C_{2.4}O_1$ to that of $C_{5.5}O_1$ for the ion-irradiated PET surface [56].

The detailed analysis of photoelectron peak shapes is reported in Figs 1 and 2 for PHMS and PET, respectively.

In particular Fig. 1, showing C_{1s} and Si_{2p} photoelectron peaks for PHMS surfaces, indicates that both plasma and ion irradiation induce a significant loss of the original methyl groups and the formation of new oxidized carbon functionalities. In particular, the C_{1s} peak for the untreated PHMS surface has been fitted using only two components, a main one centred at 284.6 ± 0.2 eV BE (C_I) and assigned to $>C–Si$ bonds, and the second one centred at approx. 286.6 ± 0.2 eV BE

Table 1.
Surface atomic compositions of the various PHMS and PET samples

Sample	O_{1s} (at%)	C_{1s} (at%)	Si_{2p} (at%)
Untreated PHMS	56.0	22.5	21.5
Plasma-treated PHMS	65.6	9.8	24.6
50 keV Ar^+-irradiated PHMS	67.8	12.0	20.2
Untreated PET	28.0	72.0	–
Plasma-treated PET	33.5	66.5	–
50 keV Ar^+-irradiated PET	14.7	85.3	–

(C_{II}), assigned to >C–OH and >C–O–C (groups belonging either to the terminal polymer groups or to solvent residues [60]). The plasma-treated PHMS exhibits the same components as the untreated one, with a strong reduction of C_I with respect to C_{II} component. In contrast, the C_{1s} peak of ion-irradiated PHMS exhibits two new components, centred at 286.1±0.2 eV (C_{III}) and 288.0±0.2 eV (C_{IV}), and assigned, respectively, to the formation of C–O–Si linkages and >C=O groups [31]. Furthermore, traces of newly formed >COO⁻ groups could be identified in the high BE tail of the peak, and confirmed by TOF-SIMS measurements (data not shown).

The Si_{2p} peak analysis evidences a symmetric band, which is well fitted using a single gaussian component, centred at about 102.2±0.2 eV BE (Si_1) for the untreated PHMS and at approx. 103.6±0.2 eV (Si_2) for both plasma and ion-irradiated PHMS. In agreement with the literature [31, 60], the Si_1 and Si_2 components have been assigned, respectively, to [SiO_3C] clusters and to the formation of an amorphous SiO_2-like phase, predominantly formed by [SiO_4] clusters. Thus, the irradiated PHMS surface appears to be essentially composed of SiO_xC_y (H_z).

Figure 2 reports the C_{1s} and O_{1s} photoelectron peaks for PET under the very same irradiation conditions as for PHMS.

The specific functional groups for PET before irradiation (Fig. 2a-b) have been found to include C–C and C–H linkages in the phenyl rings set at 285.0 eV (C_I), C–O–C bonds at 286.4±0.2 eV (C_{II}) (with the related oxygen component at 533.5±0.2 eV (O_{II})), COO⁻ groups at 288.7±0.2 eV (C_{III}), and the shake-up, due to the $\pi \rightarrow \pi^*$ transitions in the phenyl rings, at 291.8±0.4 eV. The quantitative ratios between the described components are in perfect agreement with the literature [61].

As for plasma-treated and ion-irradiated PET surfaces, one can see clearly that the modification pattern is very different for the two kinds of treatments. In fact, the plasma-treated PET (Fig. 2c, d) exhibits C_{1s} and O_{1s} photoelectron peak shapes still similar to those of the untreated PET (Fig. 2a, b), except for the relative increase of the oxidised carbon species and a strong decrease of the shake-up, related to the $\pi \rightarrow \pi^*$ transitions, at approx. 538.5±0.2 eV BE for O_{1s} and approx. 292.0±0.2 eV BE for C_{1s}, which indicates a partial disruption of the aromatic rings. In contrast, the ion-irradiated PET surfaces (Fig. 2e, f) show drastic changes in C_{1s} and O_{1s} photoelectron peak shapes, with a strong decrease of both the oxidised species and the disappearance of the shake-up, suggesting that most of the initial phenyl rings were destroyed by the irradiation. Thus, the irradiated PET phase consists essentially of an oxidized and a hydrogenated amorphous carbon phase.

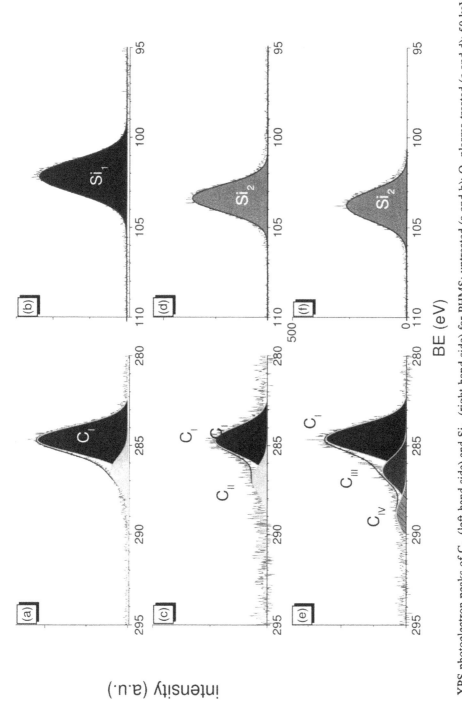

Figure 1. XPS photoelectron peaks of C_{1s} (left-hand side) and Si_{2p} (right-hand side) for PHMS: untreated (a and b); O_2-plasma-treated (c and d); 50 keV Ar^+-irradiated (e and f).

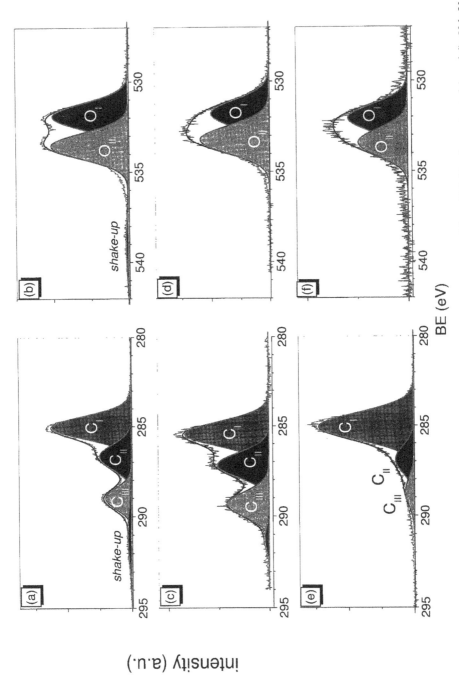

Figure 2. XPS photoelectron peaks of C_{1s} (left-hand side) and O_{1s} (right-hand side) for PET: untreated (a and b); O_2-plasma-treated (c and d); 50 keV Ar^+-irradiated (e and f).

3.1.2. Surface free energy results

The morphology at the micrometer and nanometer scale for both polymer surfaces, both before and after irradiation treatments, was found to be quite flat with roughness parameters R_q and R_a lower than 1 nm (see Table 2). Accordingly, we assume that the contact angle measurements on the various investigated surfaces are not affected by modification of roughness.

Figure 3 shows the water contact angles on PHMS and PET both before and after plasma or ion beam irradiation. The two polymers exhibit very different modification of wettability properties with ion beam and plasma treatments. In fact, while the very hydrophobic untreated PHMS surface becomes moderately hydrophobic after ion irradiation and very hydrophilic after plasma treatment (Fig. 3a), the change in water wettability of PET is very small for plasma-treated surface ($\Delta\theta{\sim}20°$) and negligible for the ion-irradiated one (Fig. 3b).

Table 3 shows the calculated surface free energy values for all the PHMS and PET surfaces. One can see that both surface modification treatments for PHMS increase the total SFE values to about three times the original value of the untreated surface, due to the increase of both dispersion (Lifshitz–van der Waals) as well as polar (Lewis acid–base) components, with a major contribution from the Lewis base term. In contrast, the modification in SFE for PET is much less marked, as expected also on the basis of the fact that the water contact angle value was nearly unchanged around a value of approx. 70° for both untreated and irradiated surfaces. Accordingly, no significant change is seen in the dispersion component and only a very small increase of the polar component is observed with irradiation.

The measured surface free energies can be related to the different radiation-induced chemical modifications for the two polymers. In fact, a parallel increase of wettability and SFE is observed for both types of treatments for PHMS. In particular, the concentration of highly polar SiO_x-related species increases along with the Lewis basic component. However, it should be noted that both plasma- and

Table 2.

R_a, R_q and R_{max} parameters for the various surfaces

Sample	R_q (nm)	R_a (nm)	R_{max} (nm)
Untreated PHMS	0.302±0.006	0.242±0.006	2.386±0.003
Plasma-treated PHMS	0.596±0.019	0.475±0.016	4.707±0.591
50 keV Ar⁺-irradiated PHMS	0.300±0.001	0.240±0.001	2.392±0.056
Untreated PET	0.213±0.011	0.169±0.008	1.817±0.210
Plasma-treated PET	0.353±0.019	0.375±0.016	5.707±0.823
50 keV Ar⁺-irradiated PET	0.338±0.023	0.255±0.016	4.536±0.909

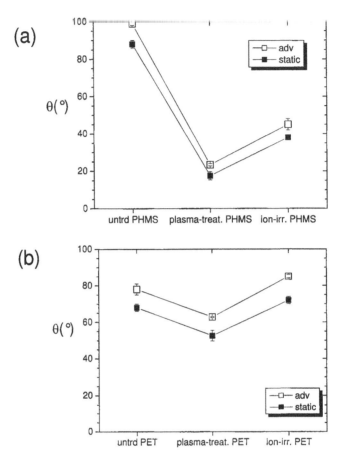

Figure 3. Static and advancing water contact angles for untreated, O_2-plasma-treated and 50 keV Ar^+-irradiated surfaces of PHMS (a) and PET (b).

Table 3.
Total surface free energy (SFE), Lifshitz–van der Waals (LW) component, Lewis acid (+) and Lewis base (−) parameters for untreated, O_2-plasma-treated and 50 keV Ar^+-irradiated PHMS and PET surfaces

Sample	SFE (mJ/m^2)	LW (mJ/m^2)	(−) (mJ/m^2)	(+) (mJ/m^2)
Untreated PHMS	23.4±1.6	22.8±0.2	8.7±0.9	0.02±0.01
Plasma-treated PHMS	61.5±2.7	39.9±0.3	43.6±1.4	2.7±0.5
50 keV Ar^+-irradiated PHMS	56.1±2.1	38.2±0.1	31.5±1.2	2.6±0.5
Untreated PET	42.8±0.6	39.6±0.3	8.5±0.2	0.3±0.05
Plasma-treated PET	47.1±4.3	39.8±0.2	25.2±1.1	0.5±0.5
50 keV Ar^+-irradiated PET	39.6±2.6	36.5±0.4	9.2±0.6	0.2±0.2

ion-beam-treated PHMS samples also undergo a considerable increase of the Lifshitz–van der Waals component.

In the case of PET, while these surfaces remain substantially hydrophobic, the SFE values are quite high due to the predominant contribution of the Lifshitz–van der Waals component, with the exception of the plasma-treated surface where a large increase of the Lewis basic component is clearly due to the increase of carboxyl and C–O–C groups.

3.2. Correlation between surface parameters and cell response

Figure 4 shows the quantitative results, from ID values derived from optical microscopy analysis, for the cell coverage of the different surfaces. It can be seen that for PHMS, while the untreated surface shows a low cytocompatibility, a dramatic increase in the number of adhered cells is obtained with both ion and plasma treatments. In contrast, the cell adhesion on PET is already good for the untreated surface, remaining practically unaffected by both ion and plasma treatments.

It should be emphasized that the adhesion before irradiation is indeed as much as 5-times higher for PET as compared to PHMS, while after ion irradiation the cell adhesion onto the PHMS surface is significantly higher than PET (by about a factor of 1.4).

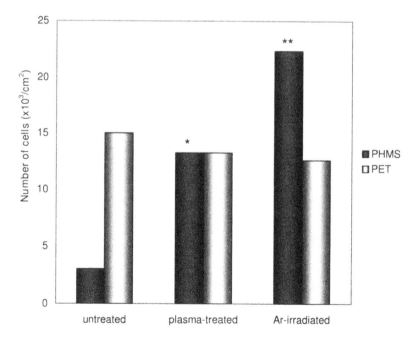

Figure 4. Cell coverage for the various PHMS and PET surfaces after 48 h of incubation. *$P < 0.05$ significance level; **$P < 0.01$ significance level.

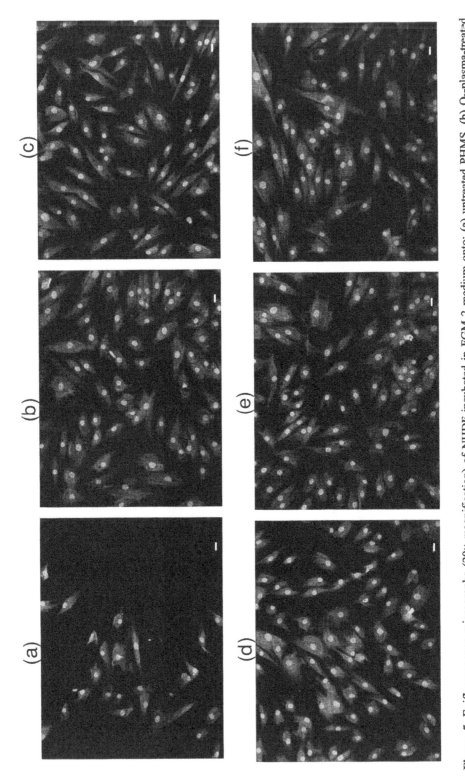

Figure 5. Epifluorescence micrographs (20× magnification) of NHDF incubated in FGM-2 medium onto: (a) untreated PHMS, (b) O₂-plasma-treated PHMS, (c) 50 keV Ar⁺-irradiated PHMS, (d) untreated PET, (e) O₂-plasma-treated PET and (f) 50 keV Ar⁺-irradiated PET. Bar = 10 μm.

Figure 5 shows the epifluorescence optical micrographs of the adhered cells onto untreated and irradiated PHMS and PET surfaces in FGM-2 medium. One can clearly see that the cells attached to untreated PHMS (Fig. 5a) show a poor viability, indicated by their predominantly rounded morphology and clustering tendency. An optimal cell adhesion and viability, i.e., elongated cell shape and alignment, is instead achieved for both plasma-treated (Fig. 5b) and 50 keV Ar$^+$-irradiated PHMS (Fig. 5c), as well as for the untreated and irradiated PET substrates. The observed cell behaviour can be discussed in terms of the surface free energy (SFE) of the various investigated surfaces.

The comparison between SFE data and cell adhesion and spreading behaviour indicates that an optimal cell response occurs for large SFE values, seemingly no less than 40 mJ/m^2 for the investigated cases. Indeed, a poor cell adhesion is observed only for untreated PHMS surface, having an SFE value of about 24 mJ/m^2, while for all the other cases, with SFE values ranging between about 40 and 60 mJ/m^2, a quite good cell adhesion behaviour is found. It should be noted that the large values of SFE for the samples showing a good cell adhesion derive from thecombination of a quite large apolar Lifshitz–van der Waals (γ^{LW}) contribution, on average 36–40 mJ/m^2, and a highly variable acid–base (γ^{AB}) component, ranging from about 3 up to 22 mJ/m^2.

However, it should be stressed that, in general, there is no linear correlation between SFE and the number of adhered cells. In fact, for the different PET samples, whose SFE values range between 39.6 and 47.1 mJ/m^2, the measured ID values are almost identical (see Fig. 4 above). On the other hand, for the higher SFE plasma-treated PHMS (61.5 mJ/m^2) the cell adhesion is lower by a factor of 1.7 with respect to Ar$^+$-irradiated PHMS, having an SFE value lower by 5.4 mJ/m^2.

This fact suggests that other factors also, unexplored in the present paper, for instance, the polar character of the surfaces, surface charge, electrical domains, topography, radicals, etc., may play important roles in prompting or preventing cell adhesion [41–53]. More experimental work is actually needed to identify the relevant parameters.

4. CONCLUSIONS

The results reported in the present paper indicate that the cell behaviour onto various untreated and plasma- or ion-beam-modified PHMS and PET surfaces can be related to surface free energy (SFE), suggesting the existence of a threshold value roughly about 40 mJ/m^2 to obtain good cell adhesion.

At the present stage of the research, the detailed mechanisms for the role of the surface free energy in the complex phenomena involved in cell–surface interactions are a matter of speculation. In particular, for high SFE surfaces as those considered here, possible mechanisms could involve either adsorption from the culture medium of a suitable protein layer for an optimal cell adhesion [55, 62], or the switching of a biochemical signalling leading to the expression of adhesion proteins directly by the cellular membranes [63].

We are also aware that SFE is just one of the possible relevant factors involved in the complex process of cell interaction with a synthetic surface. More work is being undertaken in order to understand the detailed mechanism for the cell–surface and protein–surface interactions.

Acknowledgements

The authors acknowledge the financial support of FIRB 2001-RBNE01458S "Tecnologie per la manipolazione su scala nanometrica dei materiali e loro applicazione biomedicali" and FISR "Dispositivi a Silicio per la bioelettronica" (MIUR, Rome).

REFERENCES

1. W. J. Tan, G. P. Teo, K. Liao, K. W. Leong, H. Q. Mao and V. Chan, *Biomaterials* **26**, 891-898 (2005).
2. D. Delmar-Greenberg, M. Azam Ali and K. E. Gonsalves, *J. Mater. Sci. Mater. Med.* **14**, 833-834 (2003).
3. G. Paradossi, F. Cavalieri, E. Chiessi, C. Spagnoli and M. K. Cowman, *J. Mater. Sci. Mater. Med.* **14**, 687-691 (2003).
4. S. Drotleff, U. Lungwitz, M. Breunig, A. Dennis, T. Blunk, J. Tessmar and A. Gopferich, *Eur. J. Pharm. Biopharm.* **58**, 385-407 (2004).
5. E. Binzen, A. Lendlein, S. Kelch, D. Rickert and R. P. Franke, *Clin. Hemorheol. Microcirculation* **30**, 283-288 (2004).
6. N. A. Peppas, K. M. Wood and J. O. Blanchette, *Expert Opin. Biol. Ther.* **4**, 881-887 (2004).
7. K. L. Mittal (Ed.), *Polymer Surface Modification: Relevance to Adhesion*, Vol. 3. VSP, Utrecht (2004).
8. K. L. Mittal (Ed.), *Polymer Surface Modification: Relevance to Adhesion*, Vol. 2. VSP, Utrecht (2000).
9. M. Strobel, C. S. Lyon and K. L. Mittal (Eds.), *Plasma Surface Modification of Polymers: Relevance to Adhesion*. VSP, Utrecht (1994).
10. L. J. Gerenser, *Adhesion Sci. Technol.* **7**, 1019-1026 (1993).
11. J. L. Dewez, A. Doren, Y. J. Schneider and P. G. Rouxhet, *Biomaterials* **20**, 547-559 (1999).
12. Y. Lee, S. Han, H. Lim, Y. Kim and H. Kim, *Anal. Bioanal. Chem.* **373**, 595-600 (2002).
13. R. L. Clough, *Nucl. Instrum. Metods* B **185**, 8-33 (2001).
14. Y. Ozdemir, N. Hasirci and Serbetci, *J. Mater. Sci. Mater. Med.* **13**, 1147-1152 (2002).
15. I. K. Kang, O. H. Kwon, Y. M. Lee and Y. K. Sung, *Biomaterials* **17**, 841-847 (1996).
16. I. K. Kang, S. H. Choi, D. S. Shin and S. C. Yoon, *Int. J. Biol. Macromol.* **28**, 205-212 (2001).
17. C. P. Sharma, G. Jayasree and P. P. Najeeb, *J. Biomater. Appl.* **2**, 205-218 (1987).
18. E. Piskin, *J. Biomater. Sci. Polym. Edn.* **4**, 45-60 (1992).
19. D. J. Balazs, K. Triandafillu, P. Wood, Y. Chevolot, C. van Delden, H. Harms, C. Hollenstein and H. J. Mathieu, *Biomaterials* **25**, 2139-2151 (2004).
20. Q. Zhang, C. Wang, Y. Babukutty, T. Ohyama, M. Kogoma and M. Kodama, *J. Biomed. Mater. Res.* **60**, 502-509 (2002).
21. J. S. Bae, E. J. Seo and I. K. Kang, *Biomaterials* **20**, 529-537 (1999).
22. S. H. Hsu and W. C. Chen, *Biomaterials* **21**, 359-367 (2000).
23. J. H. Lee, J. W. Park and H. B. Lee, *Biomaterials* **12**, 443-448 (1991).
24. R. B. Patterson, A. Messier and R. F. Valentini, *ASAIO J.* **41**, M625-M629 (1995).

25. D. J. Li, F. Z. Cui and H. Q. Gu, *Biomaterials* **20**, 1889-1896 (1999).
26. L. Bacakova, V. Mares, M. G. Bottone, C. Pellicciari, V. Lisa and V. Svorcik, *Gen. Physiol. Biophys.* **18**, 1-53 (1999).
27. L. Bacakova, V. Svorcik, V. Rybka, I. Micek, V. Hnatowicz, V. Lisa and F. Kocourek, *Biomaterials* **17**, 1121-1126 (1996).
28. H. Sato, H. Tsuji, S. Ikeda, N. Ikemoto, J. Ishikawa and S. Nishimoto, *J. Biomed. Mater. Res.* **44**, 22-30 (1999).
29. J. S. Lee, M. Kaibara, M. Iwaki, H. Sasabe, Y. Suzuki and M. Kusakabe, *Biomaterials* **14**, 958-960 (1993).
30. G. Xu, Y. Hibino, Y. Suzuki, K. Kurotobi, M. Osada, M. Iwaki, M. Kaibara, M. Tanihara and Y. Imanishi, *Colloids Surfaces B: Biointerfaces* **19**, 237-247 (2000).
31. C. Satriano, E. Conte and G. Marletta, *Langmuir* **17**, 2243-2250 (2001).
32. B. Pignataro, E. Conte, A. Scandurra and G. Marletta, *Biomaterials* **18**, 1461-1470 (1997).
33. J. S. Lee, M. Kaibara, M. Iwaki, H. Sasabe, Y. Suzuki and M. Kusakabe, *Biomaterials* **14**, 958-960 (1993).
34. L. Bačáková, V. Mareš, V. Lisá and V. Švorčik, *Biomaterials* **21**, 1173-1179 (2000).
35. L. Bacakova, V. Mares, M. G. Bottone, C. Pellicciari, V. Lisa and V. Svorcik, *J. Biomed. Mater. Res.* **49**, 369-379 (2000).
36. M. Kaibara, H. Iwata, H. Wada, Y. Kawamoto, M. Iwaki and Y. Suzuki, *J. Biomed. Mater. Res.* **31**, 429-435 (1996).
37. J. H. Lee, J. W. Park and H. B. Lee, *Biomaterials* **12**, 443-448 (1991).
38. P. van der Valk, A. W. van Pelt, H. J. Busscher, H. P. de Jong, C. R. Wildevuur and J. Arends, *J. Biomed. Mater. Res.* **17**, 807-817 (1983).
39. T. I. Okpalugo, A. A. Ogwu, P. D. Maguire and J. A. McLaughlin, *Biomaterials* **25**, 239-245 (2004).
40. M. Kaibara, H. Iwata, H. Wada, Y. Kawamoto, M. Iwaki and Y. Suzuki, *J. Biomed. Mater. Res.* **31**, 429-435 (1996).
41. C. Satriano, S. Carnazza, S. Guglielmino and G. Marletta, *Nucl. Instrum. Methods* **B 208**, 287-293 (2003).
42. K. E. Schmalenberg, H. M. Buettner and K. E. Uhrich, *Biomaterials* **25**, 1851-1857 (2004).
43. A. Okumura, M. Goto, T. Goto, M. Yoshinari, S. Masuko, T. Katsuki and T. Tanaka, *Biomaterials* **22**, 2263-2271 (2001).
44. T. O. Collier, J. M. Anderson, W. G. Brodbeck, T. Barber and K. E. Healy, *J. Biomed. Mater. Res.* **69A**, 644-650 (2004).
45. J. P. Fisher, Z. Lalani, C. M. Bossano, E. M. Brey, N. Demian, C. M. Johnston, D. Dean, J. A. Jansen, M. E. Wong and A. G. Mikos, *J. Biomed. Mater. Res.* **68A**, 428-438 (2004).
46. G. Csucs, R. Michel, J. W. Lussi, M. Textor and G. Danuser, *Biomaterials* **24**, 1713-1720 (2003).
47. N. Q. Balaban, U. S. Schwarz, D. Riveline, P. Goichberg, G. Tzur, I. Sabanay, D. Mahalu, S. Safran, A. Bershadsky, L. Addadi and B. Geiger, *Nature Cell Biol.* **3**, 466-472 (2001).
48. Y. Ito, *Biomaterials* **20**, 2333-2342 (1999).
49. C. Xu , F. Yang, S. Wang and S. Ramakrishna, *J. Biomed. Mater. Res.* **71A**, 154-161 (2004).
50. R. J. Vance, D. C. Miller, A. Thapa, K. M. Haberstroh, and T. J. Webster, *Biomaterials* **25**, 2095-2103 (2004).
51. T. W. Chung, D. Z. Liu, S. Y. Wang and S. S. Wang, *Biomaterials* **24**, 4655-4661 (2003).
52. C. Satriano, N. Spinella, M. Manso, A. Licciardello, F. Rossi and G. Marletta, *Mater. Sci. Eng.* **C 23**, 779-786 (2003).
53. N. Kataoka, K. Iwaki, K. Hashimoto, S. Mochizuki, Y. Ogasawara, M. Sato, K. Tsujioka and F. Kajiya, *Proc. Natl. Acad. Sci. USA* **99**, 15638-15643 (2002).
54. G. Marletta and C. Satriano, in: *Frontiers in Multifunctional Integrated Nanosystems*, E. Buzaneva and P. Scharff (Eds.), pp. 71-94. Kluwer, Dordrecht (2004).
55. C. Satriano, G. Marletta, S. Carnazza and S. Guglielmino, *J. Mater. Sci. Mater. Med.* **14**, 663-670 (2003).

56. C. Satriano, S. Carnazza, A. Licciardello, S. Guglielmino and G. Marletta, *J. Vacuum Sci. Technol.* **A 21**, 1145-1151 (2003).
57. C. J. van Oss, M. K. Chaudhury and R. J. Good, *Chem. Rev.* **88**, 927-941 (1988).
58. J. Behnisch, A. Holländer and H. Zimmermann, *J. Appl. Polym. Sci.* **49**, 117-124 (1993).
59. C. Satriano, S. Carnazza, S. Guglielmino and G. Marletta, *Langmuir* **18**, 9469-9475 (2002).
60. C. Satriano, G. Marletta and E. Conte, *Nucl. Instr. Methods* **B 148**, 1079-1084 (1999).
61. G. Beamson and D. Briggs, *High Resolution XPS of Organic Polymers*, pp. 174-175. Wiley, London (1992).
62. P. Billsten, U. Carlsson, B. H. Jonsson, G. Olofsson, F. Hook and H. Elwing, *Langmuir* **15**, 6395-6399 (1999).
63. N. Faucheux, R. Schweiss, K. Lützow, C. Werner and T. Groth, *Biomaterials* **25**, 2721- 2730 (2004).

Contact Angle, Wettability and Adhesion, Vol. 4, pp. 487–499
Ed. K.L. Mittal
© VSP 2006

Favorable effects of acid–base sites on polymer coatings in the kinetics of oil removal

LAURENCE BOULANGÉ-PETERMANN,[1,*] BERNARD LUQUET[2] and MAUD VALETTE-TAINTURIER[2]

[1] *Ugine&Alz, Centre de Recherche d'Isbergues, BP 15, 62330 Isbergues, France*
[2] *Arcelor Flat Carbon, Centre d'Etudes et de Développement, BP 30109, 60761 Montataire, France*

Abstract—Our aim was to distinguish different surface parameters such as topography and surface energy which are likely to play roles in the cleaning kinetics of soiled polymer coatings. We selected polyurethane (PU), poly(ethylene terephthalate) (PET) and fluoro-epoxy (FE) coatings and tested these in an as-received condition and after an accelerated aging in water at 40°C for 1000 h. Two PU samples with similar surface compositions but different topographies were used. The surface free energy derived from contact angle measurements discriminates acid–base (PU) and monopolar basic (PET) surfaces from an apolar surface (fluoro-epoxy coating). The cleanability of soiled polymer coatings strongly depends on the surface chemistry. Acidic and basic sites, as described by van Oss, Chaudhury and Good theory, are favorable for oil removal using a cleaning solution containing anionic and non-ionic surfactants. The apolar surface is more difficult to clean. The surface topography was found to be a minor parameter in our experimental conditions. Therefore, the adaptation of the two-liquid phase system for oil removal gave the same trends as the ones obtained using a laminar flow cell. Lastly, the study of surface hydration by Environmental Scanning Electron Microscopy (ESEM) showed that many small water droplets condense on the acid–base PU coating; whereas on apolar FE coating, the droplets preferentially germinate only on surface defects.

Keywords: ESEM; surface hydration; polymer coating; fouling; cleaning.

1. INTRODUCTION

Carbon steels covered by polymer coatings are commonly used in hygienic applications and, more particularly, for the wall panel in cold storage rooms in the food industry, hospitals and in public buildings [1]. In such cold rooms, foodstuffs are only stored for a short time and during their handling, some splashing can occur on the wall-panel surfaces. This surface fouling can then constitute a favorable environment for further bacterial development. As a consequence, proper cleaning is essential [2–4].

---

*To whom correspondence should be addressed. Tel.: (33-3) 2163-5604; Fax: (33-3) 2163-2056; e-mail: laurence.boulange@ugine-alz.arcelor.com

Various organic coatings have been proposed for the wall-panel manufacturing. Numerous studies have been reported on corrosion resistance and chemical inertness of these coatings [5–7], but only a few investigations dealing with the easy-to-clean properties have been performed on steel coatings, assessed in as-received condition and after water aging [8]. Actually, the coating properties can be modified after exposure to UV, cleaning products, or high relative humidity. In practice, a coating degradation due to blister formation is very often observed [5]. This is the reason why the materials must be assessed both in an as-received condition and after an accelerated aging test in water at 40°C for 1000 h.

Our aim here was to distinguish different surface parameters, such as topography, surface energy and surface chemistry, which are likely to play roles in the cleaning kinetics of soiled polymer coatings. We selected polyurethane (PU), poly(ethylene terephthalate) (PET) and fluoro-epoxy coatings and tested these in an as-received condition, as well as after accelerated aging in water at 40°C for 1000 h. Two samples of PU with similar surface compositions but different topographies were used. We also developed new methods for surface characterization such as the observation of water droplets germination on polymer coatings by environmental scanning electron microscopy (ESEM).

2. MATERIALS AND METHODS

2.1. Selection of materials

Galvanized steel sheets coated with polymer coatings were selected as the solids in this study. The polymer coatings were applied to steel surfaces according to the following procedure [9]. At the first stage of activation and passivation, an electrolytic degreasing was performed in hot alkaline solutions. The medium contained caustic soda, sodium ortho-phosphate and non-ionic surfactants at 90°C with a current density of 10 A/dm² [10]. A first coating, named primer, was then applied with a roller to a thickness of 5–30 μm to provide anti-corrosive and adhesion properties. The coated steel was heated in an oven at 250°C and then cooled to room temperature. A second coating, named topcoat, was applied with a coating machine (roll-coat). The second coating thickness was from 20 to 200 μm. As previously described, the coated steel was heated in an oven and then cooled.

The following coatings (topcoats) were studied: polyurethane (PU), poly(ethylene terephthalate) (PET) and fluoro-epoxy (FE). Two PU samples (PU1 and PU2) with similar surface compositions but different topographies were selected.

Prior to contact angle measurements and soiling experiments, the coated surfaces were soaked in 95% ethanol for 10 min, air dried and then soaked in an alkaline detergent (Procter & Gamble, Neuilly sur Seine, France) at a concentration of 1.2% (v/v), which corresponds to pH 8.2 at 20°C. The cleaning solution contained anionic surfactants (less than 5%) and non-ionic surfactants between 5 and 15% (w/w). The sample was rinsed in distilled water and dried with a tissue paper. Henceforth, coatings such prepared will be called fresh samples. To simulate coating aging, the samples were immersed in water at 40°C for 1000 h.

2.2. Solid surface characterization

The coating surfaces were characterized in terms of surface free energy and topography. Surface hydration was also observed by Environmental Scanning Electron Microscopy (ESEM).

2.2.1. Solid surface free energy, γ_S

Contact angles were measured on coating surfaces (S) by the sessile drop technique using diiodomethane, formamide and water (L). The surface tension and its components for liquids are given in Table 1. The contact angle was directly measured with a Krüss G-10 goniometer. The coating surface free energy, γ, was determined from contact angle measurements by the following equation [11, 12]:

$$\gamma_L(\cos\theta+1) = 2\left(\gamma_S^{LW}\gamma_L^{LW}\right)^{1/2} + 2\left(\gamma_S^-\gamma_L^+\right)^{1/2} + 2\left(\gamma_S^+\gamma_L^-\right)^{1/2} \tag{1}$$

where γ^{LW} denotes the apolar Lifshitz–van der Waals component, γ^+ the acidic parameter (electron acceptor) and γ^- the basic parameter (electron donor). The acid–base component (γ^{AB}) is given by

$$\gamma^{AB} = 2\left(\gamma^+\gamma^-\right)^{1/2} \tag{2}$$

The coating surface free energy is expressed in mJ/m². The surface free energy was measured on both fresh and aged coatings.

Table 1.
Liquid surface tension and its components (in mJ/m²) [11]

Liquid	γ_L	γ_L^{LW}	γ_L^+	γ_L^-	γ_L^{AB}
Water	72.8	21.8	25.5	25.5	51
Formamide	58	39	2.28	39.6	19
Diiodomethane	50.8	50.8	0	0	0

2.2.2. Coating surface topography

The topographic parameters measured were the arithmetic average roughness (S_a), the maximum peak-to-valley height (S_t) and the skewness S_{sk} expressed in μm [13]. These parameters were deduced from optical profiler scans of size 100×100 μm² using the Surfvision software.

2.2.3. Environmental scanning electron microscopy (ESEM)

Hydration and dehydration cycles on PU1 and FE were performed in an ESEM (Philips XL30 LaB$_6$). The acceleration voltage was 30 kV. The sample (size 5×5 mm^2) was placed on a Philips PW6750 Peltier cooled specimen stage. The first water droplets appeared by condensation on the sample surface at a pressure between 5 and 5.8 Torr and a temperature between 2 and 3°C.

2.3. Study of oil removal

2.3.1. Cleanability in dynamic conditions

The surface cleaning was carried out in a laminar flow cell [14]. Prior to cleaning, the coatings were soiled by spraying oil with a brush soaked in the oil. The nutritious sunflower oil (Lesieur, France) used is mainly composed of fatty acids (13%), monounsaturated fatty acid (oleic acid) (22%) and polyunsaturated acids (65%).

The coated surfaces thus obtained were completely covered by small droplets of oil with a diameter not exceeding 1 µm. The sample size was 13 cm². The oil quantity deposited by the brush was 8.5±1.5 mg which was controlled by weighing the sample on a precision scale (Mettler AE 240). This leads to an oil density of 0.70±0.15 mg/cm².

The laminar flow cell consisted of two flat plates between which a commercial detergent (Procter & Gamble, Neuilly sur Seine, France) at a concentration of 1.2% (v/v) circulated and the bottom plate was the sample to be cleaned. The cleaning solution contained anionic surfactants (less than 5%) and non-ionic surfactants between 5 and 15% (w/w). The cleaning was monitored in 20-s intervals at 20°C. There was no intermediate water rinse. Between every step of cleaning in the laminar flow cell, the sample was dried at 20°C at 0% relative humidity and weighed. Based on the weight difference, it was possible to evaluate the quantity of oil removed by the cleaning procedure where w_0 is the initial weight of the soiled surface by the nutritious oil and w_t the weight measured after cleaning sequences at time t. The cleaning kinetics were studied on all fresh coating materials, as well as on aged coatings. The experiment was repeated four times. An ANOVA test was performed with Statlets® software version 2.1 (StatPoint, Orlean, VA, USA).

2.3.2. Cleanability in static conditions

We adapted the two-liquid phase system initially proposed by Schultz et al. [15, 16] to measure the variations of interfacial tensions between the cleaning solution and the oil deposit on polymer coatings. First, the coated steels were horizontally immersed in the oil. Contact angles were recorded at the triple point (coating–cleaning solution–oil interface) as a function of time, for times between 0 and 60 s. The measurements were carried out on fresh PU1, PET and FE, as well as on aged coatings.

3. RESULTS

3.1. Physical chemistry of polymer coatings

As indicated by the roughness parameters S_a, S_t and S_{sk}, the coatings studied displayed smooth surfaces, except for the material referred as FE (Table 2). The PU coatings display similar surface composition but their topography differs. On PU1, the skewness (S_{sk}) is positive attesting some peaks on the surface, whereas on PU2 it is negative indicating some holes in the surface. These values are in accord with previous data [4, 8]. The PET coating possesses $C=O–O$, $O–CH_2$, $CH_2=CH–CO–O$ and $C–N$ functions, whereas PU coatings contain $C=O–O$, $C=O–O–CH_2$ and $O–CH_2$ functions. Lastly, the FE surface is mainly composed of $C=C$, CH_2 and CH_3 functions.

As shown in Table 2, water contact angles measured on fresh coatings vary from $68\pm2°$ to $100\pm3°$. On PU and FE coatings, the contact angles decreased after water aging. Nevertheless, on PET there was no significant difference in contact angle values on fresh coating or after aging in water.

The LW component varied from 26 to 42 mJ/m². The highest values were obtained on PET whereas the lowest were measured on FE. Intermediate values of $\gamma_S^{LW} = 30$ mJ/m² were calculated on PU coatings. Fluoro-epoxy coating can be considered as apolar, as only a small γ_S^- value, around 1.7 mJ/m², is obtained.

Table 2.
Main topographic characteristics measured on fresh polymer coatings

Coating	S_a (μm)	S_t (μm)	S_{sk}	H_2O	$HCONH_2$	CH_2I_2	γ_S^{LW}	γ_S^+	γ_S^-	γ_S^{AB}
PU1 fresh	0.03	1.54	0.07	81 (4)	73 (3)	57 (4)	30	0	13.1	0
PU1 after aging	0.06	0.66	0.20	71 (4)	58 (2)	52 (4)	33	0.18	14.5	3.2
PU2 fresh	0.04	0.51	-0.10	85 (2)	70 (3)	57 (2)	30	0	7.1	0
PU2 after aging	0.02	0.38	-0.08	68 (2)	49 (3)	40 (2)	40	0.3	12.7	4
PET fresh	0.04	0.86	2.74	78 (3)	60 (2)	39 (2)	40	0	8.6	0
PET after aging	0.04	0.94	1.55	78 (2)	58 (2)	36 (2)	42	0	7.6	0
FE fresh	0.69	8.25	1.50	100 (3)	81 (2)	62 (3)	27	0	1.7	0
FE after aging	0.89	11.3	1.42	90 (4)	75 (4)	64 (3)	26	0.03	5.3	0.8

S_a is the arithmetic average roughness, S_t the maximum peak-to-valley height and S_{sk} the skewness and the results of contact angle measurements on these coatings (expressed in degrees) and their surface free energy and its components (in mJ/m²). The error limits are shown in parentheses.

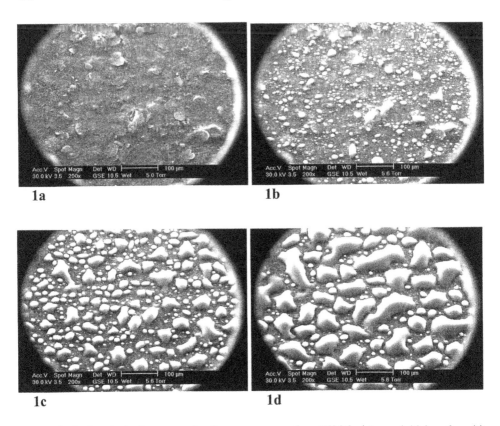

1a

1b

1c

1d

Figure 1. Surface hydration on apolar fluoro-epoxy coating. ESEM pictures: initial surface (a), germination of first water droplets (b), growth of water droplets on the FE surface defects (c) and final condensation (d).

Fresh PU and PET coatings display a monopolar basic character as γ_S^- varies from 7.1 to 13.1 mJ/m².

After water aging, the surface energy components of PET do not change whereas PU and FE coatings display acid–base character. The acidic parameter (γ_S^+) values are, respectively, 0.18, 0.3 and 0.03 mJ/m² on aged PU1, PU2 and FE. After water aging, a small acid–base component was obtained on the initial apolar FE. This difference in the surface energy could be attributed to water uptake in the first few layers of the coating [8].

To summarize, three types of coatings can be distinguished: (i) surface with strong apolar character and an irregular topography such as fresh FE, (ii) surface with a monopolar basic character such fresh PU and PET coatings, (iii) surface with acid–base properties such as PU and FE after water aging. In the study of surface hydration by ESEM, we selected two coatings for which the surface energy components varied strongly after aging. The ESEM observations confirmed

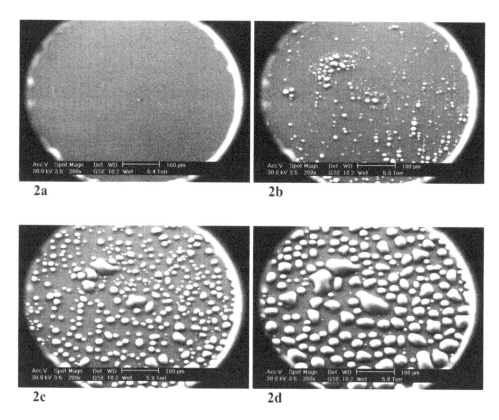

Figure 2. Surface hydration on monopolar basic PU1 coating. ESEM pictures: initial surface (a), germination of many small water droplets (b), water condensation (c) and formation of many water droplets (d).

the rough surface and the presence of peaks on FE coating (Fig. 1a). A soon as the pressure increases in the ESEM chamber, some water droplets will condense on the coatings. On FE material (Fig. 1b), the droplets preferentially germinate on the peaks. Then, in the second step, the droplets grow on the surface and more particularly on the defects (Fig. 1c) and finally they coalesce (Fig. 1d). On the other hand, the PU1 surface is smoother (Fig. 2a), but still many small water droplets germinate on PU1 (Fig. 2b). At higher pressure in the ESEM chamber, the water condensation is more homogeneous (Fig. 2c) and only a few water droplets finally condense (Fig. 2d).

3.2. Cleaning

As illustrated in Fig. 3a, the PU and PET coatings are easier to clean from the oil than the FE coating. After 40 s of cleaning in the laminar flow cell, the percentages of oil removed by the flow of cleaning solution were 72±10%, 81±4% and

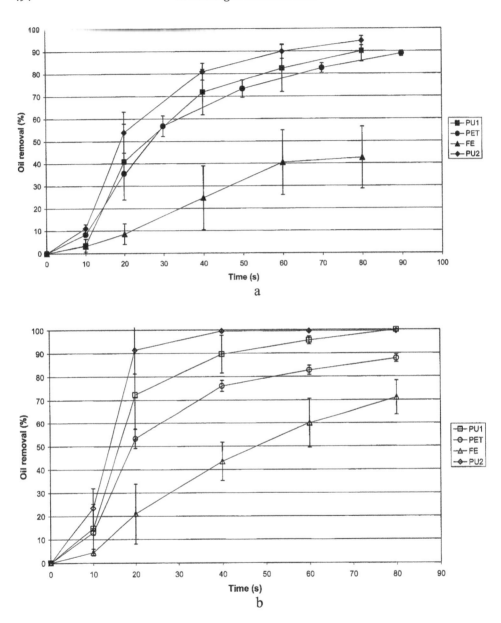

Figure 3. Cleaning kinetics for oil removal from polymer coatings tested fresh (a) and after aging in water (b). The bars represent the standard deviation.

65±5% for PU1, PU2 and PET, respectively. In the same experimental conditions, only 25±14% was removed from the FE coating ($P = 0.003$).

After aging in water, the cleaning kinetics were faster on PU coatings (Fig. 3b). The oil quantities removed after 40 s of cleaning were 90±8% and 100±0% on

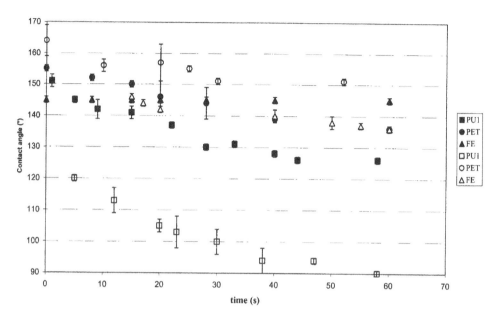

Figure 4. Evolution of contact angle as a function of time in a two-liquid phase system where the solid is initially immersed in the oil and a droplet of cleaning solution is then deposited on the coating tested fresh and after aging in water. The closed symbols correspond to fresh coatings whereas open symbols correspond to aged coatings.

PU1 and PU2 coatings, respectively. The cleaning kinetics was slightly improved on aged FE. Indeed, after 40 s, 43±8% of the oil was removed.

In the adaptation of the two-liquid phase system, the highest oil displacement by the droplet of cleaning solution was observed on aged PU1 and the contact angle was 90° after 60 s. Intermediate variations in contact angle values were recorded on fresh PU1 and PET coatings, and these were between 180° and 120° (Fig. 4). Lastly, only a small displacement of oil was seen on FE. On the whole, a similar classification as from studies in the laminar flow cell of coating surfaces was obtained here as the largest oil displacement by droplets of cleaning solution was observed on the aged PU1, followed by PET and then FE.

The study of surface cleaning in a dynamic system or in static conditions leads to the same order in the coating performance. By using these two modes, it is possible to distinguish the physico-chemical effect of surfactants from the mechanical effect induced by the laminar flow on the coating surface.

4. DISCUSSION

Our aim was to distinguish different surface parameters such as topography, surface energy and chemistry which are likely to play roles in the cleanability of soiled polymer coatings.

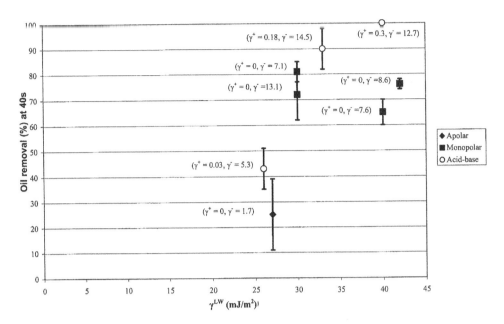

Figure 5. Percentage of oil removal after 40 s of cleaning in the laminar flow cell as a function of the Lifshitz–van der Waals component of the coating surface energy. The acidic and basic parameters of every coating material are also shown in parentheses in the figure. All the surface free energy components are expressed in mJ/m².

As shown previously, the coatings under assessment display variations in topography and surface energy components. Also, water aging can induce some modifications in the surface energy. This was clearly observed on PU and FE coatings, as both coatings displayed acidic sites after water aging. These modifications could be attributed to water adsorption on these surfaces. Indeed, the surface properties of the coating play a role in surface hydration. On PU coating surface, there are some acid–base sites on the surface; this can help germination of many small water droplets on the surface. On the other hand, on the apolar material (FE), the germination of water occurs preferentially on the surface defects and leads to the formation of larger droplets.

There is a relationship between the surface energetic characteristics of coatings and the "ease of cleaning" property. To correlate this, we represented the percentage of oil removed by the flow of cleaning solution after 40 s as a function of the LW component of the surface free energy of the coating (Fig. 5). We have also shown in this figure the acidic and basic parameters of each coating. The ease of cleaning is expressed here in terms of the percentage of oil removed after 40 s. We chose this particular timing because the scatter in the measurements is smaller than in the beginning of the cleaning as observed in Fig. 3. Lastly, all the coatings under assessment display variations in the LW component, acidic and/or basic parameters as well as topography. To distinguish the topography effect from the surface chemistry, we compared the experimental data where only one surface parameter varies, whereas the others remained constant.

As mentioned previously, the worst coating to clean was FE ($\gamma_S^{LW} = 27$ mJ/m², $\gamma_S^+ = 0$ mJ/m², $\gamma_S^- = 1.7$ mJ/m²). On aged FE ($\gamma_S^{LW} = 26$ mJ/m², $\gamma_S^+ = 0.03$ mJ/m², $\gamma_S^- = 5.3$ mJ/m²), the percentage of oil removed is twice as high. Favorable effects of acid–base interactions could explain this improvement in the cleaning properties as both surfaces have similar topographies. In addition, if we consider FE ($\gamma_S^{LW} = 27$ mJ/m², $\gamma_S^+ = 0$ mJ/m², $\gamma_S^- = 1.7$ mJ/m²), PU1 ($\gamma_S^{LW} = 30$ mJ/m², $\gamma_S^+ = 0$ mJ/m², $\gamma_S^- = 13.1$ mJ/m²) and PU2 ($\gamma_S^{LW} = 30$ mJ/m², $\gamma_S^+ = 0$ mJ/m², $\gamma_S^- = 7.1$ mJ/m²), the percentages of oil removed will be three times as high on these monopolar basic surfaces than on apolar FE. Nevertheless, the PU surfaces are smoother than the FE and we still cannot conclude definitively about the possible role of the topography. However, if we compared both PU monopolar basic surfaces where surface energetics is similar but the topography differs, we did not see a great difference in cleaning performance. Therefore, we can conclude that here the topography is not a first-order parameter in the cleaning performance.

The comparison of both fresh PU coatings with fresh PET ($\gamma_S^{LW} = 40$ mJ/m², $\gamma_S^+ = 0$ mJ/m², $\gamma_S^- = 8.6$ mJ/m²) and after water aging ($\gamma_S^{LW} = 42$ mJ/m², $\gamma_S^+ = 0$ mJ/m², $\gamma_S^- = 7.6$ mJ/m²) did not reveal any significant variation in the cleaning performance. Thus, in our experimental conditions, an increase in γ_S^{LW} value by about 10 mJ/m² did not strongly affect the degree of cleaning. Lastly, if we compare two pairs of coatings PU1 ($\gamma_S^{LW} = 30$ mJ/m², $\gamma_S^+ = 0$ mJ/m², $\gamma_S^- = 13.1$ mJ/m²) with aged PU1 ($\gamma_S^{LW} = 33$ mJ/m², $\gamma_S^+ = 0.18$ mJ/m², $\gamma_S^- = 14.5$ mJ/m²) and PET ($\gamma_S^{LW} = 40$ mJ/m², $\gamma_S^+ = 0$ mJ/m², $\gamma_S^- = 8.6$ mJ/m²) with aged PU2 ($\gamma_S^{LW} = 40$ mJ/m², $\gamma_S^+ = 0.3$ mJ/m², $\gamma_S^- = 12.7$ mJ/m²), the presence of acidic sites on the coatings appear to be favorable for cleaning. There is a 20% gain in the oil removal on acid–base surfaces in comparison with monopolar basic ones.

On the whole, acid–base interactions play an important role in the cleaning of surface coatings; the topography is a secondary parameter. Last, in our experimental conditions, variations of 10 mJ/m² in γ_S^{LW} values do not strongly affect the oil removal. These conclusions are in accord with the model proposed by van Oss and Costanzo [17, 18], where the authors used pure hexadecane to soil the surfaces and solutions of pure surfactants to clean them. Both anionic and non-ionic molecules display a strong monopolar basic character for the head group of the molecule and a LW character for the tail. Indeed for sodium dodecyl sulfate (SDS) for the head-group, $\gamma^{LW} = 34.6$ mJ/m², $\gamma^+ = 0$ mJ/m², $\gamma^- = 46$ mJ/m² and for the tail, $\gamma^{LW} = 23.8$ mJ/m², $\gamma^+ = \gamma^- = 0$ mJ/m². As for a non-ionic molecule, such as poly(ethylene oxide), this molecule is a monopolar Lewis base with $\gamma^{LW} = 43$ mJ/m², $\gamma^+ = 0$ mJ/m², $\gamma^- = 64$ mJ/m². In our work, we used a commercial formulation of cleaning product containing anionic and non-ionic surfactants where it was not possible to calculate correct values of γ_L^+ and γ_L^-. Therefore, the model proposed by van Oss [12] is shown to be applicable here, in a practical application, on industrial polymer coatings and commercial detergent solutions.

5. CONCLUSIONS

The cleanability of soiled polymer coatings (in our case by oil) strongly depends on the coating surface chemistry. Acidic and basic sites, as described by the van Oss, Chaudhury and Good theory [11], are favorable for the oil removal by a cleaning solution containing anionic and non-ionic surfactants. This is in agreement with the model proposed by van Oss [12, 17]. Apolar surfaces are more difficult to clean. The surface topography was found to be a second-order parameter in our experimental conditions. The adaptation of the two-liquid phase system to the oil removal study gave the same trends as the ones obtained by using a laminar flow cell. By using these two cleaning methods, it is possible to distinguish the physico-chemical effect of surfactants from the mechanical effect induced by the laminar flow.

Finally, water aging can change the surface chemistry and, consequently, the ease of cleaning.

REFERENCES

1. Z. W. Wicks Jr., F. N. Jones and S. P. Pappas (Eds.), *Organic Coatings Science and Technology, Vol 2: Applications, Properties and Performance*. Wiley Interscience, New York, NY (1992).
2. E. A. Zottola, *Food Technol.* **48**, 107 (1994).
3. L. Boulangé-Petermann, *Biofouling* **10**, 275 (1996).
4. L. Boulangé-Petermann, E. Robine, S. Ritoux and B. Cromières, *J. Adhesion Sci. Technol.* **18**, 213 (2004).
5. E. Deflorian, L. Fedrizzi and P. L. Bonora, in: *Organic Coatings for Corrosion Control*, G. P. Bierwagen (Ed.), pp. 92–105. American Chemical Society, Washington, DC (1998).
6. R. Lambourne, in: *Paint and Surface Coatings: Theory and Practice, second edition*. R. Lambourne and T. A. Strivens (Eds.), pp. 658–693. Woodhead, Cambridge (1993).
7. G. R. Hayward, in: *Paint and Surface Coatings, Theory and Practice, second edition*. R. Lambourne and T. A. Strivens (Eds.), pp. 725–766. Woodhead, Cambridge (1993).
8. L. Boulangé-Petermann, C. Debacq, P. Poiret and B. Cromières, in: *Contact Angle, Wettability and Adhesion, Vol 3*, K. L. Mittal (Ed.), pp. 501–519. VSP, Utrecht (2003).
9. G. R. Pilcher, *Eur. Coat. J.* **3**, 62 (1999).
10. C. Bonnebat, R. Hellouin, F. Carrara and C. Fatrez, in: *The Book of Steel*, G. Béranger, G. Henry and G. Sanz (Eds.), pp. 673–712. Lavoisier, Secaucus, NJ (1994).

11. C. J. van Oss, M. K. Chaudhury and R. J. Good, *Chem. Rev.* **88**, 927 (1988).
12. C. J. van Oss, *Interfacial Forces in Aqueous Media*. Marcel Dekker, New York, NY (1994).
13. E. S. Adelmawla, M. M. Koura, T. M. A. Maksoud, I. M. Elewa and H. H. Soliman, *J. Mater. Proc. Technol.* **123**, 133 (2002).
14. I. H. Pratt-Terpstra, A. H. Weerkamp and H. J. Busscher, *J. Gen. Microbiol.* **133**, 3199 (1987).
15. J. Schultz, K. Tsutsumi and J. B. Donnet, *J. Colloid Interface Sci.* **59**, 272 (1977).
16. J. Schultz, K. Tsutsumi and J. B. Donnet, *J. Colloid Interface Sci.* **59**, 277 (1977).
17. C. J. van Oss and P. M. Costanzo, *J. Adhesion Sci. Technol.* **6**, 477 (1992).
18. C. J. van Oss and P. M. Costanzo, in: *Contact Angle, Wettability and Adhesion*, K. L. Mittal (Ed.), pp. 879–889. VSP, Utrecht (1993).

Contact Angle, Wettability and Adhesion, Vol. 4, pp. 501–514
Ed. K.L. Mittal
© VSP 2006

Analysis of adhesion contact of human skin *in vivo*

C. PAILLER-MATTEI,[1*] R. VARGIOLU[1] and H. ZAHOUANI[1, 2]

[1]*Laboratoire de Tribologie et Dynamique des Systèmes, UMR-CNRS 5513, Ecole Centrale de Lyon, 36 Avenue Guy de Collongue, 69134 Ecully cedex, France*
[2]*Ecole Nationale d'Ingénieurs de Saint-Etienne,58 rue Jean Parot, 42100 Saint-Etienne, France*

Abstract—Recent work in the literature based on wettability has shown both hydrophobic and hydrophilic characteristics of the surface of the human skin depending on the quantity of lipids or sebum in the skin surface areas. In this work, we compare the adhesion energy determined by the wettability method and the contact mechanics method which we developed. The contact mechanics method is based on a micro-indentation test with a low load (20–30 mN). The evolution of adhesion according to various experimental parameters such as the normal load, contact time, indentation speed and liquid treatment to skin was studied. For the liquid treatment applied to skin, our approach allows to follow variations in physico-chemical and skin mechanical properties as a function of time after the application.

Keywords: Indentation test; adhesion; wettability; human skin.

1. INTRODUCTION

Human skin is a living and complex visco-elastic material, composed of several heterogeneous and anisotropic tissue layers. The different layers from the surface to the bulk are: the stratum corneum (thickness 10–20 μm), the epidermis (50–100 μm) and the dermis (800–1200 μm). The skin consists of a network of collagen and interspersed elastic fibres, all covered by an epidermal layer of partially keratinized cells that are progressively dehydrated toward the outer surface or stratum corneum.

The surface of the skin is not smooth; it has a specific topography which has a physiological and sometimes a pathological significance [1]. The mechanical properties of skin are very similar to the mechanical properties of a soft elastomer. They depend on body zone, temperature, humidity, age, sex, etc. It is possible to modify skin properties (for example, stiffness, S or viscosity, η) with diverse skin treatments as a cosmetic or dermatological treatment.

---

*To whom correspondence should be addressed. Tel.: (33-4) 7218-6291; Fax: (33-4) 7843-3383; e-mail: Cyril.Pailler-Mattei@ec-lyon.fr

The physico-chemical and adhesion properties of the skin surface are controlled by a complex mixture of lipids which are secreted by mammalian sebaceous glands, and form a liquid film over the skin surface. In general, the thickness of the skin surface film is between 0.5 and 5 µm, depending on the body zone [2].

The surface free energy components of the forearm zone have been determined by measuring the contact angles before and after ether treatment [3]. The volar forearm zone was selected as a "dry skin" because of its low density of sebaceous orifices, less than 50 cm^{-2}. It was shown that ether treatment produced a significant increase in contact angles [4]. Contact angles for diiodomethane increased from 34° to 48° and from 41° to 55° on the forehead and forearm, respectively. The forehead surface became hydrophobic (θ_{water}=84°) and the forearm surface more hydrophobic (θ_{water}=101°). The forearm has a hydrophobic or weakly monopolar basic surface and the forehead has a hydrophilic or strongly monopolar basic surface. Both epidermal lipids and sebum were characterised by their salient electron-donor component which reduces their interfacial tension with water. Consequently, they increase skin wettability by water [4].

Our aim in this context was to study the skin adhesion behaviour using the contact mechanics approach. The contact mechanics approach is based on indentation tests. First of all, to validate our skin indentation device, we analysed the adhesion energy between two identical soft elastomers, with known mechanical and physico-chemical properties, by comparing their interfacial energy when they were immersed in two different media (air and water) [5]. The results obtained showed that our indentation device was able to measure the adhesion energy between two solids. Then the skin adhesion behavior was analysed on the inner forearm surface. We studied the variation of the skin/indenter adhesion energy as functions of the normal load, the indentation speed and the nature of the indenter material. To conclude we studied the variation of skin adhesion energy after application of a liquid treatment on the skin surface. This technique allowed to measure the modification of the skin surface physico-chemical properties as a function of time after liquid treatment application.

2. MATERIALS AND METHODS

The experimental device developed to study adhesion forces and mechanical properties of human skin *in vivo* uses an indentation test with a light load on the internal forearm. The presence of adhesion forces during the indentation cycle modifies the contact area and the Hertzian pressure field in the contact zone [6]. The indentation system employs a fully automated, highly sensitive configuration based on the measurement of load versus penetration depth between a spherical indenter and a human inner forearm. The automated system controls the applied normal load, penetration depth and offers a wide range of normal velocities. The

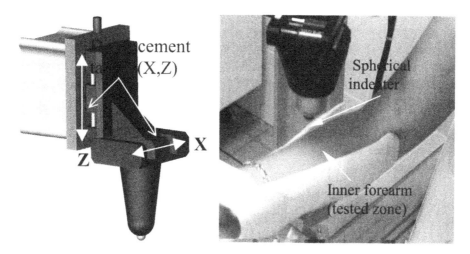

Figure 1. Skin tribometer device. Left: schematic representation of the indentation device; right: zone of the indentation tests.

displacement of the indenter is controlled by a displacement sensor (Fig. 1). The maximum displacement during the loading–unloading cycle is 1.5 cm with a resolution of 10 μm. The normal load resolution is obtained by the high precision of the mechanical probe. The load range for an indentation is 1–100 mN. In the loading–unloading cycle (hysteresis cycle), not only the elastic component but also viscous properties can be assessed. When the given normal pressure is achieved, during the unloading phase, the indenter returns to the starting position with the same velocity.

Some important quantities such as the maximum load $F_{N_{max}}$, maximum displacement δ_{max}, adhesion force F_{ad} or the slope of the initial portion of the unloading curve, $S = \dfrac{dF_N}{d\delta}$ are determined from the indentation curve (Fig. 2). S is the same as the force per unit distance, and so it is also known as the elastic stiffness of the contact. For any axisymmetric indenter the reduced Young's modulus $E*$ is obtained from [7, 8]:

$$E* = \frac{\sqrt{\pi}}{2} \frac{S}{\sqrt{A}} \tag{1}$$

where S is the stiffness of the contact and A the projected contact area. The contact area, A, between the indenter and the material is calculated with the adhesion elastic theory [9].

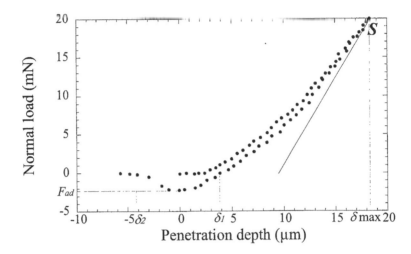

Figure 2. Indentation curve for hard plane elastomer with a spherical steel indenter. δ_1 and δ_2 are, respectively, the displacement for $F_N=0$ and the displacement at the end of indentation test. The test was carried out at indentation speed $V=20$ µm/s.

3. VALIDATION OF THE INDENTATION DEVICE FOR THE ADHESION MEASUREMENT

To validate the skin adhesion measurement by the indentation test, we used an elastomer. The idea is to compare two different values of interfacial energy $\gamma^i_{M_1/M_2}$ between the same two elastomers (M_1) and (M_2) in two different media (i=air or water) (Fig. 3). The elastomers have different reduced Young's moduli: $E_1^* \gg E_2^*$.

First of all, the values of surface energies of M_1 and M_2 in the two media were determined by the wettability tests (Table 1) (the Owens–Wendt method) [4].

We note that the surface energies of M_1 and M_2 are the same in both air and water and, thus, the physico-chemical properties of two these solid surfaces should be similar.

The adhesion energy w (or w_{123}) between two solid surfaces immersed in a third medium is given by [5] (Fig. 3):

$$w_{123} = \gamma_{13} + \gamma_{23} - \gamma_{12} = w \qquad (2)$$

γ_{13} surface energy ((M_1)/medium (air or water)),

γ_{23} surface energy ((M_2)/medium (air or water)),

γ_{12} interfacial energy ((M_1)/ (M_2)).

The w_{123} value can be positive (attraction between (M_1) and (M_2)) or negative (repulsion between (M_1) and (M_2)).

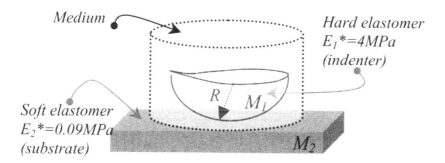

Figure 3. Experimental test to calculate the interfacial energy between M_1 and M_2. The values of reduced Young's modulus are determined by the indentation test. M_2 is a semi-infinite substrate and M_1 is the spherical elastomer indenter.

Table 1.
Surface energy (in mJ/m²) of elastomers in two different media (air, water) with the Owens–Wendt method

	Air	Water
Elastomer (M_1)	$\gamma_{M_1} = 22.7$	$\gamma_{M_1} = 55$
Elastomer (M_2)	$\gamma_{M_2} = 23$	$\gamma_{M_2} = 54$

According to relation (2) the value of γ_{12} is independent of the third medium, so:

$$\gamma_{M_1/M_2}^{air} = \gamma_{M_1/air} + \gamma_{M_2/air} - W_{air} \tag{3}$$

$$\gamma_{M_1/M_2}^{water} = \gamma_{M_1/water} + \gamma_{M_2/water} - W_{water} \tag{4}$$

and

$$\gamma_{M_1/M_2}^{air} = \gamma_{M_1/M_2}^{water} \tag{5}$$

Finally we obtain:

$$\gamma_{M_1/air} + \gamma_{M_2/air} - W_{air} = \gamma_{M_1/water} + \gamma_{M_2/water} - W_{water} \tag{6}$$

where w_{air} and w_{water} are the adhesion energies in air and water, respectively.

The values of adhesion energies (w_{air} and w_{water}) can be found with two different methods. First, it can be obtained by the indentation test with the theoretical contact mechanics method, i.e., the JKR theory [10]. The JKR theory takes into account surface forces only inside the contact zone and neglects surface forces outside of it. The effect of adhesion forces is to increase the contact area between the indenter and the material. For a JKR contact, there is a sudden separation of surfaces when the normal applied load is less than $-\dfrac{3}{2}\pi R w$.

For a sphere/plane contact, the JKR theory gives the following results:
Adhesion force:

$$F_{ad} = -\frac{3}{2}\,\pi\,R\,w \tag{7}$$

$$w = -\frac{2}{3}\frac{F_{ad}}{\pi\,R} \tag{8}$$

Contact radius:

$$a = \sqrt[3]{\frac{3}{4}\frac{R}{E^*}\,(F_N + 2F_{ad} + 2\sqrt{F_{ad}\,(F_N + F_{ad})})} \tag{9}$$

where F_N is normal load, F_{ad}, adhesion force, R, radius of curvature of indenter (R=6.35 mm), E^*, reduced Young's modulus $\dfrac{1}{E^*} = \dfrac{1-v_1^{\,2}}{E_1} + \dfrac{1-v_2^{\,2}}{E_2}$ and E_1 and E_2 are, respectively, the Young's moduli of the indenter (210 GPa for steel indenter and 4 MPa for elastomeric indenter) and the substrate. In our study $E_1 \gg E_2$; thus, the reduced Young's modulus is $\dfrac{1}{E^*} = \dfrac{1-v^2}{E_2}$, where v is the Poisson's ratio. Moreover, we can measure the experimental value of adhesion energy from the work of adhesion, W_{ad}, defined as:

$W_{ad} = -\int_{\delta_2}^{\delta_1} F_N(\delta)\big|_{unloading}\,\mathrm{d}\delta$, where δ is the penetration depth in the indentation

curve (Fig. 4). To obtain the value of experimental adhesion energy in J/m² (or mJ/m²), we divide the work of adhesion by the normal contact area $A_0 = \pi a^2\big|_{F_N=0}$ for F_N=0.

We note that the values of adhesion force and the work of adhesion are higher for the contact in water medium than for the contact in the air medium (Fig. 4).

From the interfacial energy values obtained (Table 2), we can state that our indentation device is able to measure correctly the adhesion energy between two solids. In fact, we note that the interfacial energy measured in two different media and with two different methods are similar (the difference between all interfacial energy values measured is less than 8%), which is expected as the interfacial energy does not depend on the medium.

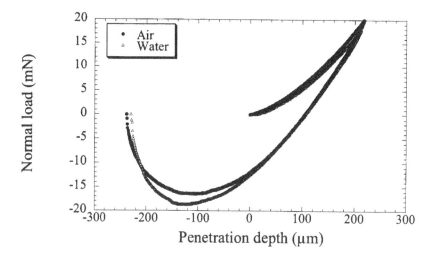

Figure 4. Indentation curves in two different media between M_1 and M_2. The test was carried out at indentation speed V=10 μm/s.

Table 2.
Values of interfacial energy in two different media obtained by two different methods (theoretical and experimental). The theoretical adhesion energy is calculated with relation (8)

Medium	Adhesion force F_{ad} (mN)	Theoretical adhesion energy w (mJ/m²)	Interfacial energy $\gamma^i_{M_1/M_2}$ (mJ/m²)	Experimental work of adhesion W_{ad} (nJ)	Experimental adhesion energy, w_{exp} (mJ/m²)	Interfacial energy $\gamma^i_{M_1/M_2}$ (mJ/m²)
Air	16.5	551	505	3900	539	493
Water	18.2	608	498	4500	583	473

4. STUDY OF SKIN ADHESION PROPERTIES

In our study, in order to reduce the effects of ageing, temperature and humidity, the tests were performed on the forearm zone of 10 Caucasian women about 30 years old. The tests were conducted at approximately constant temperature 22°C<T<24°C and constant relative humidity 20%<RH<30%. The tests were realised with a smooth steel spherical indenter with a radius of curvature R=6.35 mm. The skin reduced Young's modulus was determined from the slope of the unloading curve [11] and was found to be equal to $E^* = 9.5$ kPa ± 2 kPa.

4.1. Effect of a lipidic surface film on skin adhesion behaviour

First of all, to understand the effect of a lipidic film on skin surface we placed a water droplet (3 μl) on dry skin (without lipidic film) and on normal skin (with lipidic film). To dry the human skin we used ether. The lipidic film on the skin surface changes the skin physico-chemical properties. Normal skin is a hydrophilic surface (Fig. 5a) and the dry skin is a hydrophobic surface (Fig. 5b).

The indentation tests made on these different skins (normal and dry) show the importance of lipidic film on skin adhesion (Fig. 6). Adhesion aspects of human skin are principally determined by the thin lipidic film on the skin surface, because we see that the skin without lipidic film has no adhesion.

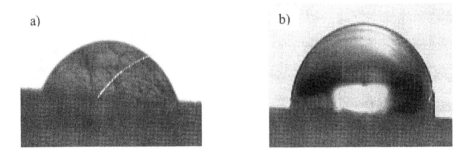

Figure 5. Water drop profiles on human skin. (a): normal skin, (b): dry skin.

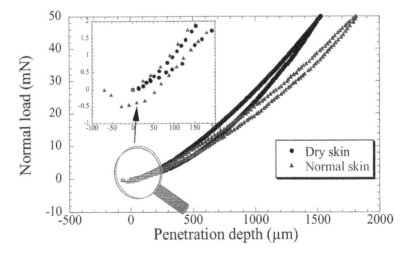

Figure 6. Indentation curves for normal and dry skins. Indentation speed V=400 μm/s.

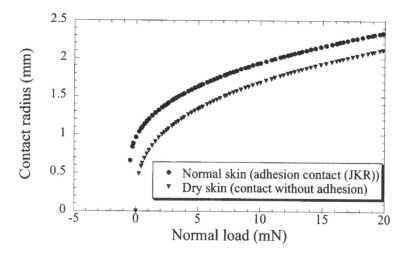

Figure 7. Variation of contact radius for normal skin and dry skin. Indentation speed V=400 μm/s.

To understand the effect of adhesion force on indenter/skin contact, we analysed the variation of skin contact radius as a function of normal load applied
(Fig. 7). The skin adhesion forces alter the contact zone between the indenter and
the skin compared to the purely elastic contact. We notice that the contact radius
for an adhesion contact is larger than for a contact without adhesion (Fig. 7).

4.2. Normal load effect on skin adhesion forces

In the indentation test case, the adhesion force measured between the indenter and
the skin is the adhesion force for a sphere/plane elastic contact. For normal load
less than 40 mN the sphere/plane elastic contact hypothesis is maintained, because the adhesion force stays constant and the parameter β is more than 85% (β
= 11% for F_N= 20 mN and β = 13% for F_N= 40 mN). The parameter β gives information on the elastic behaviour of skin. It is defined as (Fig. 2):

$$\beta = \int_0^{\delta_{max}} F_{N|loading}\,d\delta - \int_{\delta_1}^{\delta_{max}} F_{N|unloading}\,d\delta\,.$$ For β higher than 85%, the viscous

component of the skin can be neglected and we can consider skin essentially as an
elastic material. For the normal load higher than 40 mN, the adhesion force rises
and β is less than 85% (β = 18% for F_N= 80 mN). As a consequence, the elastic
sphere/plane contact hypothesis cannot be applied, because the viscous component of skin is not negligible and because the sphere/plane contact geometry between indenter and skin is not obeyed.

Finally we remark that the skin adhesion energy is constant and independent of
normal load for normal load less than 40 mN (Table 3).

Table 3.
Values of adhesion force and adhesion energy of skin

Normal load (mN)	Adhesion force (mN)	Theoretical adhesion energy (mJ/m²)
20	0.52	17.5
40	0.54	18
80	1.15	38

4.3. Indentation speed effect on skin adhesion forces

For the human skin, the thin lipidic surface film is responsible for its adhesion behaviour (Fig. 6). In general, for a solid/solid contact with a thin interfacial liquid
film, the adhesion force is influenced by the interfacial film. In this case, adhesion
force is given by:

$$F_{ad} = F_W + F_H + F_S,$$ (10)

where F_W is the wettability force, F_H the hydrodynamic force and F_S the
static force.

The hydrodynamic force F_H is given [12] as:

$$F_H = \eta \, L \, \dot{D},$$ (11)

where η is the viscosity of the thin film, L the geometric parameter [12] and
$\dot{D}$, the separation speed between solids.

The nature of the thin liquid film, and especially its viscosity, is a major parameter to determine the adhesion force. Moreover, the separation speed between
the solids is an important parameter in the hydrodynamic force expression. To
minimize the thin lipidic film effect, it is necessary to minimize the product $\eta \dot{D}$.

We cannot minimize the term $\eta \dot{D}$ by decreasing the film viscosity, because we
cannot control the film viscosity, it is constant, so the only parameter that we can
control is the separation speed between the solids.

The tests carried out at different indentation speeds confirm that the adhesion
force F_{ad} is influenced by indentation speed (Table 4). The faster the indentation
speed, the more difficult the separation between the indenter and the skin. The
adhesion force increases with the indentation speed for indentation speed higher
than 400 µm/s. As a consequence, to measure correctly the steel indenter/skin adhesion energy it is necessary to use a low indentation speed so as not to be influenced by the hydrodynamic force (Table 4). However, we note that the value of
adhesion energy for indentation speed equal to 800 µm/s is not very different
from the value for low indentation speed. It is due to the low thickness of the lipidic film in the inner forearm.

Table 4.
Values of adhesion force and adhesion energy for different indentation speeds

Indentation speed (μm/s)	Adhesion force (mN)	Theoretical adhesion energy (mJ/m²)
200	0.51	17
400	0.51	17
800	0.68	22.8

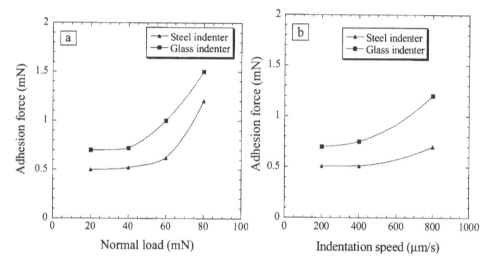

Figure 8. (a) Variation of adhesion force versus normal load for steel and glass indenters, for indentation speed V=400 μm/s. (b) Variation of adhesion force *versus* indentation speed for steel and glass indenters, for normal load F_N=20 mN.

4.4. Comparison between glass and steel indenters on skin adhesion forces

A spherical glass indenter and a spherical steel indenter were used to analyse the effect of the nature of indenter material on adhesion force. Both indenters had a radius of curvature equal to R = 6.35 mm. The glass indenter had a reduced Young's modulus of E^*=70 GPa.

Comparative indentation tests were carried out for two kinds of indenters (steel and glass). The adhesion force is plotted *versus* the normal load at constant indentation speed (Fig. 8a) or *versus* the indentation speed for constant normal load (Fig. 8b). Figure 8 clearly demonstrates that the adhesion force increases with both normal load and indentation speed, for both indenters. However, the adhesion force is always higher for the glass indenter than for the steel indenter, which is due to the difference in surface energies of the two materials. From these experiments, the influence of the surface energy on adhesion force can be noted. The nature of the contact plays a major role in adhesion phenomena.

The experimental adhesion energy can be calculated to confirm this conclusion. We have, $w_{steel} = 18 \text{mJ}/\text{m}^2$ and $w_{glass} = 24 \text{mJ}/\text{m}^2$ for the normal load of 20 mN and an indentation speed of 400 µm/s, for example.

4.5. Effect of hydration on skin adhesion and mechanical properties

In this part, we analyse the effect of hydration on the skin adhesion forces and skin bulk properties (contact stiffness) (Fig. 9). With a micro-syringe, 0.5 ml distilled water was placed on the skin surface and after 5 min the adhesion force was measured.

The main effect of the water is to increase the adhesion force between the indenter and the skin just after the water has been applied (Fig. 9) and to change the stiffness characteristics of the skin surface. A few minutes after the application of distilled water, the thickness of the surface film on the skin increases and this thickness has an effect on the adhesion force value. The indentation test shows a large increase in contact stiffness just 5 min after liquid application and after 5 min the stiffness slightly decreases but stays higher than the initial stiffness value.

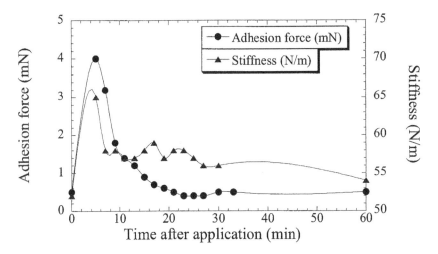

Figure 9. Variation of adhesion force and contact stiffness as a function of time after liquid water treatment. Normal load $F_N=20$ mN and indentation speed $V=400$ µm/s.

Table 5.
Adhesion and wetting forces for the normal and hydrated skin

	Normal skin	Hydrated skin
Adhesion force F_{ad} (mN)	0.5	4
Wetting force $F_W=4\pi R\gamma_{LV}$ (mN/m)	–	5.8

According to Fig. 9 the adhesion force measured between the indenter and the skin with distilled water is higher (5 min after liquid application) than the natural adhesion force of normal skin. The variation of adhesion force *versus* time after application of distilled water provides some information concerning modification of the skin physico-chemical properties. In fact, the time necessary for the skin to attain its initial condition corresponds to the time necessary for the skin to assimi-late the distilled water, which provides some information about the "action time" of the water on the skin. The surface film effect persists for about 20 min for this treatment.

Both the adhesion force and contact stiffness reach initial conditions about 1 h after treatment application. Finally, our indentation device allows to measure both the surface and the bulk skin properties at the same time and in real time, which can be very interesting to understand the cutaneous treatment effect on skin.

Hydration of skin is a complex phenomenon influenced by intrinsic (i.e., age, anatomical site) and extrinsic (i.e., ambient humidity, chemical exposure) factors [13]. The adhesion force, F_{ad}, measured for each test is given in Table 5. It is as-sumed that the liquid drop placed at the skin-sphere interface creates a meniscus. So the adhesion forces are due to the wetting forces, F_W, which are due to the Laplace pressure:

$$F_W = 4\pi R \gamma_{LV} \cos\theta, \tag{12}$$

where R is the sphere radius, γ_{LV} is the surface tension of the test liquid and θ the contact angle of the liquid with the skin surface.

Considering the scale of our measurements, we can consider that the liquid fully wets the surface of the skin, so $\theta = 0°$; then:

$$F_W = 4\pi R \gamma_{LV} \tag{13}$$

The comparison of the adhesion force, F_{ad}, measured (4 mN) for hydrated skin and the wetting force F_W obtained with pure water (5.8 mN) indicates partial wet-ting ($\theta > 0$) on skin surfaces, or more probably an impure water $\gamma_{LV} = 50.2$ mN/m.

Finally, comparison of the adhesion force and the maximum wetting force $F_W = 4\pi R \gamma_{LV}$ explains the increase of adhesion with hydrated surfaces.

5. CONCLUSIONS

The indentation system developed in this work makes it possible to characterize the effect of the skin surface film on its adhesion behaviour. To validate our skin indentation device, we compared the wettability test and a contact mechanics method to determine the interfacial energy between two elastomer solids. The re-sults showed that our device was able to measure the adhesion energy between two solids with the contact mechanics method.

Measurements of the skin adhesion force are influenced by experimental conditions, such as normal load and indentation speed. In fact, the skin adhesion force increases with normal load as well as indentation speed. Therefore, to accurately measure skin adhesion properties, the tests were carried out under controlled experimental conditions: low normal load (less than 40 mN) and low indentation speed (less than 500 μm/s). Moreover, it has been shown that the adhesion force between the indenter and the skin is sensitive to the nature of the indenter. When a liquid is applied on the skin surface, it interacts with the skin's lipidic surface and considerably modifies the skin adhesion behaviour, as well as skin mechanical properties (contact stiffness). This study showed that both skin adhesion forces and skin mechanical properties were greatly increased after a liquid (water) was applied to the skin. Comparison of the adhesion forces and the maximum wetting forces explains the increase of adhesion with hydrated surfaces. The hydration of skin has a temporary effect, which depends on the chemical composition of the liquid and the initial state of the skin's lipidic film. Finally, the thickness of the cutaneous film increases after hydration, and the nature of the contact is greatly influenced by the thickness of this liquid film.

REFERENCES

1. K. Hashimoto, *Int. J. Dermatol.* **13**, 357 (1974).
2. H. M. Sheu, S. C. Chao and T. W. Wong, *Br. J. Dermatol.* **14**, 385 (1999).
3. A. Mavon, H. Zahouani, D. Redoules, P. Agache, Y. Gall and Ph. Humbert, *Colloids Surfaces B: Biointerfaces* **8**, 147 (1997).
4. A. Mavon, D. Redoules, P. Humbert, P. Agache and Y. Gall, *Colloids Surfaces B: Biointerfaces* **10**, 243 (1998).
5. J. N. Israelachvili, *Intermolecular and Surface Forces*, 2nd edn. Academic Press, San Diego, CA (1992).
6. M. Barquins, *Wear* **158**, 87 (1992).
7. M. Nastasi, *Mechanical Properties and Deformation Behavior of Materials Having Ultra-fine Microstructures*. Kluwer, Dordrecht (1993).
8. J. L. Loubet, J. M. Georges, O. Marchesini and G. Meille, *J. Tribol.* **106**, 43 (1984).
9. D. Maugis, *J. Colloid Interface Sci.* **150**, 243 (1992).
10. K.-L. Johnson, *Contact Mechanics*, Cambridge University Press, Cambridge (2001).
11. I. N. Sneddon, *Int. J. Eng. Sci.* **3**, 47 (1965).
12. M. J. Matthewson, *Phil. Magn. A* **57**, 207 (1988).
13. J. M. Georges, S. Millot, J. L. Loubet and A. Tonck, *J. Chem. Phys.* **98**, 7345 (1993).

Contact Angle, Wettability and Adhesion, Vol. 4, pp. 515–529
Ed. K.L. Mittal
© VSP 2006

Influence of surface properties of metal leadframes on mold compound–metal adhesion in semiconductor packaging

TANWEER AHSAN*

Henkel Corporation, 15350 Barranca Parkway, Irvine, CA, USA

Abstract—Adhesion strength of a novolac epoxy-based mold compound to nickel leadframes was investigated. Surface energetics and wettabilities of untreated, thermally treated and Ar-plasma-etched leadframes were studied *via* the contact-angle approach. Surprisingly, both the most (plasma-treated) and the least (untreated) wettable nickel surfaces, as ascertained by the water contact angle, offered very low adhesion to the mold compound. Only thermally treated leadframes with modest wettability gave high adhesion. X-Ray photoelectron spectroscopy (XPS) data and scanning electron microscopy (SEM) micrographs suggest that the surface oxide layer thickness and its integrity on thermally treated leadframes are the most important factors in determining the adhesion of mold compounds to nickel leadframe surfaces. The study shows that a thermally treated nickel surface with an oxide thickness of 3.5–5.0 nm and free of organic contamination is most likely to offer high adhesion to epoxy mold compounds.

Keywords: Nickel oxide layer; wettability; surface energy; adhesion strength; plastic packaging; leadframes.

1. INTRODUCTION

The majority of adhesion science practitioners would agree that adhesion, despite being a complex phenomenon, might broadly be explained in terms of two major mechanisms, namely, chemical adhesion and mechanical adhesion. The former encompasses of all types of bond-making processes such as chemisorption, physisorption, acid–base interactions, hydrogen bonding and electrostatic interactions. On the other hand, the mechanical adhesion is related to the physical penetration and entrapment of polymer chains into the rugosity of the solid surfaces. Both of these mechanisms together are able to explain the adhesion of different materials to various substrates provided that the substrate surface is free of contamination and environmental influences such as heat, humidity and mechanical stress, have not played a major factor [1].

---

*Tel.: (1-949) 585-2037; Fax: (1-949) 789-2595; e-mail: Tanweer.Ahsan@us.henkel.com

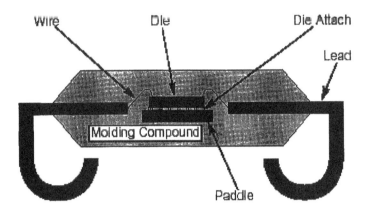

Figure 1. Schematic representation of a plastic encapsulated chip package.

As far as plastic packaging of integrated circuits (ICs) is concerned, the adhesion of epoxy-based molding compounds appears to be mainly via hydrogen bonding with some finite contribution from physical entrapment of the polymer on rough surfaces. Adhesion of mold compounds to leadframe metals such as Cu, Ni, Cu/Ni, Cu/Ni/Pd and Cu/Ni/Pd/Au appears to be very much affected by the surface energetics of these leadframes (Ref. [2] and references therein). A leadframe is a metallic portion of the device package that completes the electrical connection path from the die (chip) to ancillary hybrid circuit elements of the card assembly. A thermoset molding compound encapsulates the entire IC so that the outer ends of the leads extend beyond the plastic perimeter as shown in Fig. 1. It is desirable to relate the adhesion of mold compounds to the surface properties of the IC leadframes in order to understand their effect on delamination and adhesion when the encapsulated packages are subjected to standard JEDEC (Joint Electron Device Engineering Council) moisture stress levels followed by high temperature (approx. 260°C) solder reflow simulation [3].

Mold compounds are generally composed of epoxy resins, hardeners and fused silica. These ingredients constitute about 90–95% of the formulation. In addition, adhesion promoters such as silanes, mold release agents such as waxes, catalysts or accelerators for chain initiation reaction, pigments such as coloring agents and flame-retardants such as metal oxides (for "green" compounds) may also be present in a typical formulation. Flame retardancy is an essential requirement for polymers and plastics in the electronics industry. Environmental concerns for the use of lead and the impending strict legislations in Europe and Japan for lead-free soldering have placed great demands on mold compound manufacturers. The materials are not only expected to be green compounds (free of Br and Sb) but also must withstand high reflow temperatures (260°C or higher) often encountered in lead-free soldering conditions [2]. This is not an easy task. Specialized resin chemistry and proprietary curing methodologies are used to accomplish ever-demanding requirements of the electronics industry for package integrity and adhesion retention after exposure to high humidity and temperatures. Typical accelerated aging tests (recommended by JEDEC standards) conducted in laboratories involve moisture uptake at 85°C / 85% relative humidity (RH) for 168 h followed by exposure to 260°C reflow temperature to simulate the soldering conditions of the printed circuit boards. This is carried out to induce rapid failure of the plastic-encapsulated devices. Various physical (pull-force measurements), visual (C-mode scanning acoustic microscopy) and parametric electrical tests are carried out on molded parts to observe the extent of degradation of adhesion and the resulting electrical performance of the chip before and after accelerated aging. A stronger adhesion of mold compounds to metallic surfaces is a pre-requisite for better performance of the packaged device. Therefore, it is important to study the adhesion properties of mold compounds to various metallic surfaces.

In this paper we explore the surface properties of metal leadframes, such as copper and nickel, and attempt to relate the adhesion strength of a mold compound to the surface energetics and surface chemistry. The surface energetics was explored *via* the contact-angle technique while the adhesion strength of the molded metallic frames was assessed *via* pull-force measurements. Surface energetics of the leadframes and mold compound adhesion were determined after various pre-treatments such as thermal and plasma given to leadframes. X-Ray photoelectron spectroscopy (XPS) was used to determine the surface elemental composition of metallic surfaces.

2. SURFACE ENERGETICS: THEORETICAL CONSIDERATION

Surface energy determination of flat solid surfaces *via* contact-angle measurements is now a reasonably established technique. There are several equations that are currently in use. The interpretation and the meaning of several terms in these equations and their impact on solid surface energetics are of much debate among various theoreticians and practitioners [4, 5]. Despite all the controversies, most adhesion scientists agree on a common point that the contact angle data have, at least, some merit in understanding the surface property (chemistry) of the materials under investigation [6]. It is widely accepted that all the theories, namely, the Owens–Wendt geometric means approach [7], Wu's simple harmonic mean approach [8] and van Oss, Chaudhury and Good's (vOCG) acid–base approach [9] offer a systematic way of probing the surfaces. In all cases the data obtained are comparable, at least, in a relative sense [6]. Although, it is a rather difficult task to assign a single numerical value of surface energy to a material or specific polymeric system, the surface energy data are very helpful to make relative comparisons of the same material after the specimens have gone through some physico-chemical changes.

Amazingly, the basic idea proposed by Young almost 200 years ago [10] is still valid and is the basis of all the older and contemporary theories that relate contact angles of liquids to the solid surface energetics.

Young's equation (1) describing the contact of a liquid to a flat solid surface is expressed as

$$\cos \theta = (\gamma_{sv} - \gamma_{sl}) / \gamma_{lv} \tag{1}$$

Dupre's equation (2) relates the work of adhesion (W_{sl}) to surface energies of solid and liquid and interfacial energy as [11, 12]

$$W_{sl} = \gamma_{sv} + \gamma_{lv} - \gamma_{sl} \tag{2}$$

These two equations are combined to give the work of adhesion of a liquid on a solid surface and is called the Young–Dupre equation:

$$W_{sl} = \gamma_{lv} (1 + \cos \theta) \tag{3}$$

According to Girifalco and Good [13], W_{sl} can be estimated as

$$W_{sl} = (W_{ss} \cdot W_{ll})^{\frac{1}{2}} \tag{4}$$

Assuming $W_{ss} = 2\gamma_{sv}$ and $W_{ll} = 2\gamma_{lv}$ and introducing equation (4) in equation (3), we obtain

$$(\cos \theta + 1) \gamma_{lv} = 2 (\gamma_{sv} \cdot \gamma_{lv})^{\frac{1}{2}} \tag{5}$$

Fowkes [14] suggested that total surface energy of a material is composed of dispersion and polar interactions

$$\gamma = \gamma^d + \gamma^p \tag{6}$$

Following these considerations Owens and Wendt proposed the following equation to estimate the surface energetics of solid surfaces

$$(\cos \theta + 1) \gamma_{lv} = 2 (\gamma^d_{sv} \cdot \gamma^d_{lv})^{\frac{1}{2}} + 2 (\gamma^p_{sv} \cdot \gamma^p_{lv})^{\frac{1}{2}} \tag{7}$$

Without going into too much theoretical background the simple harmonic (SH) method of determining the surface energy, proposed by Wu, is as follows:

$$\gamma_{lv} (1 + \cos \theta) = [4 (\gamma^d_{sv} \cdot \gamma^d_{lv})/(\gamma^d_{sv} + \gamma^d_{lv}) + 4 (\gamma^p_{sv} \cdot \gamma^p_{lv})/(\gamma^p_{sv} + \gamma^p_{lv})] \tag{8}$$

Finally, the vOCG theory, expressing the surface energetics in terms of acid–base interactions and dispersion forces, is as follows:

$$\gamma_{lv} (1 + \cos \theta) = 2 (\gamma^{LW}_{sv} \cdot \gamma^{LW}_{lv})^{\frac{1}{2}} + 2 (\gamma^+_{sv} \cdot \gamma^-_{lv})^{\frac{1}{2}} + 2 (\gamma^-_{sv} \cdot \gamma^+_{lv})^{\frac{1}{2}} \tag{9}$$

We have utilized all of the above approaches to understand the surface energetics of metal leadframes and the influence of the resulting surface properties on adhesion of mold compounds. In this report we present comparisons of materials using Wu's simple harmonic means approach only because it offered practically straightforward comparative data.

Table 1.
Description of nickel frames

Leadframes	Finish	Comments
(1)	Standard Ni on Cu Ni thickness = 1–2 μm	
(2)	Standard Ni on Cu Ni thickness = 0.50 μm	
(3)	High Surface Area (HSA) Ni on Cu Ni thickness = 0.50 μm	Proprietary HSA Ni finish
(4)	HSA Ni deposition *via* sulfamate process on Cu Ni thickness = 0.5 μm	HSA Ni and sulfamate Ni on Cu

3. MATERIALS AND METHODS

Tab pull leadframes were obtained from Texas Instruments (Attleboro, MA, USA) with various metal finishes. The leadframes were made of copper base metal with 0.5–2-μm-thick nickel coatings. Some of the surfaces were textured to give so-called "high surface area surfaces" by various proprietary plating processes. Table 1 describes the various nickel frames used in this study. The mold compound was Henkel's "green" (free of bromine and antimony) novolac-based epoxy material with 80 ± 1 wt% silica filler.

3.1. Contact-angle measurements

Advancing contact angles of methylene iodide, water and ethylene glycol on metal surfaces were measured using a First Ten Angstroms (FTA) goniometer. Average contact angles of the liquids were determined by placing 3–4 drops of each liquid at various spots on the leadframes by a syringe equipped with a 27-gauge needle. Contact angle data were used to determine the surface energies of metal leadframes by using a software developed by FTA.

3.2. Adhesion measurement of molding compound

Molding compound was molded on pull frames using a press operated at 177°C. The molding compound thickness was 1–2 mm. The molded frames were post-cured at 175°C for 4 h before adhesion measurement.

Pull adhesion tests were performed using an Instron tester Model 4444 with a load cell of about 182 kg. A typical molded tab pull frame is shown in Fig. 2. The lower grip of the tester held the two metallic portions, A and B, of the tab. The holes in these leads helped in anchoring the molded compound firmly. The metallic portion C embedded in the molding compound was held at the other end of the tester. Portion C was removable at an appropriate pull force. The crosshead speed for the tester was 0.127 cm/min [15]. Pull force as recorded was the

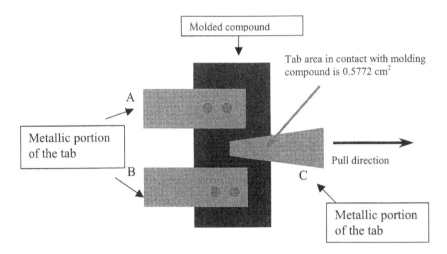

Figure 2. Schematic representation of a single molded tab frame.

maximum force needed to pull the lead C away from the molding compound. The total surface area of the lead C in contact with the molding compound was 0.5772 cm^2.

To explore the adhesion properties of oxidized nickel frames, heat treatment was given to tab pull frames in an IR reflow oven prior to molding. The oven was programmed to run at an average rate of $46 \pm 2°C/min$, peaking at $260 \pm 2°C$ for about 10–15 s before a gradual cool down to about 50°C. The total thermal treatment time was for 6.5 ± 0.10 min. Frames were placed in a desiccator for further equilibration to room temperature before the test specimens were fabricated by following standard transfer molding techniques. In-mold cure times were 100–120 s at 175°C followed by a 4-h post-cure at 175°C.

Nickel leadframes were subjected to plasma etching using a Barrel Plasma System from Anatech (Union City, CA, USA). The chamber was purged with dry nitrogen prior to charging with argon at 300 W under a pressure of 0.3–0.4 Torr. Specimens were etched for 5 min and then placed in a desiccator prior to transfer molding.

3.3. X-ray photoelectron spectroscopy (XPS) and Auger electron spectroscopy (AES)

XPS analysis was performed at Texas Instruments (TI, Attleboro, MA, USA), as well as at the University of Connecticut. Analytical work was carried out at two different places to obtain a better feel for the variations in the absolute values and the corresponding differences between the untreated and heat treated frames. XPS conditions at the University of Connecticut were as follows:

The X-ray source was monochromatic aluminum at 15 kV. The analysis region was 3 mm^2 for XPS. The pressure was 10^{-9} Torr for all experiments. Depth profiles of oxide layers on nickel surfaces were performed with AES using an argon ion sputter gun. Depth *versus* sputter time was calculated using the measured sputter rate of a tantalum oxide standard. The sputter rate of Ni in these measurements was calculated using the published relative sputter rates of tantalum oxide and pure nickel. The spot size in the Auger analysis beam was less than 1 μm^2 at 3 kV electron energy.

The Auger microprobe used at TI had electron energy of 5 kV and the spot size of the beam was larger than 100 nm^2. Spectra were taken at a chamber pressure of 10^{-9} Torr, while sputtering was carried out at 10^{-8} Torr. Tantalum oxide sputtering method was used for depth profiling.

4. RESULTS AND DISCUSSION

4.1. Mold compound–metal adhesion and surface energetics

Mold compound adhesion to various metallic surfaces within the context of semiconductor packaging (encapsulation) has been well documented [16–18]. Adhesion to copper has been studied in depth and the effect of controlled oxidation is recognized [18]. Copper is an easily oxidizable material and, therefore, precautions are often taken to protect its surface from excessive oxidation to avoid adverse effect on adhesion. The brief study of adhesion to copper in the present work has indicated that both the surface energy and adhesion increase with heat treatment. However, after a small increase in both properties with the pre-treatment temperature the adhesion is significantly reduced (Fig. 3). Over-oxidation of copper exceeds the maximum critical thickness of the oxide layer that promotes adhesion; therefore, a cohesive failure in the oxide layer is observed. This is consistent with the work cited in the literature [18].

Our main interest in this work was to explore the adhesion of mold compounds to nickel surfaces. Since nickel is less easily oxidized than copper, there are some intrinsic difficulties associated with the adhesion of novolac-based molding compounds to nickel. The adhesion data in Table 2 show that all the untreated leadframes give extremely poor adhesion with the mold compound, which is consistent with the earlier studies [19]. However, the thermally treated leadframes offered a substantially higher adhesion, especially, the higher surface area non-sulfamate process (Ni 3), offering the highest adhesion of all. It is likely that the high surface area specimen could have provided an "interlocking" effect to the mold compound. The surface energies of untreated and heat-treated frames are reported in Table 3. In our analysis of the surface energies we used all the three approaches as mentioned in Section 2. Although the numerical data differed for different approaches, we observed a systematic correlation between the various approaches. It is interesting to observe that the polar components of the thermally treated frames are almost 2–3-times higher than those of the corresponding untreated frames (Table 3). Although there is no equivalent correlation with the

Table 2.
Tab adhesion force (N) of mold compound to various untreated and thermally treated nickel frames

	Untreated frames	Thermally-treated frames
Ni (1)	9.8	182.3 ± 22
Ni (2)	2.7	191.2 ± 26
Standard 0.5 µm Ni on Cu		
Ni (3)	10.7	244 ± 75
0.5 µm high surface area Ni on Cu		
Ni (4)	7.1	107 ± 9
HSA-sulf Ni on Cu		
Ni 0.5 µm		

Table 3.
Surface energy and its components (in mJ/m^2) of nickel frames using Wu's method

	Untreated frames			Heat-treated frames		
	Total	Dispersion	Polar	Total	Dispersion	Polar
Ni (1)	44.8	39.4	5.4	60.6	43.5	17.1
Ni (2)	44.7	38.2	6.5	58.4	44.5	13.9
Ni (3)	45.8	40.8	4.9	62.3	45.7	16.6
Ni (4)	49.5	40.8	8.7	64.7	45.1	19.6

adhesion data, one conclusion appears to be certain, i.e., higher surface energy promotes higher adhesion. Higher surface energies (polarities) could result due to the presence of oxygenated species developed on metallic surfaces upon thermal treatments. These species are known to promote higher adhesion with the epoxy-based mold compounds [2].

4.2. XPS data

Table 4 presents surface atomic composition of a nickel frame with various pre-treatments. It is evident that thermally-treated frames have more oxygenated species than either of the other two samples (untreated and Ar-plasma etched). Higher levels of carbon on untreated and plasma-etched samples are unexpected but, in case of the latter, could be due to the contamination from the plasma chamber. Slightly higher levels of Si shown by the heat treated frames are definitely due to contamination from the oven. Table 5 shows the thickness of the nickel oxide layer determined by Auger electron spectroscopy. It is interesting to note that although the absolute values of the thickness are different for the two laboratories, the corresponding differences (1.65 ± 0.05 nm) between the untreated and heat-treated frames are very similar. Overall, the thickness of the oxide layer on thermally treated frames is 1.5–2-times higher than those on untreated frames.

Table 4.
XPS surface elemental composition (at%) of treated nickel surfaces

	O	Ni	C	Si
Untreated	29.9	55.5	14.5	00
Ar-Plasma etched	20.2	57.3	17.7	00
Heated at 260°C in IR reflow oven	41.1	51.1	3.1	4.9

Table 5.
Oxide thickness (nm) on nickel surfaces with various pre-treatments

Pre-treatment of frames	Data from source A	Data from source B
Untreated frames	1.9	3.5
Ar-plasma for 5 min	0.9	
Thermal treatment at 300°C	3.5	5.2
Ar-plasma + 300°C	4.0	

A, Texas Instruments, Attleboro, MA, USA; B, Dr. Kurz, University of Connecticut.

4.3. Wettability and adhesion

It is commonly believed that good wettability is a pre-requisite for adhesion enhancement [20]. The significance of both surface energy and surface topography
on wettability and, thereby, on adhesion has been well emphasized [21]. Table 6
shows various pre-treatments given to a nickel frame to induce wettability im-
provements, while Table 7 compares the surface energies of the frames. It is
noteworthy that the total surface energies (and the polar components) of the
plasma-etched frames are much higher than those of thermally-treated frame (Fig.
3). All three liquids showed wettability improvement for all treatments. Let us
consider the interaction of water drops with the surface. The untreated frame has a
water contact angle of 87° but the surface becomes more wettable with all pre-
treatments. Water contact angle on the heat-treated frame is approx. 57°, but a
drastic reduction in the contact angle (approx. 28°) is observed when the frame is
etched with Ar plasma. Intuitively, one would relate the increased wettability of a
surface by a polar liquid to be a consequence of polar moieties on the surface that
would improve adhesion of polar polymeric materials such as epoxy. We were
surprised to observe that only the thermal treatment to the metal leadframes
brought about the strongest adhesion to the mold compound. None of the other
treatment sequences, despite inducing improved wettabilities, offered any significant improvement to adhesion as compared to the untreated specimens. This apparent discrepancy between wettability and adhesion demands further attention.
One ponders as to what is the significance of water wettability of the plasma-
etched specimens and what makes the water (polar liquid) wet the surface and yet
the "implied" high surface energy specimen does not bring about any improve-

Table 6.
Contact angles of liquids (in degrees) and tab adhesion (N) of mold compound to nickel leadframe
(Ni 1) after various pre-treatments

Pre-treatment	Water	Methylene iodide	Ethylene glycol	Adhesion force (N)
Untreated	87	41.6	59.6	13
Heated at 260°C in IR reflow oven	56.7	28.3	37.3	222 ± 22
Heated at 260°C in IR reflow oven + Ar plasma	28.5	31.0	31.8	13
Ar plasma	27.8	29.3	30.5	18

Note: Contact angles are accurate to ±1–2 degrees.

Table 7.
Surface energy (mJ/m^2) of nickel leadframe 1 after various pre-treatments using Wu's method

Heat treated			Ar plasma treated			Heat treated followed by Ar plasma treatment		
Total	Dispersion	Polar	Total	Dispersion	Polar	Total	Dispersion	Polar
63.5	45.1	18.5	76.6	44.7	31.9	75.9	44.0	31.9

ment to the adhesion of mold compound to the metal surface. Perhaps the partial answer may come from examining the surface oxide layer. SEM micrographs of the untreated and treated specimens are shown in Fig. 4. The untreated surface (Fig. 4A) shows some evidence of oxide formation but the structure does not appear to be a homogeneous one. There are gaps in the form of ridges and streaks on the surface. The thickness of this layer was estimated by Auger analysis to be in the range of 2–3 nm. It appears that there are micro-voids small enough to prevent the "spreading" of water drops ($\theta = 87°$) on the surface. An alternative explanation to the poor wettability of the untreated surface may also be related to the presence of organic contamination. On the other hand, the heat-treated specimen (Fig. 4B) appears to have a denser oxide layer covering the whole surface rather homogeneously. The thickness of this oxide layer was estimated to be in the range of 3–5 nm. If it is assumed that the oxide is also slightly porous then it is conceivable that such a surface could give a modest decrease in the contact angle as was observed ($\theta = 57°$). The most astounding behavior of the surface was observed with the argon plasma-etched frame and the one that was initially heat-treated and then subjected to plasma etching. Both of these specimens had apparently very high water wettability (θ approx. 28°), despite surface contamination of Si and C from the plasma chamber. Both leadframes offered insignificant adhesion as compared to the heat-treated frames. The SEM micrographs (Fig. 4C and 4D) of

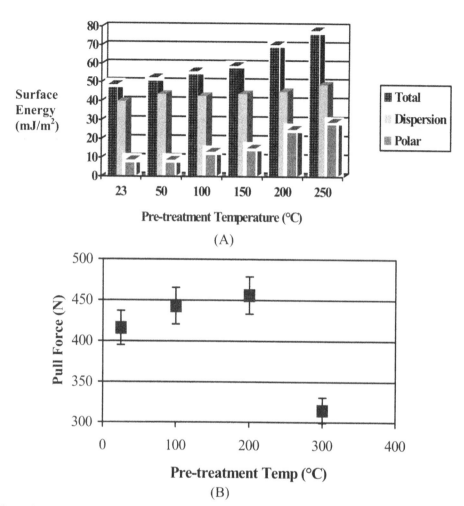

Figure 3. (A) Variation of copper surface energetics with pre-treatment temperature. (B) Tab adhesion pull force against pre-treatment temperature of encapsulated copper leadframes.

these specimens show surface oxide layers that appear less dense with isolated "broken" areas. It is likely that this type of surface promoted the "spreading" or "absorption" of the water droplets, thereby giving significantly lower contact angles. No attempts were made to measure the thickness of this oxide layer. It appears that the apparent surface energy enhancement, as measured by the contact angle approach, was as a result of the micro-roughness or spontaneous spreading of the liquids and not due to the change in the chemistry of the surface. However, in the absence of atomic force microscopy (AFM) data a complete picture of the surface roughness cannot be ascertained with confidence. Despite this, it is conceivable that Ar-plasma etching process reduced the homogeneity of the surface oxide layer, diminishing its integrity as well as the critical thickness that was

T. Ahsan

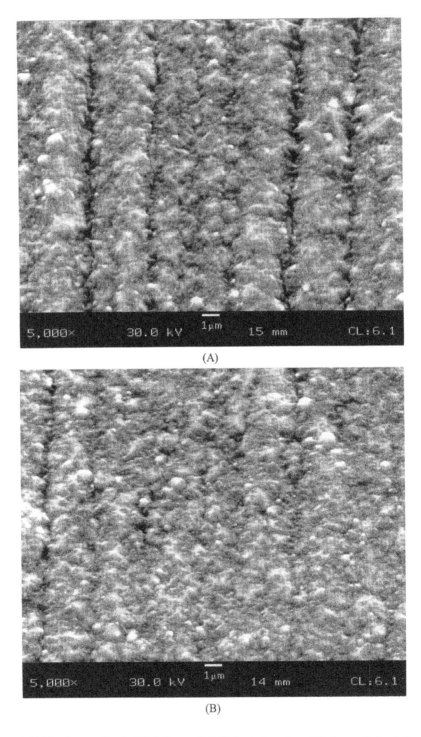

Figure 4. SEM micrographs of nickel frames. (A) Untreated surface; (B) thermally-treated surface; (C) Ar-plasma-etched surface; (D) thermally treated and then Ar plasma etched surface.

Figure 4. (Continued).

beneficial for enhanced adhesion. In view of the recent report by Park *et al.* that the oxidation of Ni under dry conditions leads to the formation of nickel oxide with surface –OH groups [22], it is not surprising that thermally-treated frames provided the highest adhesion to the novolac-based mold compound.

5. CONCLUSIONS

Surface energetics and wettability of nickel metal leadframes were related to the novolac-based mold compound adhesion. Various pre-treatments to the metal frames brought about varying degrees of surface wettabilities. Modestly wettable, thermally-treated frames offered the highest adhesion, while the most wettable surface obtained by Ar-plasma etching did not offer any improvement in adhesion over that obtained on untreated frames. These observations are unusual and perhaps specific to a nickel surface because other metals, such as copper, offer good adhesion after argon plasma etching.

XPS data and SEM micrographs revealed the thickness and the integrity of the nickel surface oxide layers. Estimated thickness of the oxide layer offering the strongest adhesion was found to be in the range of 3.5–5.2 nm. Nickel surface with homogeneous surface oxide developed by thermal treatment and free of organic contamination is most likely to offer high adhesion to the novolac-based mold compounds.

Acknowledgements

The author appreciates technical help by the R&D staff at Olean facility. Dr. Donald Abbott of Texas Instruments is thanked for help with the Auger spectroscopy. Dr. Kurz of The University of Connecticut is thanked for his help with the XPS data. Finally, the management at Henkel Technologies is acknowledged for their permission to publish this work.

Symbols used:

γ_{sv}, surface energy of solid in equilibrium with the liquid vapor;

γ_{sl}, interfacial energy between solid and liquid;

γ_{LV}, surface energy of liquid in equilibrium with its saturated vapor;

W_{sl}, work of adhesion between solid and liquid;

γ^d, dispersion component of surface energy;

γ^p, polar component of surface energy;

γ^+, acid parameter of surface energy;

γ^-, base parameter of surface energy;

γ, total surface energy;

θ, contact angle of liquid with solid;

LW, Lifshitz–van der Waals;
d,   dispersion;
p,   polar.

REFERENCES

1. S. Pignataro, in: *Mittal Festschrift on Adhesion Science and Technology*, W. J. van Ooij and H. R. Anderson Jr. (Eds.), pp. 147–159. VSP, Utrecht (1998).
2. D. Abbott, T. Ahsan, A. A. Gallo and C. Bischof, *Proc. IPCSMEM Council APEX, LF2-1*, pp. 1–11 (2001).
3. Joint Industry Standard, *Moisture/Reflow Sensitivity Classification for Non-hermetic Solid State Surface Mount Devices*. IPC/JEDEC J-STD-020B (2002).
4. K. L. Mittal (Ed.), *Contact Angle, Wettability and Adhesion*, Vol. 3, VSP, Utrecht (2003).
5. K. L. Mittal (Ed.), *Acid-Base Interactions: Relevance to Adhesion Science and Technology*. Vol. 2. VSP, Utrecht (2000).
6. C. Della Volpe, D. Maniglio and S. Siboni, in: *Contact Angle, Wettability and Adhesion*, Vol. 2, K. L. Mittal (Ed.), pp. 45–71. VSP, Utrecht (2002).
7. D. K. Owens and R. C. Wendt, *J. Appl. Polym. Sci.* **13**, 1740 (1969).
8. S. Wu and K. J. Brozowski, *J. Colloid Interface Sci.* **37**, 686 (1971).
9. C. J. van Oss, *Interfacial Forces in Aqueous Media*, Marcel Dekker, New York, NY (1994).
10. T. Young, *Phil. Trans. Roy. Soc.* **95**, 65 (1805).
11. A. Duprè, *Theorie Mecanique de la Chaleur*, pp. 367–370. Gauthier-Villars, Paris (1869).
12. J. T. Davies and E. K. Rideal, *Interfacial Phenomena*. Academic Press, New York, NY (1963).
13. R. J. Good, in: *Contact Angle, Wettability and Adhesion*, K. L. Mittal (Ed.), pp. 3–36. VSP, Utrecht (1993).
14. F. M. Fowkes, *J. Phys. Chem.* **67**, 2538 (1963).
15. C. S. Bischof, *Proc. IEEE* **45**, 827–834 (1995).
16. K. Cho, E. C. Cho, *J. Adhesion Sci. Technol.* **14**, 1333 (2000).
17. K. Cho and E. C. Cho and C. E. Park, *J. Adhesion Sci. Technol.* **15**, 439 (2001).
18. M. Lebbai, J.-K. Kim and W. K. Szeto, *J. Adhesion Sci. Technol.* **17**, 1543 (2003).
19. A. A. Gallo and D. C. Abbott, *Proc. 9th Annual Pan-Pacific Microelectronics Symposium*, p. 275 (2004).
20. K. L. Mittal, in: *Adhesion Science and Technology*, L.-H. Lee (Ed.), Part A, pp. 129–168. Plenum Press, New York, NY (1975).
21. D. E. Packham, *Int. J. Adhesion Adhesives* **23**, 437–448 (2003).
22. M. B. Chan-Park, J. Gao and A. H. L. Koo, *J. Adhesion Sci. Technol.* **17**, 1979 (2003).

Printed and bound by CPI Group (UK) Ltd, Croydon, CR0 4YY

23/10/2024

01778246-0004